T0183751

Lecture Notes in Computer Science 11234

Commenced Publication in 1973
Founding and Former Series Editors:
Gerhard Goos, Juris Hartmanis, and Jan van Leeuwen

Editorial Board

More information about this series at http://www.springer.com/series/7409

Hakim Hacid · Wojciech Cellary
Hua Wang · Hye-Young Paik
Rui Zhou (Eds.)

Web Information Systems Engineering – WISE 2018

19th International Conference
Dubai, United Arab Emirates, November 12–15, 2018
Proceedings, Part II

 Springer

Editors
Hakim Hacid ⓘ
Zayed University
Dubai, United Arab Emirates

Wojciech Cellary ⓘ
Poznan University of Economics
Poznan, Poland

Hua Wang ⓘ
University of Victoria
Footscray, VIC, Australia

Hye-Young Paik ⓘ
University of New South Wales
Sydney, NSW, Australia

Rui Zhou ⓘ
Swinburne University of Technology
Hawthorn, VIC, Australia

ISSN 0302-9743   ISSN 1611-3349 (electronic)
Lecture Notes in Computer Science
ISBN 978-3-030-02924-1   ISBN 978-3-030-02925-8   (eBook)
https://doi.org/10.1007/978-3-030-02925-8

Library of Congress Control Number: 2018958517

LNCS Sublibrary: SL3 – Information Systems and Applications, incl. Internet/Web, and HCI

This Springer imprint is published by the registered company Springer Nature Switzerland AG
The registered company address is: Gewerbestrasse 11, 6330 Cham, Switzerland

Preface

Welcome to the proceedings of the 19th International Conference on Web Information Systems Engineering (WISE 2018), held in Dubai, UAE, during November 12–15, 2018. The series of WISE conferences aims to provide an international forum for researchers, professionals, and industrial practitioners to share their knowledge in the rapidly growing area of Web technologies, methodologies, and applications. The first WISE event took place in Hong Kong, SAR China (2000). Then the trip continued to Kyoto, Japan (2001); Singapore (2002); Rome, Italy (2003); Brisbane, Australia (2004); New York, USA (2005); Wuhan, China (2006); Nancy, France (2007); Auckland, New Zealand (2008); Poznan, Poland (2009); Hong Kong, SAR China (2010); Sydney, Australia (2011); Paphos, Cyprus (2012); Nanjing, China (2013); Thessaloniki, Greece (2014); Miami, USA (2015); Shanghai, China (2016); Puschino, Russia (2017); and this year, WISE 2018 was held in Dubai, UAE, supported by Zayed University.

A total of 209 research papers were submitted to the conference for consideration, and each paper was reviewed by at least three reviewers. Finally, 48 submissions were selected as regular papers (with an acceptance rate of 23% approximately), plus 21 as short papers. The research papers cover the areas of blockchain, security and privacy, social networks, microblog data analysis, graph data, information extraction, text mining, recommender systems, medical data analysis, Web services, cloud computing, data stream, distributed computing, data mining techniques, entity linkage and semantics, Web applications, and data mining applications.

In addition to regular and short papers, the WISE 2018 program also featured four workshops: (1) the 5th WISE Workshop on data quality and trust in big data (QUAT 2018); (2) International Workshop on Edge-Based Computing for Next-Generation Wireless Networks; (3) the Third International Workshop on Information Security and Privacy for Mobile Cloud Computing, Web, and Internet of Things (ISCW 2018); (4) The 1st International Workshop on Cloud Computing Economic Impacts. This year's tutorial program included: (1) Text Mining for Social Media; (2) Towards Privacy-Preserving Identity and Access Management Systems for Web Developers; and (3) From Data Lakes to Knowledge Lakes: The Age of Big Data Analytics.

We also wish to take this opportunity to thank the general co-chairs, Prof. Zakaria Maamar and Prof. Marek Rusinkiewicz; the program co-chairs, Prof. Hakim Hacid, Prof. Wojciech Cellary, and Prof. Hua Wang; the workshop co-chairs, Prof. Michael Sheng and Prof. Tetsuya Yoshida; the tutorial and panel chair, Dr. Hye-Young Helen Paik; the sponsor chair, Dr. Fatma Taher; the finance chair, Prof. Hakim Hacid; the local arrangements co-chairs, Dr. Andrew Leonce, Prof. Huwida Saeed, and Prof. Emad Bataineh; the publication chair, Dr. Rui Zhou; the publicity co-chairs, Dr. Dickson Chiu, Dr. Reda Bouadjenek, and Dr. Vanilson Burégio; the website co-chairs, Mr. Emir Ugljanin and Mr. Emerson Bautista; and the WISE Steering Committee representative, Prof. Yanchun Zhang.

We would like to sincerely thank our keynote speakers:

- Professor Athman Bouguettaya, Professor and Head of School of Information Technologies, University of Sydney, Sydney, Australia
- Professor A. Min Tjoa, Institute of Information Systems Engineering TU Wien, Vienna University of Technology, Vienna, Austria
- Professor Mike P. Papazoglou, Executive Director of the European Research Institute in Services Science (ERISS), University of Tilburg, Tilburg, The Netherlands

In addition, special thanks are due to the members of the international Program Committee and the external reviewers for a rigorous and robust reviewing process. We are also grateful to Zayed University, UAE, Springer Nature Switzerland AG, IBM, and the International WISE Society for supporting this conference. The WISE Organizing Committee is also grateful to the workshop organizers for their great efforts to help promote Web information system research to broader domains.

The local lead organizer, Prof. Hakim Hacid, would also like to thank the following colleagues for their support and dedication at different levels in the preparation of WISE 2018 (alphabetical order): Ahmed Alblooshi, Ahmad Al Rjoub, Alia Sulaiman, Amina El Gharroubi, Ayesha Alsuwaidi, Fatima AlMutawa, Hind AlDosari, Jarita Sebastian, Jimson Lee, Osama Nasr, Rosania Braganza, Sudeep Kumar, Zenelabdeen Alsadig, Gemma Ornedo.

We expect that the ideas that have emerged in WISE 2018 will result in the development of further innovations for the benefit of scientific, industrial, and social communities.

November 2018

Hakim Hacid
Wojciech Cellary
Hua Wang
Hye-Young Paik
Rui Zhou

Organization

General Co-chairs

Zakaria Maamar   Zayed University, UAE
Marek Rusinkiewicz   Florida Gulf Coast University, USA

Program Co-chairs

Hakim Hacid   Zayed University, UAE
Wojciech Cellary   Poznań University of Economics and Business, Poland
Hua Wang   Victoria University, Australia

Workshop Co-chairs

Michael Sheng   Macquarie University, Australia
Tetsuya Yoshida   Nara Women's University, Japan

Tutorial and Panel Chair

Hye-Young Helen Paik University of New South Wales, Australia

Sponsor Chair

Fatma Taher   Zayed University, UAE

Finance Chair

Hakim Hacid   Zayed University, UAE

Local Arrangements Co-chairs

Andrew Leonce   Zayed University, UAE
Huwida Saeed   Zayed University, UAE
Emad Bataineh   Zayed University, UAE

Publication Chair

Rui Zhou   Swinburne University of Technology, Australia

Publicity Co-chairs

Dickson Chiu   The University of Hong Kong, SAR China
Reda Bouadjenek   University of Toronto, Canada
Vanilson Burégio   Federal Rural University of Pernambuco (UFRPE), Recife,
                            Brazil

Website Co-chairs

Emir Ugljanin   State University of Novi Pazar, Serbia
Emerson Bautista   Zayed University, UAE

WISE Steering Committee Representative

Yanchun Zhang   Victoria University, Australia and Fudan University, China

Program Committee

Karl Aberer   EPFL, Switzerland
Marco Aiello   University of Stuttgart, Germany
Mohammed Eunus Ali   Bangladesh University of Engineering and Technology
                            (BUET), Bangladesh
Toshiyuki Amagasa   University of Tsukuba, Japan
Boualem Benatallah   University of New South Wales, Australia
Djamal Benslimane   Lyon 1 University, France
Mohamed Reda   The University of Melbourne, Australia
   Bouadjenek
Athman Bouguettaya   The University of Sydney, Australia
Vanilson Burégio   UFRPE, Brazil
Yi Cai   South China University of Technology, China
Bin Cao   Zhejiang University of Technology, China
Xin Cao   University of New South Wales, Australia
Jinli Cao   Latrobe University, Australia
Barbara Catania   University of Genoa, Italy
Richard Chbeir   University of Pau, France
Cindy Chen   University of Massachusetts Lowell, USA
Lisi Chen   Hong Kong Baptist University, SAR China
Lu Chen   Zhejiang University, China
Jacek Chmielewski   Poznań University of Economics and Business, Poland
Ting Deng   Beihang University, China
Hai Dong   RMIT University, Australia
Schahram Dustar   Vienna University of Technology, Austria
Nora Faci   Lyon 1 University, France
Yunjun Gao   Zhejiang University, China
Dimitrios   Swinburne University of Technology, Australia
   Georgakopoulos

Jarogniew Rykowski	Poznań University of Economics and Business, Poland
Mohamed Sellami	ISEP Paris, France
Shuo Shang	King Abdullah University of Science and Technology, KSA
Wei Shen	Nankai Univcersity, China
Yanyan Shen	Shanghai Jiao Tong University, China
Yulong Shen	Xidian University, China
Abdullatif Shikfa	Qatar University, Qatar
Yain-Whar Si	University of Macau, SAR China
Shaoxu Song	Tsinghua University, China
Weiwei Sun	Fudan University, China
Reima Vesa Suomi	University of Turku, Finland
Stefan Tai	TU Berlin, Germany
Xiaohui Tao	The University of Southern Queensland, Australia
Dimitri Theodoratos	New Jersey Institute of Technology, USA
Yicheng Tu	University of South Florida, USA
Leong Hou U	University of Macau, SAR China
Athena Vakali	Aristotle University of Thessaloniki, Greece
De Wang	Georgia Institute of Technology, USA
Junhu Wang	Griffith University, Australia
Lizhen Wang	Yunnan University, China
Xin Wang	Tianjin University, China
Adam Wójtowicz	Poznań University of Economics and Business, Poland
Mingjun Xiao	University of Science and Technology of China, China
Haoran Xie	The Education University of Hong Kong, SAR China
Hayato Yamana	Waseda University, Japan
Yanfang Ye	West Virginia University, USA
Xun Yi	RMIT University, Australia
Hongzhi Yin	The University of Queensland, Australia
Jianming Yong	University of Southern Queensland, Australia
Tetsuya Yoshida	Nara Women's University, Japan
Ge Yu	Northeastern University, China
Jeffrey Xu Yu	Chinese University of Hong Kong, SAR China
Chao Zhang	University of Illinois at Urbana-Champaign, USA
Detian Zhang	Jiangnan University, China
Ji Zhang	The University of Southern Queensland, Australia
Xiangliang Zhang	King Abdullah University of Science and Technology, KSA
Yanchun Zhang	Victoria University, Australia
Lei Zhao	Soochow University, China
Xiangmin Zhou	RMIT University, Australia
Xingquan Zhu	Florida Atlantic University, USA

Additional Reviewers

Abdallah Lakhdari
Ali Hamdi Fergani Ali
Anastasia Douma
Andrei Kelarev
Anila Butt
Balaji Vasan Srinivasan
Bing Huang
Brian Setz
Bulat Nasrulin
Ch. Md. Rakin Haider
Chen Zhan
Demetris Paschalides
Di Yao
Dimitrios Papamartzivanos
Dinesh Pandey
Elio Mansour
Fan Liu
Filippos Giannakas
Gang Chen
Gang Ren
Hao Wu
Hongxu Chen
Ildar Nurgaliev
Jeff Ansah
Jeffery Ansah
Joe Tekli
Karam Bou Chaaya
Le Sun
Liandeng Su

Lili Sun
Luis Sanchez Giraldo
Marios Anagnostopoulos
Masoud Salehpour
Md Saddam Hossain Mukta
Md Saiful Islam
Md Zahidul Islam
Mohammed Bahutair
Panayiotis Smeros
Qinyong Wang
Saisai Ma
Sha Lu
Shahriar Badsha
Shi Zhi
Shiv Kumar Saini
Siuly Siuly
Sunav Choudhary
Sven Hartmann
Tam Nguyen
Thang Duong
Tugrulcan Elmas
Weiqing Wang
Weiyi Huang
Xinghao Li
Xu Yang
Xuechao Yang
Yingnan Shi
Zacharias Georgiou

Contents – Part II

Web Services and Cloud Computing

Data Stream and Distributed Computing

Data Mining Techniques

Entity Linkage and Semantics

Web Applications

Data Mining Applications

Contents – Part I

Social Network and Security

Social Network

Microblog Data Analysis

Graph Data

Information Extraction

Text Mining

Recommender Systems

SARFM: A Sentiment-Aware Review Feature Mapping Approach for Cross-Domain Recommendation

Yang Xu, Zhaohui Peng, Yupeng Hu, and Xiaoguang Hong[✉]

School of Computer Science and Technology, Shandong University, Jinan,
People's Republic of China
{xuyang0211, pzh, huyupeng, hxg}@sdu.edu.cn

Abstract. Cross-domain algorithms which aim to transfer knowledge available in the source domains to the target domain are gradually becoming more attractive as an effective approach to help improve quality of recommendations and to alleviate the problems of cold-start and data sparsity in recommendation systems. However, existing works on cross-domain algorithm mostly consider ratings, tags and the text information like reviews, and don't take advantage of the sentiments implicated in the reviews efficiently, especially the negative sentiment information which is easy to be weakened during the process of transferring. In this paper, we propose a sentiment-aware review feature mapping framework for cross-domain recommendation, called SARFM. The proposed SARFM framework applies deep learning algorithm SDAE (Stacked Denoising Autoencoders) to model the Sentiment-Aware Review Feature (SARF) of users, and transfers SARF via a multi-layer perceptron to capture the nonlinear mapping function across domains. We evaluate and compare our framework on a set of Amazon datasets. Extensive experiments on each cross-domain recommendation scenarios are conducted to prove the high accuracy of our proposed SARFM framework.

Keywords: Cross-domain recommendation · Sentiment-aware review feature
Stacked denoising autoencoders

1 Introduction

Cross-domain recommendation systems are gradually becoming more attractive as a practical approach to improve quality of recommendations and to alleviate cold-start problem, especially in small and sparse datasets. These algorithms mine knowledge on users and items in a source domain to improve the quality of the recommendations in a target domain. They can also provide joint recommendations for items belonging to different domains by the linking information among these domains [1]. Most existing works about cross domain recommendation tend to aggregate knowledge from different domains from the perspective of explicitly specified common information [2–4] or transferring latent features [5, 8, 9, 13]. However, the aggregated knowledge are merely based on ratings, tags, or the text information like reviews, and don't take advantage of the sentiments implicated in the reviews efficiently. After watching a popular film,

© Springer Nature Switzerland AG 2018
H. Hacid et al. (Eds.): WISE 2018, LNCS 11234, pp. 3–18, 2018.
https://doi.org/10.1007/978-3-030-02925-8_1

using a novel electronic product or playing a video game, users often rate them and submit reviews to share their feelings, which could convey fairly rich sentiment information.

In this paper, we investigate two key issues of review-based cross-domain recommendation:

(1) *How to effectively transfer user's sentiment information?*

For all we know, existing cross-domain recommendation algorithms which utilize user reviews didn't take full advantage of the sentiment information of these reviews. They implement knowledge transfer by mixing positive and negative reviews together, which will weaken and even lose some sentiment information of the users, especially the negative sentiment. For instance, a user may deeply care about the plot of a novel, and he made positive comments on the plots of some novels in the domain of electronic book (source domain) while made negative comments on the plots of some other novels. If we transfer the knowledge gained from user reviews from the source domain to the target domain without distinguishing the sentiment polarity of these reviews, some latent factors such as "plot", "positive sentiment" and "negative sentiment" of the reviews will be mixed up as "users' feature" to be transferred to the target domain. In the domain of movie (target domain), a movie with poor plots, namely the movie whose latent factors "plot" and "negative sentiment" take higher weight, will produce a match with the users' feature transferred to the target domain. Nevertheless, the user may not be fond of this movie.

(2) *Which model should be leveraged to extract sentiment features of reviews, topic model or deep learning model?*

Digging the latent sentiment feature of user reviews is the key of improving the performance of recommendation algorithm. Review-based recommendation algorithm tend to employ the topic model such as LDA (Latent Dirichlet Allocation) [10] to extract latent features. Each latent feature (topic) extracted by LDA is associated with a set of key words. Thus, we can acquire the interpretable recommendation results through matching topic distributions of users and products. On the other hand, deep learning is adept in learning deep latent features automatically. The deep learning model SDAE (Stacked Denoising Autoencoder) [7] is capable of extracting the deep latent feature of review information and owns lower model complexity compared with the most widely used deep learning algorithms, CNN (Convolutional Neural Network) [22] and RNN (Recurrent Neural Network) [23]. However, the latent features extracted by SDAE are not interpretable, which means SDAE based algorithm is hard to provide a natural interpretation for the recommendation result.

In this paper, we give our answers to the above two questions and thus propose a new cross-domain recommendation framework called SARFM. Under SARFM, we can effectively identify the semantic orientation of user reviews and extract the sentiment-aware review feature (SARF). To achieve the goal of transferring knowledge, we propose a multi-layer perceptron based mapping method to transfer sentiment information of users from source domain to target domain. Through transferring SARF of users, the SARFM method gets a superior performance in cross-domain recommendation.

To summarize, the major contributions of this paper are as follows:

- We consider sentiment information in the cross-domain recommendation problem and transfer positive and negative sentiment-aware review feature from source domain to target domain respectively.
- A novel cross-domain recommendation framework was brought forward in this paper, together with multiple implementations of the sentiment-aware review feature extracting component in the framework.
- We systematically compare our methods with other algorithms on the Amazon dataset. The results confirm that our methods substantially improve the performance of cross-domain recommendation.

The rest of this paper is organized as follows. Section 2 reviews the related works and Sect. 3 presents some notations and the problem formulation. Section 4 introduce the modeling method of SARF. Section 5 details the mapping method of SARF and the cross-domain recommendation approach. Experiments and discussion are given in Sect. 6 and conclusions are drawn in Sect. 7.

2 Related Work

Existing works about cross-domain algorithm mostly extract domain-specific information from ratings [5], tags [2] and the text information like reviews [16]. Ren [8] proposed the PCLF model to learn the shared common rating pattern across multiple rating matrices and the domain-specific rating patterns from each domain. Fang [2] exploited the rating patterns across multiple domains by transferring the tag co-occurrence matrix information. Xin [16] exploited review text by learning a non-linear mapping on users' preferences on different topics across domains. However, these approaches don't take advantage of the sentiments implicated in the reviews efficiently.

Sentiment analysis is widely used in recommendation systems. Computing the semantic orientation of a user review has been studied by several researchers. Diao [17] built a language model component in their proposed JMARS model to capture aspects and sentiments hidden in reviews. Zhang [18] extracted explicit product features and user opinions by phrase-level sentiment analysis on user reviews to generate explainable recommendation results. Li [19] proposed a SUIT model to simultaneously utilize the textual topic and user item factors for sentiment analysis. But all these algorithms don't employ sentiment analysis on cross-domain recommendation task.

Deep learning is usually applied on image recognition and natural language processing. However, in recent years, the attempt to apply deep learning on recommendation system has widely emerged. Wang [25] proposed a hierarchical Bayesian model that uses a deep learning model to obtain content features and a traditional CF model to address the rating information. He [12] proposed a deep learning-based recommendation framework in which users and items are represented via the one-hot encoding of their ID. In this paper, we employ deep learning model SDAE to extract sentiment-aware review feature and mapping it to target domain for the sake of making cross-domain recommendations.

3 Preliminaries

In this section, we first introduce some frequently used notations. Then, we formulate the cold-start cross-domain recommendation problem and present the SARFM framework to solve the problem.

3.1 Notations

Objects to be recommended in the cross-domain recommendation system are referred to as *items*. Let $\mathcal{U} = \{u_1, u_2, \ldots, u_{|\mathcal{U}|}\}$ denote the set of common users in both domains and $\mathcal{I}_S = \{i_1, i_2, \ldots, i_{|\mathcal{I}_S|}\}$, $\mathcal{I}_T = \{\iota_1, \iota_2, \ldots, \iota_{|\mathcal{I}_T|}\}$ are the sets of items (e.g. movies, books, or electronics) in source domain and in target domain respectively. The user review dataset is represented as $R_{S,U} = \{r_{u_1}, r_{u_2}, \ldots, r_{u_{|\mathcal{U}|}}\}$ in source domain and $R_{T,U}$ in target domain, where r_{u_i} is all of the reviews of user u_i in the corresponding domain. Similarly, we let $R_{T,I} = \{r_{\iota_1}, r_{\iota_2}, \ldots, r_{\iota_{|\mathcal{I}_T|}}\}$ denote the item review dataset in target domain, where r_{ι_j} is all of the reviews which item ι_j acquired in target domain.

In the SARFM, the sentiment analysis algorithm is employed on the review datasets mentioned above to divide them into corresponding positive review datasets (e.g. $R_{S,U}^{pos}$, $R_{T,U}^{pos}$ and $R_{T,I}^{pos}$) and negative review datasets (e.g. $R_{S,U}^{neg}$, $R_{T,U}^{neg}$ and $R_{T,I}^{neg}$). $F_{S,U}^{pos} = \left(f_{S,u_1}^{pos}, f_{S,u_2}^{pos}, \ldots, f_{S,u_{|U|}}^{pos}\right)^T$ represents the positive review feature matrix of common users in source domain, while $F_{T,U}^{neg} = \left(f_{T,u_1}^{neg}, f_{T,u_2}^{neg}, \ldots, f_{T,u_{|U|}}^{neg}\right)^T$ denotes the negative review feature matrix of common users in the target domain, which is similar to $F_{S,U}^{neg}$ and $F_{T,U}^{pos}$. In addition, $F_I^{pos} = \left\{f_{\iota_1}^{pos}, f_{\iota_2}^{pos}, \ldots, f_{\iota_{|I|}}^{pos}\right\}$ denotes the positive review feature matrix of all items in target domain, while $F_I^{neg} = \left\{f_{\iota_1}^{neg}, f_{\iota_2}^{neg}, \ldots, f_{\iota_{|I|}}^{neg}\right\}$ denotes the negative review feature matrix of all items in target domain. For a cold-start user u' in target domain, $f_{S,u'}^{pos} \left(f_{S,u'}^{neg}\right)$ denotes the positive (negative) review feature of user u' in source domain, and $\hat{f}_{T,u'}^{pos} \left(\hat{f}_{T,u'}^{neg}\right)$ represents the affine positive (negative) review feature of user u' in target domain.

3.2 Problem Formulation

Given two domains which share the same users U. Users appearing in only one domain can be regarded as the cold-start users U' in the other domain. Without loss of generality, one domain is referred to as the source domain and the other as the target domain. The most common cross-domain recommendation approaches focus on transferring information based on ratings, tags and reviews from source domain to target domain, without accounting for sentiment information implicated the reviews sufficiently.

We are tackling cross-domain recommendation task for cold-start users by modeling the Sentiment-Aware Review Feature (SARF) of users and transferring them from source domain to target domain. To achieve this purpose, we propose a cross-domain recommendation framework called SARFM. This framework contains three major steps, i.e., sentiment-aware review feature extracting, cross-domain mapping and cross-domain recommendation, as illustrated in Fig. 1.

In the first step, we apply a SO-CAL [24] based sentiment analysis algorithm to analyze the Semantic Orientation (SO) of each sentence of user reviews in both domains. Then, the original review datasets of both domains are divided into corresponding positive review datasets and negative review datasets respectively. Next, we employ the deep learning model SDAE (Stacked Denoising Autoencoder) on the sentiment tagged datasets to extract the sentiment-aware review feature of users. In addition, we designed an LDA [10] based method to extract SARF of users, which is used to compare the effectiveness with SDAE. In the second step, we model the cross-domain relationships of users through a mapping function based on Multi-Layer Perceptron (MLP) [6]. We assume that there is an underlying mapping relationship between the user's SARFs of the source and target domains, and further use a mapping function to capture this relationship. Finally, in the third step, we make recommendation for cold-start user in the target domain. We can get an affine SARF for cold-start user in the target domain, with the SARF learned for him/her in the source domain and the MLP-based mapping functions. In the rest of this paper, we will introduce each step of SARFM in details.

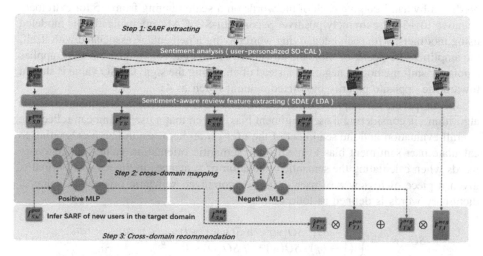

Fig. 1. Overview of the SARFM cross-domain recommendation framework

4　Sentiment-Aware Review Feature Extracting

As discussion in the previous section, in order to transfer sentiment-aware review pattern in the source domains to the target domain, the first phase of SARFM is to model the sentiment-aware review feature of common users in both domains. The key challenge is how to extract the user's focus on the item and the positive or negative emotions expressing in the user reviews. To address this problem, we propose a sentiment-aware review feature extracting method based on sentence-level sentiment analysis approach and SDAE (or LDA).

4.1　Sentiment Analysis

The sentiment analysis problem in SARFM can be formulated as follows: Given a set of reviews R, a sentiment classification algorithm classifies each sentence of review $r \in R$ into one of the two sets, positive set R^{pos} and negative set R^{neg}. Each entry of R^{pos} and R^{neg} is an (user/item, {sentence}, SO value) triplet, where {sentence} is a set of sentences with clear semantic orientation ("positive" in R^{pos} or "negative" in R^{neg}) extracted from the raw review data of the user/item, and SO value is the average of semantic orientation values of sentences in {sentence}. For this purpose, we analyze the semantic orientation of each sentence of user reviews based on the state-of-the-art sentiment analysis algorithm SO-CAL described in [24].

The Semantic Orientation CALculator (SO-CAL) is a lexicon-based sentiment analysis approach which mainly consists of three characteristics. First, SO-CAL uses four kinds of manual basic dictionary (adjective, noun, verb, and adverb), which were produced by hand-tagging each of the words on a scale ranging from -5 for extremely negative to $+5$ for extremely positive. Second, in SO-CAL, intensification is modeled using modifiers, with each intensifying word having a percentage associated with itself, the amplifiers are positive, while the downtoners are negative. Finally, SO-CAL applies a polarity shift method of negation, instead of changing the sign, the SO value is shifted toward the opposite polarity by a fixed amount (such as 4).

In this paper, we improve the SO-CAL to a user-personalized sentiment analysis algorithm via considering a user sentiment bias. Given that a user rating can reflect the overall evaluation and the sentiment of the user on an item, we utilize user's rating to calculate user sentiment bias to modify the semantic orientation values of dictionary words when calculating the semantic orientation of sentences in the review. Formally, given a piece of review r of user u, user-personalized semantic orientation value of dictionary words is defined as follows:

$$SO_u(w) = \begin{cases} (1 + b_u) \cdot SO(w) & if\ SO(w) > 0 \\ (1 - b_u) \cdot SO(w) & if\ SO(w) < 0 \end{cases}, b_{u,r} = \frac{rat_{u,r} - \overline{rat}_u}{\overline{rat}_u} \tag{1}$$

where $SO_u(w)$ is user-personalized semantic orientation value of dictionary word w for user u and $SO(w)$ represents the initial semantic orientation value of w. $b_{u,r}$ denotes the user sentiment bias for review r, $rat_{u,r}$ is the corresponding rating of review r and the average rating of user u is represented as $\overline{rat}_u$.

Since the SO calculation procedure in SO-CAL is not the focus of this paper, we refer the readers to the related literature [24] for more details. The effect of sentiment analysis on SARFM will be reported later in the experimental section.

We employ user-personalized SO-CAL method on original review sets $R_{S,U}, R_{T,U}, R_{T,I}$ to divide them into positive review subsets $R_{S,U}^{pos}, R_{T,U}^{pos}, R_{T,I}^{pos}$ and negative review subsets $R_{S,U}^{neg}, R_{T,U}^{neg}, R_{T,I}^{neg}$ on the sentence-level respectively.

4.2 Sentiment-Aware Review Feature Extracting

Sentiment-Aware Review Feature (SARF) indicates the user's focus on the item and the positive or negative emotions expressed in the reviews. In this paper, we apply the deep learning model SDAE to extract SARF. In addition, the LDA based extracting method is also described in this paper which is designed as another implementation of our recommendation framework.

Stacked Denoising Autoencoder. DAE, namely denoising autoencoder, which is shown in Fig. 2 (a), consists of an encoder and a decoder. The initial input x is converted into its corrupted version x' by means of a stochastic mapping $x' \sim qD(x'|x)$. The encoder $e(\cdot)$ takes the given corrupted input x' and maps it to its hidden representation $e(x')$, while the decoder $d(\cdot)$ maps this hidden representation back to a reconstructed version of $d(e(x'))$. Then, a denoising autoencoder is trained to reconstruct the original input x by minimizing loss $\mathcal{L}(x, d(e(x')))$. Nevertheless, existing literatures have shown that multiple layers stacked together can generate rich representations in hidden layers [20, 21]. SDAE [7] is a feedforward neural network which stacks multiple DAEs in order to form a deep learning framework, as shown in Fig. 2 (b). Each layer's output in SDAE is the input of its next layer. SDAE utilizes greedy layer-wise training strategy to train each layer in the network successively, and then pre-train the whole deep learning network. Z_c is the clean input matrix and Z_o represents the noise-corrupted input matrix. Z_L is the middle layer of the model. Let λ represents regular parameter and $\|\cdot\|_F$ denotes Frobenius norm. Then the training model of SDAE is formulated as follow:

$$\min_{\{w_l\},\{b_l\}} \|Z_c - Z_L\|_F^2 + \lambda \sum_l \left(\|w_l\|_F^2 + \|b_l\|_F^2 \right) \qquad (2)$$

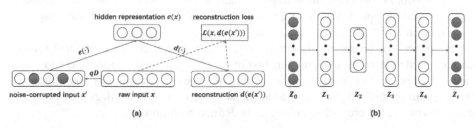

Fig. 2. (a) The architecture of DAE (b) A 2-layer SDAE with $L = 4$

SARF Extracting. In our framework, we need to learn positive (negative) review feature extraction model $SDAE_S^{pos}$ and $SDAE_T^{pos}$ ($SDAE_S^{neg}$ and $SDAE_T^{neg}$) in both domains respectively. Then we will introduce the learning method of $SDAE_S^{pos}$, that is similar to $SDAE_T^{pos}$, $SDAE_S^{neg}$ and $SDAE_T^{neg}$.

As described in Sect. 3.1, each entry of $R_{S,U}^{pos}$ is an (user, {sentence}, SO value) triplet. In our formulation, each {sentence} is considered as a "document" of the user and "corpus" is defined as the collection of {sentence}s in $R_{S,U}^{pos}$. We employ TF-IDF method to construct the "document-word" matrix $X_{S,U}^{pos}$ as the clean input of the $SDAE_S^{pos}$ model and generate noise-corrupted input $X_{S,U}^{\prime pos}$ by choosing 5 percent of matrix values randomly and setting them to 0, while the others are left unmodified. In the $SDAE_S^{pos}$ model, given the total number of layers is L, the positive review feature vector is generated from the $L/2$ layer, which is denoted as $f_{S,u}$ for user u in source domain.

Similarly, $SDAE_T^{pos}$, $SDAE_S^{neg}$ and $SDAE_T^{neg}$ are trained on sentiment classified review datasets $R_{T,U}^{pos}$, $R_{S,U}^{neg}$ and $R_{T,U}^{neg}$ respectively. Then we can get the positive and negative review feature matrix in both domains. In this paper, the SARF represents as a set of positive and negative review feature vectors, such as $SARF_{S,u} = \left(f_{S,u}^{pos}, f_{S,u}^{neg} \right)$, which is the sentiment-aware review feature of user u in source domain.

In addition, LDA can also be used to extract SARF under the above formulation. In the LDA version of our framework, we employ LDA on each sentiment classified review dataset shown in Fig. 1. Then we can get the positive and the negative topic distributions of each "document" which can be represented as SARF in both domains and the per-topic word distributions $\beta_{T_U}^{pos}$, $\beta_{T_U}^{neg}$, $\beta_{T_I}^{pos}$ and $\beta_{T_I}^{neg}$ in target domain.

5   Sentiment-Aware Review Feature Mapping and Cross-Domain Recommendation

5.1   Sentiment-Aware Review Feature Mapping

In this paper, we utilize an MLP-based method to tackle the SARF mapping problem, as shown in Fig. 1. To avoid mutual interference between positive and negative review features during the process of knowledge transfer, two MLP models, the positive MLP model and the negative MLP model, were employed to map the two parts of $SARF = (f^{pos}, f^{neg})$ from source domain to target domain respectively. Next, we will introduce the proposed mapping algorithm under f^{pos} mapping scenario, and the mapping algorithm under f^{neg} mapping scenario is similar.

In our proposed mapping algorithm, only the common users with sufficient review data are used to learn the mapping function in order to guarantee its robustness to noise caused by review data sparsity and imbalance of review distribution in both domains. We use entropy and statistical method to measure the cross-domain degree of common users. Formally, the cross-domain degree is defined as follows:

$$c(u) = \left(-q_{u,s} \log_2 q_{u,s} - q_{u,t} \log_2 q_{u,t}\right) \frac{\mathcal{N}(u,s) + \mathcal{N}(u,t)}{\sum_{u_i \in U_c} \mathcal{N}(u_i, s) + \mathcal{N}(u_i, t)}$$

$$where \quad q_{u,s} = \frac{\mathcal{N}(u,s)}{\mathcal{N}(u,s) + \mathcal{N}(u,t)}, \quad q_{u,t} = \frac{\mathcal{N}(u,t)}{\mathcal{N}(u,s) + \mathcal{N}(u,t)} \tag{3}$$

$\mathcal{N}(u,s)$ is the number of reviews in source domain of common user u, and $\mathcal{N}(u,t)$ is that in target domain. U_c denotes the set of common users between both domains. Intuitively, the previous part of $c(u)$ is conditional entropy which indicates the distribution of the interactions of the user u in both domains. The latter part of $c(u)$ measures the proportion of u's reviews in both domain. The common users with $c(u) > threshold\gamma$ are selected to learn the mapping function.

Let $\theta^S = \{\theta_1^S, \theta_2^S, \ldots, \theta_N^S\}$ denotes the set of f^{pos} s in the source domain, and $\theta^T = \{\theta_1^T, \theta_2^T, \ldots, \theta_N^T\}$ represents the set of f^{pos} s in the target domain. N is the number of common users in both domains. Under the MLP model setting, we formulate the f^{pos} mapping problem as: Given N training instance (θ_i^S, θ_i^T), $\theta_i^S, \theta_i^T \in R^M$, $(i = 1, 2, \ldots, N)$, where $\theta_i^S = (\theta_{i1}^S, \theta_{i2}^S, \ldots, \theta_{iM}^S)$ is the f^{pos} of common user u_i in the source domain and $\theta_i^T = (\theta_{i1}^T, \theta_{i2}^T, \ldots, \theta_{iM}^T)$ is that in the target domain, our task is to learn an MLP mapping function to map the f^{pos} from the source domain to the target domain.

In a feedforward MLP model, the output o_{ik} is formulated as

$$y_{ik} = \sum_{j=1}^{L'} c_{jk} a_j, \quad o_{ik} = g(y_{ik}) \tag{4}$$

where c_{jk} represents the weight of the j'th input of the output layer neuron k and L' is the number of hidden neurons in each hidden layer. $g(y)$ is the activation function of the output layer, which is set to be the softmax function in this study. a_j denotes the j'th hidden neuron activation of lower hidden layer, which can be defined as

$$y_j = \sum_{p=1}^{P} w_{pj} a_p, \quad a_j = f(y_j) \tag{5}$$

where w_{pj} is the weight of the p'th input of the hidden layer neuron j (the hidden bias can be included in the input weights) and a_p is the input θ_{ip}^S or the p'th hidden neuron activation of the lower hidden layer. P represents the number of inputs or neurons in the lower layer. $f(y)$ is the hidden layer activation function, which is set to be the ReLU function in this study. Considering that input and output of MLP model in this study are feature vectors or topic distributions, the error between θ_i^T and $o_i = (o_{i1}, o_{i2}, \ldots, o_{iM})$ is measured by Kullback-Leibler divergence:

$$E = \sum_{i=1}^{N} \sum_{k=1}^{M} o_{ik} \log \frac{o_{ik}}{\theta_{ik}^{T}} \tag{6}$$

To obtain the MLP mapping function, we utilize stochastic gradient descent to learn the parameters. We refresh the parameters of the MLP by looping through the training instances. The back-propagation algorithm is adopted to calculate the gradients of the parameters. The iterative process is stopped until the model converges, thus we can get the positive MLP mapping function $\mathcal{F}_{pos}(\cdot; \Theta_p)$, where Θ_p is its weights set. Similarly, we can learn the negative MLP mapping function $\mathcal{F}_{neg}(\cdot; \Theta_n)$.

5.2   Cross-Domain Recommendation

For a cold-start user in target domain, we do not have sufficient information to estimate his/her preference directly in target domain. However, we can get the affine SARF for him/her in target domain, with the SARF learned in source domain and the learned MLPs. In this section, we will introduce how to predict the cold-start user's preference on the items by using SARF in target domain.

Given a cold-start user u' in the target domain, we can extract user u''s $SARF_{S,u'} = \left(f_{S,u'}^{pos}, f_{S,u'}^{neg} \right)$ from $R_{S,U}^{pos}$ and $R_{S,U}^{neg}$ in the source domain. Then the affine $\widehat{SARF}_{T,u'} = \left(\hat{f}_{T,u'}^{pos}, \hat{f}_{T,u'}^{neg} \right)$ can be obtained by the following equations:

$$\hat{f}_{T,u'}^{pos} = \mathcal{F}_{pos}\left(f_{S,u'}^{pos}; \Theta_p \right), \quad \hat{f}_{T,u'}^{neg} = \mathcal{F}_{neg}\left(f_{S,u'}^{neg}; \Theta_n \right) \tag{7}$$

Since $R_{T,U}^{pos}$ and $R_{T,I}^{pos}$ share the same review data in target domain, we can extract positive review feature matrix of the items in target domain F_I^{pos} by applying the learned $SDAE_T^{pos}$ on $R_{T,I}^{pos}$. Similarly, we can get F_I^{neg} by employing the learned $SDAE_T^{neg}$ on $R_{T,I}^{neg}$. Then the predicted emotional preference of cold-start user u' to item $\imath_j \in \mathcal{I}_T$ in the target domain can be calculated as:

$$e_{u',\imath_j} = \sigma\left(OS_{S,u'}^{pos} \cdot \hat{f}_{T,u'}^{pos} \cdot f_{\imath_j}^{pos\,T} + OS_{T,\imath_j}^{neg} \cdot \hat{f}_{T,u'}^{neg} \cdot f_{\imath_j}^{neg\,T} \right) \tag{8}$$

where $OS_{S,u'}^{pos}$ and $OS_{T,\imath_j}^{neg}$ are the average OS values of "document" described in Sect. 3.1, and $\sigma(\cdot)$ is the sigmoid function.

For the LDA version of our framework, we calculate the similarity matrixes of latent topic spaces of user review and item review in the target domain, which are defined as:

$$SIM^{pos} = \left\{ sim_{i,j}^{pos} \right\}, \; where \; sim_{i,j}^{pos} = \cos\left(\beta_{T_{U,i}}^{pos}, \beta_{T_{I,j}}^{pos} \right), \; i, j = 1, 2, \ldots, M \tag{9}$$

$$SIM^{neg} = \left\{ sim_{i,j}^{neg} \right\}, \; where \; sim_{i,j}^{neg} = \cos\left(\beta_{T_{U,i}}^{neg}, \beta_{T_{I,j}}^{neg} \right), \; i, j = 1, 2, \ldots, M \tag{10}$$

where $\beta_{T_{U,i}}$ $\left(\beta_{T_{I,i}}\right)$ represents the topic-word distribution of topic i, which is in the latent topic space of user (item) review in the target domain. $\cos(\cdot)$ denotes the cosine similarity of two topic-word distributions.

Then the predicted emotional preference of cold-start user u' to item $\imath_j \in \mathcal{I}_T$ in the target domain is calculated as:

$$e_{u',\imath_j} = \sigma\left(OS_{S,u'}^{pos} \cdot \hat{f}_{T,u'}^{pos} \cdot SIM^{pos} \cdot f_{\imath_j}^{pos\,T} + OS_{T,\imath_j}^{neg} \cdot \hat{f}_{T,u'}^{neg} \cdot SIM^{neg} \cdot f_{\imath_j}^{pos\,T}\right) \quad (11)$$

Finally, the overall preference of cold-start user u' to item $\imath_j$ can be represented as a linear combination of a cross-domain average-value method and an emotional preference component as:

$$S(u', \imath_j) = b_{\mathcal{I}_T} + b_{u'} + b_{\imath_j} + e_{u',\imath_j} \quad (12)$$

where $b_{\mathcal{I}_T}$ denotes the overall average ratings of all items in the target domain. The parameter $b_{u'}$ is the user rating bias in the source domain and $b_{\imath_j}$ is the item rating bias in the target domain.

6 Experiments

6.1 Experimental Settings

Data Description. We employ Amazon cross-domain dataset [11] in our experiment. This dataset contains product reviews and star ratings with 5-star scale from Amazon, including 142.8 million reviews spanning May 1996 - July 2014. We select the top three domains with the most widely used in previous studies to employ in our cross-domain experiment. The global statistics of these domains used in our experiments are shown in Table 1.

Table 1. Characteristics of datasets

Dataset	Books	Electronics	Movies & TV
# of Users	603,668	192,403	123,960
# of Items	367,982	63,001	50,052
# of Reviews	8,898,041	1,689,188	1,697,533
Density	0.004%	0.014%	0.027%

Experiment Setup. The domains in the Amazon dataset only have user overlaps. Thus, we evaluate the validity and efficiency of SARFM on the cross-domain recommendation task under the user overlap scenario. We randomly remove all the rating information of a certain proportion of users in the target domain and take them as cross-domain cold-start users for making recommendation. For the sake of stringency of the experiments, we set different proportions of cold-start users, namely, $\phi = 30\%, 50\%$ and 70%. Moreover, we repeatedly sample users for 10 times to generate different sets to balance the effect of different sets of cold-start users on the final recommendation results and report the average results. Dimension of latent feature used in SDAE and the number of topics in LDA are set as $M = 20, 50$ and 100. For the mapping function, we set the structure of MLP as three hidden layers with $2M$ nodes in each hidden layer.

Compared Methods. We examine the performance of the proposed SARFM framework by comparing it with the following baseline methods:

- SD_AVG: SD_AVG is a single-domain average-value method, which predicts overall preference by global average rating.
- CD_AVG: CD_AVG is a cross-domain average-value method described in Sect. 4.2. It predicts overall preference by the following equation: $r_{ui} = b_T + b_u + b_i$ where b_T is the overall average ratings of all items in the target domain, b_u denotes the user rating bias in source domain and b_i represents item bias in target domain.
- CMF: This is a cross-domain recommendation method proposed in [14]. In CMF, the latent factors of users are shared between source domain and target domain.
- MF-MLP: This is a cross-domain recommendation framework based on MF and MLP, which is proposed by [15]. In our experiments, the structure of the MLP is set as three-hidden layer, and the number of nodes in the hidden layer is set as $2M$.
- SARFM_SDAE, SARFM_LDA: These are two different implementations of our framework proposed in this paper.
- RFM_SADE, RFM_LDA: These are two cut versions of our framework without sentiment analysis.

Evaluation Metric
We adopt the metrics of Root Mean Square Error (RMSE) and Mean Absolute Error (MAE) to evaluate our method. They are defined as:

$$RMSE = \sqrt{\sum_{r_{ui} \in \mathcal{O}_{test}} \frac{(\hat{r}_{ui} - r_{ui})^2}{|\mathcal{O}_{test}|}}, \quad MAE = \frac{1}{|\mathcal{O}_{test}|} \sum_{r_{ui} \in \mathcal{O}_{test}} |\hat{r}_{ui} - r_{ui}| \tag{13}$$

where $\mathcal{O}_{test}$ is the set of test ratings, r_{ui} denotes an observed rating in the test set, and $\hat{r}_{ui}$ represents the predictive value of r_{ui}. $|\mathcal{O}_{test}|$ is the number of test ratings.

6.2   Performance Comparison

Recommendation Performance. Experimental results of MAE and RMSE on the two pair of domains "Books-Movies & TV" and "Electronics-Movies & TV" are presented in Tables 2, 3, 4, and 5, respectively. The domain "Books" and "Electronics" are chosen as the target domain because they are much sparser compared with "Movies & TV".

Table 2. Recommendation performance in terms of MAE on the "Books-Movies & TV"

DIM	ϕ	SD_AVG	CD_AVE	CMF	MF_MLP	SARMF_LDA	SARFM_SDAE
K = 20	30%	0.8767	0.8655	0.8616	0.8446	0.8161	0.8155
	50%	0.8948	0.8663	0.8904	0.8459	0.8382	0.8254
	70%	0.9150	0.8763	0.9101	0.8809	0.8637	0.8485
K = 50	30%	0.8767	0.8655	0.8441	0.8442	0.8338	0.8108
	50%	0.8948	0.8663	0.8843	0.8530	0.8459	0.8252
	70%	0.9150	0.8763	0.8881	0.8790	0.8636	0.8270
K = 100	30%	0.8767	0.8655	0.8254	0.8440	0.8156	0.8111
	50%	0.8948	0.8663	0.8523	0.8687	0.8459	0.8254
	70%	0.9150	0.8763	0.8990	0.8940	0.8635	0.8278

Table 3. Recommendation performance in terms of RMSE on the "Books-Movies & TV"

DIM	ϕ	SD_AVG	CD_AVE	CMF	MF_MLP	SARMF_LDA	SARFM_SDAE
K = 20	30%	1.1440	1.1177	1.0962	1.0885	1.0849	1.0419
	50%	1.1645	1.1400	1.1267	1.1074	1.0894	1.0476
	70%	1.1735	1.1654	1.1361	1.1531	1.1141	1.0585
K = 50	30%	1.1440	1.1177	1.0858	1.0849	1.0849	1.0369
	50%	1.1645	1.1400	1.1275	1.1253	1.0945	1.0474
	70%	1.1735	1.1654	1.1560	1.1621	1.1141	1.0570
K = 100	30%	1.1440	1.1177	1.1296	1.1238	1.0756	1.0420
	50%	1.1645	1.1400	1.1513	1.1506	1.0847	1.0476
	70%	1.1735	1.1654	1.1607	1.1571	1.1076	1.0679

Table 4. Recommendation performance in terms of MAE on the "Electronics-Movies & TV"

DIM	ϕ	SD_AVG	CD_AVE	CMF	MF_MLP	SARMF_LDA	SARFM_SDAE
K = 20	30%	0.9459	0.8626	0.8594	0.8584	0.8509	0.8356
	50%	0.9529	0.9064	0.8960	0.8864	0.8637	0.8619
	70%	0.9733	0.9201	0.9175	0.9023	0.8660	0.8640
K = 50	30%	0.9459	0.8626	0.8599	0.8605	0.8504	0.8346
	50%	0.9529	0.9064	0.8829	0.8820	0.8630	0.8610
	70%	0.9733	0.9201	0.9005	0.8945	0.8658	0.8639
K = 100	30%	0.9459	0.8626	0.8509	0.8625	0.8500	0.8347
	50%	0.9529	0.9064	0.8840	0.9012	0.8623	0.8610
	70%	0.9733	0.9201	0.9060	0.9119	0.8654	0.8642

We respectively evaluate all the methods under different K and ϕ in both cross-domain scenarios. From these tables, we can see that the proposed SARFM_SDAE outperforms all baseline models in terms of both MAE and RMSE metrics. With the proportion of cold-start users becoming higher, the performance of single domain

Table 5. Recommendation performance in terms of RMSE on the "Electronics-Movies & TV"

DIM	ϕ	SD_AVG	CD_AVE	CMF	MF_MLP	SARMF_LDA	SARFM_SDAE
K = 20	30%	1.2408	1.1603	1.1425	1.1437	1.1391	1.1237
	50%	1.2654	1.2022	1.1794	1.1564	1.1536	1.1495
	70%	1.2948	1.2169	1.1833	1.1781	1.1659	1.1533
K = 50	30%	1.2408	1.1603	1.1601	1.1572	1.1392	1.1234
	50%	1.2654	1.2022	1.1835	1.1807	1.1533	1.1490
	70%	1.2948	1.2169	1.1731	1.2068	1.1656	1.1528
K = 100	30%	1.2408	1.1603	1.1401	1.1592	1.1390	1.1234
	50%	1.2654	1.2022	1.1708	1.1914	1.1532	1.1494
	70%	1.2948	1.2169	1.1838	1.2159	1.1656	1.1529

method SD_AVG will become progressively worse while the cross-domain methods keep satisfactory results, which shows the effectiveness of knowledge transfer. Compared with CMF and MF_MLP, our method gets an improvement of 5% to 10% both in RMSE and MAE. These results demonstrate that our framework is more suitable for making recommendations to cold-start users compared to other cross-domain baseline methods, especially in the dataset with high sparsity. Moreover, SARFM_SDAE performs better than CD_AVG especially in higher ϕ, which demonstrates that the SARF transferred from the source domain is highly effective. For the proposed SARFM_SDAE and SARFM_LDA methods, the optimal values of K are nearly 50 and 100 respectively. In the following experiments, K are set to the optimal values by preliminary tests.

Effect of Semantic Information. We removed the sentiment analysis component from our framework to get two incomplete versions of RFM_SDAE and RFM_LDA, which extract review features from raw review dataset directly. As shown in Fig. 3, both SARFM_SDAE and SARFM_LDA perform much better than RFM_SDAE and RFM_LDA under different ϕ in both cross-domain scenarios, which verifies the effectiveness of our method in the aspect of utilizing semantic information.

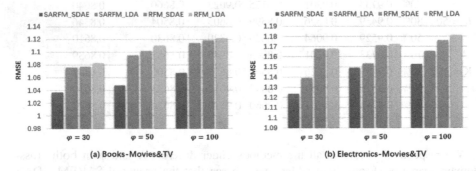

Fig. 3. Effect of semantic information on recommendation performance

Topic Model vs. Deep learning Model. We designed two implementations of our framework in this paper, SARFM_SDAE and SARFM_LDA. For the SARFM_LDA, the interpretable recommendation results can be gotten by matching topic distributions β_{T_i} and β_{T_U} in the target domain. While the SARFM_SDAE is hard to provide a natural interpretation for the recommendation result because of the latent feature extraction. However, as shown in Fig. 3, SARFM_SDAE outperforms SARFM_LDA under each of configurations, which verifies its effectiveness on extracting the deep latent feature of sentiment information.

7 Conclusions

The user reviews contain plenty of semantic information. We proposed a novel framework for cross domain recommendation that establishes linkages between the source and target domains by using sentiment-aware review features of users. In this paper, a user-personalized sentiment analysis algorithm is designed to analyze the Semantic Orientation of each review in both domains. We employed SDAE and MLP based mapping method to model user's SARF and mapped it to the target domain to make recommendations for cold-start users. In different scenarios, experiments convincingly demonstrate that the proposed SARFM framework can significantly improve the quality of cross-domain recommendation and SARFs extracted from reviews are important links between each domain.

Acknowledgments. This work is supported by NSF of Shandong, China (Nos. ZR2017MF065, ZR2018MF014), the Science and Technology Development Plan Project of Shandong, China (No. 2016GGX101034).

References

1. Cremonesi, P., Tripodi, A., Turrin, R.: Cross-domain recommender systems. In: ICDM 2012, pp. 496–503 (2012)
2. Fang, Z., Gao, S., Li, B., Li, J.: Cross-domain recommendation via tag matrix transfer. In: ICDM 2015, pp. 1235–1240 (2015)
3. Chen, W., Hsu, W., Lee, M.L.: Making recommendations from multiple domains. In: KDD 2013, pp. 892–900 (2013)
4. Yang, D., He, J., Qin, H., Xiao, Y., Wang, W.: A graph-based recommendation across heterogeneous domains. In: CIKM 2015, pp. 463–472 (2015)
5. Li, B., Yang, Q., Xue, X.: Transfer learning for collaborative filtering via a rating-matrix generative model. In: ICML 2009, pp. 617–624 (2009)
6. Ruck, D.W., Rogers, S.K., Kabrisky, M., Oxley, M.E., Suter, B.W.: The multilayer perceptron as an approximation to a Bayes optimal discriminant function. IEEE Trans. Neural Netw. 1(4), 296–298 (1990)
7. Vincent, P., Larochelle, H., Lajoie, I., et al.: Stacked denoising autoencoders: learning useful representations in a deep network with a local denoising criterion. J. Mach. Learn. Res. 11(12), 3371–3408 (2010)

8. Ren, S., Gao, S., Liao, J., Guo, J.: Improving cross-domain recommendation through probabilistic cluster-level latent factor model. In: AAAI 2015, pp. 4200–4201 (2015)
9. Gao, S., Luo, H., Chen, D., Li, S., Gallinari, P., Guo, J.: Cross-domain recommendation via cluster-level latent factor model. In: Blockeel, H., Kersting, K., Nijssen, S., Železný, F. (eds.) ECML PKDD 2013. LNCS (LNAI), vol. 8189, pp. 161–176. Springer, Heidelberg (2013). https://doi.org/10.1007/978-3-642-40991-2_11
10. Blei, D.M., Ng, A.Y., Jordan, M.I.: Latent Dirichlet allocation. J. Mach. Learn. Res. **3**, 993–1022 (2003)
11. He, R., Mcauley, J.: Ups and downs: modeling the visual evolution of fashion trends with one-class collaborative filtering. In: WWW 2016, pp. 507–517 (2016)
12. He, X., Liao, et al.: Neural collaborative filtering. In: WWW 2017. pp. 173–182 (2017)
13. Pan, W., Liu, N.N., Xiang, E.W., Yang, Q.: Transfer learning to predict missing ratings via heterogeneous user feedbacks. In: AAAI 2011, pp. 2318–2323 (2011)
14. Singh, A.P., Kumar, G., Gupta, R.: Relational learning via collective matrix factorization. In: KDD 2008, pp. 650–658 (2008)
15. Man, T., Shen, H., Jin, X., Cheng, X.: Cross-domain recommendation: an embedding and mapping approach. In: IJCAI 2017, pp. 2464–2470 (2017)
16. Xin, X., Liu, Z., Lin, C., Huang, H., Wei, X., Guo, P.: Cross-domain collaborative filtering with review text. In: IJCAI 2015, pp. 1827–1833 (2015)
17. Diao, Q., Qiu, M., Wu, C.Y., Smola, A.J., et al.: Jointly modeling aspects, ratings and sentiments for movie recommendation (JMARS). In: KDD 2014, pp. 193–202 (2014)
18. Zhang, Y., Lai, G., Zhang, M., et al.: Explicit factor models for explainable recommendation based on phrase-level sentiment analysis. In: SIGIR 2014, pp. 83–92 (2014)
19. Li, F., Wang, S., Liu, S., Zhang, M.: SUIT: a supervised user-item based topic model for sentiment analysis. In: AAAI 2014, pp. 1636–1642 (2014)
20. Rifai, S., Vincent, P., Muller, X., Glorot, X., Bengio, Y.: Contractive auto-encoders: explicit invariance during feature extraction. In: ICML 2011, pp. 833–840 (2011)
21. Chen, M., Xu, Z., Weinberger, K., Sha, F.: Marginalized denoising autoencoders for domain adaptation. Computer Science (2012)
22. Krizhevsky, A., Sutskever, I., Hinton, G.E.: Imagenet classification with deep convolutional neural networks. In: NIPS 2012, pp. 1097–1105 (2012)
23. Graves, A., Fernández, S., et al.: Connectionist temporal classification: labelling unsegmented sequence data with recurrent neural networks. In: ICML, pp. 369–376 (2006)
24. Taboada, M., Brooke, J., Tofiloski, M., et al.: Lexicon-based methods for sentiment analysis. Comput. Linguist. **37**(2), 267–307 (2011)
25. Wang, H., Wang, N., Yeung, D. Y.: Collaborative deep learning for recommender systems. In: SIGKDD 2015, pp. 1235–1244 (2015)

Integrating Collaborative Filtering and Association Rule Mining for Market Basket Recommendation

Feiran Wang[1,2], Yiping Wen[1(✉)], Jinjun Chen[1,2], and Buqing Cao[1]

[1] Key Laboratory of Knowledge Processing and Networked Manufacture,
Hunan University of Science and Technology, Xiangtan, China
ypwen_0@qq.com
[2] Swinburne Data Science Research Institute,
Swinburne University of Technology, Melbourne, Australia

Abstract. This paper proposes a market basket recommendation algorithm based on association rules and collaborative filtering. It solves the problem that traditional association rule recommendation algorithms cannot generate association rules from cold commodity items under big data environment. An implicit semantic model based on historical transaction data of all users is constructed to represent potential features of commodities and measure similarities among commodities. The missing unknown elements in the implicit semantic model are complemented by the least square method. Association rules on unpopular commodities are obtained by the similarity of the commodities, thereby improving the recommendation accuracy. Experiments with real supermarket sales data demonstrate its effectiveness.

Keywords: Recommendation · Association rules · Big data

1 Introduction

The increasing number of customers and their purchase activities in supermarkets have produced a vast amount of precious historical data. In order to increase sales and keep the business competitive, retail managers often hope to exploit the enormous amount of accumulated customer transaction data to assist in discovering relationships of purchases between different commodities [1]. Specifically, an association rule mining algorithm, e.g. the Apriori [2], can be used directly on the data to find frequent itemsets based on historical customer transactions.

The market basket recommendation approach applies association rule mining algorithms to find association between co-purchased items and then generates item recommendation based on the strength of the association between items [3]. Moreover, an improved Apriori algorithm is applied in Taobao online dress shop recommendation system solving the problem of the mass data that are continuously generated [4]. In web service recommendation, a framework for Web personalization based on association rules is presented, contiguous and non-contiguous sequential patterns discovered from

© Springer Nature Switzerland AG 2018
H. Hacid et al. (Eds.): WISE 2018, LNCS 11234, pp. 19–34, 2018.
https://doi.org/10.1007/978-3-030-02925-8_2

Web usage data [5]. How to obtain accurate recommendation results through association rule mining is a major focus in the field of recommendation algorithm research.

Sahoo, Das and Goswami [6] propose the algorithms to generate the utility based non-redundant association rules and methods for reconstructing all association rules. Al-Maolegi and Arkok [7] propose an improved algorithm called M-Apriori, which can not scan the entire database again when judging whether the k + 1-itemsets are frequent, thereby improving the efficiency of the algorithm. Valle et al. [8]. propose a methodology for market basket analysis based on minimum spanning trees, which complements the search for significant association rules among the vast set of rules that usually characterize such an analysis. Le, Lauw and Fang. [9] propose a factor-based model called CBFM that can tap deeper into the relationship between basket items. However, the above researches did not consider that the unbalanced of the number of hot items and cold items. Because the rule-based recommendation method only by support degree to recommend items, it is difficult to make accurate recommendations in situations where hot and cold items are extremely unbalanced.

Collaborative filtering algorithms [10, 11] and content-based recommendation algorithms [12, 13] have also achieved good results in other fields. However, it is difficult to automatically obtain the feature description of the product in the shopping basket recommendation. We also lack user information, and behavioral features such as favorites, dislikes, and browsing time. Some scholars have also tried to use the idea of association rules to optimize collaborative filtering algorithms to reduce the effect of sparse data on the results [14, 15]. However, it does not change the fact that collaborative filtering algorithms rely on rich data types to achieve accurate item recommendation. This method is not suitable for market basket recommendations.

This paper proposes an algorithm that integrating collaborative filtering and association rule mining. It uses the idea of collaborative filtering algorithm to solve the problem that the rule-based recommendation method cannot generate the association rule of the unpopular items, thereby improving the quality of the recommendation result.

2   Problem Statement

Market basket recommendation often is used in supermarket business. After discussing with the data analysis team of a local supermarket, if we want to reach an accurate shopping basket recommendation, we need to solve two problems:

1. **Single data type**: Supermarket business differs from e-commerce in that the data types owned by merchants are very simple and lack user personal information, preferences, and ratings data. The only reliable data in the market basket recommendation is transaction data.
2. **Single recommendation result**: Under the circumstance of sparse data, the number of hot and cold commodities are extremely unbalanced. The relevance of hot commodities in the recommendation system is too high, and almost all of the commodities are associated with hot items. Finally, the recommendation system

will recommend these hot commodities too frequently, and a single recommendation result will reduce the accuracy of the recommendation.

In view of these challenges, an approach Integrating Collaborative Filtering and Association Rule algorithm (CFAR) is proposed, which uses the idea of collaborative filtering to establish an implicit semantic model, and supplements the missing association rules in hot commodities through similarity. CFAR has the potential to balance the recommended weights of hot and cold commodities and improve the quality of recommender systems.

In order to examine our approach, we conduct our experiments on a real-world dataset which comes from a local supermarket group and involves dozens of supermarket stores. We believe that this dataset reflects the fact that market basket recommendations have the problem of data sparseness. Experiments based on this dataset can better reflect the actual performance of the market basket recommendation algorithm.

3   Integrating Collaborative Filtering and Association Rule Algorithm

3.1   Association Rules Recommended Basic Concepts

The concept of association rules is widely used in the recommendation algorithm. The recommendation algorithm based on association rules can summarize the correlation between the items from the user's behavior patterns and establish association rules [16]. Association rule algorithm is most widely used in market basket recommendation. It can dig out the potential association relationship between commodities and commodities according to the information of massive users' purchasing behavior, and organizes the relationship into effective association rules.

First, let $I = \{i_1, i_2, \cdots, i_m\}$ be a finite set of items, where each item i_l, $1 \leq l \leq m$. Let D be the task relevant database composed of utility table and the transaction table $T = \{t_1, t_2, \ldots t_n\}$, containing a set of transactions, where each transaction $t_d \in I$, $1 \leq d \leq n$, in the database be associated with a unique identifier, say t_{id}. A subset $X \subseteq T$ is called an itemset, if X contains k distinct items $\{i_1, i_2, \cdots, i_k\}$, where $i_l \in I$, $1 \leq l \leq m$, called a k-itemset.

An association rule is an implication of the form $X \rightarrow Y$, where $X \subset I$, $Y \subset I$ and $X \cap Y = \emptyset$. Association rules need to be limited by the degree of support and confidence. The support degree refers to the percentage of the support number and the total number of records corresponding to the front or back of a certain rule, $sup(X) = |\{T : X \subseteq T, T \in D\}|/|D|$. The support of $X \rightarrow Y$ is:

$$sup(X \rightarrow Y) = \frac{|\{T : X \cup Y \subseteq T, T \in D\}|}{|D|} \tag{1}$$

Confidence represents the ratio of the number of transactions that contain both X and Y to the number of transactions that contain X, or the probability that Y occurring with X is present:

$$conf(X \rightarrow Y) = \frac{sup(X \cup Y)}{sup(X)} \tag{2}$$

Then we need to set the minimum of support and confidence, denoted as *minsup* and *minconf* respectively. $X \rightarrow Y$ is called frequent item set, where $sup(X \rightarrow Y) > minsup$. $X \rightarrow Y$ is called strong association rules, where $sup(X \rightarrow Y) > minsup$, and $conf(X \rightarrow Y) > minconf$.

In the process of recommendation, the item already purchased by the user is used as the pre-rule item, and the product associated with the item following the rule is recommended to the user through the strong association rule excavated from the transaction table T to complete the association recommendation.

3.2   Algorithm Introduction

In reality, only a very small number of items have a high degree of support, and the rest of the items are unpopular commodities. Especially in the era of big data, because service providers can provide massive commodities, the number of unpopular items is even more serious. And the value and profit of cold commodities are often more than hot goods, and they are more valued by merchants. However, the traditional algorithm cannot overcome the problem of recommendation diversity because of the principle of association rules.

The CFAR algorithm classifies association rules into two categories: primary and secondary association rules. In the traditional association rule algorithm, primary association rules are called strong association rules. They are the patterns of the purchase history of the user's history, and the correlations between the commodities are summed up. The primary association rule filters the element items with minimal support to ensure that the frequent itemsets generated satisfy the minimum support.

Secondary association rules are proposed by the CFAR algorithm against the shortcomings of the existing association rule recommendation algorithm. The basic idea of the CFAR algorithm to obtain the secondary association rules is: using the idea of collaborative filtering algorithm to create a hidden semantic model to describe the attributes of each commodity, and to search for similar commodities through the similarity of attributes. Finally, the secondary association rules are established based on the relationship of similar commodities to improve the accuracy of recommendations for cold commodities.

Primary Association Rules
The CFAR algorithm uses the FP-Tree method to generate frequent itemsets, which is an improved method for mining frequent itemsets [1]. Instead of generating candidate sets, it uses frequent pattern growth to mine frequent item sets. The CFAR algorithm requires only two scans of the transaction library. The first scan results in an element entry that satisfies the minimum support. The second scan adds the read entry set to a path. If the path does not exist, a new entry is created.

Suppose assuming that there is a transaction table for a total of 6 transaction records:

1	2	3	4	5	6
a, b, c, d	a, b	c, d	a, c	a	b, c

In the first pass scan, the elements that satisfy the minimum support are: a, b, and c, where $minSup = 0.5$.

In the second pass, we traverse the set of items that satisfy the minimum support and build a set of frequent items. Starting from the root node, the transactions are read in turn, and each time a new transaction is read in, the element items are added to generate a subtree (Fig. 1).

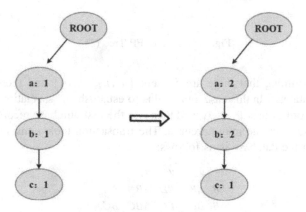

Fig. 1. Generate FP-Tree (1)

The second-level node is the frequent appearing element item. For example, a:2, it means the item appears twice in the transaction set. Finally, all the transactions that meet the conditions are added to generate a complete FP-Tree (Fig. 2).

In practical applications, the transaction database is often extremely large. The traditional algorithm for mining frequent itemsets requires multiple scans of the transaction database. For example, the Apriori algorithm needs to repeatedly scan the transaction database during the generation of the candidate set, which greatly increases the overhead of the computer hard disk IO [2]. In the CFAR algorithm, all frequent item sets are saved in the FP-Tree structure. To generate such a tree structure, only the transaction database needs to be scanned twice, which reduces the disk IO overhead and improves the efficiency of the algorithm.

Finally, the main association rule that meets the minimum confidence is filtered from the frequent itemsets. In the CFAR algorithm, the main association rule is an important basis for recommending products to users.

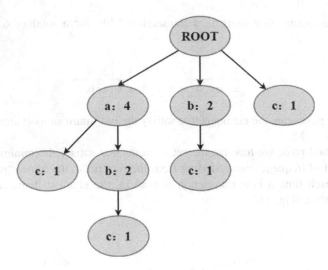

Fig. 2. Generate FP-Tree (2)

Example 1. Assuming that there are 5 users $\{u_1, u_2, \ldots, u_5\}$ and their 14 purchase records in the database. In this case, our goal is to establish an association rule based on the purchase record of user $u_1 - u_4$, and then use the first purchase record of user u_5 to predict the product that he may purchase. The transaction table consisting of 14 purchase records in the database is as follows:

u_1	u_2	u_3	u_4	u_5
AB	AB	BD	AB	AE
AB	BC	BC	ABC	BC
D	ABE	DE	AD	

In the case of considering only two frequent itemsets, we can generate frequent itemsets $\{A, B\}$ and $\{B, C\}$, where $minsup = 0.2$. The four association rules can be generated with these two frequent itemsets, with the following confidence levels:

$A \rightarrow B$	0.857
$B \rightarrow A$	0.667
$B \rightarrow C$	0.334
$C \rightarrow B$	1

The main association rules that satisfy the conditions are only $A \rightarrow B$ and $C \rightarrow B$, where $minconf = 0.8$. According to the main association rule $A \rightarrow B$, the commodity B can be recommended to the user in this case with the purchase record $\{A, E\}$ being known. The second purchase record of user u_5 shows that recommended commodity B is the correct solution.

However, the primary association rule cannot be recommended through the historical behavior of the user u_5 having purchased the commodity E because it is the cold commodity and it does not belong any frequent item sets. Since the recommendation through only the main association rules is a poor method, the CFAR algorithm proposes the secondary association rules to solve this problem.

Secondary Association Rules

Secondary association rule is obtained through the idea of a collaborative filtering algorithm. First, a hidden semantic model is constructed by using transaction data, so that the properties of the commodity can be interpreted by the low-dimensional space vector. Then calculate the similarity of the commodities based on the degree of similarity in the model. Finally, secondary association rules are established by the relationship of the most similar K items.

$(U_1, U_2, U_3, \ldots, U_M)$ represents M users, and $(V_1, V_2, V_3, \ldots, V_N)$ represents N items. The matrix $A_{(M \times N)}$ records commodity features, in which the number of times commodity V_j is purchased by user U_i is recorded as feature a_{ij}. If the same or similar user purchases V_i and t V_j, the two commodities are considered to have a high degree of similarity.

Users	Goods				
	V_1	V_2	V_3	...	V_N
U_1	0	10	6	...	2
U_2	3	0	0	...	0
...	...	...	...	...	...
U_M	0	0	4	...	1

However, the number of unpopular merchandise purchases is relatively small, resulting in an overly sparse eigenvector, and the final similar commodities may not be accurate. First, we need to map a user's characteristics into a low-dimensional vector U_i, similarly mapping a commodity's characteristics to the same dimension vector V_j, then the user's similarity to the commodities can use the inner items of the two vectors UV^T to indicate.

$$A \approx UV^T \tag{3}$$

We map all users and commodities into a k-dimensional space. In this k-dimensional space, the characteristics of the items are abstracted into k hidden features. Assuming that matrix A is approximately low rank, then matrix A can be approximated by the product of two small matrices $U_{(m \times k)}$ and $V_{(n \times k)}$. A matrix is decomposed into two submatrix of dimension reduction effect can be achieved, where $k \ll (m, n)$. Using Frobenius Norm to quantify errors generated by matrix U and matrix V reconstruction matrix A.

$$\left\| A - UV^T \right\|_F^2 \tag{4}$$

This reconstruction error contains unknowns because the large number of unknown elements in matrix A is what we need to predict. We convert the problem into an optimization problem, considering only the reconstruction error of the known data. Its optimization goal is to minimize the loss function L, where λ is a regularization coefficient:

$$L(U, V) = \sum_{(i,j) \in A} \left(a_{ij} - u_i^T v_j\right)^2 + \lambda\left(|u_i|^2 + |u_j|^2\right) \tag{5}$$

The original optimization problem can be transformed into a convex and detachable problem by selecting one of the user feature matrix U and the commodity feature matrix V. First fix the matrix V and then find the partial derivative of the loss function L on u_i. Let the derivative equal 0 to get:

$$u_i = \left(V^T V + \lambda I\right)^{-1} V^T a_i \tag{6}$$

The same holds true for fixed matrix U, which gives:

$$v_j = \left(U^T U + \lambda I\right)^{-1} U^T a_j \tag{7}$$

Where a_i is the i-th row of matrix A, a_j is the j-th column of A, and matrix I is the $k \times k$ unit matrix.

We use a random number less than 1 to initialize the matrix V, find U from the formula (6), and recalculate and overwrite the matrix V based on the calculated U and formula (7). Repeating the above two steps until you reach the set maximum number of iterations. At this point, the loss function L is approximated to the minimum, and calculating the inner item of the matrix U and the matrix V, the matrix A' is finally obtained.

We can calculate the similarity between the items by the matrix A'. Assuming that x and y are two different commodity feature vectors from the matrix A', the method of computing the similarity of two commodities through the cosine similarity is as follows:

$$C(x,y) = \frac{xy}{\|x\|^2 \times \|y\|^2} = \frac{\sum x_i y_i}{\sqrt{\sum x_i^2}\sqrt{\sum y_i^2}} \tag{8}$$

We calculate the similarity of any two items and find the most similar item for each item. We establish an association rule $x \to y$ if item y is the most similar to item x, and there is no association rule that uses x as an antecedent. Association rules generated by this method are defined as the secondary association rules. If there is a secondary association rule $x \to y$, then its weight $R(x \to y)$ is the cosine similarity between two commodities:

$$R(x \to y) = C(x, y) \tag{9}$$

According to the primary and secondary association rules, we recommend K items to each user. If there are less than K items in the user history record that satisfy the primary association rule, the recommendation is made with the highest weight association rule until the K items are recommended.

Example 2. In Example 1, we recommend the item B to the user u_5 through the primary association rule. Since the merchandise has been recommended based on the merchandise A and the merchandise E is not included in any master association rule, the master association rule cannot recommend more merchandise to the user, where $K \geq 2$. At this point, the CFAR algorithm can generate additional secondary association rules to recommend more items.

We establish a hidden semantic collaborative filtering model, which generates an item feature matrix based on training set data. The matrix generated in this case is dense:

	A	B	C	D	E
u_1	2	2	0	1	0
u_2	2	3	1	0	1
u_3	0	2	1	2	1

It can be seen from this matrix that there is a similar relationship between the characteristics of commodity C and commodity E, and they have potential relationships. According to the degree of similarity established sub-association rule $E \rightarrow C$, commodity C can be additionally recommended to the user.

4 Empirical Evaluation

4.1 Experimental Context

The experiments is based on the distributed computing platform called Hadoop which was developed by Apache, it supports data-intensive distributed applications, and it has the high fault tolerance, high throughput and can be deployed on the cheap heterogeneous hardware. Programming is implemented on the platform Spark, the Spark is an computing engine which is specially designed for large-scale data processing of fast general-purpose computing engine, enabling the distribution of memory data sets, it was a iterative MapReduce method which can better applied to data mining and machine learning needs. The following table is the specific configuration of the components used in the experiment (Table 1):

The data used in this experiment is the transaction data and membership data for a number of supermarket stores in a local supermarket group from November 2014 to November 2015. We named the data set Sales, which includes 50 million transaction records. And it involved 350,000 users and 40,000 items. We have removed invalid items and cheating users in the Sales dataset.

We sort the items in the Sales dataset in descending order of sales volume to get the following graph (Fig. 3):

Table 1. Cluster configuration

Components	Configuration
Hadoop edition	Hadoop 2.6.0
Spark edition	Spark 2.1.0
Operating system	CentOS 6.5 64 bit
Network bandwidth	1000 M
Switches	Huawei S5720-36C-EI, Huawei S5700-24TP-SI
Slave*2 (high)	Intel(R) Xeon(R) E5-2640 V3 Five-core processor*2, 64g memory, 2 TB hard disk
Slave *9 (low)	Intel(R) Core(TM) I5 6500 Quad-core processor, 8g memory, 1 TB hard disk
Master	Intel(R) Xeon(R) E5-2640 V3 Five-core processor*2, 64g memory, 2 TB hard disk

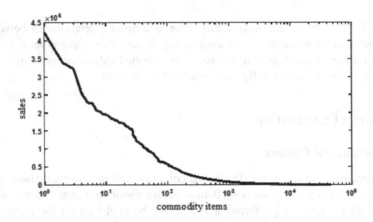

Fig. 3. The numble of commodities sales

From the above figure, we can verify that the Sales dataset is Pareto-distributed. Of the more than 40,000 items in the Sales data set, the sales volume of the top 100 items sold accounted for nearly 90% of the sales volume for the year. We can determine that the proportion of cold items and hot items is very unbalanced. The recommendation algorithm based on association rules will focus on recommending the top 100 items for sale. It lacks personalization and its accuracy is not high.

And the Sales dataset has a very sparse degree. The following table compares the Sales dataset with the MovieLens dataset, which is often used by the recommendation algorithm (Table 2):

In this paper, 80% of the data collected from the Sales data set is used as training set, which can be used for mining association rules to make recommendations. The remaining 20% of the data is used as a test set to verify the accuracy of the recommended algorithm.

Table 2. Data set basic properties

Dataset name	Sparse degree	Data size
MovieLens 100 K dataset	5.88%	100,000
MovieLens 1 M dataset	4.17%	1,000,000
MovieLens 20 M dataset	1.39%	20,000,000
Sales dataset	0.21%	30,000,000

4.2 Performance Evaluation

The traditional recommendation evaluation indicators generally include: *Precision* 、 *Coverage* and *F-measure* [17]. *Precision* is the percentage of items recommended by the user in the recommended item. *Coverage* is the percentage of items purchased by the user in the recommended item. *F-measure* has averaged the two and is an excellent indicator of the recommendation results. Assume that the set of user purchase items is *UP*, the set of system recommendation items is *RP*, and the definition of traditional recommendation indicators is as follows:

$$Precision = \frac{|UP \cap RP|}{|RP|} \tag{10}$$

$$Coverage = \frac{|UP \cap RP|}{|UP|} \tag{11}$$

$$F - measure = \frac{2 \times Coverage \times Precision}{Coverage + Precision} \tag{12}$$

However, in the traditional recommendation indicators, the increase in *Precision* will inevitably reduce the *Coverage*. The increase in *Coverage* will also affect the accuracy of *Precision*. And the above traditional indicators have no way to test the degree of personalization of the market basket recommendation algorithm. In order to verify the diversity of the proposed method on the hot-cold items with huge differences in the number of samples, this paper refers to the literature [18] and proposes a new index for shopping basket recommendation: recommended coverage(*ReCover*):

$$ReCover = \frac{|R_u|}{|I|} \tag{13}$$

Where *I* represents a set of all merchandise items, and R_u represents a set of all merchandise items recommended by the recommendation algorithm to the user. The higher the recommended coverage rate, the higher the degree of personalization of the recommended algorithm and the recommendation of more unpopular items. We think that if the two recommendation algorithms are close to the *F-measure* index, the algorithm with higher *ReCover* value can recommend more abundant commodities.

The experiment is mainly to verify the accuracy and diversity of CFAR algorithm in market basket recommendation. We compared it with the traditional association rule recommendation algorithm (AR). First, the frequent itemsets are mined using AR algorithm with different *minsup* (Fig. 4).

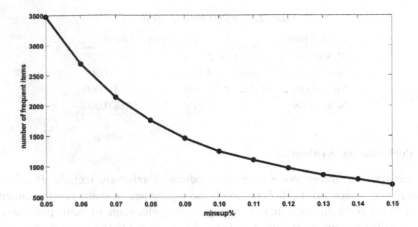

Fig. 4. The number of frequent itemsets under different support

As can be seen from the figure above, the number of frequent item sets generated depends on the minimum support value. In the process of increasing the minimum support from 0.05% to 0.15%, the number of frequent itemsets is reduced from 3476 to 705. There are a total of 40,000 items in the database, and the coverage rate of association rules generated by AR is far from enough. CFAR can generate 80,000 association rules (including primary and secondary association rules) regardless of the minimum support.

The parameter K represents the quantity of recommended products for each user. The following figure shows the average result of the *F-measure* for each user in the case of different parameters K in the CFAR algorithm (Fig. 5):

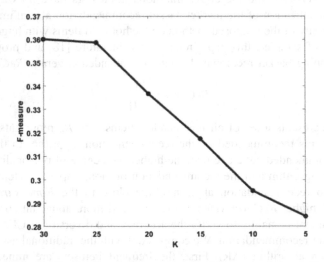

Fig. 5. The results of F-measure under different parameters K

Unlike e-commerce recommendations, it is not advisable to recommend too many products to users based on the actual market basket recommendation. And K's continued growth from 25 has little effect on the F-measure. However, as K began to decrease from 25, the quality of the shopping basket recommendation dropped rapidly. Finally we choose $K = 25$, at this time $F\text{-}measure = 0.3587$ (Fig. 6).

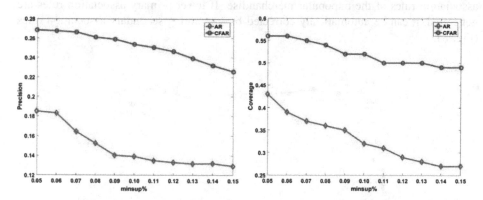

Fig. 6. Results of Precision & Coverage under different minimum support

In order to verify the superiority of CFAR, it compares with the traditional AR algorithm under different minimum support degrees. The comparison indexes include: *Precision*, *Coverage*, and *F-measure*.

As can be seen from the above figure, CFAR is superior to the traditional AR algorithm in all indicators. And in the process of changing the minimum support, CFAR has less fluctuation than AR. Because CFAR fills in missing association rules, it can maintain better results with fewer frequent itemsets (Fig. 7).

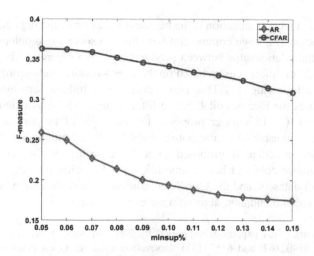

Fig. 7. Results of F-measure under different minimum support

The above figure shows the results of the recommended coverage rate under different minimum support degrees. It can be seen that the recommendation of the traditional AR algorithm is concentrated in the hot commodity items, and the relatively unpopular commodity items cannot be recommended. After increasing minimum support, AR generates fewer frequent itemsets and recommended coverage becomes lower. The CFAR will consider the similarity of the merchandise and supplement the association rules of the unpopular merchandise. If fewer primary association rules are generated, it can be automatically corrected by generating secondary association rules (Fig. 8).

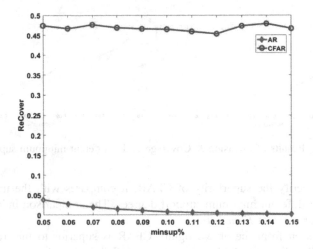

Fig. 8. The results of ReCover under different minimum support

5 Conclusion

The market basket recommendation is the business behavior of the big data era and can provide businesses with huge commercial benefits. To solve the problem that how to explores potential relationships between products and then recommends products that may be of interest to different users based on the user's historical shopping records, this article does the following: (1) This paper proposes a hidden semantic association model, which uses the idea of collaborative filtering to mine the associations between commodity items. (2) This paper proposes the concepts of primary and secondary association rules and establishes association rules for all commodities. (3) The concept of recommended coverage is proposed to test the ability of the recommendation algorithm to balance cold and hot commodities. (4) Experiments were conducted on real supermarket datasets, and the CFAR recommendation results were better than the traditional AR recommendation algorithm on each indicator.

Acknowledgements. This paper was supported by National Natural Science Fund of China (No. 61772193, 61402167, and 61572187), Innovation Platform Open Foundation of Hunan

Provincial Education Department of China (No. 17K033), Hunan Provincial Natural Science Foundation of China (No. 2017JJ2139, 2017JJ4036, and 2016JJ2056), and the Key projects of Research Fund in Hunan Provincial Education Department of China (No. 15A064).

References

1. Han, J., Kamber, M.: Data mining: concepts and techniques. Data Min. Concepts Models Methods Algorithms Second Ed. **5**(4), 1–18 (2011)
2. Agrawal, R., Imieliński, T., Swami, A.: Mining association rules between sets of items in large databases. In: ACM SIGMOD International Conference on Management of Data, pp. 207–216. ACM (1993)
3. Sarwar, B., Karypis G., Konstan, J., et al.: Analysis of recommendation algorithms for e-commerce. In: Proceedings of the 2nd ACM conference on Electronic commerce, pp. 158–167.. ACM (2000)
4. Guo, Y., Wang, M., Li, X.: Application of an improved Apriori algorithm in a mobile e-commerce recommendation system. Ind. Manage. Data Syst. **117**(2), 287–303 (2017)
5. Nakagawa, M., Mobasher, B.: Impact of site characteristics on recommendation models based on association rules and sequential patterns. In: Proceedings of the IJCAI, vol. 3, pp. 1–10 (2003)
6. Sahoo, J., Das, A.K., Goswami, A.: An efficient approach for mining association rules from high utility itemsets. Expert Syst. Appl. **42**(13), 5754–5778 (2015)
7. Al-Maolegi, M., Arkok, B.: An improved Apriori algorithm for association rules. arXiv preprint arXiv:1403.3948 (2014)
8. Valle, M.A., Ruz, G.A., Morrás, R.: Market basket analysis: complementing association rules with minimum spanning trees. Expert Syst. Appl. **97**, 146–162 (2018)
9. Le, D.T., Lauw, H.W., Fang, Y.: Basket-sensitive personalized item recommendation. In: Twenty-Sixth International Joint Conference on Artificial Intelligence, pp. 2060–2066, (2017)
10. Melville, P., Mooney, R.J., Nagarajan, R.: Content-boosted collaborative filtering for improved recommendations. In: Aaai/iaai, pp. 187–192 (2002)
11. Choi, K., Yoo, D., Kim, G., et al.: A hybrid online-product recommendation system: Combining implicit rating-based collaborative filtering and sequential pattern analysis. Electron. Commer. Res. Appl. **11**(4), 309–317 (2012)
12. Popescul, A., Pennock, D.M., Lawrence, S.: Probabilistic models for unified collaborative and content-based recommendation in sparse-data environments. In: Proceedings of the Seventeenth Conference on Uncertainty in Artificial Intelligence, pp. 437–444. Morgan Kaufmann Publishers Inc., (2001)
13. Pazzani, M.J., Billsus, D.: Content-based recommendation systems. In: Brusilovsky, P., Kobsa, A., Nejdl, W. (eds.) The Adaptive Web. LNCS, vol. 4321, pp. 325–341. Springer, Heidelberg (2007). https://doi.org/10.1007/978-3-540-72079-9_10
14. Sandvig, J.J., Mobasher, B., Burke, R.: Robustness of collaborative recommendation based on association rule mining. In: Proceedings of the 2007 ACM conference on Recommender systems. ACM, pp. 105–112 (2007)
15. Najafabadi, M.K., Mahrin, M.N., Chuprat, S., et al.: Improving the accuracy of collaborative filtering recommendations using clustering and association rules mining on implicit data. Comput. Hum. Behav. **67**, 113–128 (2017)
16. Lin, W., Alvarez, S.A., Ruiz, C.: Efficient adaptive-support association rule mining for recommender systems. Data Min. Knowl. Discov. **6**(1), 83–105 (2002)

17. Mobasher, B., Dai, H., Luo, T., et al.: Improving the effectiveness of collaborative filtering on anonymous web usage data. In: Proceedings of the IJCAI 2001 Workshop on Intelligent Techniques for Web Personalization (ITWP01), pp. 53–61 (2001)
18. Si, X., Sun, M.: Tag-LDA for scalable real-time tag recommendation. J. Comput. Inf. Syst. **6** (1), 23–31 (2009)
19. Wang, D.L., Yu, G., Bao, Y.: An approach of association rules mining with maximal nonblank for recommendation. J. Softw. **15**(8), 1182–1188 (2004)
20. He, M., Liu, W.S., Zhang, J.: Association rules recommendation algorithm supporting recommendation nonempty. J. Commun. **38**(10), 18–25 (2017)

Unified User and Item Representation Learning for Joint Recommendation in Social Network

Jiali Yang[1], Zhixu Li[1(✉)], Hongzhi Yin[2], Pengpeng Zhao[1], An Liu[1], Zhigang Chen[3], and Lei Zhao[1]

[1] School of Computer Science and Technology, Soochow University, Suzhou, China
zhixuli@suda.edu.cn
[2] School of Information Technology and Electrical Engineering, University of Queensland, Brisbane, Australia
[3] IFLYTEK Co.Ltd., Hefei, China

Abstract. Friend and item recommendation in online social networks is a vital task, which benefits for both users and platform providers. However, extreme sparsity of user-user matrix and user-item matrix issue create severe challenges, causing collaborative filtering methods to degrade significantly in their recommendation performance. Moreover, the factors those affect users' preference for items and friends are complex in social networks. For example, users may be influenced by their friends in addition to their own preferences when they choose items. To tackle these problems, we first construct two implicit graphs of users according to the users' shared neighbours in friendship network and the users' common interested items in interest network to ease data sparsity issue. Then we stand on recent advances in embedding learning techniques and propose a unified graph-based embedding model, called UGE. UGE learns two implicit representations for each user from implicit graphs, so that users can be represented as two weighted implicit representations which reflect the influence of friendship and interest. The weights and items' representation can be learnt from explicit friendship network and interest network mutually. Experimental results on real-world datasets demonstrate the effectiveness of the proposed approach.

Keywords: Social network · Joint recommendation
Graph embedding

1 Introduction

Online Social Networks (OSNs) such as Instagram, Facebook and Weibo have been growing exponentially in recent years, where users can share opinions and make friends with each other. To improve user experience and attract more users, OSNs often adopt recommendation system [1,2] to suggest potential items and friends those match users' preference.

© Springer Nature Switzerland AG 2018
H. Hacid et al. (Eds.): WISE 2018, LNCS 11234, pp. 35–50, 2018.
https://doi.org/10.1007/978-3-030-02925-8_3

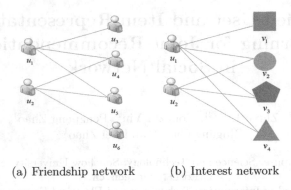

(a) Friendship network   (b) Interest network

Fig. 1. Two Types of Explicit Relations in OSNs

Unlike traditional recommendation system, such as movie recommendation which only considers the interaction relation between users and items, OSNs recommendation typically involves two types of explicit relations as illustrated in Fig. 1, which are user-user relations and user-item interactions. Thus, besides traditional item recommendation, OSNs recommendation has one more recommendation task, i.e., friend recommendation. As a result, OSNs recommendation suffers from at least two more challenges as listed below:

- **Interactive Tasks in Reciprocal Causation**. The two recommendation tasks in OSNs are usually in reciprocal causation. On one hand, people choose items not only according to their own interest but also referring to their friends especially when they have no clear preference. On the other hand, people don't necessarily make friends with those they really know in real life, they can also build relationships with those people who have similar interests in OSNs.
- **Severer Data Sparsity Problem**. Compared with traditional recommendation, OSNs recommendation faces severer data sparsity problem in both user-user relations and user-item interactions. Due to the existence of two types of relation, users in OSNs may easily ignore one of them. That is, except for those users who are inactive originally, we need to consider these biased users. Thus we meet a more serious data sparsity problem in friend recommendation or item recommendation in OSNs.

For item recommendation in OSNs, a lot of work indeed combine social relations and users' preference for items to boost recommendation performance [2,3]. But the effectiveness of methods are based on adequate extra social relations. For friend recommendation in OSNs, friendships are usually built by making use of social relations only [4]. Nevertheless, how to utilize user-item interactions to help building friendships remains a challenge given the data sparsity and reciprocal causation problems in OSNs recommendation. Besides, few work focus on joint friend and item recommendation because it involves two type of relations' modeling. Even though co-factorization methods suggest that users share latent features in rating space and the social space, they can not be directly applied to user link predictions due to the indirect modeling process of social relations [5]. Recently, Crossfire [6] realizes the cross media joint friend and item recommendation. They don't consider a robust user representation although the users exist in two relationships. Moreover, most recommendation systems are based on matrix factorization and its variants [3,7,8]. These methods do work well in state-of-art models but are easily exposed to problems of high-dimensional and massive calculation in most instances.

Recently, graph embedding methods have been studied and proved to be effective in capturing latent semantics of how items interact with each other in network structure [9,10], it has the advantage of low storage space and computational efficiency compared with traditional embedding methods. This technique is also effectively used for recommendation systems. For instance, Xie et.al. [11] use graph-based metric embedding and joint four embedding models for POI (point of interest) recommendation. PNRL [12] realize friend prediction by utilizing graph embedding technique in network structure. However, all these work only consider the recommendation of a single relationship.

Inspired by the above work, we study a new unified graph-based embedding model to jointly learn user and item representation. To ease the data sparsity, we construct two implicit graphs from friendship network and interest network. For example, in Fig. 1, user u_1 and u_2 are not linked in two graphs, but they hold common friends and items. We build two implicit relations for u_1 and u_2, which enrich source data greatly. We first map the user in implicit graphs to low-dimensional Euclidean latent space, then a robust user embedding is represented by weighted implicit representations of itself. To account for the reciprocal causation problem, the weights are learnt by training friendship network and interest network. After training, the two implicit representations and corresponding weights of each user can be obtained. The whole model is efficiently trained through the back-propagation algorithm.

The main contributions of our work can be summarized as follows:

- We propose a novel Unified Graph-based Embedding (UGE) model which stands on recent efficient graph embedding technique to realize joint friend and item recommendation in OSNs.
- We utilize implicit graphs constructed according to neighbours of users in interest network and friendship network to ease data sparsity problem, then we adopt the mutually training method to learn robust user embedding and item embedding jointly from two explicit relations.
- We conduct experiments on two real-world datasets, and the experimental results prove the effectiveness of our proposed model for friend and item recommendation.

The rest of the paper is organized as follows. In Sect. 2, we briefly review some related work. Section 3 details our proposed unified graph-based embedding method. In the end, we report experimental results in Sect. 4 and conclude our work in Sect. 5.

2   Related Work

Item Recommendation in OSNs. Most existing work believe that user social relation information can provide better item recommendation results [2,3,5,8]. In [13], the authors interpret one user's final rating decision as the balance between this user's own taste and his/her trusted users' favors. Following this view, Ma et al. [5] propose a general matrix factorization framework with two social regularization to improve the recommendation systems. Besides seeking suggestions from users' local friends, Tang et al. [14] think users may also refer to those people with high global reputations and they take global social context into account. Although these work boost item recommendation effectively, they did not consider the situation that both user ratings and social relations are sparse.

Friend Recommendation in OSNs. Friend recommendation is another important task in OSNs [6]. Most existing work on friend recommendation are concentrated on extracting information on social relations. Ding et al. [1] utilize the user's adjacency relationship to improve recommendation accuracy on the basis of convolutional neural network and Bayesian personalized ranking learning algorithm. PNRL [12] realizes friend link prediction by learning node representation of users in social network structure. However, friendships in OSNs not only include influence-oriented relations but also interested-oriented relations. That is, users those are never associated in social relations can be friends if they share same interests significantly. Nevertheless the work that combines two factors remains relatively limited.

Joint Recommendation in OSNs. Joint friend and item recommendation is faced up with the problem of complexity of two relations' modeling. Singh et al. [7] propose a collective matrix factorization method to factor user-item interaction matrix and user social relations matrix to predict links of users for friends and items concurrently. CrossFire [6] realizes cross-platform joint recommendation. Nevertheless, it fails to deal with the severe data imbalance issue due to the sparseness of OSN data.

Embedding Methods. Embedding method is a type of feature learning technology which can map item to a low dimensional representation. Matrix factorization [15,16] is one of the most typical methods for embedding. Recently, graph embedding methods have been studied and proved to be effective in capturing latent semantics of how items interact with each other in network structure [9,10]. PME [17] is a general embedding method that captures multi-dimensional information on heterogeneous networks to realize link prediction, which can be applied on arbitrary network. This technique is also effectively used for recommendation systems. Xie et.al. [11] use graph-based metric embedding and joint four embedding models for POI (point of interest) recommendation. Yin et.al. [18] embed all the observed relations among five entities such as users, events and so on. They all use extra associated content and contextual information to address the cold-start issue effectively. In our method, we try to use simple

Table 1. Notations

Symbols	Description
U, V	The set of users and items
G_{uu}, G_{uv}	User-user friendship graph and user-item interaction graph
G_{u^s}, G_{u^p}	User implicit social relation graph and preference relation graph
E_{uu}, E_{uv}, E^s, E^p	The sets of edges in G_{uu}, G_{uv}, G_{u^s} and G_{u^p}
u^s, u^p	Embedding of users in G_{u^s} and G_{u^p} respectively
γ^s, γ^p	Implicit social weight and implicit preference weight
u, v	Final embedding of users and items
M_{uu}, M_{uv}	User-user map matrix and user-item map matrix

friendships and interactions and their implicit information to extract user's and item's low-dimensional embedding representation.

3 Unified Graph-Based Embedding Model

In this section, we first introduce the notations and definitions in this paper, and then present our proposed Unified Graph-based Embedding (UGE) model and the corresponding optimization method.

3.1 Problem Formulation

For ease of presentation, we define the key data structures and notations used in this paper. Table 1 lists them.

Definition 1 (User-User Friendship Graph). *A user-user friendship graph, denote as $G_{uu} = (U, E_{uu})$, where U is a set of users in OSN and E_{uu} is a set of edges between the users those have friendships.*

Definition 2 (User-Item Interaction Graph). *A user-item interaction graph, denote as $G_{uv} = (U \cup V, E_{uv})$, where $U \cup V$ is a set of users and items in OSN and E_{uv} is a set of edges between the users and items those have interactions.*

Definition 3 (First-order Proximity). *The first-order proximity between a pair of nodes (a, b) in a network is the local pairwise proximity between two nodes, i.e. the weight of edge e_{ab}.*

In OSNs, the first-order proximity usually reflects the similarity of two nodes. However, the links observed are only a small proportion in a real-world network, and the rest missing [19]. Many users share a similar interest in the social network do not link up such as user u_1 and user u_2 in Fig. 1. Therefore, we define the second-order proximity to complement the first-order proximity, which can ease data sparsity problem significantly.

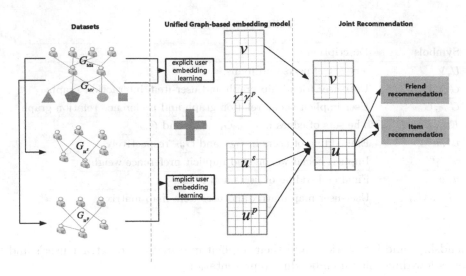

Fig. 2. Overview of the Proposed UGE Model.

Definition 4 (Second-order Proximity). *The second-order proximity between a pair of nodes (u, v) in a network is the similarity between their neighborhood network structures.*

Definition 5 (User Implicit Social Relation Graph). *A user implicit social relation graph, denote as $G_{u^s} = (U, E^s)$, encodes the implicit relationships between users according to their friends. For each edge e_{ab} in E^s, its weight is their second-order proximity in user-user friendship graph.*

Definition 6 (User Implicit Preference Relation Graph). *A user implicit preference relation graph, denote as $G_{u^p} = (U, E^p)$, encodes the implicit relationships between users according to their preference for items. For each edge e_{ab} in E^p, its weight is their second-order proximity in user-item interaction graph.*

Definition 7 (Joint Recommendation Problem). *Given a user-user graph $G_{uu} = (U, E_{uu})$ and a user-item graph $G_{uv} = (U \cup V, E_{uv})$ on OSNs, where U is the set of users, V is the set of items, E_{uu} and E_{uv} are the set of edges in G_{uu} and G_{uv}, we aim to convert users in U and items in V to a d-dimensional space, such that we can use users' and items' d-dimensional embeddings to make friend and item recommendations for target users.*

3.2 Model Description

We represent this section from modeling implicit relation graph, modeling explicit relation graph and parameter optimization. The overview of our proposed model is illustrated in Fig. 2.

Modeling Implicit Relation Graph. In our problem, there are two type of explicit relations for each user including user-user friendship relation and user-item interaction relation. We construct two implicit graphs from these two graphs separately. Intuitively, for two nodes of the same type, if there exists a path between them, there should be certain implicit relation between them. Unfortunately, counting the paths between two nodes has a rather high complexity of an exponential order. To ease the problem, we only consider the second-order proximity. We define the social second-order proximity between two users according to their common friends [20] as follows:

$$w_{ij}^s = \frac{\sum_{u_k \in U} w_{ik} w_{jk}}{\sum_{u_k \in U} w_{ik}} \tag{1}$$

where w_{ij} is the weight of edge $e_{u_i u_j}$ in E_{uu} which equals to 1 when u_j is a friend of u_i, otherwise it equals to 0. Similarly, the second-order proximity between two users according their common items is defined as follows:

$$w_{ij}^p = \frac{\sum_{u_k \in V} w_{ik} w_{jk}}{\sum_{v_k \in V} w_{ik}} \tag{2}$$

where w_{ij} is the weight of edge $e_{u_i v_j}$ in E_{uv} which equals to the u_i's normalized rating to v_j.

In this way, we get two implicit graphs. Above methods define implicit relations are directed and obviously this setting is reasonable. The trust relationship between users is usually unequal. Naturally, users those have real links in G_{uu} usually have links in implicit relation graphs, hence implicit relation graphs not only preserve the primary information to a large extent, but also add abundant new information. Next, we will introduce two view-specific representation u^s and u^p to preserve the structure information encoded in two implicit graphs separately. Given graphs $G_{u^s} = (U, E^s)$ and $G_{u^p} = (U, E^p)$, for an edge (u_i, u_j), we define the probability $p_s(u_j|u_i)$ and $p_p(u_j|u_i)$ as follows [10]:

$$p_s(u_j|u_i) = \frac{exp(c_j{}^T u_i^s)}{\sum_{c_k \in U} exp(c_k{}^T u_i^s)}, \quad p_p(u_j|u_i) = \frac{exp(c_j{}^T u_i^p)}{\sum_{c_k \in U} exp(c_k{}^T u_i^p)} \tag{3}$$

where c_j is a context representations of user u_j. In our approach, the context representations are shared in two implicit graphs, so that the two implicit graph-specific nodes representations can be located in the same semantic space. To preserve the weight of edge $e_{u_i u_j}$, we make the conditional distribution $p_s(\cdot|u_i)$ and $p_p(\cdot|u_i)$ close to their empirical distribution $\hat{p}_s(\cdot|u_i)$ and $\hat{p}_p(\cdot|u_i)$ separately. The empirical distribution is defined as $\hat{p}_s(u_j|u_i) = w_{ij}^s/d_i^s$ and $\hat{p}_p(u_j|u_i) = w_{ij}^p/d_i^p$, w_{ij}^s and w_{ij}^p are defined as Eq. 1 and 2, d_i^s and d_i^p are out-degree of user u_i in implicit social graph and implicit preference graph, We minimize the KL-divergence between the estimated neighbour distribution $p_s(\cdot|u_i)$, $p_p(\cdot|u_i)$ and the empirical neighbor distribution $\hat{p}_s(\cdot|u_i)$, $\hat{p}_p(\cdot|u_i)$ [9,10]. After some simplification, we obtain the following objective function for two implicit graphs:

$$O_s = - \sum_{(u_i,u_j) \in E^s} w_{ij}^s log p_s(u_j|u_i), \quad O_p = - \sum_{(u_i,u_j) \in E^p} w_{ij}^p log p_p(u_j|u_i) \tag{4}$$

Directly optimizing the above objective is computationally expensive because it needs to sum over the entire set of users. To address this problem, a negative sampling technique [21] is adopted, which modify the conditional probability in Eq. 4 as follows:

$$log\sigma(\boldsymbol{c_j}^T\boldsymbol{u_i^s}) + \sum_{k=1}^{K} E_{u_k \sim P_{neg}^s(u)}[log\sigma(-\boldsymbol{c_k}^T\boldsymbol{u_i^s})] \qquad (5)$$

$$log\sigma(\boldsymbol{c_j}^T\boldsymbol{u_i^p}) + \sum_{k=1}^{K} E_{u_k \sim P_{neg}^p(u)}[log\sigma(-\boldsymbol{c_k}^T\boldsymbol{u_i^p})] \qquad (6)$$

where $\sigma(x) = 1/(1 + exp(-x))$ is the sigmoid function, K is the number of negative edges. $P_{neg}^s(u) \propto d_u^{s\,3/4}$ and $P_{neg}^p(u) \propto d_u^{p\,3/4}$, among them d_u^s and d_u^p are out-degree of user u in two implicit graphs separately.

By minimizing the objective function 4, the $\boldsymbol{u^s}$ and $\boldsymbol{u^p}$ can preserve the information encoded in two implicit graphs. To integrate these two types of information. We propose a unified user representation based on the weighted implicit representations. To be specific, the unified user representation can be modeled as follows:

$$\boldsymbol{u_i} = \lambda_i^s\boldsymbol{u_i^s} + \lambda_i^p\boldsymbol{u_i^p} \qquad (7)$$

where λ_i^s and λ_i^p are the weight of two implicit graphs assigned by user u_i. For normalization, we introduce non-negative latent variables γ_i^s and γ_i^p to indicate how much information contained in two implicit graphs for user u_i. As a result, λ_i^s and λ_i^p can be defined as:

$$\lambda_i^s = \frac{\gamma_i^s}{\gamma_i^s + \gamma_i^p}, \quad \lambda_i^p = \frac{\gamma_i^p}{\gamma_i^s + \gamma_i^p} \qquad (8)$$

As more information the implicit graph contains, more important the user is, we try to assign different weights to distinguish implicit graphs. Therefore, we introduce the following regularization term:

$$O_r = \sum_{i=1}^{|U|}(\lambda_i^s \left\| \boldsymbol{u_i^s} - \boldsymbol{u_i} \right\|_2^2 + \lambda_i^p \left\| \boldsymbol{u_i^p} - \boldsymbol{u_i} \right\|_2^2) \qquad (9)$$

where $\|\cdot\|_2$ is the Euclidean norm of a vector. By integrating above objectives, the final objective of the implicit relations' modeling can be summarized below:

$$O_{im} = O_s + O_p + \eta O_r \qquad (10)$$

Modeling Explicit Relation Graph. In explicit graphs, the probability of user u_i choosing user u_j as friend i.e. $p(u_j|u_i)$ can be defined similar to the implicit relations graph. However, to differentiate implicit and explicit relation between users, we introduce a harmonious matrix M_{uu}. Hence the $p(u_j|u_i)$ is redefined as:

$$p_{uu}(u_j|u_i) = \frac{exp(\boldsymbol{u_j}^T M_{uu} \boldsymbol{u_i})}{\sum_{u_k \in U} exp(\boldsymbol{u_k}^T M_{uu} \boldsymbol{u_i})} \tag{11}$$

Similarly, the objective of user-user friendship graph is:

$$O_{uu} = - \sum_{(u_i,u_j) \epsilon E_{uu}} w_{ij} log p_{uu}(u_j|u_i) \tag{12}$$

where $p_{uu}(u_j|u_i)$ is simplified as:

$$log\sigma(\boldsymbol{u_j}^T M_{uu} \boldsymbol{u_i}) + \sum_{k=1}^{K} E_{u_k \sim P_{neg}^{uu}(u)}[log\sigma(-\boldsymbol{u_k}^T M_{uu} \boldsymbol{u_i})] \tag{13}$$

Because the user embedding is represented as Eq. 7, its learning can be transferred to learn corresponding weights $\gamma_i^s(\gamma_j^s)$ and $\gamma_i^p(\gamma_j^p)$ indirectly by minimizing the Eq. 12.

We could define $p_{uv}(v_j|u_i)$ by replacing $\boldsymbol{u_j}$ and M_{uu} with $\boldsymbol{v_j}$ and M_{uv} respectively in Eq. 13 to represent the probability of user u_i choosing item v_j in user-item interaction graph. Here, M_{uv} is aimed to differentiate the semantic space between users and items. Then O_{uv} can be calculated as:

$$O_{uv} = - \sum_{(u_i,u_j) \epsilon E_{uv}} w_{ij} log p_{uv}(v_j|u_i) \tag{14}$$

By minimizing O_{uv}, we can learn the weight γ_i^s and γ_i^p for user u_i and item v_j's embedding. To sum up above, the overall objective of our approach is summarized below:

$$O = O_{im} + O_{uu} + O_{uv} \tag{15}$$

We introduce the optimization strategy of this objective in the next section.

3.3   Parameter Optimization

To optimize the model, we utilize the coordinate gradient descent algorithm [22] and the back-propagation algorithm. Specifically, in each iteration, we first sample a set of edges from the implicit relation graph to update implicit graph-specific user embedding and user context embedding. For each sample in implicit graphs, the gradients w.r.t $\Theta_1 = \{\boldsymbol{u^s}, \boldsymbol{u^p}, \boldsymbol{c}\}$ are calculated as follows:

$$\frac{\partial O}{\partial \Theta_1} = \frac{\partial O_{im}}{\partial \Theta_1} \tag{16}$$

Then, we infer and update the voting weights for user nodes and item embeddings with explicit graphs. For each sample in explicit graphs, the gradients w.r.t $\Theta_2 = \{\gamma^s, \gamma^p, v, M_{uu}, M_{uv}\}$ are calculated as follows:

$$\frac{\partial O}{\partial \Theta_2} = \frac{\partial O_{uu}}{\partial \Theta_2} + \frac{\partial O_{uv}}{\partial \Theta_2} \tag{17}$$

The detailed calculation of gradients will not be introduced due to the limited space. In conclusion, the training algorithm is shown as Algorithm 1.

Algorithm 1: Optimization Algorithm of UGE Model

Input:
    G_{u^s}, G_{u^p}, G_{uu}, G_{uv}, K, η, T_1 and T_2

Output:
    u, v, M_{uu} and M_{uv};

1: **while** not converge **do**
2:     **while** $smp1 < T_1$ **do**
3:         Randomly pick up an implicit graph, denoted as G_{u^l};
4:         Sample (u_i, u_j) from G_{u^l} and K negative edges, update u_i^l, c_j and c_k;
5:     **end while**
6:     **while** $smp2 < T_2$ **do**
7:         Sample (u_i, u_j) from E_{uu} and K negative edges, update γ_i^s, γ_i^p, γ_j^s, γ_j^p,
        γ_k^s, γ_k^p and M_{uu}, then update u_i, u_j and u_k;
8:         Sample (u_j, v_j) from E_{uv} and K negative edges, update γ_i^s, γ_i^p, v_j, v_k and
        M_{uv}, then update u_i;
9:     **end while**
10: **end while**

---

4 Experimental Evaluation

In this section, we conduct experiments on two real-world datasets to answer the following questions: (1) Is our model able to improve friend and item recommendation by exploiting implicit and explicit relations simultaneously? and (2) How effective our model is to ease data sparsity problem?

4.1 Experimental Settings

Datasets. In the experiments, we use two datasets, i.e. Ciao and Epinions[1] to evaluate our UGE model. The statistics of the datasets is shown in Table 2.

Evaluation Metrics. We randomly select 80% of user-item interactions for each user as training set, the same method is applied on user-friend relations. In training set, the last 10% is divided for tuning parameters. And, we use the

---

[1] https://www.cse.msu.edu/~tangjili/trust.html.

Table 2. Description of Our Datasets

Statistics	Ciao	Epinions
Number of users	7375	49290
Number of items	105114	139738
Number of user-item interactions	284086	664824
Number of friendship links	111781	487181

commonly used metric recently to verify the experiment result. We define *hit@k* for a single test case as either the value 1, if the ground truth friend or item appears in the top-k results, or 0, if otherwise. The *Accuracy@k* is defined by averaging over all test cases:

$$Accuracy@k = \frac{\#hit@k}{|D_{test}|} \tag{18}$$

where $\#hit@k$ denotes the number of hits in the whole test set, and $|D_{test}|$ is the number of the test cases.

Parameter Settings. Based on past experience, we set the embedding dimension d as 50, the initial learning rate is set as 0.025. Besides, we investigate the sensitivity of η, K, T_1 and T_2 in our model. The performance w.r.t. η and K is shown in Fig. 3. From the result, we can see that the UGE performs best when the η is 0.02 and the value of K has little effect on performance in terms of *Accuracy@10* when K is in {2,4,6,8}. As a result, we set η to 0.02 and K to 2 in our final model.

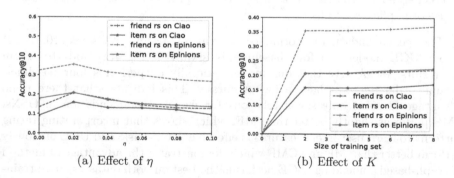

(a) Effect of η (b) Effect of K

Fig. 3. Effect of Parameters

The performance on Ciao in terms of *Accuracy@10* with different number of samples T_1 and T_2 is shown in Table 3. From the results, we can see that the performance of friend recommendation and item recommendation can reach the highest and keep stable when T_1 and T_2 are both greater or equal to 5 million,

Table 3. Effect of T_1 and T_2 On Ciao

(a) Accuracy@10 of friend rs

$T_2(10k)$ / $T_1(10k)$	400	450	500	550
400	0.147	0.165	0.194	0.198
450	0.154	0.185	0.216	0.218
500	0.189	0.204	**0.235**	0.234
550	0.205	0.216	0.234	0.235

(b) Accuracy@10 of item rs

$T_2(10k)$ / $T_1(10k)$	400	450	500	550
400	0.110	0.138	0.162	0.168
450	0.118	0.147	0.170	0.169
500	0.128	0.168	**0.174**	0.173
550	0.165	0.179	0.174	0.174

so we set them to 5 million. Similar experiments are implemented on Epinions and we set T_1 to 6 million, T_2 to 6.5 million.

4.2 Friend Recommendation

In this subsection, we check whether the proposed method can improve the performance of friend recommendation. We compare our UGE model with the following methods those parameters are tuned well.

- **Adamic/Adar:** This a metric for quantifying the similarity between two nodes in a network based on common neighbours [19].
- **MF:** Matrix Factorization method [16] factorizes the users' adjacent matrix into two low rank latent matrices and predicts the links by the matrix reconstructed by them.
- **CMF:** Collective matrix Factorization [7] is a matrix factorization model that utilizes users' social relation matrix and user-item preference matrix.
- **UGE-basic:** UGE-basic is a variant of our proposed method that makes no use of implicit relations.

The recommendation performance in terms of $Accuracy@k(k = 5, 10, 15, 20)$ of our UGE model and four baselines is shown in Fig. 4(a) and (b). From the results, we observe that our model generally outperforms four baselines. Adamic/Adar and MF have a low Accuracy. This indicates that there is no advantage in using only social relations for friend recommendation in OSNs. CMF performs slightly better than MF, which shows that incorporating rating patterns to learn user latent features is effective. UGE-basic and UGE obviously perform better than MF and CMF which demonstrates the advantage of method of graph-based embedding. UGE performs the best on both datasets on all training set. There are two reasons for this result, firstly UGE utilizes the implicit relations which can learn more from source data; secondly the iterative training method that learns the weights from friendship network and interest network can well capture the users' bias.

To evaluate the effectiveness of our model in alleviating data sparsity, we compare the performance of UGE and the best baseline UGE-basic. We randomly extract 20%, 40%, 60% and 80% train cases from whole dataset. The

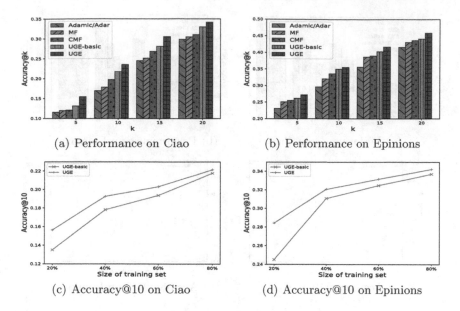

(a) Performance on Ciao (b) Performance on Epinions

(c) Accuracy@10 on Ciao (d) Accuracy@10 on Epinions

Fig. 4. Effectiveness Evaluation for Friend Recommendation.

Accuracy@10 of two methods are shown in Fig. 4(c) and (d). It can be clearly seen that in both datasets, UGE performs much better than UGE-basic at 20% setting and its advantage gradually decreases with the increase of training data. To be specific, UGE gain 15.8%, 8.1%, 4.8%, 1.7% and 16%, 3.2%, 2.1%, 1.5% relative improvement compared with UGE-basic in terms of four settings in two datasets respectively, which indicates that exploiting implicit relations can effectively ease data sparsity problem.

4.3 Item Recommendation

In this subsection, we check whether the proposed UGE model can improve the performance of item predictions. We compare our UGE model with the methods of MF, CMF, SoRec and UGE-basic. Among them, the descriptions of MF, CMF and UGE-basic are same as those in Sect. 4.2, besides, SoRec is introduced as follows:

- **SoRec:** This method [8] performs a co-factorization in user-rating matrix and user-user degree centrality relation confidence matrix by sharing same user latent factor through a probabilistic matrix factorization model.

Figure 5(a) and (b) show the results of our model and four baselines for item recommendation. MF still has no advantage. By incorporating social relations, CMF and SoRec have increased 16.37% and 36.31% on average compared to MF. It means that exploiting social relations can help improve item recommendation

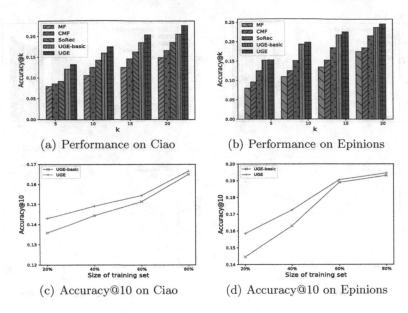

(a) Performance on Ciao   (b) Performance on Epinions

(c) Accuracy@10 on Ciao   (d) Accuracy@10 on Epinions

Fig. 5. Effectiveness Evaluation for Item Recommendation.

UGE-basic and UGE are still the best two methods. UGE gains average 20.89% and 6.1% higher performance than SoRec and UGE-basic.

Similarly, we compare UGE with best baseline UGE-basic to test our model's robustness for item recommendation. As shown in Fig. 5(a) and (b), we can see that as training ratio increases, the change of UGE in terms of *Accuracy@10* is more gradual than UGE-basic. This indicates that UGE is less sensitive to training data size and thus can better handle data sparsity problem.

To sum up, we conclude from the experiments that (1) the proposed model significantly improves both friend and item recommendation performance; and (2) our model is robust and it can ease data sparsity problem effectively.

5   Conclusion

In this paper, we propose a Unified Graph-based Embedding (UGE) model to predict users' social links and preference for items. We construct two implicit graphs to capture the second-order proximity of users and ease data sparsity problem. In addition, a mutual optimization method can learn user-user friendship relations and user-item interaction relations directly so that the user and item embedding can be obtained. Experimental results on two datasets show that our method outperforms baselines regarding both friend and item recommendation. However, the training process of UGE is complicated and need careful adjustment. Moreover, how to perform item and friend recommendations in a streaming way for more practical use in real world cases still to be solved. We will study these problems in the future work.

Acknowledgment. This research is partially supported by National Natural Science Foundation of China (Grant No. 61632016, 61402313, 61472263), and the Natural Science Research Project of Jiangsu Higher Education Institution (No. 17KJA520003).

References

1. Ding, D., Zhang, M., Li, S.Y., Tang, J., Chen, X., Zhou, Z.H: BayDNN: Friend recommendation with Bayesian personalized ranking deep neural network, pp. 1479–1488 (2017)
2. Tang, J., Xia, H., Liu, H.: Social recommendation: a review. Soc. Netw. Anal. Min. **3**(4), 1113–1133 (2013)
3. Tang, J., et al.: Recommendation with social dimensions. In: AAAI, pp. 251–257 (2016)
4. Siyao, H., Yan, X.: Friend recommendation of microblog in classification framework: using multiple social behavior features. In: International Conference on Behavior, Economic and Social Computing, pp. 1–6 (2015)
5. Ma, H., Zhou, D., Liu, C., Lyu, M.R., King, I.: Recommender systems with social regularization, pp. 287–296 (2011)
6. Shu, K., Wang, S., Tang, J., Wang, Y., Liu, H.: Crossfire: cross media joint friend and item recommendations. In: WSDM (2018)
7. Singh, A.P., Gordon, G.J.: Relational learning via collective matrix factorization. In: KDD, pp. 650–658 (2008)
8. Ma, H., Yang, H., Lyu, M.R., King, I.: Sorec: social recommendation using probabilistic matrix factorization. Comput. Intell. **28**(3), 931–940 (2008)
9. Tang, J., Qu, M., Mei, Q.: PTE: predictive text embedding through large-scale heterogeneous text networks, pp. 1165–1174 (2015)
10. Tang, J., Qu, M., Wang, M., Zhang, M., Yan, J., Mei, Q.: Line: large-scale information network embedding. In: WWW (2015)
11. Xie, M., Yin, H., Wang, H., Xu, F., Chen, W., Wang, S.: Learning graph-based POI embedding for location-based recommendation. In: CIKM, pp. 15–24. ACM (2016)
12. Wang, Z., Chen, C., Li, W.: Predictive network representation learning for link prediction. In: SIGIR, pp. 969–972 (2017)
13. Ma, H., King, I., Lyu, M.R.: Learning to recommend with social trust ensemble. In: SIGIR, pp. 203–210 (2009)
14. Tang, J., Hu, X., Gao, H., Liu, H: Exploiting local and global social context for recommendation. In: IJCAI, pp. 2712–2718 (2013)
15. Yin, H., Chen, H., Sun, X., Wang, H., Wang, Y., Nguyen, Q.V.H.: SPTF: a scalable probabilistic tensor factorization model for semantic-aware behavior prediction. In: ICDM, pp. 585–594 (2017)
16. Menon, A.K., Elkan, C.: Link prediction via matrix factorization. In: Gunopulos, D., Hofmann, T., Malerba, D., Vazirgiannis, M. (eds.) ECML PKDD 2011. LNCS (LNAI), vol. 6912, pp. 437–452. Springer, Heidelberg (2011). https://doi.org/10.1007/978-3-642-23783-6_28
17. Chen, H., Yin, H., Wang, W., Wang, H., Nguyen, Q.V.H., Li, X.: PME: projected metric embedding on heterogeneous networks for link prediction. In: SIDKDD (2018)
18. Yin, H., Zou, L., Nguyen, Q.V.H., Huang, Z., Zhou, X.: Joint event-partner recommendation in event-based social networks. In: ICDE (2018)

19. Liben-Nowell, D., Kleinberg, J.: The link prediction problem for social networks. In: CIKM, pp. 556–559 (2003)
20. Deng, H., Lyu, M.R., King, I.: A generalized co-hits algorithm and its application to bipartite graphs. In: SIGKDD, pp. 239–248 (2009)
21. Mikolov, T., Sutskever, I., Chen, K., Corrado, G., Dean, J.: Distributed representations of words and phrases and their compositionality, vol. 26, pp. 3111–3119 (2013)
22. Wright, S.J.: Coordinate descent algorithms. Math. Program. 151(1), 3–34 (2015)

Geographical Proximity Boosted Recommendation Algorithms for Real Estate

Yonghong Yu[1(✉)], Can Wang[2], Li Zhang[3], Rong Gao[4], and Hua Wang[5]

[1] College of Tongda, Nanjing University of Posts and Telecommunications,
Nanjing, China
yuyh@njupt.edu.cn

[2] School of Information and Communication Technology, Griffith University,
Gold Coast, Australia
canwang613@gmail.com

[3] Department of Computer and Information Sciences, Northumbria University,
Newcastle upon Tyne, UK
li.zhang@northumbria.ac.uk

[4] School of Computer Science, Hubei University of Technology, Wuhan, China

[5] Institute for Sustainable Industries and Liveable Cities, Victoria University,
Melbourne, Australia

Abstract. China's real estate sector has become the major force for the rapid growth of China's economy. There is a great demand for the real estate applications to provide users with their personalized property recommendations to alleviate information overloading. Unlike the recommendation problems in traditional domains, the real estate recommendation has its unique characteristics: users' preferences are significantly affected by the locations (e.g. school district housing) and prices of those properties. In this paper, we propose two geographical proximity boosted real estate recommendation models. We capture the relations between the latent feature vectors of real estate items by utilizing the average-based and individual-based geographical regularization terms. Both terms are integrated with the weighted regularized matrix factorization framework to model users' implicit feedback behaviors. Experimental results on a real-world data set show that our proposed real estate recommendation algorithms outperform the traditional methods. Sensitivity analysis is also carried out to demonstrate the effectiveness of our models.

Keywords: Real estate · Geographical proximity
Weighted regularized matrix factorization · Recommender systems

1 Introduction

In recent years, China's real estate market is explosively growing and has become the major force for the large increase of China's economy. For instance, in 2015,

© Springer Nature Switzerland AG 2018
H. Hacid et al. (Eds.): WISE 2018, LNCS 11234, pp. 51–66, 2018.
https://doi.org/10.1007/978-3-030-02925-8_4

the real estate sector contributed 6.1 percent to the overall GDP in China. This contribution continued to increase and reached 6.5% in 2016 [17]. In order to facilitate the real estate purchase and sale process, many applications that focus on the real estate information emerge on Internet, such as SouFun[1], Fang[2] and House365[3] etc. In these applications, the service providers supply users with various kinds of real estate features, including location, price, floor area ratio, and etc. With such information, users may choose suitable real estate items which meet their demands.

It is, however, still difficult for users to find their most relevant and interested properties, because there are a large number of real estate items available. That is to say, users usually suffer from the serious information overload problem. An effective solution is to build a real estate recommendation system. It discovers users' hidden preferences and provides the personalized real estate information for users by analyzing their historical activities and real estate features. For instance, a real estate recommendation system may suggest a list of school district housings to users who want to provide high-quality education for their children. Hence, a real estate recommendation system plays an important role in addressing the issue of real estate information overloading.

Recommendation systems [1] have been extensively studied in industry and academic. They are widely employed by many e-commerce web sites, such as Amazon, Youtube, Netflix, LinkedIn, and etc. By treating real estate properties as "items" (e.g., movies, music, and book) in traditional recommender systems, it is intuitive to apply the traditional item recommendation algorithms for real estate. Differ from the traditional recommendation, however, the real estate recommendation has several unique characteristics: (1) Each real estate item has a geographical feature. Users' purchase decisions are dramatically affected by the geographical positions of properties. For example, users often prefer the real estate items that are closer to towns and have convenient transport link. Alternatively, some parents prefer the school district housings with high-quality schools around, because they want to provide better education for their children; (2) It does not necessarily need to consider the price factor, when users make ratings for movies and music. By contrast, the property price is a key factor that influences users' final purchase decisions. Most existing research work focuses on the movie recommendation [2,14], the product recommendation [8,20,21], the point-of-interest recommendation [3,18,19,22], app recommendation [12], and task recommendation [15], and etc. However, there is few research work that addresses the real estate recommendation, due to those unique features above.

In this paper, therefore, we propose two geographical proximity boosted real estate recommendation models. Based on the empirical analysis on a real world data set from House365, we assume that the closer the geographical locations of two real estate items, the larger the similarity between them. We then propose two geographical regularization terms: average-based and individual-based,

---

[1] http://china.soufun.com/.
[2] http://www.fang.com/.
[3] http://www.house365.com/.

which are integrated with the weighted regularized matrix factorization to model users' implicit feedback behaviors. The average-based term constrains the latent feature vector of a property with the weighted average of its neighbors' latent feature vectors. In contrast, the individual-based term takes the individual difference between properties into account, and associates the real estate items having relatively large distance with different latent feature vectors. To the best of our knowledge, this is the first work that focuses on the real estate recommendation. In summary, the contributions of this paper are listed as follows:

- We propose two real estate recommendation algorithms, assuming that users are generally interested in the real estate items that have relatively large geographical proximity.
- We capture the relations between the latent feature vectors of properties by applying two different geographical regularization terms, which are derived from the geographical proximity. Both terms are combined with the weighted regularized matrix factorization framework to model users' click behaviors.
- We conduct extensive experiments on a real world data set from House365 to evaluate our methods. Experimental results show that our proposed algorithms for real estate recommendations are superior to traditional methods.

2 Related Work

Collaborative filtering (CF) methods [1] are widely used techniques for building recommender systems and have achieved great success in e-commerce. As the most popular approach among various CF methods, matrix factorization methods [6,9] have attracted a lot of attentions due to their effectiveness and efficiency in dealing with a very large scale user-item rating matrix. The basic assumption of matrix factorization is that only a few latent factors contribute to the preferences of users and characteristics of items. Typical matrix factorization approaches include NMF [7], PMF [9] and SVD++ [5]. Those matrix factorization based recommendation algorithms generally learn latent feature vectors of users and items from users' explicit feedback, i.e., users' ratings on items. However, explicit feedback may not always be available since it is difficult to collect. In contrast, in most situations, users' preferences are usually captured by their implicit behaviors, such as clicks, bookmarks and purchases. Although implicit feedback is relatively easy to collect, it only contains positive instances. Negative instances and missing values are mixed together, which makes the matrix factorization methods fail to learn the latent feature vectors of users and items. Collaborative filtering with implicit feedback is referred as the One-Class Collaborative Filtering (OCCF) [4,10]. To solve OCCF problem, Pan et al. [10] and Hu et al. [4] proposed a Weighted Regularized Matrix Factorization (WRMF) method. Rendle et al. [13] modeled the rankings of feedback and proposed a Bayesian Personalized Ranking (BPR) criterion for recommendation systems with implicit feedback. Pan et al. [11] extended BPR and proposed Group Bayesian Personalized Ranking (GBPR), via introducing richer interactions among users. GBPR aggregates the features of similar users in groups to reduce sampling uncertainty. Recently, Zhao et al. [23] proposed Social Bayesian Personalized Ranking (SBPR) model, which integrates social connections with users' implicit feedback to estimate users' rankings of items.

As users implicitly express their preferences for real estate items, i.e., via clicking on real estate item. The real estate recommendation can be regarded as OCCF problem. The direct idea of making real estate recommendation is to adopt traditional recommendation algorithms targeting at the OCCF problem. However, these traditional recommendation algorithms do not take the unique characteristics of real estate recommendation into account, such as geographical position and price factors etc. Unlike them, in this paper, we focus on real estate recommendation problem and exploit geographical proximity to boost the performance of real estate recommendation systems.

In addition, the task of real estate recommendation is related to but differs from the point-of-interest (POI) recommendation in location-based social networks [3,18,19,22], which leverages social influence, temporal influence and geographical influence to improve POIs recommendation systems. Unlike the POIs recommendation task, social influence has no direct effect on the real estate recommendation since the social networks are not available in real estate Internet applications. Moreover, the adoption of geographical influence is based on the fact that users' check-in activities require physical interactions between users and POIs, while the online access of real estate items do not require physical interactions between users and real estate items. Hence, our proposed geographical proximity boosted real estate recommendation approaches are different from POIs recommendation algorithms integrated with geographical influence.

3   Preliminary Knowledge

3.1   Problem Definition

In typical real estate recommendation systems, there are two types of entities: the set of M users $U = \{u_1, u_2, ..., u_M\}$ and the set of N real estate items $H = \{h_1, h_2, ..., h_N\}$. Besides a unique identifier, each real estate item may contain additional description information, such as district, price, and geographical location. We simplify "real estate item" as "item". Users' clicks are formed as a user-item click frequency matrix C. Each entry c_{ui} of C represents the number of clicks on real estate item i by user u. The number of clicks reflects users' preferences for certain real estate items. The set of real estate items clicked by user u is denoted as H_u. In general, the click frequency matrix C is extremely sparse since users are usually interested in a small portion of real estate items. For example, in the House365 dataset used in our experiments, 97.98% entries are missed. In this paper, we use "real estate" and "house" interchangeably.

The goal of real estate recommender system is to learn users' hidden preferences based on users' click history and the properties of real estate items, and then provide users with ranked lists of real estate items that users may be interested in.

3.2 Matrix Factorization

Owing to the effectiveness and efficiency in dealing with the large scale user-item rating matrix R, matrix factorization [6] is widely employed in recommender systems. The underlying assumption of matrix factorization is that only a few latent factors contribute to the preferences of users and characteristics of items. Matrix factorization decomposes the user-item rating matrix R into two low-rank latent feature matrices $P \in \mathbb{R}^{K \times M}$ and $Q \in \mathbb{R}^{K \times N}$, where $K \ll \min\{M, N\}$, and then approximates the rating matrix R using the product of P and Q.

Matrix factorization learns P and Q by minimizing the following objective function:

$$\mathcal{L}^{MF} = \min_{P,Q} \frac{1}{2} \sum_{(u,i) \in \Omega} (r_{ui} - p_u^T q_i)^2 + \frac{\lambda_1}{2} \|P\|_F^2 + \frac{\lambda_2}{2} \|Q\|_F^2, \tag{1}$$

where p_u and q_i denote the K-dimensional user and item feature vectors, respectively. $\| \cdot \|_F^2$ is the Frobenius norm, and Ω indicates the set of the (u, i) pairs for known ratings. λ_1 and λ_2 are regularization parameters. Matrix factorization usually utilizes stochastic gradient descent (SGD) to seek a local minimum of the above objective function.

3.3 Weighted Regularized Matrix Factorization

The above matrix factorization is designed for recommendation scenarios with explicit feedback, such as ratings on movies or products. By contrast, weighted regularized matrix factorization (WRMF) [10] is suitable for scenarios with implicit feedback, such as clicks, bookmarks and purchases etc. In order to solve the OCCF problem, WRMF treats all missing entries as negative instances and assigns varying confidence to positive and negative instances. Formally, the confidence level of entry c_{ui} is defined as:

$$w_{ui} = \begin{cases} 1 + \alpha \times c_{ui}, & if \ c_{ui} > 0 \\ 1, & otherwise, \end{cases} \tag{2}$$

where α is the weight parameter, which controls the confidence level of the positive instances. According to definition of confidence level, w_{ui} encodes the confidence of users' preferences. Based on all the weights of positive and negative instances, the objective function of WRMF is summarized as:

$$\mathcal{L}^{WRMF} = \min_{P,Q} \frac{1}{2} \sum_{u=1}^{M} \sum_{i=1}^{N} w_{ui}(r_{ui} - p_u^T q_i)^2 + \frac{\lambda_1}{2} \|P\|_F^2 + \frac{\lambda_2}{2} \|Q\|_F^2 \tag{3}$$

It should be noted that: in Eq. (1), r_{ui} indicates the value of rating on item i given by user u ($r_{ui} \in \{0, 1, 2, 3, 4, 5\}$). But, in Eq. (3), r_{ui} is the binarized value of c_{ui}, which indicates whether user u has clicked real estate item i, i.e., $r_{ui} \in \{0, 1\}$. Similar to matrix factorization, stochastic gradient descent or gradient descent algorithms are often used to seek the local optimal solution of Eq. (3).

4   Our Real Estate Recommendation Algorithms

We first empirically analyze the characteristics of House365 dataset used in our experiments, and then describe the details of our proposed geographical proximity boosted real estate recommendation algorithms.

4.1   Empirical Analysis

The dataset used in this paper is crawled from House365, which is an important information platform for China's real estate market. House365 provides various services related to housing, including sale, leasing and decoration etc., for the residents of Nanjing in China as well as its surrounding residents. We collect users' click records about new houses from 18 Jun. 2015 to 18 Sep. 2015. Meanwhile, the property information of new houses, such as name, address, district, and price etc., are also collected along with users' click records. In House365 dataset, users who have clicked fewer than 5 new houses are removed. In addition, in order to collect the accurate geographical position for each house, we apply Baidu map API[4] to retrieve their longitude and latitude based on the name and address properties of house. After data cleaning and preprocessing, there are 4,682 users and 1,098 real estate items in House365 dataset. The user-item click matrix contains 102,587 observations, and users clicked 21.9 new houses on average.

In this paper, users' click records from Jun. 18, 2015 to Sep. 6, 2015 are extracted as training data set and the remaining click records are regarded as testing set.

The number of new houses in each region is depicted in Fig. 1(a). As shown in Fig. 1(a), there are 18 regions in House365 data set. New houses mainly locate in main city zone (e.g., Gulou, Xuanwu, Jianye and Qinhuai districts) and rapidly developing zone (e.g., Jiangning, Pukou, Liuhe and Qixia districts). And Jiangning district has the largest number of new houses.

To analyze users' preferences on the house districts, Fig. 1(b) plots the distribution of user ratios in terms of the number of visited districts in the cleaned data set. From Fig. 1(b), we can observe that most of users are interested in a small portion of districts. For example, the number of districts visited by around 73% users is less than or equal to 4. In the original data set, around 75% users have 4 interested regions at most. This observation implies that real estate recommendation system do not need to consider all districts when generating house recommendations.

In order to further analyze users' click behaviors, we randomly select three users from House365, and then mark the geographical positions of houses visited by these three users on Baidu map. Figure 2(a), (b) and (c) represent the geographical position distributions of houses visited by the small-scale, middle-scale and large-scale users in terms of the number of visited houses, respectively. From Fig. 2, we can see that the major portion of houses visited by the small-scale user locate in Pukou district, which is a less-developed region with relatively lower

---

[4] http://lbsyun.baidu.com.

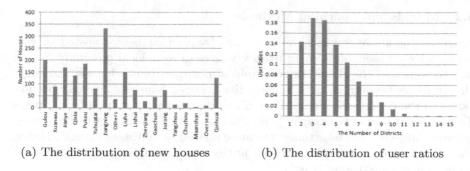

(a) The distribution of new houses   (b) The distribution of user ratios

Fig. 1. The distributions of users and new houses

house price. In addition, the middle-scale user is mainly interested in houses located in the main city zones, which are developed regions with higher house price. The houses visited by the large-scale user mainly locate at the south of the Yangtze River, especially within the regions of Jiangning district, which is rapidly developing recent years.

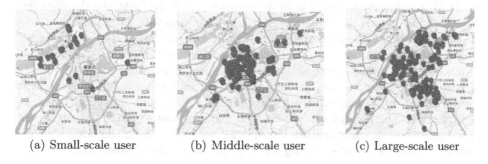

(a) Small-scale user   (b) Middle-scale user   (c) Large-scale user

Fig. 2. The geographical position distribution of houses visited by randomly selected users

According to the above analysis, we conclude the following observations: (1) the distribution of new houses is imbalanced, and most of new houses are located in the fast growing districts; (2) each user has his own interest in districts, and the average number of visited districts per user is very small; (3) the houses visited by a random user are usually clustered into several groups. Moreover, houses located in the same group generally have relatively large geographical proximity. In other words, users usually share more common interests in houses with larger geographical proximity. Hence, if two houses are similar in terms of geographical proximity, it is more possible that users show similar click behaviors on them.

4.2 Models and Learning Methods

Based on the above empirical analysis, we observe that houses visited by randomly selected users often share relatively large geographical proximity. So, we may assume that the closer the geographical positions of two houses, the larger the similarity between them. The similarity between houses is inversely proportional to the distances between them. Similar conclusion has been found in the Tobler's First Law of Geography [16], which reports that "Everything is related to everything else, but near things are more related than distant things". Hence, the First Law of Geography can also justify our assumption: near houses are more similar than distant houses.

According to the geographical positions of two houses, i.e., the longitudes and latitudes, we define the similarity between houses h_i and h_j as follows:

$$sim(i,j) = e^{-\frac{||x_i - x_j||^2}{\sigma^2}}, \tag{4}$$

where x_i and x_j indicate the geographic coordinate vectors of house h_i and h_j, respectively. The geographic coordinate vector is encoded as $[longitude, latitude]$, retrieved from Baidu map API. σ is a constant, which is set as 0.01 in our experiments. In Eq. (4), the similarity between two houses $sim(i,j)$ is in inverse proportion to the distance $||x_i - x_j||^2$. As the the distance $||x_i - x_j||^2$ increases, the similarity $sim(i,j)$ monotonically decreases.

Based on the above assumption, we incorporate the geographical proximity of real estate items into weighted regularized matrix factorization framework to model users' implicit feedback behaviors. Specifically, we propose two kinds of regularization terms, i.e., average-based and individual-based geographical regularization terms, to constrain the process of weighted regularized matrix factorization. The goal of using these two geographical regularization terms is to accurately learn latent feature vectors of houses by making the latent feature vectors as close as possible, if they share large geographical proximity.

Average-Based Geographical Regularization. As mentioned above, houses located in the same group generally have relatively large geographical proximity. Hence, the latent feature vector of a house is similar to those of nearby houses. Based on this intuition, we define an average-based geographical regularization term Reg^A as follows:

$$Reg^A = \frac{\lambda_g}{2} \sum_{i=1}^{N} \left\| q_i - \frac{\sum_{f \in N(i)} sim(i,f) q_f}{\sum_{f \in N(i)} sim(i,f)} \right\|_F^2, \tag{5}$$

where λ_g is the parameter which controls the effect of average-based geographical regularization term, $N(i)$ indicates the set of most similar neighbors of house h_i in terms of geographical proximity. In our experiments, we define a similarity threshold δ to form the set of most similar neighbors, i.e., $N(i) = \{f | sim(i,f) \geq \delta, h_f \in H\}$.

The average-based geographical regularization term Reg^A minimizes the distances between the latent feature vectors of target houses and the weighted average of latent feature vectors of their most similar neighbors. In fact, Reg^A makes the latent feature vector of a house as close as possible to the weighted average of neighbors' latent feature vectors. For example, suppose that q_i refers to the latent feature vector of house h_i, and the most similar neighbor set of house h_i is $N(i)$, then the distance between q_i and $\frac{\sum_{f \in N(i)} sim(i,f) q_f}{\sum_{f \in N(i)} sim(i,f)}$, i.e., the weighted average feature vector of all neighbors in $N(i)$, should be small.

Incorporating average-based geographical regularization term Reg^A into weighted regularization matrix factorization results in the following objective function:

$$\mathcal{L}^A = \min_{P,Q} \frac{1}{2} \sum_{u=1}^{M} \sum_{i=1}^{N} w_{ui} \left(r_{ui} - p_u^T q_i \right)^2 + \frac{\lambda_1}{2} \|P\|_F^2 + \frac{\lambda_2}{2} \|Q\|_F^2 + Reg^A. \quad (6)$$

We use the stochastic gradient descent algorithm to seek the local optimal solution of the objective function $\mathcal{L}^A$. The derivatives of objection function $\mathcal{L}^A$ regarding to p_u and q_i are computed as follows:

$$\frac{\partial \mathcal{L}^A}{\partial p_u} = \sum_{i=1}^{N} w_{ui}(p_u^T q_i - r_{ui})q_i + \lambda_1 p_u,$$

$$\frac{\partial \mathcal{L}^A}{\partial q_i} = \sum_{u=1}^{M} w_{ui}(p_u^T q_i - r_{ui})p_u + \lambda_2 q_i + \lambda_g \left(q_i - \frac{\sum_{f \in N(i)} sim(i,f) q_f}{\sum_{f \in N(i)} sim(i,f)} \right) \quad (7)$$

$$- \lambda_g \sum_{g \in N^-(i)} \frac{sim(i,g) \left(q_g - \frac{\sum_{f \in N(g)} sim(g,f) q_f}{\sum_{f \in N(g)} sim(g,f)} \right)}{\sum_{f \in N(g)} sim(g,f)}.$$

where $N^-(i)$ indicates the set of houses that take house h_i as their neighbors.

Individual-Based Geographical Regularization. The average-based geographical regularization term Reg^A makes the latent feature vector of a house close to the weighted average of feature vectors of its most similar neighbors. This method ignores the difference between relatively distant houses, and can not capture the individual characteristics of houses. For example, suppose that house h_i and h_j are neighbors, but the distance between them is relatively large. Hence, the similarity between h_i's latent feature vector and h_j's latent feature vector should be small. However, Reg^A makes their latent feature vectors close to the same weighted average latent feature vector. Consequently, houses h_i and h_j have similar latent feature vectors, leading to inaccurate learning of their hidden characteristics. To deal with this problem, we propose another geographical regularization term, called individual-based geographical regularization term Reg^I, which considers individual difference between houses, especially for distant ones. The individual-based geographical regularization term Reg^I is formulated as follows:

$$Reg^I = \frac{\lambda_g}{2} \sum_{i=1}^{N} \sum_{f \in N(i)} sim(i,f) \|q_i - q_f\|_F^2. \tag{8}$$

A small value of $sim(i,f)$ indicates that houses h_i and h_f are distant, so the difference between their latent feature vectors should be relatively large, and vice versa.

After incorporating individual-based geographical regularization term Reg^I into weighted regularization matrix factorization, we have the following objective function:

$$\mathcal{L}^I = \min_{P,Q} \frac{1}{2} \sum_{u=1}^{M} \sum_{i=1}^{N} w_{ui}(r_{ui} - p_u^T q_i)^2 + \frac{\lambda_1}{2} \|P\|_F^2 + \frac{\lambda_2}{2} \|Q\|_F^2 + Reg^I. \tag{9}$$

Similar to the way to solve the objective function $\mathcal{L}^A$, the stochastic gradient descent algorithm is applied to seek the local optimal solution of the objective function $\mathcal{L}^I$. The derivatives of objection function $\mathcal{L}^I$ with respect to p_u and q_i are computed as:

$$\frac{\partial \mathcal{L}^I}{\partial p_u} = \sum_{i=1}^{N} w_{ui}(p_u^T q_i - r_{ui})q_i + \lambda_1 p_u,$$

$$\frac{\partial \mathcal{L}^I}{\partial q_i} = \sum_{u=1}^{M} w_{ui}(p_u^T q_i - r_{ui})p_u + \lambda_2 q_i + \lambda_g \sum_{f \in N(i)} sim(i,f)(q_i - q_f) \tag{10}$$

$$+ \lambda_g \sum_{g \in N^-(i)} sim(i,g)(q_i - q_g).$$

5   Experiments

In this section, we conduct several experiments on a real data set to evaluate the performance of our proposed real estate recommendation methods. The statistics of data set used in our experiments are presented in Sect. 4.1.

5.1   Evaluation Metrics

The real estate recommendation algorithms provide each target user with a ranked list of real estate items. We employ two widely used ranking metrics to evaluate the performance of real estate recommendation algorithms, i.e., Precision@k and Recall@k, where k is the length of ranked recommendation list of real estate. For both metrics, we set $k = 5, 10, 20$ to evaluate the performance in our experiments.

5.2 Baseline Approaches

In order to evaluate the effectiveness of our proposed real estate methods, we choose the following state-of-art approaches as baselines.

- UserKNN: This method is the user-based collaborative filtering [2]. We use the cosine similarity to compute the similarity between users based on users' clicks.
- ItemKNN: This method is the item-based collaborative filtering [14]. We use the cosine similarity to compute the similarity between real estate items.
- PMF: PMF [9] can be viewed as a probabilistic extension of the SVD model. We learn user and item latent feature vectors from user-real estate click frequency matrix.
- SVD++: This method is proposed by Koren [5]. SVD++ extends the basic matrix factorization method by exploiting both explicit and implicit feedback of users to provide recommendations.
- WRMF: This method is the weighted regularization matrix factorization [10], and addresses the OCCF problem by assigning different confidence values to positive and negative instances.
- WRMF-AG: Our proposed WRMF-AG is described in Sect. 4.2, and incorporates the average-based geographical regularization term into the framework of WRMF.
- WRMF-IG: Our proposed WRMF-IG is also presented in Sect. 4.2, and incorporates the individual-based geographical regularization term into the framework of WRMF.

We set the parameters of each approach as its default setting to get the best performance. In UserKNN and ItemKNN, the number of most similar neighbors are set to be 250 and 120, respectively. In PMF, we set λ_U and λ_V to be 0.01, and the dimension of latent feature vector K to be 140. In SVD++, λ is set to be 0.01, and the dimension of latent feature vector K is equal to 80. In WRMF, $\lambda = 0.01$, $\alpha = 1$ and $K = 30$. In WRMF-AG, $\lambda_1 = 0.01$, $\lambda_2 = 0.01$, $\lambda_g = 0.5$, $\alpha = 2$ and $K = 60$. In WRMF-IG, $\lambda_1 = 0.01$, $\lambda_2 = 0.01$, $\lambda_g = 0.5$, $\alpha = 3$ and $K = 60$. In both WRMF-AG and WRMF-IG, we set the similarity threshold δ to be 0.6. Finally, we set the learning rate η involved in the stochastic gradient descent algorithm to be 0.0001.

5.3 Performance Comparison

Table 1 reports the real estate recommendation quality of all compared methods on House365 data set.

From Table 1, we have the following observations: (1) UserKNN performs better than ItemKNN. A possible reason is that the similarity between real estate items is less accurate than the similarity between users in the contexts of data sparsity, as well as the number of users is larger than the number of real estate items. In fact, ItemKNN performs the worst among all the compared methods. (2) SVD++ achieves better performance than PMF, which indicates

Table 1. Performance comparison on House365

Metric	UserKNN	ItemKNN	PMF	SVD++	WRMF	WRMF-AG	WRMF-IG
Precision@5	0.0515	0.0019	0.0421	0.0469	0.0906	0.0920	**0.0952**
Precision@10	0.0446	0.0020	0.0393	0.0399	0.0772	0.0786	**0.0812**
Precision@20	0.0223	0.0010	0.0196	0.0199	0.0386	0.0393	**0.0406**
Recall@5	0.0585	0.0018	0.0516	0.0607	0.1236	0.1285	**0.1330**
Recall@10	0.0982	0.0041	0.0968	0.1033	0.2064	0.2096	**0.2189**
Recall@20	0.0982	0.0041	0.0968	0.1033	0.2064	0.2096	**0.2189**

that simultaneously considering explicit and implicit influence of feedback is better than only considering the explicit influence of feedback. Meanwhile, the overall performance of PMF and SVD++ is worse than that of UserKNN. This is because that both of PMF and SVD++ take users' clicks on real estate items as ratings in traditional recommender systems. In those systems, there are both positive and negative instances. However, only positive instances are included in the click records of real estate recommender systems. Consequently, PMF and SVD++ can not accurately learn the latent feature vectors of users and real estate items. (3) WRMF, WRMF-AG and WRMF-IG are superior to other compared methods. This is because that WRMF, WRMF-AG and WRMF-IG take real estate recommendation as OCCF problem, i.e., they learn the latent feature vectors by assigning different confidence levels to positive and negative instances. Hence, in comparison with PMF and SVD++, WRMF based models are more suitable for solving the real estate recommendation problem, and can achieve better performance. (4) WRMF-AG and WRMF-IG consistently outperform WRMF, which demonstrates the effectiveness of our proposed real estate recommendation algorithms. The improvement ratios of WRMF-AG and WRMF-IG over WRMF are 1.5% and 5.1% in terms of Precision@5, respectively. In terms of Recall@5, WRMF-AG and WRMF-IG improve WRMF by 4.0% and 7.6%, respectively. In addition, WRMF-IG is superior to WRMF-AG. A possible reason is that WRMG-AG makes the latent feature vector of a real estate item towards the weighted average of its most similar neighbors' latent feature vectors, while ignores the difference between relatively distant real estate items.

5.4 Sensitivity Analysis

Impact of Parameter K. The dimension of latent feature vector K is an important parameter that affects the performance of real estate recommendation algorithms. In this section, we conduct a group experiments to evaluate the sensitivity of parameter K. We vary the value of K from 10 to 100 with the step 10, and observe the changes of recommendation quality. In this group of experiments, we set $\lambda_g = 0.5, \alpha = 2$ in WRMF-AG, and $\lambda_g = 0.5, \alpha = 3$ in WRMF-IG. The experimental results of WRMF-AG are shown in Fig. 3. The

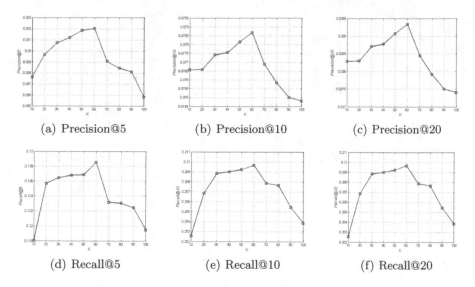

(a) Precision@5   (b) Precision@10   (c) Precision@20

(d) Recall@5   (e) Recall@10   (f) Recall@20

Fig. 3. The Impact of K on WRMF-AG

results of WRMF-IG show similar trends and thus they are omitted due to space limitation.

As demonstrated in both Fig. 3, the parameter K significantly affects the recommendation quality of real estate recommendation algorithms. As the values of K increase, the Precision@5 and Recall@5 first increase, which indicates that the recommendation quality boosts. Then, Precision@5 and Recall@5 drop down after K reaches a certain threshold. Other metrics show similar trends. This observation indicates that the large value of K does not necessarily result in the improvement of real estate recommendation systems. A possible reason is that: a large value of K makes WRMF have more latent factors to characterize users and real estate items, leading to a more powerful representation ability for WRMF. But, it may also introduce noises into WRMF model with more latent factors. This observation is consistent with the underlying assumption of matrix factorization [6], i.e., only a few latent factors contribute to the preferences of users and characteristics of items. On House365 data set, WRMF-AG and WRMF-IG achieve their best performance when K is around 60.

Impact of λ_g. The parameter λ_g is another important parameter in our proposed methods. It controls the weights of geographical proximity in learning the latent feature vectors of users and real estate items. In this section, we vary λ_g within the range of $[0.1, 1]$, and observe the sensitivity of λ_g. We set $K = 60, \alpha = 2$ in WRMF-AG, and $K = 60, \alpha = 3$ in WRMF-IG. Due to space limitation, we only report the experimental results of WRMF-AG in Fig. 4. And the experimental results of WRMF-IG show similar trends.

We can see that our proposed real estate approaches are highly sensitive to the values of λ_g. Precision@5 and Recall@5 first move upwards with the increase

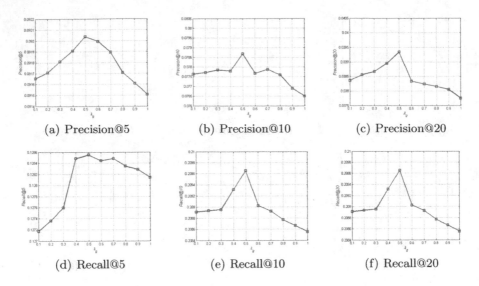

(a) Precision@5　　(b) Precision@10　　(c) Precision@20

(d) Recall@5　　(e) Recall@10　　(f) Recall@20

Fig. 4. The Impact of λ_g on WRMF-AG

of λ_g, and then begin to drop down as the value of λ_g increases. Other evaluation metrics show similar trends with Precision@5 and Recall@5. Both WRMF-AG and WRMF-IG achieve the best performance when λ_g is around 0.5.

6　Conclusion

China's real estate sector has become the major force for the growth of China's economy. It is urgent for real estate internet applications to provide users with personalized real estate information, and address the real estate information overload problem. In this paper, we proposed two geographical proximity boosted real estate recommendation methods, which incorporate the geographical proximity into the framework of the weighted regularized matrix factorization to model users' implicit feedback behaviors. The relations of latent feature vectors of real estate items are captured by the geographical regularization terms derived from geographical proximity among real estate items. Experimental results on real world data set show the superiority of our proposed methods.

The study of real estate recommendation has just begun, and we only apply geographical proximity to improve the performance of real estate recommendation systems. The price of real estate item and whether a real estate item is school district housing are important factors that affect users' final decisions. However, it is difficult to collect the prices of real estate items and the information related to school district housing due to privacy preserving. Hence, inferring the prices of real estate items and school district housing information to improve the existing real estate recommendation algorithms would be an interesting direction in our future research.

Acknowledgments. The authors would like to acknowledge the support for this work from the Natural Science Foundation of the Higher Education Institutions of Jiangsu Province (Grant No. 17KJB520028), NUPTSF (Grant No. NY217114) and Qing Lan Project of Jiangsu Province.

References

1. Adomavicius, G., Tuzhilin, A.: Toward the next generation of recommender systems: a survey of the state-of-the-art and possible extensions. IEEE Trans. Knowl. Data Eng. **17**(6), 734–749 (2005)
2. Breese, J.S., Heckerman, D., Kadie, C.: Empirical analysis of predictive algorithms for collaborative filtering. In: UAI, pp. 43–52. Morgan Kaufmann Publishers Inc. (1998)
3. Gao, R., Li, J., Li, X., Song, C., Zhou, Y.: A personalized point-of-interest recommendation model via fusion of geo-social information. Neurocomputing **273**, 159–170 (2018)
4. Hu, Y., Koren, Y., Volinsky, C.: Collaborative filtering for implicit feedback datasets. In: ICDM, pp. 263–272. IEEE (2008)
5. Koren, Y.: Factorization meets the neighborhood: a multifaceted collaborative filtering model. In: KDD, pp. 426–434. ACM (2008)
6. Koren, Y., Bell, R., Volinsky, C.: Matrix factorization techniques for recommender systems. Computer **8**, 30–37 (2009)
7. Lee, D.D., Seung, H.S.: Algorithms for non-negative matrix factorization. In: NIPS, pp. 556–562 (2001)
8. Linden, G., Smith, B., York, J.: Amazon.com recommendations: item-to-item collaborative filtering. IEEE Internet Comput. **7**(1), 76–80 (2003)
9. Mnih, A., Salakhutdinov, R.: Probabilistic matrix factorization. In: NIPS, pp. 1257–1264 (2007)
10. Pan, R., et al.: One-class collaborative filtering. In: ICDM, pp. 502–511. IEEE (2008)
11. Pan, W., Chen, L.: GBPR: group preference based bayesian personalized ranking for one-class collaborative filtering. In: IJCAI, vol. 13, pp. 2691–2697 (2013)
12. Peng, M., Zeng, G., Sun, Z., Huang, J., Wang, H., Tian, G.: Personalized app recommendation based on app permissions. World Wide Web **21**(1), 89–104 (2018)
13. Rendle, S., Freudenthaler, C., Gantner, Z., Schmidt-Thieme, L.: BPR: Bayesian personalized ranking from implicit feedback. In: UAI, pp. 452–461. AUAI Press (2009)
14. Sarwar, B., Karypis, G., Konstan, J., Riedl, J.: Item-based collaborative filtering recommendation algorithms. In: WWW, pp. 285–295. ACM (2001)
15. Shu, J., Jia, X., Yang, K., Wang, H.: Privacy-preserving task recommendation services for crowdsourcing. IEEE Trans. Serv. Comput. (2018)
16. Tobler, W.R.: A computer movie simulating urban growth in the detroit region. Econ. geogr. **46**, 234–240 (1970)
17. Wu, Y.: Real estate's contribution to GDP falling. http://timmurphy.org/2009/07/22/line-spacing-in-latex-documents/. Accessed 1 Nov 2017
18. Yin, H., Cui, B., Zhou, X., Wang, W., Huang, Z., Sadiq, S.: Joint modeling of user check-in behaviors for real-time point-of-interest recommendation. TOIS **35**(2), 1–44 (2016)

19. Yu, Y., Chen, X.: A survey of point-of-interest recommendation in location-based social networks. In: Workshops at the Twenty-Ninth AAAI Conference on Artificial Intelligence (2015)
20. Yu, Y., Gao, Y., Wang, H., Wang, R.: Joint user knowledge and matrix factorization for recommender systems. In: Cellary, W., Mokbel, M.F., Wang, J., Wang, H., Zhou, R., Zhang, Y. (eds.) WISE 2016. LNCS, vol. 10041, pp. 77–91. Springer, Cham (2016). https://doi.org/10.1007/978-3-319-48740-3_6
21. Yu, Y., Gao, Y., Wang, H., Wang, R.: Joint user knowledge and matrix factorization for recommender systems. World Wide Web **21**(4), 1141–1163 (2018)
22. Yu, Y., Wang, H., Sun, S., Gao, Y.: Exploiting location significance and user authority for point-of-interest recommendation. In: Kim, J., Shim, K., Cao, L., Lee, J.-G., Lin, X., Moon, Y.-S. (eds.) PAKDD 2017. LNCS (LNAI), vol. 10235, pp. 119–130. Springer, Cham (2017). https://doi.org/10.1007/978-3-319-57529-2_10
23. Zhao, T., McAuley, J., King, I.: Leveraging social connections to improve personalized ranking for collaborative filtering. In: CIKM, pp. 261–270. ACM (2014)

Cross-domain Recommendation with Consistent Knowledge Transfer by Subspace Alignment

Qian Zhang, Jie Lu$^{(\boxtimes)}$, Dianshuang Wu, and Guangquan Zhang

Decision Systems and e-Service Intelligence Laboratory,
Center for Artificial Intelligence, Faculty of Engineering and Information Technology,
University of Technology, Sydney, Australia
{qian.zhang-1,jie.lu,dianshuang.wu,guangquan.zhang}@uts.edu.au

Abstract. Recommender systems have drawn great attention from both academic area and practical websites. One challenging and common problem in many recommendation methods is data sparsity, due to the limited number of observed user interaction with the products/services. Cross-domain recommender systems are developed to tackle this problem through transferring knowledge from a source domain with relatively abundant data to the target domain with scarce data. Existing cross-domain recommendation methods assume that similar user groups have similar tastes on similar item groups but ignore the divergence between the source and target domains, resulting in decrease in accuracy. In this paper, we propose a cross-domain recommendation method transferring consistent group-level knowledge through aligning the source subspace with the target one. Through subspace alignment, the discrepancy caused by the domain-shift is reduced and the knowledge shared local top-n recommendation via refined item-user bi-clustering two domains is ensured to be consistent. Experiments are conducted on five real-world datasets in three categories: movies, books and music. The results for nine cross-domain recommendation tasks show that our proposed method has improved the accuracy compared with five benchmarks.

Keywords: Recommender systems
Cross-domain recommender systems · Knowledge transfer
Collaborative filtering

1 Introduction

Recommender systems have been in existence for more than twenty years with wide application [15]. They are now an indispensable part of Internet websites such as Amazon.com, YouTube, Netflix, Yahoo, Facebook, Last.fm and Meetup. With great success and promising future, recommender systems are developed to provide users with more accurate and various options. User demands of diverse

© Springer Nature Switzerland AG 2018
H. Hacid et al. (Eds.): WISE 2018, LNCS 11234, pp. 67–82, 2018.
https://doi.org/10.1007/978-3-030-02925-8_5

recommendation have prompted recommender systems to expand from single-domain to multi-domain [1]. The mining of correlation between several domains can benefit every single domain, meanwhile possibly discovering user preferences that cannot be found with single domain data. Further, exploiting several domains together provides a way to solve the data sparsity problem, which is a common and challenging issue in lots of existing recommendation methods. For example, one user may have few records in a book category in an online review and rating system, but a lot of movie ratings available. The abundance of data in another domain can assist the recommendation in a specific target domain. All of these have brought about the increasing research on cross-domain recommender systems.

Cross-domain recommender systems are developed to solve the data sparsity problem taking advantage of data in multiple domains. These systems aim to extract knowledge from domains that contain relatively rich data and adapt it to the target domain where data is insufficient. Two different types of cross-domain recommender systems have been developed. Some methods connect multiple domains through auxiliary information rather than preference data [17]. It is assumed that some side information on users/items is available, either user generated information [7], social information [9] or item attributes [19]. On the other hand, some methods focus on preference data which are the most commonly collected data on e-commerce or online rating websites. This type of systems is designed in various ways according to the overlap of users and items [16] or the form of the data [8]. Due to the privacy issue, the user IDs are usually de-identified and items from two domains are not always the same. Some previous research tried to find the linkage through user display name [13] or user spatial behavior [2]. In this paper, we focus on cross-domain recommender systems where users and items have no intersections. In this situation, the basic assumption of cross-domain recommendation similar to the basic assumption of collaborative filtering [19]: A group of users tend to rate a group items similarly as another group of users implies similar preference to another group of items implies similar attributes. The knowledge shared by these two domains is group-level rating pattern.

Though cross-domain recommender systems have attracted lots of attention and efforts from academia, they still suffer the "negative transfer" problem [14]. The main reason is that data collected from two correlated domains are probably from two related but different distributions. Thus, domain shift is an issue that must be taken into account and carefully handled in cross-domain recommender systems. Without adaptation to domain shift, cross-domain recommender systems are likely to fail to provide useful and accurate recommendation in the target domain where data are not sufficient [3]. Most existing methods on cross-domain recommendation ignore the domain shifts and fail to extract consistent knowledge shared by two or multiple domains. For example, codebook transfer (CBT) clusters users and items into groups and extracts group-level knowledge as a "codebook" [11]. Later, a probabilistic model named rating matrix generative model (RMGM) is extended from CBT, relaxing the hard group membership to soft membership [12]. These two methods cannot ensure that knowledge extracted from two different domains is consistent, and the effectiveness of knowledge transfer is not guaranteed.

In this paper, we investigate how to eliminate the domain shift and improve the accuracy of cross-domain recommender systems. Domain adaptation is a transfer learning technique that can deal with the shift between distributions of data from the source domain and the target domain. Subspace alignment is one of the promising domain adaptation approaches to reduce the discrepancy between two domains by moving source subspace basis closer to the target subspace, especially in unsupervised setting as we conduct group-level knowledge transfer where no label is available. Before shared group-level knowledge is extracted from two domains, the source subspace and the target subspace are aligned so that the consistency of knowledge extracted is ensured. Thus, we propose a cross-domain recommendation method with consistent knowledge transfer by subspace alignment (CKTSA). The main contributions of this paper are as follows:

1. A cross-domain recommendation method CKTSA is developed where subspace alignment is used to match the source subspaces and the target subspaces to ensure the consistency of knowledge extracted.
2. The proposed method CKTSA is evaluated on five real-world datasets with nine cross-domain recommendation tasks compared with five other non-transfer or cross-domain recommendation methods. The results show that our proposed method has superior advantages in providing accurate recommendation in sparse data especially when source domain data has divergence with target domain data.

The rest of the paper is organized as follows. Section 2 gives some preliminary and a formal definition of the problem. Section 3 describes our method using subspace alignment to ensure the consistent knowledge transfer in cross-domain recommendation. In Sect. 4, we present our experiments on five real-world datasets spanning three categories of data. Finally, in Sect. 5, conclusion is provided with some future directions of this research.

2   Preliminary and Problem Formation

In this section, before presenting our proposed method, recommendation by tri-factorization is briefly introduced. The problem targeted in this paper is also formally formulated.

2.1   Recommendation by Tri-Factorization

Matrix factorization projects both users and items onto the same latent space so that they are comparable, and through their inner products reconstructs the rating matrix [10]. Similarly, the rating matrix $X \in \mathbb{R}^{M \times N}$ (bold letters represent matrixes) can be factorized into three matrixes (suppose there are M users and N items). Users and items are mapped to different latent spaces and the two spaces are mapped together through a mediate interaction matrix, as in [5]: $X = USV^T$, where $U \in \mathbb{R}^{M \times K}$ is user feature matrix, representing users clustered into K groups, $V \in \mathbb{R}^{N \times L}$ is item feature matrix, representing items clustered into L groups and $S \in \mathbb{R}^{K \times L}$ is the group preference matrix, i.e. the group-level knowledge. Thus, $\Theta = \{U, S, V\}$ are the parameters used to predict the ratings and provide a recommendation.

To calculate the missing values, the user-item rating matrix is reconstructed through $\hat{X} = USV^T$. Tri-factorization of X minimizes the loss function $\mathcal{L}(X, USV^T)$, which measures the error of prediction. Since X is usually sparse, the loss function is in a weighted form as follows:

$$\mathcal{L}(X, USV^T) = \|I \circ (X - USV^T)\|_F \tag{1}$$

where I is the rating indicator matrix the same size of X representing whether the rating in X is observed or not, $I_{ij} \in \{0, 1\}$. $I_{ij} = 1$ indicates that the rating is observed and $I_{ij} = 0$ otherwise and $\circ$ denotes the Hadamard product of matrixes. The tri-factorization is:

$$\min_{U, S, V} \mathcal{L}(X, USV^T)$$
$$\text{s. t. } U > 0, S > 0, V > 0$$

2.2 Problem Formulation

In this problem setting, the users/items have no correspondence across the domains and are treated as completely different users/items. We assume that explicit rating data are available for both the source and target domains. Formally, the problem is defined as:

Given a source rating matrix $X_s \in \mathbb{R}^{M_s \times N_s}$ and a target rating matrix $X_t \in \mathbb{R}^{M_t \times N_t}$, our goal is to develop a recommendation method aiming to help recommendation tasks in the target domain predict the rating $\hat{X}_t = U_t S_t V_t^T$ using knowledge in the source rating matrix X_s and $\Theta_s = \{U_s, S_s, V_s\}$, where $\mathcal{U}_s \cap \mathcal{U}_t = \emptyset$ and $\mathcal{I}_s \cap \mathcal{I}_t = \emptyset$. $\mathcal{U}_s$ and $\mathcal{I}_s$ represent the user set and item set in the source domain, while $\mathcal{U}_t$ and $\mathcal{I}_t$ represent the user set and item set in the target domain.

3   Cross-Domain Recommendation with Consistent Knowledge Transfer

In this section, our proposed CKTSA method is presented beginning with an overview of the method procedure containing five steps. Each of the five steps is then explained in detail.

3.1 The Method Overview

The proposed method CKTSA uses subspace alignment to ensure the knowledge extracted from the source domain is consistent with that in the target domain. The procedure consists five steps, as show in Fig. 1. (1) Users and items are clustered separately and user feature matrixes and item feature matrixes are obtained; (2) Subspace alignment is conducted to move the source subspace closer to the target domain; (3) Consistent knowledge is extracted since source subspaces is aligned to the target subspace and domain discrepancy is eliminated; (4) Feature representation is regulated in the target domain to retain domain specific characteristics; (5) Recommendation in the target domain.

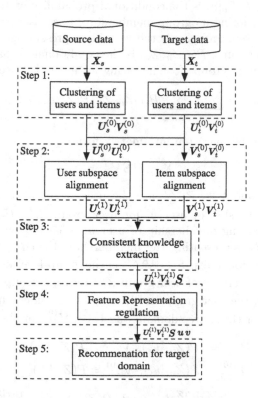

Fig. 1. The CKTSA method procedure.

3.2 The CKTSA Method

Our proposed method consists five steps.

Step 1: Clustering Users and Items in both Domains. We choose the Flexible Mixture Model (FMM) to cluster the users and items separately [18], since this method allows both users and items to fall into multiple groups with different memberships. This fits to the situation that users may have various preferences and items may have diverse content. The same clustering procedure is used for both the source domain and the target domain; however, for simplicity, we have only provided the description for one domain.

Suppose users are clustered into K user groups $\{Z_u^{(1)}, \ldots, Z_u^{(K)}\}$, while items are clustered into L item groups $\{Z_v^{(1)}, \ldots, Z_v^{(L)}\}$. Z_u and Z_v are two latent variables that denote the user and item groups respectively. $P(Z_u|u)$ is the conditional probability of a user belonging to a user group, denoting the group membership of the user; $P(Z_v|v)$ is the conditional probability of an item belonging to an item group, denoting its group membership. Each user group has a rating preference for each item group. r is the variable representing the preference of user groups to item groups. $P(r|Z_u, Z_v)$ is the conditional probability of r given user group Z_u and item group Z_v. The rating for a coupled user-item pair is:

$$R(u, v) = \sum_r r \sum_{Z_u, Z_v} P(r|Z_u, Z_v) P(Z_u|u) P(Z_v|v) \tag{2}$$

Equation (2) can be rewritten into matrix form:

$$\boldsymbol{X} = \boldsymbol{USV}^T \tag{3}$$

where $\boldsymbol{U} \in \mathbb{R}^{M \times L}$ and $\boldsymbol{V} \in \mathbb{R}^{N \times L}$ are the user and item feature matrix. U_{ij} represents the membership of user u_i for user group $Z_u^{(j)}$. U_{i*} is the ith row of matrix U representing membership of user u_i to each group. U_{*j} is the jth column of matrix $\boldsymbol{U}$ representing the membership of each user to user group $Z_u^{(j)}$. The same goes for items. $\boldsymbol{S} \in \mathbb{R}^{K \times L}$ is the group-level knowledge matrix. S_{ij} represents the preference of user group $Z_u^{(i)}$ for item group $Z_v^{(j)}$.

After clustering, the user group and item group membership matrixes $\boldsymbol{U}_s^{(0)}$, $\boldsymbol{V}_s^{(0)}$ are acquired for the source domain and $\boldsymbol{U}_t^{(0)}$, $\boldsymbol{V}_t^{(0)}$ for the target domain.

$$\boldsymbol{U}_s^{(0)} = P(Z_{u_s}|u_s), \ \boldsymbol{V}_s^{(0)} = P(Z_{v_s}|v_s) \tag{4}$$

$$\boldsymbol{U}_t^{(0)} = P(Z_{u_t}|u_t), \ \boldsymbol{V}_t^{(0)} = P(Z_{v_t}|v_t) \tag{5}$$

where $P(Z_u|u) = \frac{P(u|Z_u)P(Z_u)}{\sum_{Z_u} P(u|Z_u)P(Z_u)}$ and $P(Z_v|v) = \frac{P(v|Z_v)P(Z_v)}{\sum_{Z_v} P(v|Z_v)P(Z_v)}$. Five parameters $P(u|Z_u)$, $P(v|Z_v)$, $P(r|Z_u, Z_v)$, $P(Z_u)$ and $P(Z_v)$ are learned from the FMM. An expectation maximization (EM) algorithm is used to learn FMM. Suppose there are H ratings in user-item rating matrix represented as $\{(u_1, v_1, r_1), (u_2, v_2, r_2), (u_H, v_H, r_H)\}$. In the E-step, the parameters are fixed to optimize joint posterior probability. For any user-item-rating triplet (u_h, v_h, r_h):

$$P(Z_u, Z_v|u_h, v_h, r_h) = \frac{P(r_h|Z_u, Z_v)P(u_h|Z_u)P(v_h|Z_v)P(Z_u)P(Z_v)}{\sum_{Z_u, Z_v} P(r_h|Z_u, Z_v)P(u_h|Z_u)P(v_h|Z_v)P(Z_u)P(Z_v)} \tag{6}$$

Then, in the M-step, the parameters are updated as follows:

$$P(Z_u) = \frac{\sum_h \sum_{Z_v} P(Z_u, Z_v | u_h, v_h, r_h)}{H} \tag{7}$$

$$P(Z_v) = \frac{\sum_h \sum_{Z_u} P(Z_u, Z_v | u_h, v_h, r_h)}{H} \tag{8}$$

$$P(u|Z_u) = \frac{\sum_{u_h=u} \sum_{Z_v} P(Z_u, Z_v | u_h, v_h, r_h)}{H \cdot P(Z_u)} \tag{9}$$

$$P(v|Z_v) = \frac{\sum_{v_h=v} \sum_{Z_u} P(Z_u, Z_v | u_h, v_h, r_h)}{H \cdot P(Z_v)} \tag{10}$$

$$P(r|Z_u, Z_v) = \frac{\sum_{r=r_h} P(Z_u, Z_v | u_h, v_h, r_h)}{\sum_h P(Z_u, Z_v | u_h, v_h, r_h)} \tag{11}$$

By alternatively computing the E-step and the M-step, users and items are clustered into latent groups according to the user-item rating matrix. (For details, see [18]).

Step 2: Subspace Alignment of User and Items. After users and items are clustered in both source and target domains, user feature matrixes $U_s^{(0)}$ and $U_t^{(0)}$ and item feature matrixes $V_s^{(0)}$ and $V_t^{(0)}$ are obtained. It is possible to cluster users in source and target domains both in K groups to make users both lie in K-dimensional space, but in fact the user feature matrixes are from different marginal distributions. The divergence in data distributions need to be eliminated before conducting knowledge extraction. We use subspace alignment in this paper to learn new representations of users and items so that they are in the same subspace coordinate system.

To better handle the characteristics of data distributions, we firstly use the function of Z-score to normalize the original representations of users in both domains. Then, we use principal component analysis (PCA) to extract d eigenvectors corresponding to the largest d eigenvalues which are treated as the basis of source and target spaces, denoted as D_s^u and D_t^u for user subspaces and D_s^v and D_t^v for item subspaces. For subspace alignment, we align the subspace basis according to [6]. User and item transition matrixes T^u and T^v are learned through optimizing:

$$\mathcal{L}(T^u) = ||D_s^u T^u - D_t^u||_F$$
$$\min_{T^u} \mathcal{L}(T^u) \tag{12}$$
$$\mathcal{L}(T^v) = ||D_s^v T^v - D_t^v||_F$$
$$\min_{T^v} \mathcal{L}(T^v) \tag{13}$$

Closed forms of optimal T^u and T^v are as follows (details refer to [6]):

$$T^u = (D_s^u)^T D_t^u \tag{14}$$
$$T^v = (D_s^v)^T D_t^v \tag{15}$$

Once the transition matrixes are obtained, the user and item subspaces can be aligned to the same one. How the subspace alignment is done is summarized in Algorithm 1. There are two advantages using the subspace alignment: (1) Compared with other domain adaptation methods that directly project source and target domain data to a shared common subspace, subspace alignment is not only limited to the shared features but able to exploit more correlations between two domains; (2) Compared with other domain adaptation methods that model domain shift and learn new representations through a large number of subspaces, subspace alignment uses a linear transition function that is simple and powerful. These two advantages fit well in our cross-domain recommendation problem setting. The first advantage contributes to the potential diverse demand of users while the second advantage meets users' requirement on recommender systems to respond quickly and provide timely support.

Algorithm 1. User and Item Subspace Alignment

Input:
    $U_s^{(0)}$, $U_t^{(0)}$, the source and target user feature matrix
    $V_s^{(0)}$, $V_t^{(0)}$, the source and target item feature matrix
Output:
    $U_s^{(1)}$, $U_t^{(1)}$, the aligned source and target user feature matrix
    $V_s^{(1)}$, $V_t^{(1)}$, the aligned source and target item feature matrix
1: $D_s^u \leftarrow PCA(f_{zs}(U_s^0))$, $D_t^u \leftarrow PCA(f_{zs}(U_t^0))$
   $D_s^v \leftarrow PCA(f_{zs}(V_s^0))$, $D_t^v \leftarrow PCA(f_{zs}(V_t^0))$
2: $T^u \leftarrow (D_s^u)^T D_t^u$, $T^v \leftarrow (D_s^v)^T D_t^v$
3: $U_s^{(1)} \leftarrow U_s^{(0)} D_s^u T^u$, $U_t^{(1)} \leftarrow U_t^{(0)} D_t^u$
   $V_s^{(1)} \leftarrow V_s^{(0)} D_s^v T^v$, $V_t^{(1)} \leftarrow V_t^{(0)} D_t^v$
4: Normalize $U_s^{(1)}$, $U_t^{(1)}$, $V_s^{(1)}$ and $V_t^{(1)}$ to $[0, 1]$
5: **return** $U_s^{(1)}$, $U_t^{(1)}$, $V_s^{(1)}$, $V_t^{(1)}$

Step 3: Consistent Knowledge Extraction. After the domain adaptation, $U_s^{(1)}$, $U_t^{(1)}$ and $V_s^{(1)}$, $V_t^{(1)}$ aligned to the same subspaces. Once the source subspace is aligned to the target subspace, the recommender systems learned from the source and target domains will share the same group-level knowledge matrix S.

Consistent knowledge S is obtained by maximizing the approximation of the available data in both the source rating matrix and the target rating matrix. To qualify the approximation, Frobenius norm is used as a measure between the original rating matrix and the approximation. We list the cost function here and more details can be found in [20]:

$$J_s(S) = \frac{1}{M_s N_s}\|I_s \circ (X_s - U_s^{(1)}S(V_s^{(1)})^T)\|_F + \frac{1}{M_t N_t}\|I_t \circ (X_t - U_t^{(1)}S(V_t^{(1)})^T)\|_F + \frac{\lambda}{2KL}\|S\|_F \quad (16)$$

where I_s is an indicator matrix for X_s, if $(I_s)_{ij} = 1$, then $(X_s)_{ij} \neq 0$ and $(I_s)_{ij} = 0$, otherwise. The same applies to I_t for X_t. $\circ$ is an entry-wise product,

λ is the parameter for regularization. Finally, consistent knowledge is learned through the following optimization problem:

$$\min_{S} J_s(S)$$
$$\text{s.t. } S > 0$$

Step 4: Feature Representation Regulation. In our problem setting, some domain-specific characteristics are embedded in the small amount of available data in the target rating matrix. To reveal these idiosyncrasies of the target domain, we amend feature representations of the target rating matrix to make the model fit better to the task in target rating matrix. The representation regulation is achieved through an optimization problem. The cost function is:

$$J_r(u,v) = \|I_t \circ (X_t - U_t^{(1)} u S (V_t^{(1)} v)^T)\|_F \tag{17}$$

The tuning factors can be learned through optimizing

$$\min_{u,v} J_r(u,v)$$
$$\text{s.t. } u \geq 0, v \geq 0$$

The optimization problem is solved by alternatively estimating tuning factors u and v. For more details, see [20].

Step 5: Recommendation in Target Domain. The recommendation in target domain is given by Eq. (18).

$$\hat{X}_t = (U_t^{(1)} u) S (V_t^{(1)} v)^T \tag{18}$$

where $\hat{X}_t$ is the reconstructed user-item rating matrix for prediction, u, v are user and item tuning factors for target domain, S is the consistent knowledge, $U_s^{(1)}$, $U_t^{(1)}$ are user and item feature matrixes for the target domain after subspace alignment obtained from Algorithm 1.

4 Experiments

In this section, the proposed method CKTSA is evaluated. First, the datasets and evaluation metrics used are introduced in Sect. 4.1, followed by experimental settings and the baseline methods 4.2. The results of the experiments are presented in Sect. 4.3.

4.1 Datasets and Evaluation Metrics

To test our proposed method, we need to choose data from similar data so that transfer learning is meaningful, but divergence still exists between the source

Table 1. Statistical information on the original datasets

	Each Movie	Movielens 1M	Library thing	Amazon book	Yahoo music_1	Yahoo music_2
#user	72916	6040	7279	8026324	200000	200000
#item	1628	3900	37232	2330066	136736	136736
#rating	2811983	1000209	749401	22507155	78344627	78742463
Sparsity	97.63%	95.75%	99.72%	99.99%	99.71%	99.71%
Range	0–1	1–5	0.5–5	1–5	1–5	1–5

and target domains. Similar to [20], we choose movie, book and music as three categories for experiments. Our experiments comprise nine cross-domain recommendation tasks with all the combinations of the three categories. Five real-world datasets were used: EachMovie[1], Movielens1M[2], LibraryThing[3], Amazon Book[4] and YahooMusic[5]. Each is publicly available and has been used to test recommender systems in a variety of scenarios for recommender systems in single domain. But tests on these dataset in this novel cross-domain setting are lacking. For AmazonBooks, we removed all users who had given exactly the same rating for every book, as these data are not effective for constructing a recommender system [20]. EachMovie and LibraryThing was normalized to the range of $\{1, 2, 3, 4, 5\}$ before conducting experiments. The statistical information for original datasets is provided in Table 1. Across all the datasets, 1000 items that had been rated more than 10 times were randomly chosen. We then filtered out the users who had given less than a total of 20 ratings. For the source domain data, we randomly selected 500 users to be regular customers of the site. The source domain data were controlled to be more dense than the target domain data. For the target domain data, we randomly selected 200 users to be regular customers of the site, and another 300 users to be new customers. For new users, five observed ratings were given, and the rest of the ratings were used for evaluation. In the end, the rating matrixes for both the source and target domains were all 500×1000 matrixes. The details of the final datasets are summarized in Table 2. Mean absolute error (MAE) and root mean square error (RMSE) were used as the evaluation metrics:

---

[1] http://www.cs.cmu.edu/~lebanon/IR-lab/data.html#intro.
[2] http://grouplens.org/datasets/movielens/1m/.
[3] https://www.librarything.com.
[4] http://jmcauley.ucsd.edu/data/amazon/.
[5] https://webscope.sandbox.yahoo.com/catalog.php?datatype=r.

Table 2. Description of data subsets in three categories

Data type	Data source	Domain	Sparsity	Average
Movie	EachMovie	Source	96.00%	4.32
	Movielens1M	Target	98.50%	2.91
Book	LibraryThing	Source	87.43%	3.97
	AmazonBook	Target	97.87%	3.13
Music	YahooMusic_1	Source	95.70%	4.14
	YahooMusic_2	Target	97.27%	2.66

$$MAE = \sum_{u,v,X_{uv} \in Y} \frac{|\hat{X}_{uv} - X_{uv}|}{|Y|}$$

$$RMSE = \sqrt{\sum_{u,v,X_{uv} \in Y} \frac{(\hat{X}_{uv} - X_{uv})^2}{|Y|}}$$

where Y is the test set, and $|Y|$ is the number of test ratings.

4.2 Experimental Settings and Baselines

The rating average is a very important statistics of the data which we used in our experiments to represent whether data in two domains are of high similarity or not. According to Table 2, we can see a big divergence in the rating average between the source domain data and the target domain data. This fits to our problem setting in Sect. 2 that data in the source domain and the target domain are similar but divergence still exists.

Three non-transfer learning methods and two cross-domain methods were chosen as comparisons for the proposed method. The non-transfer learning methods were: Pearson's correlation coefficient (PCC) [4], FMM [18] and single value decomposition (SVD) [10]. The cross-domain methods were: CBT [11] and RMGM [12]. These two cross-domain recommendation methods are all developed without fully considering the domain-shift widely existed in data of two domains. PCC uses user-based CF, and the number of neighborhoods was set at 50. For SVD, the latent feature number was fixed at 10, the regularization factor was set to 0.015, and the learning rate was set to 0.003. For FMM, CBT and RMGM, the user group number and item group number were both set to 10. For the proposed method, CKTSA, the user feature number and the item feature number were both set to 10, and the regularization factor was set to 0.5. Further analysis of the parameters is provided in Sect. 4.3. All the methods (except for PCC) need to initialize the factorized matrix randomly, we ran 20 random initializations and report the averaged results and standard deviations.

4.3 Results

The experiment results of our proposed CKTSA compared with the other five baselines on two accuracy metrics are presented in Tables 3, 4 and 5. Overall, CKTSA has the best performance in all the nine tasks relates to three categories: movie, book and music. In sparse data settings, CKTSA can significantly increase the recommendation accuracy compared with non-transfer learning recommendation methods like PCC, FMM and SVD. These results suggest that CKTSA is an effective method to transfer knowledge that can assist recommendation in the target domain.

Also, CKTSA has better performance than other two cross-domain recommendation methods CBT and RMGM. Compared with other non-transfer learning methods, CBT fails to transfer knowledge to the target domain in most cases. The transfer learning in CBT is even negative in many of the experiments. This is because the core algorithm in CBT is very simple and contains direct group-level knowledge transfer without adjustment or adaptation to the divergence between two domains. On the other hand, RMGM achieves at least positive transfer in most of the tasks compared with CBT. But RMGM still cannot meet the requirement of cross-domain recommendation to extract consistent knowledge from a source domain which is similar but slightly different to the target domain. The comparison between CKTSA and these two cross-domain recommendation methods implies that CKTSA can transfer consistent knowledge from the source domain to the target domain. The advantage is more obvious when it comes to the situation that divergence exists between two domains.

Table 3. Prediction performance on a movie target domain

Methods		Source data	MAE	RMSE
Non-trans	PCC	-	1.2123	1.5722
	FMM	-	1.1104 ± 0.0118	1.3540 ± 0.0143
	SVD	-	1.0949 ± 0.0049	1.3540 ± 0.0074
Cross-domain	CBT	Movie	1.4587 ± 0.0176	1.7883 ± 0.0161
		Book	1.2275 ± 0.0077	1.5551 ± 0.0152
		Music	1.2892 ± 0.0239	1.6383 ± 0.0217
	RMGM	Movie	1.1896 ± 0.0170	1.4810 ± 0.0225
		Book	1.1357 ± 0.0231	1.4102 ± 0.0335
		Music	1.1337 ± 0.0191	1.4042 ± 0.0283
	CKTSA	Movie	$\mathbf{1.0273} \pm 0.0066$	$\mathbf{1.2379} \pm 0.0071$
		Book	$\mathbf{1.0293} \pm 0.0062$	$\mathbf{1.2392} \pm 0.0062$
		Music	$\mathbf{1.0275} \pm 0.0060$	$\mathbf{1.2385} \pm 0.0078$

Table 4. Prediction performance on a book target domain

Methods		Source data	MAE	RMSE
Non-trans	PCC	-	1.1802	1.4907
	FMM	-	1.0260 ± 0.0118	1.2670 ± 0.0146
	SVD	-	1.2264 ± 0.0133	1.5283 ± 0.0166
Cross-domain	CBT	Movie	1.2762 ± 0.0091	1.5918 ± 0.0085
		Book	1.0951 ± 0.0040	1.3997 ± 0.0107
		Music	1.1337 ± 0.0115	1.4588 ± 0.0113
	RMGM	Movie	1.0072 ± 0.0104	1.2418 ± 0.0155
		Book	1.0152 ± 0.0150	1.2492 ± 0.0209
		Music	1.0073 ± 0.0117	1.2429 ± 0.0159
	CKTSA	Movie	$\mathbf{0.9772} \pm \mathbf{0.0047}$	$\mathbf{1.1814} \pm \mathbf{0.0064}$
		Book	$\mathbf{0.9762} \pm \mathbf{0.0037}$	$\mathbf{1.1804} \pm \mathbf{0.0043}$
		Music	$\mathbf{0.9810} \pm \mathbf{0.0066}$	$\mathbf{1.1852} \pm \mathbf{0.0077}$

Table 5. Prediction performance on a music target domain

Methods		Source data	MAE	RMSE
Non-trans	PCC	-	1.4843	1.8539
	FMM	-	1.2959 ± 0.0211	1.5492 ± 0.0172
	SVD	-	1.4797 ± 0.0208	1.7229 ± 0.0242
Cross-domain	CBT	Movie	1.8215 ± 0.0111	2.1673 ± 0.0119
		Book	1.6315 ± 0.0092	1.9342 ± 0.0153
		Music	1.6932 ± 0.0125	2.0236 ± 0.0147
	RMGM	Movie	1.3329 ± 0.0162	1.6057 ± 0.0180
		Book	1.3306 ± 0.0190	1.6014 ± 0.0229
		Music	1.3306 ± 0.0190	1.6014 ± 0.0229
	CKTSA	Movie	$\mathbf{1.2915} \pm \mathbf{0.0101}$	$\mathbf{1.5022} \pm \mathbf{0.0131}$
		Book	$\mathbf{1.2922} \pm \mathbf{0.0149}$	$\mathbf{1.5046} \pm \mathbf{0.0149}$
		Music	$\mathbf{1.2911} \pm \mathbf{0.0107}$	$\mathbf{1.5004} \pm \mathbf{0.0119}$

4.4   Parameter Analysis

We analyzed how the parameter K, L and λ affect the performance of CKTSA. K is the number of user groups and L is the number of item groups, while λ is the trade-off parameter for the consistent knowledge extraction. Due to the space limitation, only the result of movie-to-movie cross-domain recommendation task is presented. Performance on both MAE and RSME is presented.

To analyze λ, K and L are fixed to 10. In Fig. 2, both MAE and RSME are not significantly affected by λ. To analyze K and L, λ is set to be 0.5. As for the K and L, we use grid search to analyze how these two parameters can interact

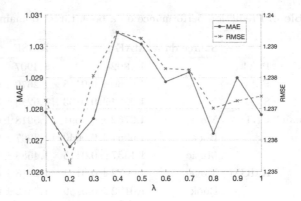

Fig. 2. Results with different settings on parameter λ.

with each other and how they can affect the performance of CKTSA in Fig. 3. In the range of 5 to 70, the higher the number, the better the performance. Since the higher K and L, the complexity of this method will significantly increase. Thus, in our experiments, we choose K and L for 10 for convenience. And we choose the same value for K and L in every tri-factorization based methods.

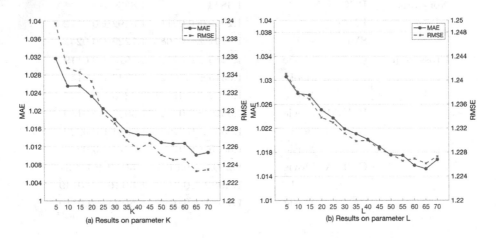

Fig. 3. Results with different settings on parameters K and L.

5   Conclusion

In this paper, we investigate the cross-domain recommendation problem, specially for two domains where domain-shifts happen. A cross-domain recommendation method called CKTSA is developed that can ensure consistent knowledge transfered from the source domain to the target domain. In the proposed method, we firstly obtained the subspace eigenvectors of both users and items in the source domain and the target domain. Through subspace alignment, the two coordinate systems are aligned to the same one. Since the subspaces are aligned, consistent knowledge can be extracted from two domains which can help to improve the accuracy of recommendation in the target domain. Extensive experiments on five real-world datasets with nine cross-domain recommendation tasks show that the proposed CKTSA achieves the best performance compared with five baselines including both non-transfer learning and cross-domain recommendation methods. In the future, we will try to develop methods that can deal with heterogeneous data in this problem setting. In this way, implicit feedback and other user behaviors can all be involved to profile the user in multiple domains.

References

1. Cantador, I., Fernández-Tobías, I., Berkovsky, S., Cremonesi, P.: Cross-domain recommender systems. In: Ricci, F., Rokach, L., Shapira, B. (eds.) Recommender Systems Handbook, pp. 919–959. Springer, Boston (2015). https://doi.org/10.1007/978-1-4899-7637-6_27
2. Chen, W., Yin, H., Wang, W., Zhao, L., Hua, W., Zhou, X.: Exploiting spatio-temporal user behaviors for user linkage. In: Proceedings of the 2017 ACM on Conference on Information and Knowledge Management, pp. 517–526. ACM (2017)
3. Cremonesi, P., Quadrana, M.: Cross-domain recommendations without overlapping data: myth or reality? In: Proceedings of the 8th ACM Conference on Recommender Systems, pp. 297–300. ACM (2014)
4. Deshpande, M., Karypis, G.: Item-based top-N recommendation algorithms. ACM Trans. Inf. Syst. **22**(1), 143–177 (2004)
5. Ding, C., Li, T., Peng, W., Park, H.: Orthogonal nonnegative matrix t-factorizations for clustering. In: Proceedings of the 19th ACM SIGKDD International Conference on Knowledge Discovery and Data Mining, pp. 126–135. ACM (2006)
6. Fernando, B., Habrard, A., Sebban, M., Tuytelaars, T.: Unsupervised visual domain adaptation using subspace alignment. In: 2013 IEEE International Conference on Computer Vision (ICCV), pp. 2960–2967. IEEE (2013)
7. Hao, P., Zhang, G., Martinez, L., Lu, J.: Regularizing knowledge transfer in recommendation with tag-inferred correlation. IEEE Trans. Cybern. (99), 1–14 (2017)
8. Hu, L., Cao, J., Xu, G., Cao, L., Gu, Z., Zhu, C.: Personalized recommendation via cross-domain triadic factorization. In: Proceedings of the 22nd International Conference on World Wide Web, pp. 595–606. ACM (2013)
9. Jiang, M., Cui, P., Chen, X., Wang, F., Zhu, W., Yang, S.: Social recommendation with cross-domain transferable knowledge. IEEE Trans. Knowl. Data Eng. **27**(11), 3084–3097 (2015)
10. Koren, Y., Bell, R., Volinsky, C.: Matrix factorization techniques for recommender systems. Computer **42**(8), 30–37 (2009)
11. Li, B., Yang, Q., Xue, X.: Can movies and books collaborate? Cross-domain collaborative filtering for sparsity reduction. In: IJCAI, vol. 9, pp. 2052–2057 (2009)
12. Li, B., Yang, Q., Xue, X.: Transfer learning for collaborative filtering via a rating-matrix generative model. In: Proceedings of the 26th Annual International Conference on Machine Learning, pp. 617–624. ACM (2009)

13. Li, Y., Peng, Y., Zhang, Z., Wu, M., Xu, Q., Yin, H.: A deep dive into user display names across social networks. Inf. Sci. **447**, 186–204 (2018)
14. Lu, J., Behbood, V., Hao, P., Zuo, H., Xue, S., Zhang, G.: Transfer learning using computational intelligence: a survey. Knowl.-Based Syst. **80**, 14–23 (2015)
15. Lu, J., Wu, D., Mao, M., Wang, W., Zhang, G.: Recommender system application developments: a survey. Decis. Support Syst. **74**, 12–32 (2015)
16. Mirbakhsh, N., Ling, C.X.: Improving top-N recommendation for cold-start users via cross-domain information. ACM Trans. Knowl. Discov. Data **9**(4), 33 (2015)
17. Shi, Y., Larson, M., Hanjalic, A.: Collaborative filtering beyond the user-item matrix: a survey of the state of the art and future challenges. ACM Comput. Surv. **47**(1), 3 (2014)
18. Si, L., Jin, R.: Flexible mixture model for collaborative filtering. In: Proceedings of the 20th International Conference on Machine Learning, pp. 704–711 (2003)
19. Singh, A.P., Gordon, G.J.: Relational learning via collective matrix factorization. In: Proceedings of the 14th ACM SIGKDD International Conference on Knowledge Discovery and Data Mining, pp. 650–658. ACM (2008)
20. Zhang, Q., Wu, D., Lu, J., Liu, F., Zhang, G.: A cross-domain recommender system with consistent information transfer. Decis. Support Syst. **104**, 49–63 (2017)

Medical Data Analysis

D-ECG: A Dynamic Framework
for Cardiac Arrhythmia Detection
from IoT-Based ECGs

Jinyuan He[1], Jia Rong[1(✉)], Le Sun[2], Hua Wang[1], Yanchun Zhang[1],
and Jiangang Ma[3]

[1] Institute of Sustainable Industries and Liveable Cities, Victoria University,
Melbourne, Australia
jinyuan.he@live.vu.edu.au, {jia.rong,hua.wang,yanchun.zhang}@vu.edu.au
[2] School of Computer and Software,
Nanjing University of Information Science and Technology, Nanjing, China
sunle2009@gmail.com
[3] College of Science and Engineering, James Cook University, Queensland, Australia
jiangang.ma@jcu.edu.au

Abstract. Cardiac arrhythmia has been identified as a type of cardiovascular diseases (CVDs) that causes approximately 12% of all deaths globally. The current progress on arrhythmia detection based on ECG recordings is facing a bottleneck for adopting single classifier and static ensemble methods. Besides, most of the work tend to use a static feature set for characterizing all types of heartbeats, which may limit the classification performance. To fill in the gap, a novel framework called D-ECG is proposed to introduce dynamic ensemble selection (DES) technique to provide accurate detection of cardiac arrhythmia. In addition, the proposed D-ECG develops a result regulator that use different features to refine the classification result from the DES technique. The results reported in this paper have shown visible improvement on the overall heartbeat classification accuracy as well as the sensitivity of disease heartbeats.

Keywords: ECG · Cardiac arrhythmia detection
Dynamic ensemble selection

1 Introduction

Cardiac arrhythmia is a type of cardiovascular diseases (CVDs) that seriously affects millions of people around the world and accounts for approximately 80% of the sudden cardiac death [46]. Critically, arrhythmia can be categorized as life-threatening and non-life-threatening arrhythmia [56]. Life-threatening arrhythmia imminently threatens patients' lives and emergency treatment is required. Non-life-threatening arrhythmia just imposes a long-term health risk to patients, but special care is still needed to avoid further deterioration of heart function.

© Springer Nature Switzerland AG 2018
H. Hacid et al. (Eds.): WISE 2018, LNCS 11234, pp. 85–99, 2018.
https://doi.org/10.1007/978-3-030-02925-8_6

ECG provides a noninvasive and inexpensive way for arrhythmia study. With a rapid growth of *Internet-of-Things* (IoT) techniques, more and more wearable ECG monitoring devices have been invented to produce high-quality ECG recordings [52]. Comparing to the traditional Holter device, the IoT-based ECG monitoring devices provide a more human-friendly way for heart status tracking. The ECG recordings are stored in the IoT cloud space, which are easy and convenient for both clinicians and patients to access. However, manual interpretation of the large amount of continuously generated ECG recordings is time-demanding and error-prone. Therefore, a computer-assisted method is needed to help analyze and interpret the ECG signals. This work aims to develop an efficient and effective method to help detect arrhythmia from the IoT-based ECG recordings.

Many researches attempts have been made to address the arrhythmia detection. Current methods are facing a bottleneck for adopting a single classifier trained with a predefined feature set [9,15,16,54], which may bias the classification results and lead to a relatively low generalization performance. This is because that the value of features vary significantly from patients to patients (even in the same patient) while a single classifier is believed to be an expert only in certain local regions of the feature space [57]. Although some ensemble methods, such as *random forest* [2] and *ensemble of support vector machine* [21], have been employed to remedy the disadvantages, the problem is only partly solved because the diversity of the traditional ensembles is relatively low. Similar to the use of a single classifier, the ensemble of classifiers are determined in the training phase. Besides, using a static set of features for classifying all types of heartbeats can limit the classification performance, as the sensitivity to certain feature varies with heartbeat types [56].

Apart from the above problems, the performance of arrhythmia detection related heavily to the training data preparation. Since the heartbeat data is naturally imbalanced, the trained classifier may have difficulties in correctly recognizing the disease heartbeats which only accounts for a small portion of the whole heartbeat data set [9]. A proper measure is demanded to eliminate the negative impacts of the imbalance [45].

To solve the above problems, this work proposes a dynamic framework called D-ECG for cardiac arrhythmia detection. The D-ECG introduces the *dynamic ensemble selection* (DES) technique to estimate a competence level of each classifier from a pool of classifiers in the training phase. In testing, the DES technique selects the most competent classifiers to predict the heartbeat label according to each heartbeat to be classified. To the best of our knowledge, this is the first time that the DES technique is used in cardiac arrhythmia detection scenario. Besides, a result regulator is creatively developed to refine the result from the previous phase. Essentially, the regulator is a classifier trained with a feature set which is different from the one used for DES training.

In summary, the contributions of this work include:

- proposing a dynamic framework named D-ECG, which first introduces the DES techniques for automatic cardiac arrhythmia detection.

- customizing a result regulator to improve the heartbeat classification performance.
- experimentally comparing the performance of various DES techniques in ECG-based heartbeats classification.
- experimentally evaluating the feasibility of D-ECG in arrhythmia detection, with the results being compared against the stat-of-the-art methods in the same field in terms of overall accuracy, sensitivity and positive predictive.

The rest of this paper is structured as follows. Section 2 introduces the IoT-based ECG monitoring system and Dynamic ensemble selection techniques, and reviews current methods in arrhythmia detection. Section 3 presents the proposed D-ECG. The experiment results and discussion are presented in Sect. 4. Section 5 concludes this paper and discusses the future work.

2   Related Work

This section firstly introduces the IoT-based ECG monitoring system and then reviews current methods in arrhythmia detection. After that, the dynamic ensemble technique is presented in detail.

2.1   IoT-Based ECG Monitoring System

An IoT-based ECG monitoring system mainly consists of three parts: an ECG sensing network, an IoT cloud, and a graphical user interface (GUI) [52]. The ECG sensing network is responsible for generating ECG recordings for patients and transmitting the produced data to the IoT cloud. The IoT cloud is mainly used for data storage and analysis. Specifically, the IoT cloud performs ECG signal interpretation and heart disease detection, such as arrhythmia detection [2, 9,15,20,28,34,56] and sleep apnea detection [2,4,9,10,23,35]. The GUI module is often used for data visualization and management.

Data security and privacy are important concerns for a IoT-based ECG monitoring system [44], since the ECG recordings stored in the cloud may suffer from various types of security threats, such as linking attacks and unauthorized access [39]. In order to ensure the safety and privacy of the ECG recordings, security mechanisms, such as anonymization [40,55] and access control [22,25,38,41–43], should be considered when designing an IoT-based ECG monitoring system.

2.2   Current Methods on Arrhythmia Detection

Since the life-threatening arrhythmia has been well studied [10], this study focus on the investigation of non-life-threatening arrhythmia (Supraventricular ectopic beat (S)) and the related ectopic heartbeats (Ventricular ectopic beat (V)). Figure 1 intuitively shows the differences between the normal heartbeats (N), supraventricular (S) and ventricular (V) ectopic beats. The N and S beats share similar shapes but not the RR-intervals. The S beats generally have a shorter

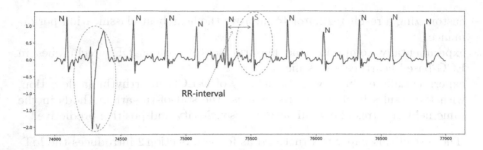

Fig. 1. A sample ECG recording that contains N, S and V heartbeats, where RR-intervals denote the time distance between two successive R peaks.

previous-RR compared to the N beats from the same patient. On the contrast, the V beats significantly differ from the N and S beats in terms of ECG shape.

Current methods on non-life-threatening arrhythmia detection mainly focused on signal features extraction and selection, and the use of various classifiers combinations. A heartbeat is normally represented by features extracted from cardiac rhythm, or time/frequency domains. Commonly used features include RR-intervals [2,9,56], samples or segments of ECG curves [32], higher-order statistics [2,17], wavelet coefficients [15,20,34], and signal energy [54]. To avoid negative impacts of noisy data, some techniques have been employed to reduce the feature space, such as the *floating sequential search* [27] and the *weighted LD model* with a forward-backward search strategy [18]. Although the extracted features in previous work have been proven to be effective in characterizing an ECG-based heartbeat, the classification performance is limited by using a static set of features for classifying all types of heartbeats.

Regarding the choice of classifiers for arrhythmia classification, the *support vector machine* (SVM) is the most widely used for its robustness, good generalization and computationally efficiency [1,13]. Besides, the *nearest neighbors* (NN) and *artificial neural networks* (ANN) are also frequently found in the literature. Other classifiers include *weighted linear discriminant* (WLD), *decision tree*, *optimum-path forest* (OPF). Nevertheless, the use of a single classifier can bias the classification and lead to a relatively low generalization performance. Although some ensemble methods have been employed to remedy the disadvantages, the problem can only be partly solved.

2.3   Dynamic Ensemble Selection

Dynamic ensemble selection (DES) is one of the promising approaches to construct a multiple classifier system (MCS). Recently, an increasing number of works have reported the superior performance of the DES over the static methods [6,11]. A DES-based system is composed of three stages: generation, selection and aggregation [6]. In the generation stage, a collection of classifiers are trained to create an accurate and diverse classifier pool. In the selection stage, an ensemble containing the most competent classifiers is selected. Finally, in the aggregation stage, the output of each classifier in the selected ensemble are aggregated to give the final decision of the system.

The core issue of DES techniques is the selection of the most competent classifiers for any testing sample [12]. The competence of a classifier in the pool is measured by its performance over a local region of the feature space where the testing sample is located. Methods for defining a local region includes clustering [26], k-nearest neighbors [36], potential function model [49,50] and decision space [7]. The criterion for measuring the performance of a base classifier can be divided as individual-based and group-based criterion. In individual-based criterion, each base classifier is independently measured by evaluation metrics such as ranking, accuracy, probabilistic, behavior [7], meta-learning [11]. In the group-based criterion, the performance of a base classifier relates to its iterations with other classifiers in the pool. For example, diversity, data handling [51] and ambiguity [19] are widely used group-based performance metrics.

Regarding the aggregation approaches, there are three main strategies for results combination: static combiner, trained combiner and dynamic weighting. The majority voting scheme is a representative static combiner, which is also commonly used in the traditional ensemble methods. In trainable combiners, the outputs of the selected based classifiers are used as the input features for another learning algorithm, such as [6,30]. In dynamic weighting, higher weight value will be allocated to the most competent classifier and then the outputs of all the weighted classifiers are aggregated to give the final decision.

Currently, the most prevalent DES techniques include DES-KL [50], DES-KNN [37], KNORA-E [24], KNORA-U [24], KNOP [7], DES-P [50], DES-RRC [49], and META-DES [11], which are discussed in Sect. 3.4.

3   Dynamic ECG Framework

The proposed D-ECG is detailedly presented in this section. The D-ECG is composed of five phases: preprocessing, feature extraction, classifier pool training, dynamic selection classification and result refinement.

3.1   ECG Data Preprocessing

IoT-collected data usually come with uncertainties [33]. Specifically, the IoT-collected ECG signals have serious background noise and baseline wandering. Baseline wandering is the effect that the base axis (X-axis) of individual heartbeats appear to move up or down rather than being straight all the time. Proper measures should be taken to eliminate the negative effect caused by the noises on cardiac arrhythmia detection.

To remove the baseline wandering, each ECG signal is processed with a 200-ms width median filter followed by a 600-ms median filter to obtain the signal baseline. The baseline is then subtracted from the raw ECG signal to get the baseline corrected data.

After that, the discrete wavelet transform is employed to remove the background noise from the baseline corrected signals. The *Daubechies-4* (DB4) mother wavelet function [14] is used because its short vanishing moment is ideal for analyzing signals with sudden changes like ECG. The coefficients of detail information (cD_x) in each frequency band are processed by a high-pass filter with a threshold value $T = \sqrt{2 * log(n)}$, where n is the length of the input signal. The clean signals are obtained by employing inverse discrete wavelet transform on the coefficients.

The clean signals are segmented to individual heartbeats by taking advantage of the labeled R peak locations. For each R peak, 90 samples (250 ms) before R peak and 144 samples (400 ms) after R peak are taken to represent a heartbeat, which is long enough to catch samples to represent the re-polarization of ventricles and short enough to exclude the neighbor heartbeats [2].

3.2   Feature Extraction

Disease heartbeats can cause disorders to heartbeat shape and heart rhythms in ECG signal. To effectively catch these anomalies, three types of features are used to represent individual heartbeats: RR-intervals, higher order statistics and wavelet coefficients.

As experimentally proven in [56], the RR-interval is one of the most indispensable features for heartbeat classification and it has great capacity to tell both the S and V beats from the normal beats. In this work, four types of RR-intervals are extracted from ECG signals: *pre_RR*, *post_RR*, *local_RR* and *global_RR* [28].

However, it should be noted that the RR-intervals can significantly vary from patient to patient. Therefore, the RR-intervals are normalized for each heartbeat in the way below:

$$nomalized\_pre\_RR = \frac{pre\_RR}{mean(ds.pre\_RR)} \tag{1}$$

$$nomalized\_post\_RR = \frac{post\_RR}{mean(ds.post\_RR)} \tag{2}$$

$$nomalized\_local\_RR = \frac{local\_RR}{mean(ds.local\_RR)} \tag{3}$$

$$nomalized\_global\_RR = \frac{global\_RR}{mean(ds.global\_RR)} \tag{4}$$

where $ds.pre\_RR$ denotes the average of all pre_RRs in the ds that the heartbeat belongs to, and so on.

Regarding the *higher order statistics* (HOS), it is reported being useful in catching subtle changes in ECG data [29]. In this work, the *skewness* (3rd order statistics) and *kurtosis* (4th order statistics) are calculated for each heartbeat. They can be mathematically defined as follows, where $X_{1...,N}$ denotes all the data samples in a signal, $\bar{X}$ is the mean and s is the standard deviation.

$$Skewness = \frac{\sum_{i=1}^{N}(X_i - \bar{X})^3/N}{s^3} \tag{5}$$

$$Kurtosis = \frac{\sum_{i=1}^{N}(X_i - \bar{X})^4/N}{s^4} - 3 \tag{6}$$

The wavelet coefficients provides both time and frequency domain information of a signal, which is claimed to be the best features of ECG signal [28]. The choice of the mother wavelet function used for coefficients extraction is crucial to the final classification performance. In this work, the *Haar* wavelet function is chosen because of its simplicity and that it has been demonstrated as the ideal wavelet for short time signal analysis [54].

3.3 Classifier Pool Training

A classifier pool contains a set of base classifiers that are trained both accurately and diversely. First of all, the SMOTEENN technique [3,8,47] is adopted to remedy the training data imbalance problem by employing to over-sample the minority classes (class S and V) to the same amount of class N.

In order to increase the diversity of the classifier pool, 6 different classifiers, including *multi-layers perceptron, support vector machine* (SVM), *linear SVM, Bayesian model with Gaussian kernel, decision tree*, and *K-nearest neighbors model*, are used in this work, which are trained using different training subsets.

A training database is split into a training set and a dynamic selection data set (DSEL), which account for 70% and 30%, respectively. Six different subsets are generated from the training set on use of 6-folds-cross-validation for base classifiers training. The DSEL data set is used for ensemble selection. In testing, a query heartbeat is calculated its local region within the scope of the DSEL data set for classifier competence measure.

3.4 Dynamic Selection Classification

In this stage, a dynamic ensemble selection technique is equipped into the framework. There are several prevalent DES techniques which have been proven their effectiveness in some classification problems. The differences between them are summarized in Table 1. Given a well-defined local region in DSEL,

- **DES-KNN** selects the top N accurate classifiers and top J diverse classifiers to compose the ensemble.
- **DES-KL** measures the competence level using

$$\sigma_{i,j} = \sum_{x_k \in DSEL} C_{src} exp(-d(x_k, x_j)^2) \tag{7}$$

where x_j is a query sample, C_{src} is the KL divergence between the uniform distribution and the vector of class supports.
- **DES-P** selects the classifiers that have a better accuracy than a random classifier into the ensemble.

– **DES-RRC** measures the competence level using

$$\sigma_{i,j} = \sum_{x_k \in DSEL} C_{src} K(x_k, x_j) \tag{8}$$

where x_j is a query sample, C_{src} denotes the source competence proposed in [48] and $K(x_k, x_j)$ is a Gausssian potential function. The classifiers that have a higher competence level than a random classifier are selected.

– **KNORA-E** selects the classifiers that correctly recognize all samples in the local region. If no base classifier is qualified, the local region shrinks.

– **KNORA-U** selects all classifiers that correctly recognize at least one sample in the local region. A majority voting scheme is used to give the final result.

– **KNOP** works similarly to KNORA-U. The difference is that KNORA-U works in the feature space, while KNOP works in the decision space [7], by which, all samples in MIT-BIH-AR are transformed into the decision space in advance.

– **META-DES** considers a base classifier as competent or incompetent by defining a set of meta features for classifier and training a meta-classifier which takes in the meta features to predict if a base classifier is competent.

Table 1. A brief comparison between prevalent DES techniques.

Technique	Local region definition	Competence measure	Reference
DES-KNN	K-NN	Accuracy & diversity	Soares et al. [37]
DES-KL	Potential function	Probabilistic	Woloszynski et al. [50]
DES-P	Potential function	Probabilistic	Woloszynski et al. [50]
DES-RRC	Potential function	Probabilistic	Woloszynski et al. [49]
KNORA-E	K-NN	Oracle	Ko et al. [24]
KNORA-U	K-NN	Oracle	Ko et al. [24]
KNOP	K-NN	Behavior	Cavalin et al. [7]
META-DES	K-NN	Meta-learning	Cruz et al. [11]

3.5 Result Refinement

A result regulator is developed in this stage. An SVM classifier is trained with features which are different with those used in the training stage. The motivation is two-folds: (1) the sensitivity to certain feature varies with heartbeat types [56]; and (2) the S beats are likely to be misrecognized as V beats.

In order to precisely catch the differences between class S and V, the RR-intervals are excluded from the feature set and merely the S and V beats are used to train the regulator. When testing, if a query heartbeat is classified as class S or V in the previous phase, it is then passed through the regulator for result refinement.

4   Experiment Result Analysis and Discussion

In this section, the effectiveness of the proposed D-ECG in cardiac arrhythmia detection is evaluated on use of the MIT-BIH-AR benchmark database. The metrics used for classifier performance evaluation are *sensitivity (Se)*, *positive predictive value (+P)* and *accuracy value (Acc)*. The DES techniques listed in Table 1 are individually evaluated. Their performances are then compared to that of single classifiers and traditional ensembles. The META-DES technique is fit into the proposed D-ECG to get the final arrhythmia detection result, which is compared against the stat-of-the-art methods.

4.1   The MIT-BIH Arrhythmia Database

The MIT-BIH arrhythmia database [31] contains 48 two-lead ambulatory ECG records from 47 patients (22 females and 25 males). Each record has approximately 30 min in length. These recordings were digitized at 360 Hz. For most of them, the first lead is *modified limb lead II* (except for the recording 114). The second lead is a pericardial lead (usually $V1$, sometimes are $V2$, $V4$ or $V5$, depending on subjects).

The Association for the Advancement of Medical Instrumentation (AAMI) categorized the heartbeats in the database into 5 super-classes: Normal (N), Supraventricular ectopic beat (S), Ventricular ectopic beat (V), Fusion beat (F) and Unknown beat (Q) [5]. Penalties would not be applied for the misclassification of class F and Q, as recommended by the AAMI standard [16].

Intra-patient and inter-patient paradigm [2,16,53,56] are two different types of paradigms concerning the use of the MIT-BIH-AR database for performance evaluation. The intra-patient paradigm separates the entire data set into a training set and a testing set merely according to the heartbeat labels, whereas inter-patient paradigm groups the heartbeats by patients and partitions the patients into a training set (DS1) and a testing set (DS2), as shown in Table 2. It has been empirically proven that the intra-patient paradigm can over optimistic the classification result by allowing training and testing heartbeats coming from the same patient [28]; therefore, in order to reveal the true performance of the proposed model and have a fair comparison with the stat-of-the-art rivals, the inter-patient paradigm is strictly followed in this work.

4.2   Comparative Analysis of DES Techniques

The effectiveness of the DES techniques in arrhythmia detection are evaluated in this subsection. They are trained on the same classifier pool as mentioned in Sect. 3.3. The used features are RR-intervals, HOS and wavelet coefficients.

In Table 3, all DES techniques have a similar performance in terms of the overall accuracy and positive predictive of class N except DES-KNN. META-DES obtains the best result in overall accuracy, sensitivity of class N and positive predictive of class S. KNORA-U achieves the best sensitivity of class N, which is approximately 6.16% higher than the second best result obtained by DES-KNN.

Table 2. Recording distributions and class proportions on DS1 and DS2.

Data set	N	S	V	F	Q	Recordings (patient ID)[a]
DS1	45808	943	3786	414	8	101, 106, 108, 109, 112, 114, 115, 116, 118, 119, 122, 124, 201, 203, 205, 207, 208, 209, 215, 220, 223, 230
DS2	44198	1836	3219	388	7	100, 103, 105, 111, 113, 117, 121, 123, 200, 202, 210, 212, 213, 214, 219, 221, 222, 228, 231, 232, 233, 234

[a] Each recording is denoted by a 3-digits number and the numbers are originally discontinuous.

Regarding the detection of class V, all DES techniques have a promising performance, with the lowest sensitivity higher than 93%. However, the corresponding positive predictive values are struggling around 60%.

Table 3. Heartbeats classification performance of DES techniques, single classifiers and traditional ensembles.

Classification technique	Classifiers	Acc	N		S		V	
			Se	+P	Se	+P	Se	+P
DES	DES-KNN	81.79	82.15	99.28	68.70	28.10	93.72	35.90
	DES-KL	90.39	91.86	98.98	65.62	32.44	94.78	61.38
	DES-P	90.42	91.87	99.01	65.56	32.46	95.22	61.44
	DES-RRC	89.40	90.81	99.01	64.37	28.58	94.81	60.40
	KNORA-E	89.43	90.74	99.04	65.11	30.72	95.65	57.85
	KNORA-U	89.97	91.10	98.77	**74.86**	35.69	93.72	59.67
	KNOP	90.54	92.02	98.88	64.37	35.56	95.62	60.00
	META-DES	**90.77**	**92.35**	98.80	63.63	**36.06**	95.00	60.80
Single classifier	SVM	81.83	81.35	**99.48**	88.08	30.17	94.91	36.95
	KNN	84.22	84.84	98.50	72.01	22.01	92.73	52.89
	Perceptron	80.22	81.75	98.93	40.99	12.85	89.90	39.53
	Linear SVM	85.08	86.90	99.01	48.52	18.14	89.69	48.28
	Bayesian (Gaussian)	79.13	80.52	97.59	64.37	31.06	77.63	28.65
	Decision tree	78.37	80.38	96.56	52.74	18.37	74.37	39.95
Homogeneous ensemble [a]	E-SVM	82.76	82.87	99.35	72.92	30.89	**96.68**	36.76
	E-KNN	84.27	84.90	98.62	72.06	22.86	92.54	51.46
	E-perceptron	85.95	87.98	99.00	43.90	16.44	91.61	52.79
	E-linear SVM	86.70	88.96	98.95	46.12	18.55	88.51	52.74
	E-Gaussian Bayesian	81.58	83.89	96.58	36.66	17.90	84.28	40.19
	Forrest	87.53	89.20	98.59	56.04	33.60	90.37	45.86
Heterogeneous ensemble	Mix all	89.48	91.13	99.02	61.63	27.00	92.98	**61.70**

[a] The amount of classifiers in each homogeneous ensemble is 50.

In order to demonstrate the advantages of DES techniques, they are compared to single classifier and traditional ensembles methods. Notice that the traditional ensemble methods were categorized into homogeneous ensemble and heterogeneous ensemble. The former generates an ensemble by training a classifier with different training sets, while the later contains various types of classifiers. In traditional ensemble methods, the final decision is given based on the majority voting scheme. Obviously, the average performance of DES techniques is significantly better than that of single classifier and homogeneous ensemble methods. Surprisingly, the mixture ensemble performs closely to the average performance of DES techniques. However, since the number of base classifiers in the ensemble is small, the performance is sensitive to the choices of classifiers.

4.3   D-ECG Performance Evaluation

Since META-DES obtains the best results in overall accuracy, sensitivity of class N and positive predictive of class S, and performs closely to other DES techniques in the rest of the metrics, it is suggested to be selected as the DES component of the proposed D-ECG.

Table 4. Arrhythmia detection result of **META-DES** and **D-ECG** in DS2.

		Predicted class (META-DES)					Predicted class (D-ECG)				
		N	S	V	F	Q	N	S	V	F	Q
True class	N	40698	1921	1331	0	119	40698	2053	1199	0	119
	S	125	1116	513	0	0	125	1472	157	0	0
	V	102	58	3058	0	1	102	103	3013	0	1
	F	3	0	4	0	0	3	0	4	0	0
	Q	263	0	124	0	1	263	2	122	0	1

Table 5. Arrhythmia detection performance of before- and after-refinement.

Refinement	Acc	N		S		V	
		Se	+P	Se	+P	Se	+P
Before	90.77	92.35	98.8	63.63	36.06	95.0	60.8
After	**91.4** ↑	92.35	98.8	**83.92** ↑	**40.55** ↑	93.6	**67.03** ↑

The result coming from META-DES is eventually refined by the regulator component to give the final decision of the proposed D-ECG. Table 4 shows the arrhythmia detection results of META-DES and D-ECG on $DS2$. Table 5 summarizes the comparison results. The result shows that the regulator component has make visible improvements to the result obtained by META-DES, with overall accuracy increasing from 90.77% to 91.4%, sensitivity of class S increasing by more than 20%, positive predictive of class S and V increasing from 36.06% and 60.8% to 40.55% and 67.03%, respectively.

Table 6 summarizes the comparative results between D-ECG and the state-of-the-art methods in cardiac arrhythmia detection. The proposed D-ECG achieves the best sensitivity of both class S and V, and takes the second best place in

Table 6. Arrhythmia detection comparison between the proposed D-ECG and the stat-of-the-art rivals in DS2 of the MIT-BIH-AR database

Method	Acc	N		S		V	
		Se	+P	Se	+P	Se	+P
Proposed D-ECG	91.4	92.35	98.8	**83.92**	40.55	**93.6**	67.03
De Chazal [16]	81.9	86.9	99.2	75.9	38.5	77.7	81.9
Ye [53]	86.4	88.5	97.5	60.8	**52.3**	81.5	63.1
Zhang [56]	86.7	88.9	99.0	79.1	36.0	85.5	**92.8**
Shan [9]	**93.1**	**98.4**	95.4	29.5	38.4	70.8	85.1
Mariano [27]	78.0	78.0	99.0	76.0	41.0	83.0	88.0

overall accuracy and sensitivity of class N. Shan's model [9] obtains the highest accuracy and class N sensitivity, but fails in detection of class S, with the sensitivity of class S being merely 29.5%. Though the proposed D-ECG has a relative low positive predictive of both class S and V, it still a more appropriate choice than other listed works for cardiac arrhythmia detection from a clinical point of view. This is because in clinical environment, misclassification of a normal heartbeat would not lead to a disaster, but missing a disease heartbeat can kill.

5   Conclusion

In this work, a dynamic framework named D-ECG for automatic cardiac arrhythmia detection from IoT-based ECG recordings is proposed. The D-ECG introduces the dynamic ensemble selection techniques to improve the heartbeat classification performance. Technically, the proposed D-ECG consists of five phases: preprocessing, feature extraction, classifier pool training, dynamic selection classification and result refinement. A result regulator is customized in the last stage of D-ECG that uses different features to refine the disease heartbeats classification result from the DES technique. Eight DES techniques are evaluated in experiments. The results show that the DES techniques generally have a superior performance than single classifier and traditional ensemble methods. The final results given by the D-ECG have shown a marked improvement to the stat-of-the-art methods in cardiac arrhythmia detection in terms of overall heartbeat classification accuracy and the sensitivity of disease heartbeats.

References

1. Abawajy, J.H., Kelarev, A.V., Chowdhury, M.: Multistage approach for clustering and classification of ECG data. Comput. Methods Programs Biomed. **112**(3), 720–730 (2013)
2. Afkhami, R.G., Azarnia, G., Tinati, M.A.: Cardiac arrhythmia classification using statistical and mixture modeling features of ECG signals. Pattern Recogn. Lett. **70**, 45–51 (2016)

3. Alejo, R., Sotoca, J.M., Valdovinos, R.M., Toribio, P.: Edited nearest neighbor rule for improving neural networks classifications. In: Zhang, L., Lu, B.-L., Kwok, J. (eds.) ISNN 2010. LNCS, vol. 6063, pp. 303–310. Springer, Heidelberg (2010). https://doi.org/10.1007/978-3-642-13278-0_39

4. Alonso-Atienza, F., Morgado, E., Fernandez-Martinez, L., García-Alberola, A., Rojo-Alvarez, J.L.: Detection of life-threatening arrhythmias using feature selection and support vector machines. IEEE Trans. Biomed. Eng. **61**(3), 832–840 (2014)

5. ANSI/AAMI: Testing and reporting performance results of cardiac rhythm and ST segment measurement algorithms. In: Association for the Advancement of Medical Instrumentation - AAMI ISO EC57 (1998–2008)

6. Britto Jr., A.S., Sabourin, R., Oliveira, L.E.: Dynamic selection of classifiersa comprehensive review. Pattern Recognit. **47**(11), 3665–3680 (2014)

7. Cavalin, P.R., Sabourin, R., Suen, C.Y.: Dynamic selection approaches for multiple classifier systems. Neural Comput. Appl. **22**(3–4), 673–688 (2013)

8. Chawla, N.V., Bowyer, K.W., Hall, L.O., Kegelmeyer, W.P.: SMOTE: synthetic minority over-sampling technique. J. Artif. Intell. Res. **16**, 321–357 (2002)

9. Chen, S., Hua, W., Li, Z., Li, J., Gao, X.: Heartbeat classification using projected and dynamic features of ECG signal. Biomed. Signal Process. Control **31**, 165–173 (2017)

10. Cheng, P., Dong, X.: Life-threatening ventricular arrhythmia detection with personalized features. IEEE Access **5**, 14195–14203 (2017)

11. Cruz, R.M., Sabourin, R., Cavalcanti, G.D.: META-DES. Oracle: meta-learning and feature selection for dynamic ensemble selection. Inf. Fusion **38**, 84–103 (2017)

12. Cruz, R.M., Sabourin, R., Cavalcanti, G.D.: Dynamic classifier selection: recent advances and perspectives. Inf. Fusion **41**, 195–216 (2018)

13. Daamouche, A., Hamami, L., Alajlan, N., Melgani, F.: A wavelet optimization approach for ECG signal classification. Biomed. Signal Process. Control **7**(4), 342–349 (2012)

14. Daubechies, I.: Ten Lectures on Wavelets, vol. 61. Siam, Philadelphia (1992)

15. De Albuquerque, V.C.H., et al.: Robust automated cardiac arrhythmia detection in ECG beat signals. Neural Comput. Appl. **29**(3), 679–693 (2018)

16. De Chazal, P., O'Dwyer, M., Reilly, R.B.: Automatic classification of heartbeats using ECG morphology and heartbeat interval features. IEEE Trans. Biomed. Eng. **51**(7), 1196–1206 (2004)

17. De Lannoy, G., François, D., Delbeke, J., Verleysen, M.: Weighted conditional random fields for supervised interpatient heartbeat classification. IEEE Trans. Biomed. Eng. **59**(1), 241–247 (2012)

18. Doquire, G., De Lannoy, G., François, D., Verleysen, M.: Feature selection for interpatient supervised heart beat classification. Comput. Intell. Neurosci. **2011**, 1 (2011)

19. Dos Santos, E.M., Sabourin, R., Maupin, P.: A dynamic overproduce-and-choose strategy for the selection of classifier ensembles. Pattern Recognit. **41**(10), 2993–3009 (2008)

20. Güler, İ., Übeylı, E.D.: ECG beat classifier designed by combined neural network model. Pattern Recognit. **38**(2), 199–208 (2005)

21. Huang, H., Liu, J., Zhu, Q., Wang, R., Hu, G.: A new hierarchical method for inter-patient heartbeat classification using random projections and RR intervals. Biomed. Eng. Online **13**(1), 90 (2014)

22. Kabir, M.E., Wang, H., Bertino, E.: A role-involved purpose-based access control model. Inf. Syst. Front. **14**(3), 809–822 (2012)

23. Karthika, J., Thomas, J.M., Kizhakkethottam, J.J.: Detection of life-threatening arrhythmias using temporal, spectral and wavelet features, pp. 1–4. IEEE (2015)
24. Ko, A.H., Sabourin, R., Britto Jr., A.S.: From dynamic classifier selection to dynamic ensemble selection. Pattern Recognit. **41**(5), 1718–1731 (2008)
25. Li, M., Sun, X., Wang, H., Zhang, Y., Zhang, J.: Privacy-aware access control with trust management in web service. World Wide Web **14**(4), 407–430 (2011)
26. Lin, C., Chen, W., Qiu, C., Wu, Y., Krishnan, S., Zou, Q.: LibD3C: ensemble classifiers with a clustering and dynamic selection strategy. Neurocomputing **123**, 424–435 (2014)
27. Llamedo, M., Martínez, J.P.: Heartbeat classification using feature selection driven by database generalization criteria. IEEE Trans. Biomed. Eng. **58**(3), 616–625 (2011)
28. Luz, E.J.S., Schwartz, W.R., Cámara-Chávez, G., Menotti, D.: ECG-based heartbeat classification for arrhythmia detection: a survey. Comput. Methods Programs Biomed. **127**, 144–164 (2016)
29. Martis, R.J., Acharya, U.R., Ray, A.K., Chakraborty, C.: Application of higher order cumulants to ECG signals for the cardiac health diagnosis. In: 2011 Annual International Conference of the IEEE Engineering in Medicine and Biology Society, EMBC, pp. 1697–1700. IEEE (2011)
30. Masoudnia, S., Ebrahimpour, R.: Mixture of experts: a literature survey. Artif. Intell. Rev. **42**(2), 275–293 (2014)
31. Moody, G.B., Mark, R.G.: The impact of the MIT-BIH arrhythmia database. IEEE Eng. Med. Biol. Mag. **20**(3), 45–50 (2001)
32. Özbay, Y., Tezel, G.: A new method for classification of ECG arrhythmias using neural network with adaptive activation function. Digit. Signal Process. **20**(4), 1040–1049 (2010)
33. Qin, Y., Sheng, Q.Z., Falkner, N.J., Dustdar, S., Wang, H., Vasilakos, A.V.: When things matter: a survey on data-centric internet of things. J. Netw. Comput. Appl. **64**, 137–153 (2016)
34. Sahoo, S., Kanungo, B., Behera, S., Sabut, S.: Multiresolution wavelet transform based feature extraction and ECG classification to detect cardiac abnormalities. Measurement **108**, 55–66 (2017)
35. Sharma, H., Sharma, K.: An algorithm for sleep apnea detection from single-lead ECG using hermite basis functions. Comput. Biol. Med. **77**, 116–124 (2016)
36. Sierra, B., Lazkano, E., Irigoien, I., Jauregi, E., Mendialdua, I.: K nearest neighbor equality: giving equal chance to all existing classes. Inf. Sci. **181**(23), 5158–5168 (2011)
37. Soares, R.G., Santana, A., Canuto, A.M., de Souto, M.C.P.: Using accuracy and diversity to select classifiers to build ensembles. In: International Joint Conference on Neural Networks, IJCNN 2006, pp. 1310–1316. IEEE (2006)
38. Sun, L., Wang, H., Yong, J., Wu, G.: Semantic access control for cloud computing based on e-healthcare. In: 2012 IEEE 16th International Conference on Computer Supported Cooperative Work in Design (CSCWD), pp. 512–518. IEEE (2012)
39. Sun, X., Li, M., Wang, H., Plank, A.: An efficient hash-based algorithm for minimal k-anonymity. In: Proceedings of the thirty-first Australasian conference on Computer science-Volume 74, pp. 101–107. Australian Computer Society, Inc. (2008)
40. Sun, X., Wang, H., Li, J., Zhang, Y.: Injecting purpose and trust into data anonymisation. Comput. Secur. **30**(5), 332–345 (2011)
41. Vimalachandran, P., Wang, H., Zhang, Y., Heyward, B., Zhao, Y.: Preserving patient-centred controls in electronic health record systems: a reliance-based model

implication. In: 2017 International Conference on Orange Technologies (ICOT), pp. 37–44. IEEE (2017)

42. Vimalachandran, P., Wang, H., Zhang, Y., Zhuo, G., Kuang, H.: Cryptographic access control in electronic health record systems: a security implication. In: Bouguettaya, A., et al. (eds.) WISE 2017. LNCS, vol. 10570, pp. 540–549. Springer, Cham (2017). https://doi.org/10.1007/978-3-319-68786-5_43

43. Wang, H., Cao, J., Zhang, Y.: A flexible payment scheme and its role-based access control. IEEE Trans. knowl. Data Eng. **17**(3), 425–436 (2005)

44. Wang, H., Zhang, Z., Taleb, T.: Special issue on security and privacy of IoT. World Wide Web **21**(1), 1–6 (2018)

45. Wang, Y., Li, H., Wang, H., Zhou, B., Zhang, Y.: Multi-window based ensemble learning for classification of imbalanced streaming data. In: Wang, J., et al. (eds.) WISE 2015. LNCS, vol. 9419, pp. 78–92. Springer, Cham (2015). https://doi.org/10.1007/978-3-319-26187-4_6

46. Wikipedia contributors: Heart arrhythmia (2018). https://en.wikipedia.org/wiki/Heart_arrhythmia

47. Wilson, D.L.: Asymptotic properties of nearest neighbor rules using edited data. IEEE Trans. Syst. Man Cybern. **3**, 408–421 (1972)

48. Woloszynski, T., Kurzynski, M.: A measure of competence based on randomized reference classifier for dynamic ensemble selection. In: 2010 20th International Conference on Pattern Recognition (ICPR), pp. 4194–4197. IEEE (2010)

49. Woloszynski, T., Kurzynski, M.: A probabilistic model of classifier competence for dynamic ensemble selection. Pattern Recogn. **44**(10–11), 2656–2668 (2011)

50. Woloszynski, T., Kurzynski, M., Podsiadlo, P., Stachowiak, G.W.: A measure of competence based on random classification for dynamic ensemble selection. Inf. Fusion **13**(3), 207–213 (2012)

51. Xiao, J., Xie, L., He, C., Jiang, X.: Dynamic classifier ensemble model for customer classification with imbalanced class distribution. Expert Syst. Appl. **39**(3), 3668–3675 (2012)

52. Yang, Z., Zhou, Q., Lei, L., Zheng, K., Xiang, W.: An iot-cloud based wearable ECG monitoring system for smart healthcare. J. Med. Syst. **40**(12), 286 (2016)

53. Ye, C., Kumar, B.V., Coimbra, M.T.: Heartbeat classification using morphological and dynamic features of ECG signals. IEEE Trans. Biomed. Eng. **59**(10), 2930–2941 (2012)

54. Yu, S.N., Chen, Y.H.: Electrocardiogram beat classification based on wavelet transformation and probabilistic neural network. Pattern Recogn. Lett. **28**(10), 1142–1150 (2007)

55. Zhang, J., et al.: On efficient and robust anonymization for privacy protection on massive streaming categorical information. IEEE Trans. Dependable Secure Comput. **14**(5), 507–520 (2017)

56. Zhang, Z., Dong, J., Luo, X., Choi, K.S., Wu, X.: Heartbeat classification using disease-specific feature selection. Comput. Biol. Med. **46**, 79–89 (2014)

57. Zhu, X., Wu, X., Yang, Y.: Dynamic classifier selection for effective mining from noisy data streams. In: Fourth IEEE International Conference on Data Mining, ICDM 2004, pp. 305–312. IEEE (2004)

Jointly Predicting Affective and Mental Health Scores Using Deep Neural Networks of Visual Cues on the Web

Hung Nguyen[1], Van Nguyen[2], Thin Nguyen[1(✉)], Mark E. Larsen[3],
Bridianne O'Dea[3], Duc Thanh Nguyen[1], Trung Le[2], Dinh Phung[2],
Svetha Venkatesh[1], and Helen Christensen[3]

[1] Deakin University, Geelong, Australia
thin.nguyen@deakin.edu.au
[2] Monash University, Clayton, Australia
[3] Black Dog Institute, University of New South Wales, Sydney, Australia

Abstract. Despite the range of studies examining the relationship between mental health and social media data, not all prior studies have validated the social media markers against "ground truth", or validated psychiatric information, in general community samples. Instead, researchers have approximated psychiatric diagnosis using user statements such as "I have been diagnosed as X". Without "ground truth", the value of predictive algorithms is highly questionable and potentially harmful. In addition, for social media data, whilst linguistic features have been widely identified as strong markers of mental health disorders, little is known about non-textual features on their links with the disorders. The current work is a longitudinal study during which participants' mental health data, consisting of depression and anxiety scores, were collected fortnightly with a validated, diagnostic, clinical measure. Also, datasets with labels relevant to mental health scores, such as emotional scores, are also employed to improve the performance in prediction of mental health scores. This work introduces a deep neural network-based method integrating sub-networks on predicting affective scores and mental health outcomes from images. Experimental results have shown that in the both predictions of emotion and mental health scores, (1) deep features majorly outperform handcrafted ones and (2) the proposed network achieves better performance compared with separate networks.

Keywords: Deep learning · Social media
Visual features · Health analytics
Behavioral monitoring · Mental health

1 Introduction

By 2030 depression is forecast to be the second largest cause of disease burden worldwide [13]. A key component in reducing the prevalence of depression is

---

H. Nguyen and V. Nguyen—contributed equally

© Springer Nature Switzerland AG 2018
H. Hacid et al. (Eds.): WISE 2018, LNCS 11234, pp. 100–110, 2018.
https://doi.org/10.1007/978-3-030-02925-8_7

the 'translation' of what is already known into practice. It is accepted that progress in lowering depression prevalence over the next decade is more likely to be made by putting into practice what we already know [24] rather than by developing new treatments, with once-hopeful pharmaceutical products having "failed in clinical trials" [11]. Targeting those individuals who are currently not in treatment or not optimally in treatment has been estimated to be able to avert disease burden by 23%, and thus would represent a major advance, if achieved [1]. However, getting people into treatment is difficult. Many don't seek help, with estimates of 60% not receiving evidence-based treatments [2]. Others delay, by which time the disorder has become entrenched. Delays are typically long, and range from 6 to 14 years, depending on the condition [3,15,21]. Relapse prevention is rare, with crisis or serious and tragic outcomes possible [22].

A new, potentially game-changing idea is that data from sources that individuals generate in the normal course of their lives, can provide the basis by which risk can be detected. The use of such 'natural' data to help identify and manage illness is gaining traction. For instance, Patrick et al. urged researchers to establish a framework to allow 'natural data' to be captured [16]. Symptoms, responses to medications and other parameters, can help patients and physicians better tailor medical treatments. Much 'natural data' is lying dormant, although data scientists are being attracted in increasing number to the new science of computational social science [5]. The rapid expansion of social media sites such as blogs, Twitter, Facebook, where data potentially can be used to assess and identify risks, combined with the increasing numbers of people using Facebook and social media, make the field ripe for investigation. The use of social media to measure depressive symptoms was first described by De Choudhury and Nimrod [4,14] in a study over 12 months examining Twitter in individuals recruited via crowd sourcing [4]. Social media was able to characterize the onset of depression in individuals. These characteristics were reduced social activity, low affect expressed in content, tight networks, and content about medication and relationships.

Using a longitudinal study design, conducted from July 2014 to October 2016, the current study aims to identify the valid markers of depression and anxiety within blog content. This study will collect and analyze visual features expressed within blog content which will then be correlated with the mental health scores, as measured by validated clinical scales. The current study attempts to provide one of the first clinical validations of visual markers of mental health in social media data.

As shown in the literature, there is also a strong correlation between emotions and mental health and thus there would be beneficial to use emotion information for mental health prediction. However, it may be difficult to have datasets including the ground-truth for both emotions and mental health. For example, our dataset includes two subsets, one is associated with only emotion scores while the other is associated with only mental health scores. To exploit the benefit from both the sets and, at the same time, to perform prediction of both emotion and mental health scores, in this paper, we propose a deep neural network-based method joining independent tasks (from separate neural networks). We have verified out proposed network on various visual feature types and compared with existing approaches. Experimental results have shown that (1) deep features majorly outperform handcrafted ones in the predictions and (2) the proposed network achieves better performance compared with separate networks.

The outcomes will then enable prediction models to be established which can then be used to monitor online sites automatically, and in real time, for mental health risk. The findings may have implication in improving the delivery of precise and adaptive mental health care which utilities the Internet to reduce costs and empower patients. The value of this type of data, its transparency, and its potential for supplementing or replacing human decision making with automated algorithms represents a revolution for all healthcare systems.

The remainder of the paper is organized as follows. Section 2 reports our study setting to collect ground-truth mental health scores for participants as well as the social media data they made. It also presents our proposed approaches to extract visual features for our image data. The experimental results are reported in Section 3. Section 4 provides remarks and concludes the paper.

2 Methodology

2.1 Study Design

The basic design is a longitudinal study during which participants' online data will be captured continuously over the study period and participants' mental health data, consisting of depression and anxiety scores, will be surveyed fortnightly. Ethics was received from relevant institutions.

2.2 Participants, Recruitment, and Procedure

Recruitment took place between July 2014 and October 2016. Using a series of online adverts published on various social media channels, individuals who blogged were invited to visit the study website in which they were provided information about the study. There was no exclusion criteria, although, to be included participants must provide the URL of their blog site and self-identify as a mental health blogger.

The study procedure involved completing an online mental health assessment (via email) at baseline, and then again every fortnight, for a period of 16 weeks. These assessments were completed online using a customized online survey platform. After reading the Participant Information Sheet and Consent Information on the study website, participants were asked to provide their consent to having their mental health measured and social media data extracted fortnightly. Once consent was given, participants were asked to provide their blog-site details and complete the first online mental health assessment. Additional eight online assessments were then scheduled for fortnightly administration. Each participant received $20 in Amazon Webstore credit for their participation.

This method of recruiting participants to the trial has proved to be successful for current online trials [6,23].

2.3   Measures

Demographics were assessed using questions on age, gender, prior diagnosis of depression or anxiety from medical practitioner, and medication use for depression and anxiety. Participants were also asked to rate their overall general health as either "very good", "good", "moderate", "bad" or "very bad". Depressive symptoms were assessed using the self-report Patient Health Questionnaire (PHQ-9 [8]). This nine-item questionnaire assessed the presence of depressive symptoms in the past two weeks. Individuals were asked to rate the frequency of depressive symptoms using a four point Likert scale ranging from "none of the time" to "every day or almost every day". A total score is then calculated, ranging from 0 to 27, which can then be classified as "nil-minimal" (0–4), "mild" (5–9), "moderate" (10–14), "moderately severe" (15–19) or "severe" (20–27).

Anxiety symptoms were assessed using the self-report Generalized Anxiety Disorder Scale (GAD-7 [19]). This seven-item questionnaire assessed the presence of generalized anxiety symptoms in the past two weeks. It uses the same response scale as the PHQ-9 and participants' total scores, ranging from 0 to 21, can also be classified into "mild", "moderate" or "severe" using thresholds of 5, 10, and 15 respectively.

2.4   Social Media Data Collection and Analysis

Social media data was crawled using the publicly accessible Application Programming Interface (APIs) for each platform. In this work we used images posted by the participants in their own blogs as the potential predictors of their mental health scores. Then features for the images were learned using a machine learning approach. In particular, we propose a deep-learning based framework to capture visual features potentially capturing mental scores (patient health questionnaire (PHQ) and generalized anxiety disorder (GAD)).

2.5   Proposed Method

As shown in the literature, there is a strong correlation between emotions and mental health. In our study, dataset 1 and 2 respectively contain images annotated with mental health and emotion scores. However, each dataset is associated with only one type of scores, e.g. dataset 1 is annotated with PHQ scores and dataset 2 is annotated with Dominance scores. A straightforward approach is to develop independent predictors for each score type. However, this approach would not take any benefit from the dual impacts between emotion and mental health. In this paper, we propose a deep neural network-based method that is able to integrate independent tasks in a joint framework. In particular, our proposed neural network consists of a (fully connected) joint network which takes input directly from visual features and wires to different separate (fully connected) networks, each of which handles an independent prediction task.

Figure 1 illustrates our proposed method with two sub-tasks, e.g., prediction of PHQ score (Net 1) and Dominance score (Net 2). As shown in this figure, without using the Joint net (i.e., directly connecting Net 1 and Net 2 to the input

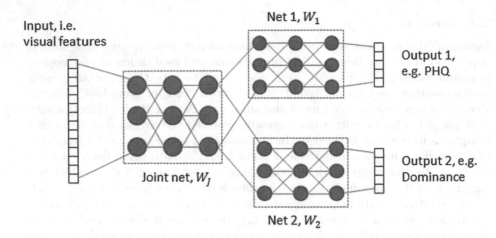

Fig. 1. Illustration of the proposed method with two sub-tasks, e.g., prediction of PHQ score (Net 1) and Dominance score (Net 2).

features), we obtain two separate networks devoted for two independent tasks. Meanwhile, the Joint net implicitly encodes the correlation between two outputs, e.g. Dominance (for emotion) and PHQ (for mental health), and transfers the information from one prediction task to the other. By engaging different independent but correlated tasks, our proposed network leverages the performance of one prediction task by other tasks. The proposed method has also been validated on various visual feature types and proved to outperform independent networks.

Let $\mathcal{L}$ be the loss of whole framework (e.g., including the Joint net and separate nets) and $\mathcal{L}_i$ be the loss of the i-th net. Let W denote the weights of the whole framework including the weights W_J of the Joint net and the weights W_i of each i-th net. We define the loss $\mathcal{L}$ as,

$$\mathcal{L} = \sum_i \alpha_i \mathcal{L}_i$$

where $\sum_i \alpha_i = 1$.

Since our network aims to predict emotion and mental health scores, each $\mathcal{L}_i$ is defined by a least squared error function. In addition, *tanh* is used as activation functions for all layers in the network.

Back-propagation then can be done as,

$$\frac{\partial \mathcal{L}}{\partial W} = \sum_i \alpha_i \left(\frac{\partial \mathcal{L}_i}{\partial W} \right),$$

Since each training sample has only one associated output (e.g., each image is associated to either a PHQ score or a Dominance score), training the framework with a training data point can be done on the corresponding sub-net. For example, with the framework shown in Fig. 1, the training sample which is associated with a PHQ score will invoke Net 1 and the one which is associated with a Dominance score will activate Net 2. In other words, $\frac{\partial \mathcal{L}}{\partial W}$ can be simplified to $\alpha \left(\frac{\partial \mathcal{L}_i}{\partial W} \right)$ and the calculation of $\frac{\partial \mathcal{L}_i}{\partial W}$ consists of the calculation of $\frac{\partial \mathcal{L}_i}{\partial W_J}$ and

$\frac{\partial \mathcal{L}_i}{\partial W_i}$. We note that our proposed network can be extended easily to adapt with more sub-tasks by adding more corresponding separate networks. In addition, as shown in our experiments, the network can also be customized to become an end-to-end system (i.e., from images to emotion/mental health scores). To train the joint network, stochastic gradient descent is adopted [10].

3 Experimental Results

3.1 Participants and Data

A total of 153 participants participated in the study (age range = 18–67 years; mean age: 29.48 years, SD: 10.29, 87.6% female). Of these, 127 (83.0%) had received a diagnosis of depression or anxiety from a medical practitioner, and 87 (56.9%) were taking medication for depression or anxiety. For depression scores at baseline, 22 were nil-minimal (14.4%), 36 mild (23.5%), 29 moderate (19.0%), 39 moderately severe (25.5%) and 27 severe (17.6%). For anxiety scores, 34 were minimal (22.2%), 47 mild (30.7%), 40 moderate (26.1%), and 32 severe (20.9%). A total of 72 (47.1%) reporting having had an anxiety attack in the two weeks prior, with the mean number of attacks in this period being 5.29 (SD:7.36, range: 1–50). At baseline, 81 (52.9%) had "thoughts that they would be better off dead, or of hurting themselves" for several days or more in the past two weeks. At baseline, 68 (44.4%) had both moderate/severe depression and anxiety, with depression and anxiety scores highly correlated (r = 0.78, p < .001).

We created a dataset, called dataset 1, by crawling all images posted by participants in their own blogs within the windows of two weeks for each mental health assessment. This dataset contains **29,340** images made by the participants during their survey time. These images are used to explain the depression and anxiety scores the participants experienced. In particular, images posted within two weeks of a survey will be assigned the GAD-7 and PHQ-9 scores assessed for the survey. Then this data will be split into a derivation cohort (80%) from which deep learning models will be trained, and a validation cohort (20%) in which the accuracy of the models will be assessed using the Root Mean Square Error (RMSE).

We used the standard dataset International Affective Picture System (IAPS) [9], called dataset 2. This dataset is provided by the NIMH Center for Emotion and Attention at the University of Florida. There are 1,196 pictures in dataset 2, each of which was rated with affective scores, consisting of pleasure, arousal, and dominance, ranging from 1 to 9. For example, for valence score, the scale is from completely happy to completely unhappy. We also divided this dataset into training (80% of the images) and testing (20%) sets.

3.2 Experimental Setup

We have experimented our proposed network with various features. In particular, hue, saturation, and value (HSV) [12,17] that capture human perception to color information were used. We also implemented the bag-of-features approach in which SURF was used for building a dictionary of visual code-words (100 words were used in our implementation). Each image was then encoded by a histogram of code words. HSV and SURF are handcrafted features; they are well-known for their potential in various applications but do require the domain knowledge of the applications.

Recently, deep convolutional neural networks (CNNs) have been used to automatically learn features. In this work, we have experimented with state-of-the-art CNNs including Very Deep Convolutional Networks (VGG) [18], Deep residual nets (ResNet) [7], and Inception [20]. The VGG was introduced by Simonyan and Zisserman [18] for Large Scale Image Recognition. This network is characterized by its simplicity, using only 3×3 convolutional layers stacked on top of each other in increasing depth. The significant improvement on the prior-art configurations can be obtained by pushing the depth to 16–19 weight layers. Reducing volume size is handled by max pooling. The two fully-connected layers are followed by a softmax classifier.

ResNet was designed based on the use of so-called micro-architecture modules (also called network-in-network architectures). The term micro-architecture refers to the set of building blocks used to construct the network. Since the first time of introduction in 2015, ResNet has become a seminal architecture for its robustness, demonstrating that extremely deep networks can be trained using standard SGD (and a reasonable initialization function) through the use of residual modules. In our work, we reform layers of ResNet as learning residual functions with reference to the layer inputs, instead of learning unreferenced functions.

Szegedy et al. [20] proposed the so-called Inception network. The main aim of this architecture is to make the inception act as a multi-level feature extractor via 1×1, 3×3, and 5×5 convolutions. This architecture has been proved to improve the utilization of the computing resources inside the network. In the network, the output of filters are stacked along the channel dimension before being fed into following layers.

We take the output of the penultimate layer of each aforementioned network to get the features of images. To VGG and Inception network, we use the output of layer before the prediction layer named "predictions". Each image going through VGG and Inception will have 4,096 features respectively. For ResNet network, the output of layer before the final layer for prediction is taken, and every image from ResNet will have 2,048 features. For features learned using deep neural networks, we adopted the corresponding network architectures and then fine-tuned the networks on our collected datasets. This manner also enables an end-to-end system (i.e. from images to emotion/mental health scores).

3.3   Prediction Performance on Mental Health Scores

To predict mental health scores from images, we used two hand-crafted features, HSV and SURF, as well as four deep learning based features, including Inception, Resnet, VGG16, and VGG19. The result is shown in Table 1. Deep features

Table 1. Performance (in RMSE) of different deep architectures for mental health scores. The best performance is in **bold**.

Features	PHQ	GAD
HSV (handcrafted)	5.62	4.76
SURF (handcrafted)	5.45	4.64
Inception (CNN)	4.95	**4.07**
Resnet (CNN)	**4.92**	4.28
VGG16 (CNN)	5.08	4.34
VGG19 (CNN)	5.24	4.33

performed better in the estimation. Among them, ResNet gained the best performance, at an RMSE of **4.92** for PHQ, which is in the range of 0–27, and an RMSE of **4.07** for GAD, which is in the range of 0–21.

Table 2. Performance (in RMSE) of different deep architectures for pleasure, arousal and dominance scores. The best performance is in **bold**.

Features	Pleasure	Arousal	Dominance
HSV (handcrafted)	1.81	1.21	**1.26**
SURF (handcrafted)	1.98	1.38	1.66
Inception (CNN)	2.09	1.55	1.64
Resnet (CNN)	**1.48**	**0.90**	1.77
VGG16 (CNN)	1.49	0.96	1.66
VGG19 (CNN)	1.53	1.01	1.71

3.4 Prediction Performance on Affective Scores

To examine the robustness of the proposed framework, we conducted the same experiments on affective scores for images. We used the standard dataset International Affective Picture System (IAPS) [9].

We again divided the dataset into training (80% of the images) and testing (20%) sets. Visual features extracted by different state-of-the-art convolutional neural networks were then used to estimate the affective scores for the images and RMSE was used to evaluate the performance. Table 2 reports the performance of the proposed framework.

For this task, in general, deep features gained better results in the prediction of Pleasure and Arousal scores for the images. In particular, similar to the mental health score prediction, ResNet again achieves the best results, with the RMSE of **1.48** and **0.90** for the estimation of Pleasure and Arousal, respectively, which are in the scale of 1–9. For Dominance score, the hand-crafted features performed better than did the deep ones. Especially, simple HSV feature is the best at the prediction, at **1.26** RMSE.

Table 3. Predictions using joint and single models. The better performance is in **bold**.

Score	Joint model	Single model	Difference
PHQ	**4.71**	5.45	0.74
GAD	**3.99**	4.64	0.65

(a) Performance (in RMSE) in mental health scores predictions with SURF features.

CNNs	Joint model	Single model	Difference
HSV	**1.13**	1.26	0.13
SURF	**1.34**	1.66	0.32
Inception	**1.16**	1.64	0.48
Resnet	**1.09**	1.77	0.68
VGG16	**1.23**	1.66	0.43
VGG19	**1.34**	1.71	0.37

(b) Performance (in RMSE) of different deep architectures in Dominance scores predictions.

3.5   Joint Network vs. Separate Networks

Experimental results also show that our proposed network outperforms separate networks. For example, for prediction of mental health scores, as shown in Table 3a, the proposed method with the Joint net, gained RMSEs of 4.71 and 3.99 for PHQ and GAD, respectively. On the other hand, the corresponding separate network (i.e., Net 1 in Fig. 1) achieved RMSEs of 5.45 and 4.64 for PHQ and GAD, respectively.

For prediction of affective scores, for example, Dominance, as reported in Table 3b, the proposed network also shows better performance compared with the corresponding separate network (i.e., Net 2 in Fig. 1) on all feature types. Possibly, hidden sharing visual features predictive of both affective and mental health scores helped improve the performance.

4   Conclusion

This paper proposes a deep neural network-based method for prediction of affective scores and mental health outcomes from images. The proposed method makes use of sub-neural networks on sub-prediction tasks. In the study, various visual features were investigated. The proposed method was thoroughly evaluated against benchmark and large-scale datasets. Experimental results revealed valid and reliable social media markers of depression, demonstrating the validity of using social media to identify the risk of depression. This allows us to develop models that can be applied to a range of social media platforms for the identification of person-specific warning signs of depression automatically and in real-time – something currently not possible. The findings have the potential to produce an "opt in" service that social media companies can offer to monitor for depression, and with further research, deliver help-seeking support and/or treatment.

References

1. Andrews, G., Issakidis, C., Sanderson, K., Corry, J., Lapsley, H.: Utilising survey data to inform public policy: comparison of the cost-effectiveness of treatment of ten mental disorders. Br. J. Psychiatry **184**(6), 526–533 (2004)
2. Burgess, P.M., Pirkis, J.E., Slade, T.N., Johnston, A.K., Meadows, G.N., Gunn, J.M.: Service use for mental health problems: findings from the 2007 national survey of mental health and wellbeing. Aust. New Zealand J. Psychiatry **43**(7), 615–623 (2009)
3. Christiana, J.M.: Duration between onset and time of obtaining initial treatment among people with anxiety and mood disorders: an international survey of members of mental health patient advocate groups. Psychol. Med. **30**(3), 693–703 (2000)
4. De Choudhury, M., Counts, S., Horvitz, E.: Social media as a measurement tool of depression in populations. In: Proceedings of the Annual ACM Web Science Conference, pp. 47–56. ACM (2013)
5. Giles, J.: Making the links. Nature **488**(7412), 448–450 (2012)
6. Gosling, J.A.: The GoodNight study - online CBT for insomnia for the indicated prevention of depression: study protocol for a randomised controlled trial. Trials **15**(1), 56 (2014)
7. He, K., Zhang, X., Ren, S., Sun, J.: Deep residual learning for image recognition. In: Proceedings of the Conference on Computer Vision and Pattern Recognition, pp. 770–778 (2015)
8. Kroenke, K., Spitzer, R.L., Williams, J.B.W.: The PHQ-9. J. Gen. Internal Med. **16**(9), 606–613 (2001)
9. Lang, P.J., Bradley, M.M., Cuthbert, B.N.: International affective picture system (IAPS): affective ratings of pictures and instruction manual. Technical report, NIMH Center for the Study of Emotion and Attention (2005)
10. LeCun, Y., Bottou, L., Bengio, Y., Haffner, P.: Gradient-based learning applied to document recognition. Proc. IEEE **86**(11), 2278–2324 (1998)
11. Ledford, H.: If depression were cancer. Nature **515**(7526), 182–184 (2014)
12. Manikonda, L., De Choudhury, M.: Modeling and understanding visual attributes of mental health disclosures in social media. In: Proceedings of the Conference on Human Factors in Computing Systems, CHI 2017, pp. 170–181 (2017)
13. Mathers, C.: The global burden of disease: 2004 update. World Health Organization (2008)
14. Nimrod, G.: Online depression communities: members' interests and perceived benefits. Health Commun. **28**(5), 425–434 (2013)
15. Olfson, M., Kessler, R.C., Berglund, P.A., Lin, E.: Psychiatric disorder onset and first treatment contact in the United States and Ontario. Am. J. Psychiatry **155**(10), 1415–1422 (1998)
16. Patrick, K.: Gaining insight from patient and person-generated real world/real time data. In: Medicine 2.0 Conference (2013)
17. Reece, A.G., Danforth, C.M.: Instagram photos reveal predictive markers of depression. EPJ Data. Sci. **6**(1), 15 (2017)

18. Simonyan, K., Zisserman, A.: Very deep convolutional networks for large-scale image recognition. In: Proceedings of the International Conference on Learning Representations (2015)
19. Spitzer, R.L., Kroenke, K., Williams, J.B.W., Löwe, B.: A brief measure for assessing generalized anxiety disorder: the GAD-7. Arch. Internal Med. **166**(10), 1092–1097 (2006)
20. Szegedy, C., et al.: Going deeper with convolutions. In: Proceedings of the Conference on Computer Vision and Pattern Recognition (2014)
21. Thompson, A., Issakidis, C., Hunt, C.: Delay to seek treatment for anxiety and mood disorders in an Australian clinical sample. Behav. Change **25**(2), 71–84 (2008)
22. Thornicroft, G., Sartorius, N.: The course and outcome of depression in different cultures: 10-year follow-up of the WHO collaborative study on the assessment of depressive disorders. Psychol. Med. **23**(4), 1023–1032 (1993)
23. van Spijker, B.A.J.: Reducing suicidal thoughts in the Australian general population through web-based self-help: study protocol for a randomized controlled trial. Trials **16**(1), 62 (2015)
24. Woolf, S.H.: The meaning of translational research and why it matters. JAMA **299**(2), 211–213 (2008)

Preserving Data Privacy and Security in Australian My Health Record System: A Quality Health Care Implication

Pasupathy Vimalachandran[1](✉), Yanchun Zhang[1], Jinli Cao[2],
Lili Sun[1], and Jianming Yong[3]

[1] Institute for Sustainable Industries and Liveable Cities, VU Research,
Victoria University, Melbourne, Australia
{Pasupathy.Vimalachandran,yanchun.zhang,
lili.sun}@live.vu.edu.au
[2] The Department of Computer Science and IT, La Trobe University,
Melbourne, Australia
j.cao@latrobe.edu.au
[3] School of Management and Enterprise, Faculty of Business, Education,
Law and Arts, University of Southern Queensland, Toowoomba, Australia
yongj@usq.edu.au

Abstract. Australian My Health Record (MyHR) system must enable efficient availability of meaningful, accurate, complete and up-to-date health data. However, the major challenge must be to ensure the security of the clinical information of the MyHR. The foremost question that remains unanswered is 'are current information security settings adequate to protect MyHR?'. To build an adequate security setup and increase the uptake of the MyHR system, it is imperative to show the MyHR is safe to use. In addressing this issue and implementing the adoption of the initiative, we determine and systematically analyse the existing threats to the system. We assess strengths of various solutions against possible threats and discuss the development and implementation process of the proposed model.

Keywords: Data security · Access control in healthcare

1 Introduction

Applying cloud computing technology in healthcare promises a range of benefits and the most one of them is Electronic Health Record (EHR) system. Healthcare is a vigorous and complex setting with a number of participants. EHRs have a variety of functionalities which include storage of health information and data, results management, order entry and management, decision support, administrative processes, reporting and population management [1, 2]. The participant needs are vary depending on the purpose of involvement. For example, in a general practice organisation environment, the participants might be; patients, general practitioners (GP), nurses, lab technicians, practice manager, reception staff and information technology (IT) professionals. These participants use the system where a patient's information is stored for

© Springer Nature Switzerland AG 2018
H. Hacid et al. (Eds.): WISE 2018, LNCS 11234, pp. 111–120, 2018.
https://doi.org/10.1007/978-3-030-02925-8_8

various purposes. For example, while a GP will access a patient's clinical data in order to make a decision in delivering healthcare, a reception staff will access the patient demographic details to book an appointment. To enable this process, access control is used [3, 9, 15]. Hence, access control of a patient's health data must be flexible considering the dynamic nature of the healthcare settings. Conversely, considering the confidentiality around health data, the access control must be stronger.

In addition, in a healthcare provider organisation level, access policies and procedures should support MyHR operations and every user role within that organisation. For instance, the access policies must limit the operation of read, write, copy and print of any sensitive health information which is MyHR. Some users need access to de-identified health data for their role requirements. Medical Researchers need health data for research purposes, however sensitive or identified data cannot be shared and therefore, de-identified data must be available [28, 31]. To enable researchers to access appropriate data, it is important to have the ability to alter access rights for some users.

Considering these all requirements of an access control mechanism, it would be difficult to assign only one access control mechanism to satisfy all needs of the organisation [8, 9, 19, 21]. Therefore we consider two-layered model as a solution to ensure the privacy and security of the MyHR system. We firstly analyse previous EHR related threats. A wide range of attacks have been documented previously in this area. Hence, it is essential to account for the different attacks that have targeted health-based databases in the past.

1.1 Motivations

Healthcare organisations are inherently complex and dynamic environments which makes it difficult for administrators to define access control policies [4–7]. MyHR user privileges are therefore often defined at a coarse level to minimise workflow inefficiencies and maximise flexibility in the management of a patient. The consequence of such practice is that MyHR systems are left vulnerable to potential abuse from insiders who are authenticated within the organisation, which ultimately can compromise patient confidentiality. Furthermore, Information Technology (IT) technical staff or the system operators who maintain the IT systems and the databases also may access patients' clinical information. This leads to a risk of intentional or unintentional leakage, despite privacy and confidentiality agreements. Such agreements do not eliminate leaks occurring, but they mitigate the risk based on the person's professional integrity [10–13]. This demonstrates that information stored in MyHR databases or cloud servers face a significant risk of exposure. This potential for internal abuse must be addressed.

1.2 EHR Related Security Breaches

Issues of confidentiality and abuse of data cause many healthcare providers to oppose the coordination of medical databases despite their potential benefits [14, 17]. The followings are some of the incidents of security breaches in relation to EHRs:

- Researchers from University of Minnesota mistakenly revealed the names of deceased kidney donors to the recipients in a survey [15, 16].
- A hacker had access to sensitive health data from an unidentified medical centre in New York and another in Holland [18, 32].
- A hacker infiltrated the University of Washington's Medical Centre computer system and stole at least 5000 cardiology and rehabilitation medicine patients records [20, 33].
- A Florida state public health worker brought home a computer disk with the names of 4000 HIV positive patients and shared the contents with two Florida newspapers [22, 34].

After review the literature of the data privacy and security in EHR systems in Sect. 2, we analyse potential attacks targeting healthcare environment, provide solutions for those attacks in Sect. 3. Section 4 discusses the current Australians' My Health Record Systems while Sect. 5 describes a proposed model. Section 6 concludes the paper and suggests future work.

2 Related Work

There are different access control strategies for EHR and EMR that have been developed in the past [22]. According to one Forrester study, 80% of data security and privacy breaches involve insiders, employees or those with internal access to an organisation, putting information at risk [23]. With health sensitive data, this risk becomes more prominent. Many researchers have proposed various resolutions to solve the security and privacy problems associated with the EMRs and EHRs. These problems mainly refer to access control. The term "access control" is simply defined as "the ability to permit or deny the use of something by someone" [24]. The key objective of access control mechanisms is to permit authorised users to manipulate data and thus maintain the privacy of data [25]. There are different access control mechanisms that have been identified in the literature review. The basic models of the access control principles are (i) Discretionary Access Control (DAC), (ii) Mandatory Access Control (MAC), (iii) Role Based Access Control (RBAC) and (iv) Purpose Based Access Control (PBAC). However, the development is not to a satisfactory degree, in order to fulfil the privacy requirements of EMRs and EHRs [26, 27].

DAC uses access restriction set by the owner and restricts access to the objects. However a user who is allowed to access an object by the owner of the object has the capability to pass on the access right to other users without the involvement of the owner of the object [28, 29]. In RBAC, each user's access right is determined based on user roles and the role-specific privileges associated with them. RBAC policy uses the need-to-know principle to assign permissions to roles and to fulfil the least privileged condition by the system administrator. However, RBAC does not integrate other access parameters or related data that are significant in allowing access to the user [30]. PBAC is based on the notion of relating data objects with purposes [8, 19]. Many researchers have identified that greater privacy preservation is possible by assigning objects with purposes [30]. PBAC leads to a great deal of complexity at the access control level.

3 Potential Attacks Analysis

The potential attacks in healthcare industry are discussed in Table 1. Every potential threat or attack that has been identified in Table 1 are discussed and analysed to determine the right solution to prevent the threat or attack in Table 2.

Table 1. Potential Attacks

Potential attack	Description	Example
Excessive privileges	Application users are granted exceed privileges that exceed the requirements of their job functions	An employee whose job requires name and address of other employees take privileges to view the salary information as well
Privilege abuse	Application users may abuse legitimate data access privileges for unauthorised purposes	A user with privileges to view an employee details may abuse that privileges to retrieve all employee records
Unauthorised privilege elevation	Attackers may take advantage of vulnerabilities in DBMS software to convert low-level access privileges to high-level access privileges	Sometimes, an attacker may take advantage of database buffer overflow vulnerability to grant administrative privileges

Table 2. Possible solutions for the attacks

Identified attack	Possible solutions	Description
Excessive privileges	Access Control Mechanism (ACM) and use view of database	Proper access control mechanism restricts privileges to minimum data access for their job designation of an organisation
Privileges Abuse	Access Control Policy (ACP) and use view of database	The Access Control Policy that not only states what data is accessible, but also how data is accessed. Also the policy must identify users who are abusing access privileges
Unauthorised privilege elevation	Access Control Mechanism, use view of database and Intrusion Prevention System (IPS)	Proper Control Mechanism can detect a user who suddenly uses an unusual SQL operation. The IPS can identify a specific documented threat within the operation

The next step would be the strengths of the solution have been assessed in order to providing health-based database privacy and security in Table 3.

Table 3. Prevention method assessment

Possible attack	Access control mechanism	Intermediate database	Cryptography
Excessive privileges	√		
Privileges Abuse	√		
Unauthorised privilege elevation	√		
Platform vulnerabilities	√		
SQL injections	√	√	
Weak authentication	√	√	
Exposure of back-up data		√	√

4 Data Privacy and Security of the MyHR Systems

My Health Record (MyHR) was created to make it easier for healthcare professionals to access important information about patients quickly, and provide the best possible medical care [15].

Patient confidentiality has been a cornerstone of medical treatment since its very beginnings. This is a key reason why healthcare providers are among the most highly trusted group of professionals. Healthcare providers can generally only collect, use and disclose health information in a patient's record in order to provide healthcare. Any uses of outside of those allowed by the MyHR Act may constitute a data breach.

However, reviewing the current settings and controls of the health, the privacy and confidentiality around the MyHR involves further to the healthcare providers. The system operator of the MyHR which manages the system or practice staff in a healthcare provider organisation, have ability to access data but they may intentionally leak patients' clinical information.

Hence the access must carefully be monitored and controlled to maintain the privacy and confidentiality of patients to mitigate this risk. The other solution would be non-sensitive information accessibility for non-clinical staff. Discussing this concept in healthcare setting adds another layer of access for researchers and non-clinical staff, in addition to an improved access control level, is imminent.

5 Proposed Model

The basic access control alone model will not provide sufficient security for a system, because once anyone enters into system, then the data can be viewed or retrieved easily.

However improved or advanced access control will provide a full protection for MyHR systems. When access control is implemented well, it adds additional security to the health database. Considering all these factors behind the security model, it is clear that advanced access control will provide more security to MyHR systems and its databases. Considering the importance of medical research, on the other hand, the researchers must have access to de-identified data sets for the research.

Therefore, there is a need for an another level, in addition to improved access control, such as the presence of an intermediate state of database with clear de-sensitive information or non-intelligent information, which is most useful for an organisation as most system users refer to databases for identifying health issues, analysis, processing documents purposes. Figure 1 illustrates the overview of the proposed model.

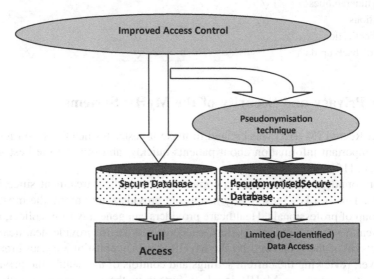

Fig. 1. The proposed framework

5.1   Layer One: An Improved Access Control Method for the MyHR

After considering several aspects of access control mechanisms through literature review, it was realized that there is a real need to put more control on this level of security. This led to develop a mechanism called "Give-Authority" which will be an ideal answer to minimise the potential for misuse or abuse of health data within a healthcare organisation.

Health sensitive and confidential data (e.g.; clinical notes/medical conditions) are stored in databases of clinical software systems. These data are susceptible to internal abuse as they can be viewed by anyone within an organisation using the internal settings. Hence this sensitive data needs to be protected from internal abuse. We have reason to believe that the "Give-Authority" method prevent these types of internal abuse.

Give-Authority: To access data through this system, an employee who has top level privilege (e.g. manager) has to give authorisation to a user to access health sensitive data. Hence, the manager keeps track of what the user does with the sensitive data. Every system user is already aware that they can be monitored once they log in the system and keeps track of what is being accessed. It is like a counter check. The responsibility and the accountability are shared, ensuring high security.

When the pairs are set-up in a organisation, the following main factors need to be considered; the physical location of the users and super users (e.g.; sharing the same office), job discipline of users (employees who are working in a similar discipline are paired) and the frequency and time an employee enters and uses the system (e.g.; an employee who needs to use the system for the whole day, all seven days a week should be paired with another employee who also uses the system for the same period rather than with an employee who only needs to access the system for a few hours in a week).

This system has its own problems:

(1) If manager is absent for a user then the user cannot enter into the system or perform his or her routine jobs.
(2) The system cannot prevent both the user and the manager as a pair deciding to abuse the data.

To overcome the first problem, a manager may be able to give permission through the Internet or networking as a future development. Alternatively doctors or other top level staff can also be considered as a manager to give permission for users to work on sensitive data. However, it is very difficult to overcome the second problem. A system monitoring facility can be developed as a part of this system to monitor the users and the managers. A system audit and/or quality improvement process is suggested.

5.2   Layer Two: Pseudonymisation Technique

Using the Pseudonymisation technique patients' personally identifiable fields including name and Healthcare Identifier (HI) within a health data record are replaced by one or more artificial (meaningless) identifiers which is called pseudonym.

With this technique, a user can still search data for relationships, however cannot capture all the value of the data. On the other hand, copying pseudonymised data is similarly pointless as they keys connecting the valuable links between the accessible pseudonym and the actual data itself are held elsewhere as shown in Table 4.

The hidden connective index data is stored in a secure destination or another PC where ordinary (or basic) users cannot access. The difference between encryption and pseudonymisation is; encryption or password permission exposes sensitive data and relationship. However, in pseudonymisation, the sensitive data is hidden and the relationships are exposed. The two key requirements for pseudonymisation are; data patterns must be maintained for linkage or analysis and personal data that will be shared, either internally or with a partner, must be hidden during the usage. Thus adopting the psudonymisation technique to the record will preserve privacy, reduce risk exposure and mitigate any potential impact of internal and external security breaches. Pseudonymisation renders stolen data effectively useless for identity theft and other fraud. This facilitates secure outsourcing and off shoring by using de-identified data to identify health records. The healthcare organisations can attain cost savings whilst significantly reducing the security concerns of using third party processors. The health software system integrators, developers and systems administrators can use de-identified data for estimating eHealth projects that work with health sensitive data,

Table 4. Basic pseudonymisation technique

Healthcare Identifier	Medication	Date	Condition	Name
8001567898761234	Insulin	01-10-2014	CD	John Smith
8008123456785000	Dapotum	05-10-2014	MH	Jane Doe
8001567898761234	Thalitone	10-10-2014	CKD	John Smith

Pseudonymisation Process

Healthcare Identifier Pseudonym	Medication	Date	Condition	Name Pseudonym
0102	Insulin	01-10-2014	CD	A12
452	Dapotum	05-10-2014	MH	B02
2712	Thalitone	10-10-2014	CKD	N17

designing and testing new systems that source health sensitive data from existing operations, and maintaining eHealth systems that manipulate sensitive data. Healthcare identifier and other health related number systems including Medicare number effectively become sensitive through their long term usage.

6 Conclusion and Future Suggestion

This paper presents two-layered model for ensuring privacy and security in the Australian My Health Record system. The first layer covers an improved access control method for healthcare workers and the second layer includes psuedonymisation technique for non-clinical staff and researchers to use de-identified data for their use purposes. With the introduction of pseudonymisation for the sensitive health information in EHR systems, indirect identifiable personal data is a long standing privacy technique to reduce the risks of identifiability.

More research is needed on the industry-wide prevalence of each type of EHR risk and the impact on health record integrity, patient safety, and quality of care. Further research is also needed on the causes of EHR-related errors and on effective strategies for preventing and correcting them.

References

1. Bosch, M. et al.: Review article: effectiveness of patient care teams and the role of clinical expertise and coordination: a literature review. Med. Care Res. Rev. (2009)
2. Kannampallil, T.G., et al.: Considering complexity in healthcare systems. J. Biomed. Inf. **44**, 943–947 (2011)

3. Wang, H., Sun, L., Bertino, E.: Building access control policy model for privacy preserving and testing policy conflicting problems. J. Comput. Syst. Sci. **80**(8), 1493–1503 (2014)

4. Zhang, Y., et al.: On secure wireless communications for IoT under eavesdropper collusion. IEEE Trans. Autom. Sci. Eng. **13**(3), 1281–1293 (2016)

5. Zhang, J., et al.: On efficient and robust anonymization for privacy protection on massive streaming categorical information. IEEE Trans. Dependable Secure Comput. **14**(5), 507–520 (2017)

6. Chin, T.: Security breach: hacker gets medical records. Am. Med. News **44**, 18–19 (2001)

7. Sun, X., et al.: Publishing anonymous survey rating data. Data Min. Knowl. Disc. **23**(3), 379–406 (2011)

8. Kabir, M.E., Wang H.: Conditional purpose based access control model for privacy protection. In: Proceedings of the Twentieth Australasian Conference on Australasian Database, vol. 92, pp. 135–142 (2009)

9. Wang, H., Sun L.: Trust-involved access control in collaborative open social networks. In: The 4thInternational Conference on Network and System Security, pp. 239–246 (2010)

10. Zhang, J., Tao, X., Wang, H.: Outlier detection from large distributed databases. World Wide Web. **17**(4), 539–568 (2014)

11. Carter, M.: Integrated electronic health records and patient privacy: possible benefits but real dangers. Med. J. Aust. **172**, 28–30 (2000)

12. Sittig, D.F., Singh, H.: Defining health information technology-related errors: new developments since to err is human. Arch. Intern Med. **171**, 1281–1284 (2011)

13. Wang, H., Zhang, Z., Taleb, T.: Special issue on security and privacy of IoT. World Wide Web **21**(1), 1–6 (2018)

14. Weir, C.R., et al.: Direct text entry in electronic progress notes. An evaluation of input errors. Methods Inf. Med. **42**, 61–67 (2003)

15. Wang, H., Cao, J., Zhang, Y.: Ticket-based service access scheme for mobile users. Aust. Comput. Sci. Commun. **24**(1), 285–292 (2002)

16. Australian Government: The eHealth consultation (2013). http://www.health.gov.au/internet/main/publishing.nsf/Content/pacd-ehealth-consultation-faqs. Accessed 15 Mar 2015

17. Shu, J., et al.: Privacy-preserving task recommendation services for crowd sourcing. IEEE Trans. Serv. Comput. (2018). https://doi.org/10.1109/TSC.2018.2791601

18. American Health Information Management Association: AHIMA Data Quality Management Model (2012)

19. Kabir, M.E., Wang, H., Bertino, E.: A role-involved purpose-based access control model. Inf. Syst. Front. **14**(3), 809–822 (2012)

20. Wang, H., Jiang, X., Kambourakis, G.: Special issue on security, privacy and trust in network-based big data. Inf. Sci. **318**(C), 48–50 (2015)

21. Wang, H., Cao, J., Zhang, Y.: Ubiquitous computing environments and its usage access control. In: Proceedings of the 1st International Conference on Scalable Information Systems, Hong Kong, p. 6 (2006)

22. Vimalachandran, P., Wang, H., Zhang, Y., Zhuo, G., Kuang, H.: Cryptographic access control in electronic health record systems: a security implication. In: Bouguettaya, A., et al. (eds.) WISE 2017. LNCS, vol. 10570, pp. 540–549. Springer, Cham (2017). https://doi.org/10.1007/978-3-319-68786-5_43

23. Sandhu, R.S., Samarati, P.: Access control: principle and practice. IEEE Commun. Mag. **32**, 40–48 (1994)

24. Wang, H., Cao, J., Zhang, Y.: A flexible payment scheme and its role-based access control. TKDE **17**(3), 425–436 (2005)

25. Li, H., et al.: Multi-window based ensemble learning for classification of imbalanced streaming data. World Wide Web **20**(6), 1507–1525 (2017)

26. Sun, X., Wang, H., Li, J., Truta, T.M.: Enhanced p-sensitive k-anonymity models for privacy preserving data publishing. Trans. Data Priv. **1**(2), 53–66 (2008)

27. Sun, L., et al.: Semantic access control for cloud computing based on e-Healthcare. In: IEEE 16th International Conference on Computer Supported Cooperative Work in Design, pp. 512–518 (2012)

28. Li, M., et al.: Privacy-aware access control with trust management in web service. World Wide Web **14**(4), 407–430 (2011)

29. Sun, X., et al,: An efficient hash-based algorithm for minimal k-anonymity. In: Proceedings of the thirty-first Australasian Conference on Computer Science, vol. 74, pp. 101–107 (2008)

30. Sun, X., et al.: Injecting purpose and trust into data anonymization. Comput. Secur. **30**(5), 332–345 (2011)

31. Sun, X., et al.: Satisfying privacy requirements before data anonymization. Comput. J. **55**(4), 422–437 (2012)

32. Mark, E., Serge, B.: A case study in access control requirements for a health information system. In: Proceedings of the Second Workshop on Australasian Information Security, Data Mining and Web Intelligence, and Software Internationalisation, vol. 32, pp. 53–61 (2004)

33. Motta, G., Furuie, S.: A contextual role-based access control authorization model for electronic patient records. IEEE Trans. Inf Technol. Biomed. **7**(3), 202–207 (2003)

34. Vimalachandran, P., et al.: The Australian PCEHR system: ensuring privacy and security through an improved access control mechanism. EAI Endorsed Trans. Scalable Inf. Syst. **3**(8), e4 (2016)

A Framework for Processing Cumulative Frequency Queries over Medical Data Streams

Ahmed Al-Shammari[1,2]($\boxtimes$), Rui Zhou[1]($\boxtimes$), Chengfei Liu[1], Mehdi Naseriparsa[1], and Bao Quoc Vo[1]

[1] Swinburne University of Technology, Melbourne, Australia
[2] University of Al-Qadisiyah, Al Diwaniyah, Iraq
{aalshammari,rzhou,cliu,mnaseriparsa,bvo}@swin.edu.au

Abstract. Medical data streams processing becomes increasingly important since it extracts critical information from a continuous flow of patient data. Various types of problems have been studied on medical data streams, such as classification, clustering, anomaly detection, etc.; however, efficient evaluation of cumulative frequency queries has not been well studied. The cumulative frequency of patients' status can play an instrumental role in monitoring the patients' health conditions. Up to now, efficiently processing cumulative frequency queries on medical data streams is still a challenging task due to the large size of the incoming data. Therefore, in this paper, we propose a novel framework for processing the cumulative frequency queries over medical data streams to support the online medical decision. The proposed framework includes two components: data summarisation and dynamic maintenance. For data summarisation, we propose a hybrid approach that combines two data structures and exploits a classification algorithm to select the more efficient data structure for computing the cumulative frequency. For dynamic maintenance, we propose an incremental maintenance approach for updating the cumulative frequencies when new data arrive. The experimental results on a real dataset demonstrate the efficiency of the proposed approach.

Keywords: Medical data streams · Cumulative frequency query
Binary indexed tree · Dynamic maintenance

1 Introduction

Data summarisation is an important data mining technique that aims to generate a concise representation of the underlying data [1,13]. The summarised data is represented in a compact form and still informative [2,9]. More recently, medical streaming data summarisation has become a useful task in healthcare information systems. That is because the data summary contains critical information for patients' health status. For instance, in the clinical bio-statistics, the cumulative

© Springer Nature Switzerland AG 2018
H. Hacid et al. (Eds.): WISE 2018, LNCS 11234, pp. 121–131, 2018.
https://doi.org/10.1007/978-3-030-02925-8_9

frequency is often used to determine the number of observations that exist above or below a specific value, or between two bounding values [5]. It may indicate the vital signs of blood pressure and electrocardiogram data. For example, consider the case where the patients data is recorded by different measurement devices. Then, these raw data is transferred to the medical centre for further analysis, and finally is presented to medical professionals. During this process, devices are designed to measure the physical status of the human body, such as blood flow, blood pressure, and electrocardiogram, in a very short period of time. Thus, to ensure the effective monitoring of patients' health status, it is necessary for the medical professionals to check the patient health-related data over time. For monitoring purposes, a doctor may submit a cumulative frequency query that includes the lower and upper bound values. Then, the system retrieves the cumulative frequency of the relevant measurements during a certain time interval for the specified range. However, processing a cumulative frequency query on medical data streams is a challenging task due to the large size of the continuous flow of patients data. For instance, to monitor a patient with a heart disease, the measurement device generates the streams very frequently (every half a minute). Thus, we face a vast amount of streaming data that are technically difficult to process. In this paper, we address some challenges for processing the cumulative frequency query on medical streaming data. First, there is a need to design an index structure that is capable of calculating the cumulative frequency efficiently. Second, there is also a need to design an incremental maintenance approach for maintaining the designed indexes in a timely manner. Therefore, we propose a new framework for processing the cumulative frequency queries on the medical streaming data. Technically, the proposed framework includes two components: (1) data summarisation, and (2) dynamic maintenance. The main contributions of this paper are summarised as follows:

- We propose a novel framework for processing the cumulative frequency queries over medical data streams. The framework includes a hybrid approach that combines two data structures and exploits a classification algorithm to minimise the computations.
- We further propose an incremental maintenance approach for updating the data summaries in a sliding time window.
- We validate the proposed approaches with experiments on a real-world dataset and demonstrate their efficiency.

The rest of the paper is organised as follows: Sect. 2 presents the problem formulation, which is followed by the proposed solution in Sect. 3. Section 4 presents the experimental results. The related work is presented in Sect. 5, and finally the paper is concluded in Sect. 6.

2   Problem Formulation

We first show how a medical data stream is modelled, and then define necessary preliminary terms. Finally, we define the problem studied in this paper formally.

Definition 1 *(Medical Data Stream). A medical data stream S is defined as a sequence of tuples $S = \langle s_1, t_1 \rangle, \langle s_2, t_2 \rangle, \ldots, \langle s_n, t_n \rangle$, where s_i is a set of data values, and $s_i.x_j$ indicates the jth measurement value for s_i on the jth dimension/symptom, and t_i is the associated timestamp of $\langle s_i, t_i \rangle$.*

Definition 2 *(Time/Tuple based Window). A window (w) is defined as a range that bounds the flow of data stream. The time-based window is specified based on the time units, whereas tuple-based window is specified based on the number of tuples.*

In this paper, we consider time-based window to manage the flow of medical data streams. Tuple-based window can be handled similarly.

Definition 3 *(Frequency). A frequency $f(s_i.x)$ is defined as the number of times that an attribute value $s_i.x_j$ appears within a specified time window w.*

Practically, in many medical applications, it is useful to check the total number of occurrences of some measurement value appearing within a doctor prespecified range so that the patients' status can be monitored. The main indicator that determines the risky and normal statuses of the patients depends on the health-related boundaries. Usually, there is a need to specify an interval and monitor patients' the cumulative frequency of a particular attribute value.

Definition 4 *(Cumulative Frequency Query). Given an interval $[l, u]$, a cumulative frequency query aims to retrieve the total of frequencies of all the values within the range $[l, u]$, which can be expressed by the following formula:*

$$cf(s_i.x) = \sum_{i=1}^{n} f(s_i.x), s_i.x \in [l, u] \tag{1}$$

Medical data streams are evolving over time. The cumulative frequency results are likely to change over time accordingly. It is desirable to provide query answers as soon as possible for doctors if their cumulative frequency queries are continuous. As a result, it is significant that the cumulative frequency summaries can be maintained incrementally when new data arrive. The maintenance of C includes inserting new data stream values, updating the cumulative frequency values in $\{cf_i, \ldots, cf_n\}$, and deleting expired data stream values. Now, we summarise the problem that is studied in this paper below:

Problem Definition. Given a medical data stream S, a set of cumulative frequency queries $Q = \{q_1, q_2, \ldots, q_n\}$, and their corresponding specified windows $W = \{w_1, w_2, \ldots, w_n\}$, the required task is to retrieve and maintain the cumulative frequencies $C = \{cf_1, cf_2, cf_3, \ldots, cf_n\}$ from the patients data for the corresponding specified time windows W efficiently.

3   The Proposed Solution

This section clarifies the technical aspects of the proposed framework. Figure 1 shows the main components of our framework. From the figure, the main components are as follows: (1) data summarisation, and (2) dynamic maintenance.

In the first component, we discuss how to process cumulative frequency queries. In the second component, we discuss how to maintain the cumulative frequency results when new stream items arrive and old stream items expire. The technical details of the data summarisation and dynamic maintenance are presented in Sects. 3.1 and 3.2 respectively.

3.1　Data Summarisation

The first component of the proposed approach is data summarisation. Computing the cumulative frequency of the data streams is one of the data summarisation processes. We propose a new approach for processing the cumulative frequency queries on medical data streams efficiently. Specifically, our approach is a combination of a Binary Indexed Tree (BIT) based approach and a baseline approach. In the design, we build a classifier to predict the better approach for processing the cumulative frequency cf of the medical data streams. Since the incoming data may involve big or small range of data, we take the value range of data streams into consideration, which can be classified into two categories: small and large. The BIT approach is employed to efficiently calculate the cumulative frequency of large query ranges, whereas the baseline approach is employed to calculate the cumulative frequency of small query ranges. In The technical details of the proposed approaches are presented in Sects. 3.1.1, 3.1.2, and 3.1.3 respectively.

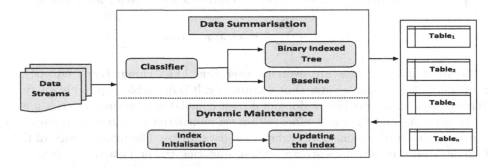

Fig. 1. The components of proposed framework

3.1.1　Baseline Approach

The baseline approach employs an array $\mathcal{A}$ to store and update the frequency of the values appearing within a specific time window on a medical data stream. For example, we assume that $\mathcal{B}$ is a bag of attribute values in a specific time window, eg., $\mathcal{B} = \{95, 120, 95, 99, 95, 100, 64, 66, 95, 77\}$, here the value frequencies stored in $\mathcal{A}$ are $f(64) = 1$, $f(66) = 1$, $\cdots$, $f(95) = 4$, $\cdots$, $f(100) = 1$. Note that, for an absent value, the frequency field in $\mathcal{A}$ is recorded as 0, eg., $f(65) = 0$. Then, to compute the cumulative frequency of a certain value range $[l, r]$, we apply a linear scan on $\mathcal{A}$ and aggregate the frequency values from $f(l)$ to $f(u)$ by using the Formula 1. Clearly, due to a linear scan on an array, this process requires $O(n)$ time complexity, where n is the possible number of values. When a data value enters or leaves the window, the corresponding frequency can be updated in $O(1)$ time. The baseline approach works better in updating time and is good when data value range is small. However, the $O(n)$ complexity makes query efficiency not satisfactory, especially, when data value range is large, eg., when data value range is enlarged by dividing data values with smaller intervals in order to provide better precision.

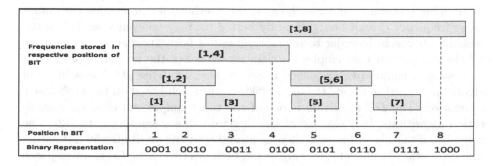

Fig. 2. The structure of BIT

3.1.2   Binary Indexed Tree Approach

Binary Indexed Tree (BIT) [7] is a data structure that accelerates the retrieval of the cumulative frequency of a certain range. It calculates the cumulative frequency in $O(\log n)$ time complexity. The core idea of BIT is that the elements of the array store the cumulative frequencies of specific ranges. Each node stores a sub-sum of frequencies, covering the number of values as some power of two. Figure 2 shows the frequencies stored in a binary indexed tree array, assuming data values start from 1. In this BIT array, bit[1] stores the frequency of elements at arr[1]; bit[2] stores the frequency of elements at arr[1] and arr[2]; bit[3] stores the frequency of elements at arr[3]; bit[4] stores the cumulative frequency of elements from arr[1] to arr[4]. As a result, given a value n, the BIT can help to calculate the cumulative frequencies of the values $[1, n]$. The binary representation of n includes at most $\lceil \log_2 n \rceil$ bits. For processing a cumulative frequency query $q = [l, u]$, the BIT can help to calculate the cumulative frequency of the lower end bit[$l-1$] (which records the values in $[1, l-1)$) and the upper end bit[u] (which records the values in $[1, u]$). And then, we take the difference bit[u] − bit[$l-1$] and get the cumulative frequencies of the range $[l, u]$. Note that, for ease of illustration, in Fig. 2, we adopt 1 as the starting value, and 1 as the increment step between two adjacent values. In practice, the starting value can be any and the increment step can be also set to a larger or smaller value, depending on the need of users. A mapping can be easily done by taking the measured value as input and using a mapping function to locate the right BIT array entry for recording the occurrence of such value.

3.1.3 Hybrid Approach

The baseline approach uses a linear scan on an array data structure to retrieve cumulative frequencies of a given user-interested value range. However, this incurs high computational cost, when the range is large and the interval increment is small. Therefore, to improve the performance, the Binary Indexed Tree (BIT) approach exploits an index to retrieve the cumulative frequency in $O(\log n)$ time. This BIT approach does not excel when the query range is small. This is because, under such cases, the overhead for processing the cumulative frequency cannot be significantly better (or may be even worse) than the baseline approach. In order to take advantage of both approaches, we propose a hybrid approach that employs a classifier to select the proper approach for processing a cumulative frequency query. A Decision Tree (DT) classification algorithm is used to decide the more efficient approach based on the probability estimation. The DT algorithm is one of the most simple and effective classification algorithms. It is capable of learning from the streaming data [12]. The core idea of this algorithm is to choose optimal splitting attributes by estimating some statistics. Technically, this algorithm includes two parts: (a) training phase and (b) testing phase. In the training phase, a decision tree is built based on the input, output and the training medical data streams. To build the model, the DT would exploit the entropy function ($E(P_1, P_2) = -\sum_{i=1}^{2} P_i \log P_i$) to decide on which input features the data will split, where P_1 is the probability that an instance from S results in the outcome 0. While, P_2 is the probability that the instance from S results in the outcome 1. We design the feature vector $\boldsymbol{v}$ that includes three main statistics as follows: the mean ($m = \frac{\sum_{i=1}^{n} s.x_i}{n}$), the variance ($r = \frac{\sum_{i=1}^{n}(s.x_i - m)^2}{n}$), and a set of cumulative frequency queries $\mathcal{Q}$.

Algorithm 1. Hybrid Approach

    Input   : Data stream S , cumulative frequency query $q = [l, u]$
    Output: Cumulative frequency cf
1 $S^{train} \leftarrow getSubset(S)$
2 $S^{test} \leftarrow S \setminus S^{train}$
3 $DT \leftarrow Train(S^{train})$
4 **if** $Test(DT)$ **then**
5       $m \leftarrow calculateAvg(s.x)$;
6       $r \leftarrow calculateVar(s.x, m)$;
7       $v \leftarrow createVector(m, r, q)$;
8       $y \leftarrow classify(DT, v)$;
9   **if** $y == 0$ **then**
10   use baseline to retrieve cf for q;
11   **else**
12   use BIT to retrieve cf for q;
13 **else**
14   go to line 1
15 **return** cf;

Algorithm 1 presents the steps of processing the cumulative frequency query using the hybrid approach. In lines 1–2, we prepare the training and testing data. In lines 5–7, we create the feature vector $\boldsymbol{v}$. In lines 9–12, we select the more efficient approach for processing the cumulative frequency query based on the the output y of the DT classifier in line 8. If y returns 0, we use the baseline approach, otherwise we use the BIT approach. In line 14, we require to use the training phase if the test is not successful. In line 15, we retrieve the cumulative frequency cf for the specified q.

3.2 Dynamic Maintenance

The second component of the proposed approach is the dynamic maintenance component. When data streams slide into a new time window w, the cumulative frequencies of BIT are to be updated incrementally. The BIT approach maintains the cumulative frequencies in two primary steps presented as follows: (1) index initialisation, and (2) index update. Firstly, we initialise the BIT index when the data streams in the first time window w arrive. To build the index, we firstly extract the distinct attribute values in the window w and compute the frequency cf of these values. Then, to include these values and their corresponding frequencies into the BIT, we use an insertion operation inside the BIT data structure. Secondly, the cumulative frequency of some of the existing attribute values may be subjected to change in a new time window. That is because more instances of a value may arrive into the window or some instances exit from the window. In order to update the index structure efficiently, we enhance the system to be capable of maintaining the index of the cumulative frequencies by using two operations that are stated as follows:

- **Increment:** this operation increments the cumulative frequency associated with the attribute value $s.x$ which is from a new instance $<s, t>$ in the current time window w_c.
- **Decrement:** this operation decrements the cumulative frequency associated with the attribute value $s.x$ which is from an old instance $<s, t>$ that was in the old window w_o but has exited from the current time window w_c.

Algorithm 2. BIT Maintenance

Input : Old window w_o, Current window w_c, Index $\mathcal{B}$
Output: Updated index $\mathcal{B}$
1 **foreach** *stream item* $<s, t> \in w_c - w_o$ **do**
2 $\quad$| $\quad$Increment: $f(s.x) = f(s.x) + 1$ in $\mathcal{B}$;
3 **foreach** *stream item* $<s, t> \in w_o - w_c$ **do**
4 $\quad$| $\quad$Decrement: $f(s.x) = f(s.x) - 1$ in $\mathcal{B}$;
5 **return** $\mathcal{B}$

Algorithm 2 presents the steps of the maintenance using the BIT approach. In lines 1–2, we update the cumulative frequency of an attribute value by using the increment operation to add in new stream item values. In lines 3–4, we update the cumulative frequency of an attribute value by using the decrement operation to remove expired stream items. In line 5, we retrieve the new index $\mathcal{B}$.

4 Experimental Results

This section highlights the experimental results for the proposed approaches as follows: (a) baseline, (b) BIT, and (c) hybrid approach. We conducted experi-

ments on a real medical dataset. We use the Cuff-Less Blood Pressure Estimation dataset[1]. The format of the dataset is Matlab's v7.3, and the total size is
3.17 GB. The dataset includes four subsets, each subset contains 3000 records.
The experiments are implemented in Java and executed by a processor Intel, Core
(i5)-3570 CPU 3.40 GHz. We verify the performance of the proposed approaches
by varying the size of streaming data and varying the query bounds.

4.1   Varying the Data Stream Size

Figure 3 presents the execution time of the proposed approaches by varying the
size of the data streams. We set the cumulative frequency query $q = [55, 75]$
for this set of experiments. We use 4 subsets of data streams with various sizes.
The order of their sizes are as follows: Subset 3 < Subset 1 < Subset 4 <
Subset 2. When the size of the data streams is set from 500 to 3000 with an
increase of 500 records at each iteration, we note that the subset 2 which has
the largest size requires a long execution time in comparison with other subsets
for all approaches. Conversely, subset 3 which has the smallest size requires a
less execution time for all approaches.

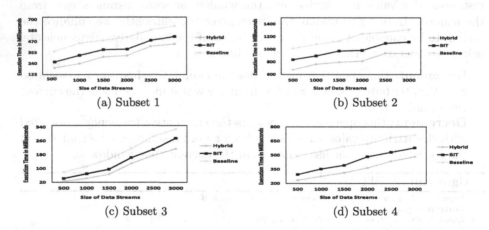

Fig. 3. Execution time for the baseline, BIT and hybrid approaches with varying the
data streams size

That's because when the size of data streams increase, all approaches scale
linearly with the execution time to retrieve the cumulative frequency q. Moreover, the hybrid approach shows the best performance in comparison with two
other approaches. That's because the hybrid approach employs the decision tree
classifier which effectively selects the more efficient approach. We can observe
that the data stream size has the minimum effect on the performance of hybrid
approach.

---

[1] https://archive.ics.uci.edu/ml/datasets/Cuff-Less+Blood+Pressure+Estimation.

4.2 Varying the Query Bounds

Figure 4 presents the execution time of the proposed approaches by varying the query bounds of in the experimented data streams. We quantify the effect of the baseline, BIT and hybrid approaches in processing the cumulative frequency of the blood pressure. The query bounds are categorised into four groups as follows: (1) small ($q = [55, 60]$), (2) medium ($q = [60, 70]$), (3) large ($q = [70, 90]$), and (4) very large ($q = [90, 130]$). From Fig. 4a, we observe that the baseline approach performance is close to BIT or in other cases it beats the BIT approach. That's because when the query bounds are small, the BIT computational overhead deteriorates its performance. Conversely, when the query bounds are larger the performance of BIT improves in comparison with the baseline approach as shown in Fig. 4c and d. We also observe that the hybrid approach outperforms the baseline and the BIT approaches in all cases. As a result, the hybrid approach is always the best approach.

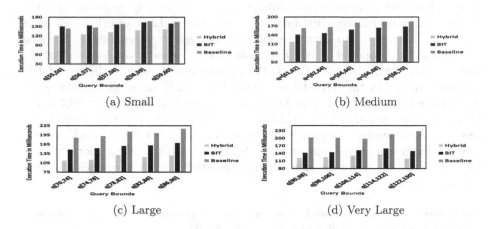

(a) Small   (b) Medium

(c) Large   (d) Very Large

Fig. 4. Execution time for the baseline, BIT and hybrid approaches with varying the query bounds

5 Related Work

This section highlights the basic findings in the studies related to the binary indexed tree and range query processing [14]. Most of the existing works have been focused on the Binary Indexed Tree (BIT). This data structure introduced by Fenwick [7] for efficiently calculating and maintaining the cumulative frequencies. Dima et al. [6] employed the BIT to solve the Range Minimum Query (RMQ) problem. Bille et al. [3] proposed two succinct models of the (BIT) for solving the partial sums problem. The first model requires $nk + O(n)$ bits of space with supporting the sum and update operations. In the second model, the optimal time for both operations is increased partially. Han et al. [8] utilized (BIT) to estimate the number of join results to help with load shedding over data streams. Mladenović et al. [11] proposed a variable neighbourhood search approach. They used the (BIT) for updating and checking of solutions in the neighbourhoods. As mentioned earlier, the tree-based index structure plays a significant role for answering the range queries [4,10]. Zhu et al. [15] proposed a new binary index, called binary obstructed tree (OB-tree) for indexing composite items in the obstructed space. The basic idea of OB-tree is to divide the obstructed space into non-obstructed sub spaces. The distinction of this paper is that we compute the cumulative frequency value between two ranges of the given query. The literature reveals that none of the proposed approaches employ the BIT for processing cumulative frequency queries in the streaming environment.

6 Conclusion

In this paper, we propose a novel framework for processing the cumulative frequency queries over medical data streams. Our proposed framework includes two main components: data summarisation and dynamic maintenance. In the first component, we proposed a hybrid approach which uses a combination of the baseline and Binary Indexed Tree (BIT) approaches to retrieve the cumulative frequency query. In the second component, the hybrid approach maintains the cumulative frequency summaries by initialising and maintaining the BIT index. Verified experimentally, the hybrid approach improves the performance of processing the cumulative frequency queries on average. For future work, we will consider further improvements on the hybrid approach to make it more effective to retrieve and maintain the continuous cumulative frequency queries over very fast data streams.

Acknowledgement. This work was partially supported by the ARC Discovery Project under Grant No. DP170104747 and DP180100212.

References

1. Abbas, A.M., Bakar, A.A., Ahmad, M.Z.: Fast dynamic clustering SOAP messages based compression and aggregation model for enhanced performance of Web services. J. Netw. Comput. Appl. **41**, 80–88 (2014)
2. Al-Shammari, A., Liu, C., Naseriparsa, M., Vo, B.Q., Anwar, T., Zhou, R.: A framework for clustering and dynamic maintenance of XML documents. In: Cong, G., Peng, W.-C., Zhang, W.E., Li, C., Sun, A. (eds.) ADMA 2017. LNCS, vol. 10604, pp. 399–412. Springer, Cham (2017). https://doi.org/10.1007/978-3-319-69179-4_28
3. Bille, P., Christiansen, A.R., Prezza, N., Skjoldjensen, F.R.: Succinct partial sums and Fenwick trees. In: Fici, G., Sciortino, M., Venturini, R. (eds.) SPIRE 2017. LNCS, vol. 10508, pp. 91–96. Springer, Cham (2017). https://doi.org/10.1007/978-3-319-67428-5_8

4. Chen, L., Gao, Y., Li, X., Jensen, C. S., Chen, G., Zheng, B.: Indexing metric uncertain data for range queries. In: Proceedings of the 2015 ACM SIGMOD International Conference on Management of Data, pp. 951–965. ACM (2015)
5. Chen, L., Gao, Y., Zhong, A., Jensen, C.S., Chen, G., Zheng, B.: Indexing metric uncertain data for range queries and range joins. VLDB J. **26**(4), 585–610 (2017)
6. Dima, M., Ceterchi, R.: Efficient range minimum queries using binary indexed trees. Olymp. Inform. **9**, 39–44 (2015)
7. Fenwick, P.M.: A new data structure for cumulative frequency tables. Softw.: Pract. Exp. **24**(3), 327–336 (1994)
8. Han, D., Xiao, C., Zhou, R., Wang, G., Huo, H., Hui, X.: Load shedding for window joins over streams. In: Yu, J.X., Kitsuregawa, M., Leong, H.V. (eds.) WAIM 2006. LNCS, vol. 4016, pp. 472–483. Springer, Heidelberg (2006). https://doi.org/10.1007/11775300_40
9. Hoplaros, D., Tari, Z., Khalil, I.: Data summarization for network traffic monitoring. J. Netw. Comput. Appl. **37**, 194–205 (2014)
10. Jung, H., Kim, Y.S., Chung, Y.D.: QR-tree: an efficient and scalable method for evaluation of continuous range queries. Inf. Sci. **274**, 156–176 (2014)
11. Mladenović, N., Urošević, D., Ilić, A.: A general variable neighborhood search for the one-commodity pickup-and-delivery travelling salesman problem. Eur. J. Oper. Res. **220**(1), 270–285 (2012)
12. Rutkowski, L., Jaworski, M., Pietruczuk, L., Duda, P.: The CART decision tree for mining data streams. Inf. Sci. **266**, 1–15 (2014)
13. Wang, C., Zhang, R., He, X., Zhou, G., Zhou, A.: Event phase extraction and summarization. In: Cellary, W., Mokbel, M.F., Wang, J., Wang, H., Zhou, R., Zhang, Y. (eds.) WISE 2016. LNCS, vol. 10041, pp. 473–488. Springer, Cham (2016). https://doi.org/10.1007/978-3-319-48740-3_35
14. Wang, Y., Meliou, A., Miklau, G.: RC-Index: diversifying answers to range queries. Proc. VLDB Endow. **11**(7), 773–786 (2018)
15. Zhu, H., Yang, X., Wang, B., Lee, W.-C.: Range-based nearest neighbor queries with complex-shaped obstacles. IEEE Trans. Knowl. Data Eng. **30**(5), 963–977 (2018)

Web Services and Cloud Computing

Knowledge-Driven Automated Web Service Composition—An EDA-Based Approach

Chen Wang[1(✉)], Hui Ma[1], Aaron Chen[1], and Sven Hartmann[2]

[1] School of Engineering and Computer Science, Victoria University of Wellington,
Wellington, New Zealand
{chen.wang,hui.ma,aaron.chen}@ecs.vuw.ac.nz
[2] Department of Informatics, Clausthal University of Technology,
Clausthal-Zellerfeld, Germany
sven.hartmann@tu-clausthal.de

Abstract. Service Oriented Architecture starts with the concept of web services, which give birth to an application of web service composition that selects and combines web services to accommodate users' complex requirements. These requirements often cover functional parts (i.e., semantic matchmaking of services' inputs and outputs) and non-functional parts (i.e., Quality of Service). Service composition is an NP-hard problem. Evolutionary Computation (EC) techniques have been successfully proposed for finding solutions with near-optimal Quality of Semantic Matchmaking (QoSM) and/or Quality of Service (QoS) using knowledge of promising solutions. Estimation of Distribution Algorithm (EDA) has been applied to semi-automated QoS-aware service composition, since it is capable of extracting knowledge of good solutions into a explicit probabilistic model. However, existing works do not support extracting knowledge for fully automated service composition that does not obeying a given workflow. In this paper, we proposed an EDA-based fully automated service composition approach to jointly optimize Quality of Semantic Matchmaking and Quality of Services. This approach is compared with a PSO-based approach that was recently proposed to solve the same problem.

Keywords: Web service composition · QoS optimization
Combinatorial optimization · EDA

1 Introduction

Web services are self-describing web-based applications, which can be deployed, discovered and invoked over the Internet by service users [3]. Because a single service often cannot completely satisfy users' complex requirements, *web service composition* is achieved by loosely coupling web services to provide added values in relation to both the functional and non-functional aspects, i.e., *Quality of Semantic Matchmaking* (QoSM) and *Quality of service* (QoS). Therefore,

© Springer Nature Switzerland AG 2018
H. Hacid et al. (Eds.): WISE 2018, LNCS 11234, pp. 135–150, 2018.
https://doi.org/10.1007/978-3-030-02925-8_10

semantic web services composition and *QoS-aware service composition* raise the interests of many researchers in optimizing QoSM and QoS respectively. *Fully automated service composition* has been a promising research field, and it constructs service workflows automatically with service selections, without strictly obeying any specific workflows [11].

Knowledge-driven Web service composition is achieved by utilizing knowledge, which is defined as useful information acquired through experience (i.e., promising service composition solutions). The knowledge can be implicit or explicit based on practical or theoretical understanding of promising solutions. By iteratively updating and utilizing the knowledge, new candidate solutions are generated until a most desired solution found.

Conventional Evolutionary Computation (EC) techniques use implicit knowledge of promising solutions to successfully achieve QoS-aware web service composition [7,12,19,24], where new candidate solutions are generated using implicit knowledge by one or more variation operators on parent individuals. For example, Genetic Algorithms produce new candidate solutions by crossover operated on two selected parent individuals. Whereas, Estimation of Distribution Algorithm (EDA) is different from most conventional EC techniques, EDA uses explicit knowledge encoded by a probabilistic model based on the distribution of a set of parent individuals, which often refers to a superior subpopulation that is made of vector-based solutions. It has been suggested in some problem domains, information revealed by the explicit knowledge, in particular, distributions and dependencies of variables in vector-based solutions, can make the search more effective and efficient [2].

Despite recent successes in other problem domains [22,23], such as arc routing and assembly flow-shop scheduling problems, EDA remains an important research question for successfully solving service composition problems. Two existing service composition approaches utilize EDA for service composition problems, but the probabilistic models in these works have no clear definition [10] or do not support fully automated service composition [9]. Therefore, opportunities still exist to further investigate the effectiveness of other models for supporting fully automated service composition.

The overall goal of this paper is to *propose an EDA-based approach for fully automated service composition* where QoS and QoSM jointly optimized. We achieve three objectives in this work, and some initial ideas have been published in a poster [21].

1. To learn explicit knowledge of solutions, the service composition problem is transferred into a permutation-based problem. To achieve that, we propose a fixed-length, vector-based indirect representation of service composition solutions. This representation enables reliable and accurate learning of the underlying probability distribution model.
2. To study the effective exploitation of the learned knowledge through two updating strategies for the probability distribution model.

3. To demonstrate the effectiveness of our EDA-based approach, we conduct experiments to compare it against a recently proposed PSO-based method [19] as a baseline.

2 The Semantic Web Service Composition Problem

A *semantic web service* (*service*, for short) is considered as a tuple $S = (I_S, O_S, QoS_S)$ where I_S is a set of service inputs that are consumed by S, O_S is a set of service outputs that are produced by S, and $QoS_S = \{t_S, c_S, r_S, a_S\}$ is a set of non-functional attributes of S. The inputs in I_S and outputs in O_S are parameters modeled through concepts in a domain-specific ontology $\mathcal{O}$. The attributes t_S, c_S, r_S, a_S refer to the response time, cost, reliability, and availability of service S, respectively, which are four commonly used QoS attributes [25].

A *service repository* $\mathcal{SR}$ is a finite collection of services supported by a common ontology $\mathcal{O}$. A *composition task* (also called *service request*) over a given $\mathcal{SR}$ is a tuple $T = (I_T, O_T)$ where I_T is a set of task inputs, and O_T is a set of task outputs. The inputs in I_T and outputs in O_T are parameters that are semantically described by concepts in the ontology $\mathcal{O}$.

Two special atomic services $Start = (\emptyset, I_T, \emptyset)$ and $End = (O_T, \emptyset, \emptyset)$ are considered for accounting for the input and output of a given composition task T, and add them to $\mathcal{SR}$. We use *matchmaking types* to describe the level of a match between outputs and inputs [8]. For concepts a, b in $\mathcal{O}$ the *matchmaking* returns *exact* if a and b are equivalent ($a \equiv b$), *plugin* if a is a sub-concept of b ($a \sqsubseteq b$), *subsume* if a is a super-concept of b ($a \sqsupseteq b$), and *fail* if none of previous matchmaking types is returned. In this paper we are only interested in *exact* and *plugin* matches for robust compositions, see [4]. As argued in [4] *plugin* matches are less preferable than *exact* matches due to the overheads associated with data processing. the semantic similarity of concepts is suggested to be considered when comparing different *plugin* matches.

A *robust causal link* [5] is a link between two matched services S and S', noted as $S \rightarrow S'$, if an output a ($a \in O_S$) of S serves as the input b ($b \in O_{S'}$) of S' satisfying either $a \equiv b$ or $a \sqsubseteq b$. For concepts a, b in $\mathcal{O}$, the *semantic similarity* $sim(a, b)$ is calculated based on the edge counting method in a taxonomy like WorldNet or an ontology [13]. One advantage of this method is simple calculation and good performance [13]. Therefore, the *matchmaking type* and *semantic similarity* of a robust causal link can be defined as follow:

$$typе_{link} = \begin{cases} 1 & \text{if } a \equiv b \ (exact \text{ match}) \\ p & \text{if } a \sqsubseteq b \ (plugin \text{ match}) \end{cases} \tag{1}$$

$$sim_{link} = sim(a, b) = \frac{2N_c}{N_a + N_b} \tag{2}$$

with a suitable parameter $p, 0 < p < 1$, and with N_a, N_b and N_c, which measure the distances from concept a, concept b, and the closest common ancestor c of a and b to the top concept of the ontology $\mathcal{O}$, respectively. However, if more than one pair of matched output and input exist from service S to service S', $type_e$ and sim_e will take on their average values.

The *QoSM* of the service composition can be obtained by aggregating over all robust causal links as follow:

$$MT = \prod_{j=1}^{m} type_{link_j} \tag{3}$$

$$SIM = \frac{1}{m} \sum_{j=1}^{m} sim_{link_j} \tag{4}$$

Formal expressions as in [6] are used to represent service compositions. The constructors •, ‖, + and * are used to denote sequential composition, parallel composition, choice, and iteration, respectively. The set of *composite service expressions* is the smallest collection SC that contains all atomic services and that is closed under sequential composition, parallel composition, choice, and iteration. That is, whenever $C_0, C_1, \ldots, C_d$ are in SC then •$(C_1, \ldots, C_d)$, ‖ $(C_1, \ldots, C_d)$, +$(C_1, \ldots, C_d)$, and *C_0 are in SC, too. Let C be a composite service expression. If C denotes an atomic service S then its QoS is given by QoS_S. Otherwise the QoS of C can be obtained inductively as summarized in Table 1. Herein, $p_1, \ldots, p_d$ with $\sum_{k=1}^{d} p_k = 1$ denote the probabilities of the different options of the choice +, while ℓ denotes the average number of iterations. Therefore, QoS of a service composition solution (i.e., A, R, T, and C) can be obtained by aggregating a_C, r_C, t_C and c_C in Table 1.

Table 1. QoS calculation for a composite service expression C

$C =$	$r_C =$	$a_C =$	$c_C =$	$t_C =$
•$(C_1, \ldots, C_d)$	$\prod_{k=1}^{d} r_{C_k}$	$\prod_{k=1}^{d} a_{C_k}$	$\sum_{k=1}^{d} c_{C_k}$	$\sum_{k=1}^{d} t_{C_k}$
‖$(C_1, \ldots, C_d)$	$\prod_{k=1}^{d} r_{C_k}$	$\prod_{k=1}^{d} a_{C_k}$	$\sum_{k=1}^{d} c_{C_k}$	$MAX\{t_{C_k} \| k \in \{1, ..., d\}\}$
+$(C_1, \ldots, C_d)$	$\prod_{k=1}^{d} p_k \cdot r_{C_k}$	$\prod_{k=1}^{d} p_k \cdot a_{C_k}$	$\sum_{k=1}^{d} p_k \cdot c_{C_k}$	$\sum_{k=1}^{d} p_k \cdot t_{C_k}$
*C_0	$r_{C_0}{}^{\ell}$	$a_{C_0}{}^{\ell}$	$\ell \cdot c_{C_0}$	$\ell \cdot t_{C_0}$

When multiple quality criteria are involved in decision making, the fitness of a solution can be defined as a weighted sum of all individual criteria using Eq. (5), assuming the preference of each quality criterion is provided by users.

$$Fitness = w_1 \hat{MT} + w_2 \hat{SIM} + w_3 \hat{A} + w_4 \hat{R} + w_5(1 - \hat{T}) + w_6(1 - \hat{C}) \tag{5}$$

with $\sum_{k=1}^{6} w_k = 1$. This objective function is defined as a *comprehensive quality model* for service composition. We can adjust the weights according to users' preferences. $\hat{MT}$, $\hat{SIM}$, $\hat{A}$, $\hat{R}$, $\hat{T}$, and $\hat{C}$ are normalized values calculated within the range from 0 to 1 using Eq. (6). To simplify the presentation we also use the notation $(Q_1, Q_2, Q_3, Q_4, Q_5, Q_6) = (MT, SIM, A, R, T, C)$. Q_1 and Q_2 have minimum value 0 and maximum value 1. The minimum and maximum value of Q_3, Q_4, Q_5, and Q_6 are calculated across all task-related candidates in the service repository $\mathcal{SR}$ using the greedy search in [7,15].

$$
\hat{Q}_k = \begin{cases} \frac{Q_k - Q_{k,min}}{Q_{k,max} - Q_{k,min}} & \text{if } k = 1, \ldots, 4 \text{ and } Q_{k,max} - Q_{k,min} \neq 0, \\ \frac{Q_{k,max} - Q_k}{Q_{k,max} - Q_{k,min}} & \text{if } k = 5, 6 \text{ and } Q_{k,max} - Q_{k,min} \neq 0, \\ 1 & \text{otherwise.} \end{cases} \tag{6}
$$

The goal of comprehensive quality-aware service composition is to maximize the objective function in Eq. (5) to find the best possible solution for a given composition task T.

3   Our EDA-Based Approach for Service Composition

In this section, we present our new EDA-based approach for fully automatic semantic web service composition. We will start with an outline in Sect. 3.1. Subsequently, we discuss two proposed ideas behind this approach: one is that a vector-based representation of service composition solutions is proposed to allow reliable and accurate learning of knowledge from promising solutions; another is that two adaptive updating methods are proposed to facilitate knowledge reuse.

The EDA strategy has been applied with some success to optimization problems where candidate solutions can be represented as permutations over a given set of elements [2]. The success, however, strongly depends on the ability to define a suitable probability distribution model for the problem domain under investigation. Service compositions are commonly represented as directed acyclic graphs (DAG). The DAG-representation of a service composition is essential, in particular, it allows an efficient computation of the quality of a service composition. One idea would be to represent a service composition as a queue of services, that is, a permutation of atomic services from the service repository $\mathcal{SR}$. Such a permutation, however, needs to be interpreted. For that, we will define a suitable mapping between DAG-representations of service composition solutions and permutations. For details see Sect. 3.2.

Moreover, to properly balance between exploration and exploitation during the evolution, we propose a general method to adaptively adjust the learned probability distribution model at every generation, resulting in the development of two specific adjusting strategies to be discussed in Sect. 3.4.

3.1   Outline of Our EDA-Based Method

Our proposed approach is outlined in Algorithm 1. To begin with, we initialize the initial population $\mathcal{P}^0$ by randomly generating m service composition solutions. In line 2, we update each individual in $\mathcal{P}^0$ with an encoded queue of service

---

Algorithm 1. Our EDA-based method for service composition.

Input   : composition task T, service repository $\mathcal{SR}$
Output: an optimal composition solution
1: Initialize $\mathcal{P}^0$ with m randomly generated solutions, each represented as a vector$_k^g$ (where $k = 1, \ldots, m$);
2: Update each solution in $\mathcal{P}^0$ with an encoded sol$_k^g$;
3: Generate $\mathcal{NHM}^0$ from the top $\frac{1}{2}$ of best solutions in $\mathcal{P}^0$;
4: Set generation counter $g \leftarrow 0$;
5: **while** $g <$ *maximum number of generations* **do**
6:      Populate $\mathcal{P}^{g+1}$ with m solutions vector$_k^{g+1}$ sampled from $\mathcal{NHM}^g$;
7:      Update each solution in $\mathcal{P}^{g+1}$ with an encoded sol$_k^{g+1}$;
8:      Generate $\mathcal{NHM}^{g+1}$ from the top $\frac{1}{2}$ of the best solutions in $\mathcal{P}^{g+1}$;
9:      Update $\mathcal{NHM}^{g+1}$ using Eq. (9);
10:     Set $g \leftarrow g + 1$;
11: Let sol^{opt} be the best solution in $\mathcal{P}^g$;

---

indexes sol_k^g. To produce sol_k^g, we firstly decode the randomly generated $vector_k^g$ into DAG-based solutions using a forward graph building technique in [16,19], during the decoding, the fitness values of each solution is calculated. Second, we encode each DAG-based solutions into sol_k^g using BFS. The details of encoding and decoding will be discussed in Sect. 3.2. In line 3, based on the fitness value, only top $\frac{1}{2}$ best solutions are used to generate NHM^g. See more details of lines 2 and 3 in Sect. 3.2. The iterative part (lines 5 to 10) will be repeated until the maximum number of generations is reached. During each iteration, NHBSA is applied to sample new solutions for the next population $\mathcal{P}^{g+1}$. We update $\mathcal{P}^{g+1}$ similarly to line 2. This next population $\mathcal{P}^{g+1}$ is then used for generating $\mathcal{NHM}^{g+1}$, and then a moving updating technique is proposed to update $\mathcal{NHM}^{g+1}$ based on $\mathcal{NHM}^g$.

In a nutshell, our proposed method introduces a vector-based representation sol$_k^g$ that requires an encoding process (see lines 2 and 7), and an updating process for NHM (see line 9). These two processes are new compared to the standard EDA strategy.

3.2   A Novel Vector-Based Representation

Herein we consider two constructors $\bullet$ and $\|$ in most automated service composition works [7,14–16,19,20], where service composition solutions are represented as Directed Acyclic Graph (DAG). We can calculate QoS easily on a DAG-based solution [19]. For example, response time T is the time of the most time-consuming path in the DAG. Given a queue of service indexes (i.e., a vector), we can decode a DAG using a forward graph building algorithm [19]. Let $\mathcal{G} = (V, E)$ be a DAG-based service composition solution from *Start* to *End*, where nodes correspond to the services and edges correspond to the robust causal links. Often, $\mathcal{G}$ does not contain all services in $\mathcal{SR}$.

Often, different queues could be decoded into identical DAG-based composition solution. These queues could leads conflicts in learning the knowledge of service positions for one composition solution. To reduce the chances of conflicts, we aim to efficiently produce a nearly unique and more reliable service queue for the identical DAG-based composition solution. Thus, we encode this DAG into vector-based solutions using BFS, since BFS is a simple algorithm that efficiently transfer a DAG into a vector. Let $[S_0, \ldots, S_t]$ be a queue of services discovered by BFS traverses on the whole $\mathcal{G}$, starting from $Start$, $[S_{t+1}, \ldots, S_{n-1}]$ be a queue of remaining services in $\mathcal{SR}$ not utilized by $\mathcal{G}$. We use $sol_k^g = [I_k^g(S_0), \ldots, I_k^g(S_t), \ldots, I_k^g(S_{n-1})]$ to represent the kth (out of m, m is population size) service composition solution, and $P(g) = [sol_0^g, \ldots, sol_k^g, \ldots, sol_{m-1}^g]$ to represent a population of solutions of generation g. $I_k^g(S_x)$, $x \in \{1, \ldots, n-1\}$, represents the index of service S_x in $\mathcal{SR}$. To summarize a process of producing encoded vector-based solutions, we outline this process in Fig. 1.

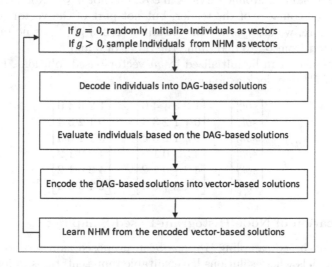

Fig. 1. A process of producing encoded vector-based solutions

Example 1. Let us consider a composition task $T = (\{a, b\}, \{i\})$ and a service repository $\mathcal{SR}$ consisting of five atomic services. $S_0 = (\{b\}, \{i\}, QoS_{S_0})$, $S_1 = (\{a\}, \{f, g\}, QoS_{S_1})$, $S_2 = (\{a, b\}, \{h\}, QoS_{S_2})$, $S_3 = (\{f, h\}, \{i\}, QoS_{S_3})$ and $S_4 = (\{a\}, \{f, g, h\}, QoS_{S_4})$. The two special services $Start = (\emptyset, \{a, b, e\}, \emptyset)$ and $End = (\{i\}, \emptyset, \emptyset)$ are defined by a given composition task T. Figure 2 illustrates the encoding process to produce an encoded solution.

As an example, take an arbitrary service index queue vector$_0^0 = [4, 1, 2, 3, 0]$. This service index queue is decoded into a DAG $\mathcal{G}_0^0$ representing a service composition that satisfies the composition task T. Afterwards $\mathcal{G}_0^0$ is encoded as a vector-based $sol_0^0 = [1, 2, 3 \mid 4, 0]$. Herein, the each position on the left side of

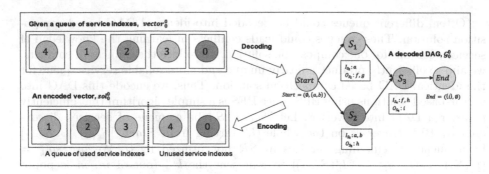

Fig. 2. An example of an encoded solution for a composition task T

| corresponds to a service discovered by BFS on $\mathcal{G}_0^0$, while the right side corresponds to the remaining atomic services in $\mathcal{SR}$, but not in $\mathcal{G}_0^0$. Note, that | is just displayed for the courtesy of the reader, but not part of the vector representation. Furthermore, we also permit the encoding $[1, 2, 3 \mid 0, 4]$, as no information can be extracted from $\mathcal{G}_0^0$ to determine the order of 0 and 4.

A population $\mathcal{P}^0$ can be initialized by m vector-based solutions. For $m = 6$, an example of $\mathcal{P}^0$ may look as follows:

$$\mathcal{P}^0 = \begin{bmatrix} sol_0^0 \\ sol_1^0 \\ sol_2^0 \\ sol_3^0 \\ sol_4^0 \\ sol_5^0 \end{bmatrix} = \begin{bmatrix} 1\,2\,3 \mid 4\,0 \\ 0 \mid 1\,2\,3\,4 \\ 0 \mid 1\,2\,3\,4 \\ 4\,3 \mid 0\,1\,2 \\ 4\,3 \mid 0\,1\,2 \\ 2\,1\,3 \mid 0\,4 \end{bmatrix} = \begin{bmatrix} 1\,2\,3\,4\,0 \\ 0\,1\,2\,3\,4 \\ 0\,1\,2\,3\,4 \\ 4\,3\,0\,1\,2 \\ 4\,3\,0\,1\,2 \\ 2\,1\,3\,0\,4 \end{bmatrix}$$

3.3   Application of Node Histogram-Based Sampling

Node histogram-based sampling [18] has been proposed as a tool for sampling probability models where solutions have suitable representations in form of permutations. Using our new vector-based representation of candidate solutions for composition tasks, we are now able to apply this tool for our problem.

The *node histogram matrix* (NHM) at generation g, denoted by $\mathcal{NHM}^g$, is an $n \times n$-matrix with entries $e_{i,j}$ as follows:

$$e_{i,j}^g = \sum_{k=0}^{m-1} \delta_{i,j}(sol_k^g) + \varepsilon \tag{7}$$

$$\delta_{i,j}(sol_k^g) = \begin{cases} 1 & \text{if } I_k^g(S_i) = j \\ 0 & \text{otherwise} \end{cases} \tag{8}$$

where $i, j = 0, 1, \ldots, n-1$, and $\varepsilon = \frac{2m}{n-1} b_{ratio}$ is a predetermined bias. Roughly speaking, entry $e_{i,j}^g$ counts the number of times that service S_i appears in position j of the service queue over all solutions in population $\mathcal{P}^g$.

Once we have computed $\mathcal{NHM}^g$, we use node histogram-based sampling [18] to sample new candidate solutions vector$_k^{g+1}$ (with $k = 1, \ldots, n$) for generation $g + 1$. Thus, in particular, the service index of each position is sampled with a random position sequence.

3.4 Strategies for Adaptive Updating of NHM

To trace the promising searching area, we attempt to select a proper learning discount rate α in Eq. (9) for updating NHM.

$$\mathcal{NHM}^{g+1} \leftarrow \left((1 - \alpha) \times e_{i,j}^g + \alpha \times e_{i,j}^{g+1}\right)_{i,j=1,\ldots,n} \qquad (9)$$

This formula defines a mechanism to compute the new $\mathcal{NHM}^{g+1}$ by updating the previous $\mathcal{NHM}^g$. Traditionally, in EDA a fixed discount rate $\alpha = 1$ is predetermined, potentially leading to premature convergence. To address this challenge, we want to propose an adaptive discount rate for NHM that changes dynamically over the different generations. In fact, NHM is increasingly concentrated at desired solutions with each and every new generation. Therefore, the impact of incorporating previous experiences, $\mathcal{NHM}^g$ would increase. Thus, a decreasing discount rate α should be assigned for every new $\mathcal{NHM}^{g+1}$. Based on this idea, we adjust the discount rate dynamically during evolution. Ideally, we choose α such that a good balance of exploration and exploitation is achieved for evolving high-quality solutions.

Our first strategy for adjusting α is is based in the information level of NHM, see Eq. (10). It chooses α based on its linear relationship to the changes in information level of NHM within a certain interval $[A, B] \subseteq [0, 1]$.

$$\alpha = \frac{\overline{\mathcal{H}(\mathcal{P}^{g+1})} - min\{\mathcal{H}(\mathcal{P}^g), \mathcal{H}(\mathcal{P}^{g+1})\}}{max\{\mathcal{H}(\mathcal{P}^g), \mathcal{H}(\mathcal{P}^{g+1})\} - min\{\mathcal{H}(\mathcal{P}^g), \mathcal{H}(\mathcal{P}^{g+1})\}} \times (B - A) + A \quad (10)$$

We measure the information level of NHM through its average entropy $\overline{\mathcal{H}(\mathcal{P}^g)}$, see Eq. (11). Therefore we call this strategy for adjusting α *entropy-based*. In general, a low knowledge level is initially represented by NMH, a high value is returned by entropy of NHM. We expect that knowledge converges at a high knowledge level with a decreasing entropy value.

$$\overline{\mathcal{H}(P(g))} = \frac{1}{n} \sum_{i=}^{n-1} \sum_{j=0}^{n-1} -\frac{e_{i,j}^g}{\sum_{j=0}^{n-1} e_{i,j}^g} \log_2 \frac{e_{i,j}^g}{\sum_{j=0}^{n-1} e_{i,j}^g} \qquad (11)$$

Herein, $min\{\mathcal{H}(\mathcal{P}^g), \mathcal{H}(\mathcal{P}^{g+1})\}$ and $max\{\mathcal{H}(\mathcal{P}^g), \mathcal{H}(\mathcal{P}^{g+1})\}$ are theoretically minimum and maximum entropy values of NHM that are calculated based on the historically found values during the run.

Besides Eq. (10), we propose another, much simpler strategy (called *linear-decrement* strategy) for adjusting α, see Eq. (12), for the purpose of comparison.

$$\alpha = \frac{g_{max} - g}{g_{max}} \times (B - A) + A \qquad (12)$$

Herein, g_{max} denotes the maximum number of generations, and g is the counter for the current generation, see Algorithm 1.

In summary, we propose two methods to adaptively tune the discount rate α for NHM. We call them *Entropy-based EDA* (E-EDA, for short) and *Linear decrement EDA* (L-EDA, for short), respectively. These two enhanced methods are expected to be less prone to premature convergence than our basic EDA-based method.

4   Experimental Evaluation

We conduct experiments to evaluate the performance of our approach. We compare EDA (i.e., EDA without utilizing any updating method) and its variations L-EDA and E-EDA against a PSO-based method that was recently proposed to solve the same service composition problem [19]. We focus on two benchmarks, WSC-08 and WSC-09 extended with QoS attributes, that have been widely employed in service composition research, e.g. in [7,12,19,24].

To assure a fair comparison in terms of the number of evaluations, the population size is set to 200, number of generations equals to 100, and b_{ratio} is 0.0002. The interval [A, B] is set to [0.2, 0.9]. We run each experiment with 30 independent repetitions. Following existing work [19,20], the weights in the fitness function Eq. (5) are set to balance the QoSM and QoS, i.e., w_1 and w_2 are set to 0.25, and w_3, w_4, w_5 and w_6 to 0.125. Following the recommendation in [4] the parameter p for the plugin match is set to 0.75. We have also conducted tests with other weights and parameters, and generally observed the same behavior.

4.1   Comparison of the Fitness

Both, WSC-08 and WSC-09, define a set of composition tasks. Table 2 shows the mean value of the solution fitness and the standard deviation over 30 repetitions.

Table 2. Mean fitness values for our approach in comparison to the baseline PSO-based approach [19] (Note: the higher the fitness the better)

Task	EDA	L-EDA	E-EDA	PSO [19]
WSC-08-1	0.50621 ± 0.0096	0.50692 ± 0.0112	0.50639 ± 0.0100	0.52216 ± 0.0044
WSC-08-2	0.61433 ± 0.0000	0.61433 ± 0.0000	0.61433 ± 0.0000	0.61355 ± 0.0030
WSC-08-3	0.45509 ± 0.0001	0.45513 ± 0.0001	0.45513 ± 0.0001	0.45415 ± 0.0005
WSC-08-4	0.46447 ± 0.0001	0.46447 ± 0.0001	0.46450 ± 0.0001	0.46451 ± 0.0001
WSC-08-5	0.46908 ± 0.0003	0.46910 ± 0.0002	0.46908 ± 0.0003	0.46863 ± 0.0011
WSC-08-6	0.47422 ± 0.0001	0.47424 ± 0.0001	0.47424 ± 0.0001	0.47326 ± 0.0006
WSC-08-7	0.48075 ± 0.0001	0.48077 ± 0.0000	0.48077 ± 0.0000	0.47900 ± 0.0005
WSC-08-8	0.46182 ± 0.0000	0.46182 ± 0.0000	0.46182 ± 0.0000	0.46156 ± 0.0003
WSC-09-1	0.56690 ± 0.0085	0.56835 ± 0.0076	0.56857 ± 0.0086	0.56944 ± 0.0089
WSC-09-2	0.47114 ± 0.0000	0.47115 ± 0.0000	0.47116 ± 0.0000	0.47110 ± 0.0003
WSC-09-3	0.55116 ± 0.0000	0.55116 ± 0.0000	0.55116 ± 0.0000	0.55109 ± 0.0003
WSC-09-4	0.47255 ± 0.0003	0.47242 ± 0.0004	0.47246 ± 0.0003	0.47129 ± 0.0008
WSC-09-5	0.47041 ± 0.0000	0.47041 ± 0.0000	0.47041 ± 0.0000	0.47008 ± 0.0003

We use an independent-sample T-test with a significance level of 5% to verify the observed differences in fitness. In particular, we use pairwise comparison to compare all methods, and then identify the top performance and its related value which is highlighted in the table. This top performance also includes those methods that consistently find the known best solutions over 30 runs, with a standard deviation of 0. The pairwise comparison results for fitness are summarized in Table 3. Herein, *win/draw/loss* shows the scores of one method compared to all the others, and displays the frequency that this method outperforms, equals or is outperformed by the competing method.

Table 3. Summary of the statistical significance tests for fitness, where each column shows the win/draw/loss score of one method against a competing one for all tasks of WSC-08 and WSC-09.

Dataset	Method	EDA	L-EDA	E-EDA	PSO [19]
WSC-08 (8 tasks)	EDA	-	3/5/0	3/5/0	1/0/7
	L-EDA	0/5/3	-	0/8/0	1/0/7
	E-EDA	0/5/3	0/8/0	-	1/0/7
	PSO [19]	**7**/0/1	**7**/0/1	**7**/0/1	-
WSC-09 (5 tasks)	EDA	-	0/5/0	1/4/0	0/2/3
	L-EDA	0/5/0	-	1/4/0	0/2/3
	E-EDA	0/4/1	0/4/1	-	0/1/4
	PSO [19]	**3**/2/0	**3**/2/0	**4**/1/0	-

Tables 2 and 3 show that the quality of solutions produced using our proposed approach compares favorable to those produced using the baseline PSO-based approach, with a single exception for task WSC 08-1. This corresponds well with our observation that our EDA-based approach is more competent at improving the quality of composite services by effectively utilizing the knowledge on the probability distribution learned through NHM.

For the different variations of our approach, the two enhanced methods, L-EDA and E-EDA outperform or are at least comparable to the basic EDA, as can be observed from the top performances in Table 2 and the scores in Table 3. Furthermore, L-EDA and E-EDA are comparable to each other in achieving competitive solutions for the data sets WSC-08 and WSC-09.

It should be emphasized that even a small improvement in terms of fitness can make a big difference in the practical use of the computed composite service. This point has been demonstrated for several example solutions analyzed in [19,20] in terms of the improvements in QoSM and QoS.

4.2   Comparison of the Execution Time

Table 4 shows the mean value of the execution time and the standard deviation over 30 repetitions. The pairwise comparison results for execution time are summarized in Table 5.

Table 4 shows that our proposed approach consistently requires less execution time than the baseline PSO-based approach. This corresponds well with our

Table 4. Mean execution time (in s) for our approach in comparison to the baseline PSO-based approach [19] (Note: the shorter the time the better)

Task	EDA	L-EDA	E-EDA	PSO [19]
WSC-08-1	56.32 ± 3.59	53.81 ± 3.88	56.62 ± 2.18	62.76 ± 33.27
WSC-08-2	36.36 ± 1.16	35.23 ± 2.06	35.91 ± 2.15	41.72 ± 22.91
WSC-08-3	7815.85 ± 2040.97	8449.16 ± 2355.53	7106.93 ± 1156.56	12152.72 ± 1971.99
WSC-08-4	36.25 ± 0.91	35.73 ± 1.53	35.96 ± 1.07	118.53 ± 29.69
WSC-08-5	410.84 ± 66.05	431.95 ± 52.27	421.86 ± 53.11	1174.28 ± 380.70
WSC-08-6	6419.64 ± 257.47	6185.60 ± 369.85	6338.24 ± 355.60	11321.95 ± 2269.06
WSC-08-7	954.36 ± 167.92	1022.54 ± 175.85	1026.04 ± 158.96	2133.10 ± 753.53
WSC-08-8	1729.35 ± 157.70	1657.01 ± 156.29	1632.35 ± 193.30	4864.01 ± 1141.94
WSC-09-1	56.69 ± 3.32	57.41 ± 2.68	57.63 ± 3.59	91.61 ± 47.36
WSC-09-2	1180.21 ± 144.15	1116.08 ± 135.96	1105.67 ± 92.66	2201.56 ± 522.42
WSC-09-3	805.82 ± 28.33	790.33 ± 23.15	803.40 ± 26.19	1298.32 ± 445.03
WSC-09-4	26741.33 ± 1464.03	26386.08 ± 2818.80	25733.08 ± 1333.06	36804.51 ± 7670.98
WSC-09-5	5861.03 ± 366.95	6006.96 ± 472.10	5892.08 ± 419.45	9556.08 ± 2194.68

Table 5. Summary of statistical significance tests for execution time, where each column shows the win/draw/loss score of one method against a competing one for all tasks of WSC-08 and WSC-09.

Dataset	Method	EDA	L-EDA	E-EDA	PSO [19]
WSC-08 (8 tasks)	EDA	-	3/5/0	2/6/0	0/0/8
	L-EDA	0/5/3	-	2/3/3	0/0/8
	E-EDA	0/6/2	3/3/2	-	0/0/8
	PSO [19]	8/0/0	8/0/0	8/0/0	-
WSC-09 (5 tasks)	EDA	-	0/5/0	2/3/0	0/0/5
	L-EDA	0/5/0	-	2/3/0	0/0/5
	E-EDA	0/3/2	0/3/2	-	0/0/5
	PSO [19]	5/0/0	5/0/0	5/0/0	-

observation that solutions evolved by our EDA-based approach are more likely to have all useful services required to build a suitable DAG placed at the very front of the service queue.

For the different variations of our approach, the two enhanced methods, L-EDA and E-EDA, require less execution time for execution for most tasks than the basic EDA. This corresponds well with our observation that the decoding process for them is usually faster. This confirms that L-EDA and E-EDA are more efficient in learning the probability distributions of high-quality solutions through NHM.

4.3   Comparison of the Convergence Rate

To explore the effectiveness of our proposed approach, we have also investigated the convergence rate over 30 independent runs. We have used WSC08-3 as an example to illustrate the performance of all the compared methods. Figure 3 shows the evolution of the mean fitness value of the best solution found along the execution time over for EDA, L-EDA and E-EDA compared against the baseline PSO-based method. We observe a significant increase in the fitness value towards the optimum until execution time 2.5e+3. In the remaining execution time, all methods tend to reach a plateau with stable improvements. In particular, all our

EDA-based methods happen to converge fast given the same execution time, and require significantly less time for execution than the baseline PSO-based method at significance level of 5%.

For the different variations of our approach, we look at a zoomed-in view of the mean fitness to observe differences between them, see Fig. 3: the enhanced methods, L-EDA and E-EDA, eventually achieve a slightly higher fitness value compared to the basic EDA. This observation matches well with our expectation, as L-EDA and E-EDA are tailored such that more exploration is performed in the beginning and more exploitation in later phases of the evolution.

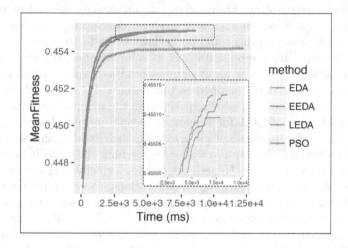

Fig. 3. Mean fitness values of best solutions over execution time

5   Related Work

To automatically construct composite services with individually or jointly optimized QoS and QoSM, AI planning and EC techniques have been widely adopted in web service composition. AI planning techniques often employ agents to plan composition works in dynamic scenarios, where combinatorial optimization is not a focus [17]. EC techniques have been investigated for achieving a global optimal in QoS and/or QoSM for web service composition. These works design effective solution representations, which always fall into two different types: *direct representations* and *indirect representations*.

GP-based approaches use tree-based representations to directly represent service composition solutions [7,12,15,20,24]. In [12], randomly initialized tree-based individuals are generated using GP by a context-free grammar, where individuals are penalized, and invalid individuals are eliminated. [24] proposes an adaptive GP-based approach for dynamically justifying crossover and mutation rates throughout the evolutionary process, where the correctness of randomly initialized tree-based individuals are ensured in the same way as those in

[12]. Combining GP with a greedy algorithm [7] is proposed to initialize valid tree-based individuals, which are transferred from a set of DAG-based solutions using an unfolding technique. A different transformation technique is investigated by [15] to present composition constructs as the functional nodes of trees. Those GP-based approaches above often suffer scalability problems in tree-based individuals because of duplicate subtrees. To overcome these shortcomings, [20] proposes a tree-like representation to eliminate the replicas of subtrees, and enables the evaluation of QoSM. Apart from these tree-based representations, GraphEvol [14] employs DAGs to directly represent and evolve service composition solutions.

The direct-representations above often rely on domain-dependent operators to explore and exploit search spaces. Therefore, developing effective operators for these direct representations could potentially bring forth some difficulties to researchers. With indirect representations in semi-automated service composition, a composition solution is always represented as a queue of services, each service in a queue is strictly mapped to one service slot of one given abstract service workflow according the service position in the queue. Two existing works consider possible uses of EDA for supporting semi-automated service composition [9,10], one work [10] does not clear explain their model, and another work [9] utilizes Restricted Boltzmann Machine for learning the explicit knowledge of promising solutions. [1] adopts Genetic Algorithm for achieving semi-automated service composition with constraints considerations. With indirect representations in fully automated service composition, a composition solution must be decoded from a sequence of services [16,19]. [16] utilizes PSO to handle large and complex search spaces, and searches for composition solutions with the best possible QoS. They propose a forward graph building algorithm to decode vector-based individuals into DAG-based composition solutions. [19] extends [16] to tackle a more complex service composition problem, where QoS and QoSM are optimized simultaneously.

6   Conclusion

In this paper, we proposed an effective and efficient EDA-based approach for the service composition problem using explicit knowledge of promising solutions. The novel vector-based representation in this work supports a reliable and accurate learning of NHM in the domain of automated service composition. In addition, two adaptive updating methods are proposed to properly balance exploitation and exploration for the searching process. Our experimental evaluation shows that EDA-based approaches are more effective and efficient compared to the PSO-based approach [19]. This demonstrates that learning the knowledge of promising composition solutions does help find near-optimal solutions. In addition, two updating methods proposed in E-EDA and L-EDA achieve reasonably good results compared to EDA.

In the future, we can investigate the influence of different intervals for our adaptive updating methods in the future, as the interval [A, B] for updating α

plays an important role for these two updating methods. Besides that, we can investigate other methods to decide α based on its non-linear relationship to the entropy of NHM for EDA. We are currently working on extending EDA-based approaches by hybridizing local search operators for improving the performances of EDA.

Acknowledgments. This work is partially supported by the New Zealand Marsden Fund with the contract numbers (VUW1510), administrated by the Royal Society of New Zealand.

References

1. Abbassi, I., Graiet, M., Gaaloul, W., Hadj-Alouane, N.B.: Genetic-based approach for ATS and SLA-aware web services composition. In: Wang, J., et al. (eds.) WISE 2015. LNCS, vol. 9418, pp. 369–383. Springer, Cham (2015). https://doi.org/10.1007/978-3-319-26190-4_25
2. Ceberio, J., Irurozki, E., Mendiburu, A., Lozano, J.A.: A review on estimation of distribution algorithms in permutation-based combinatorial optimization problems. Prog. Artif. Intell. **1**(1), 103–117 (2012)
3. Curbera, F., Nagy, W., Weerawarana, S.: Web services: why and how. In: Workshop on Object-Oriented Web Services-OOPSLA, vol. 2001 (2001)
4. Lécué, F.: Optimizing QoS-aware semantic web service composition. In: Bernstein, A., et al. (eds.) ISWC 2009. LNCS, vol. 5823, pp. 375–391. Springer, Heidelberg (2009). https://doi.org/10.1007/978-3-642-04930-9_24
5. Lécué, F., Delteil, A., Léger, A.: Optimizing causal link based web service composition. In: ECAI, pp. 45–49 (2008)
6. Ma, H., Schewe, K.D., Thalheim, B., Wang, Q.: A formal model for the interoperability of service clouds. Serv. Oriented Comput. Appl. **6**(3), 189–205 (2012)
7. Ma, H., Wang, A., Zhang, M.: A hybrid approach using genetic programming and greedy search for QoS-aware web service composition. In: Hameurlain, A., Küng, J., Wagner, R., Decker, H., Lhotska, L., Link, S. (eds.) Transactions on Large-Scale Data- and Knowledge-Centered Systems XVIII. LNCS, vol. 8980, pp. 180–205. Springer, Heidelberg (2015). https://doi.org/10.1007/978-3-662-46485-4_7
8. Paolucci, M., Kawamura, T., Payne, T.R., Sycara, K.: Semantic matching of web services capabilities. In: Horrocks, I., Hendler, J. (eds.) ISWC 2002. LNCS, vol. 2342, pp. 333–347. Springer, Heidelberg (2002). https://doi.org/10.1007/3-540-48005-6_26
9. Peng, S., Wang, H., Yu, Q.: Estimation of distribution with restricted Boltzmann machine for adaptive service composition. In: IEEE ICWS, pp. 114–121 (2017)
10. Pichanaharee, K., Senivongse, T.: QoS-based service provision schemes and plan durability in service composition. In: Meier, R., Terzis, S. (eds.) DAIS 2008. LNCS, vol. 5053, pp. 58–71. Springer, Heidelberg (2008). https://doi.org/10.1007/978-3-540-68642-2_5
11. Rao, J., Su, X.: A survey of automated web service composition methods. In: Cardoso, J., Sheth, A. (eds.) SWSWPC 2004. LNCS, vol. 3387, pp. 43–54. Springer, Heidelberg (2005). https://doi.org/10.1007/978-3-540-30581-1_5
12. Rodriguez-Mier, P., Mucientes, M., Lama, M., Couto, M.I.: Composition of web services through genetic programming. Evol. Intell. **3**(3–4), 171–186 (2010)

13. Shet, K., Acharya, U.D., et al.: A new similarity measure for taxonomy based on edge counting. arXiv preprint arXiv:1211.4709 (2012)
14. Sawczuk da Silva, A., Ma, H., Zhang, M.: GraphEvol: a graph evolution technique for web service composition. In: Chen, Q., Hameurlain, A., Toumani, F., Wagner, R., Decker, H. (eds.) DEXA 2015. LNCS, vol. 9262, pp. 134–142. Springer, Cham (2015). https://doi.org/10.1007/978-3-319-22852-5_12
15. Sawczuk da Silva, A., Ma, H., Zhang, M.: Genetic programming for QoS-aware web service composition and selection. Soft Comput. **20**, 1–17 (2016)
16. Sawczuk da Silva, A., Mei, Y., Ma, H., Zhang, M.: Particle swarm optimisation with sequence-like indirect representation for web service composition. In: Chicano, F., Hu, B., García-Sánchez, P. (eds.) EvoCOP 2016. LNCS, vol. 9595, pp. 202–218. Springer, Cham (2016). https://doi.org/10.1007/978-3-319-30698-8_14
17. Tong, H., Cao, J., Zhang, S., Li, M.: A distributed algorithm for web service composition based on service agent model. IEEE Trans. Parallel Distrib. Syst. **22**(12), 2008–2021 (2011)
18. Tsutsui, S.: A comparative study of sampling methods in node histogram models with probabilistic model-building genetic algorithms. In: IEEE International Conference on Systems, Man and Cybernetics, SMC 2006, vol. 4, pp. 3132–3137. IEEE (2006)
19. Wang, C., Ma, H., Chen, A., Hartmann, S.: Comprehensive quality-aware automated semantic web service composition. In: Peng, W., Alahakoon, D., Li, X. (eds.) AI 2017. LNCS, vol. 10400, pp. 195–207. Springer, Cham (2017). https://doi.org/10.1007/978-3-319-63004-5_16
20. Wang, C., Ma, H., Chen, A., Hartmann, S.: GP-based approach to comprehensive quality-aware automated semantic web service composition. In: Shi, Y., et al. (eds.) SEAL 2017. LNCS, vol. 10593, pp. 170–183. Springer, Cham (2017). https://doi.org/10.1007/978-3-319-68759-9_15
21. Wang, C., Ma, H., Chen, G.: EDA-based approach to comprehensive quality-aware automated semantic web service composition. In: Proceedings of the Genetic and Evolutionary Computation Conference Companion, pp. 147–148. ACM (2018)
22. Wang, J., Tang, K., Lozano, J.A., Yao, X.: Estimation of the distribution algorithm with a stochastic local search for uncertain capacitated arc routing problems. IEEE Trans. Evol. Comput. **20**(1), 96–109 (2016)
23. Wang, S.Y., Wang, L.: An estimation of distribution algorithm-based memetic algorithm for the distributed assembly permutation flow-shop scheduling problem. IEEE Trans. Syst. **46**(1), 139–149 (2016)
24. Yu, Y., Ma, H., Zhang, M.: An adaptive genetic programming approach to QoS-aware web services composition. In: IEEE CEC, pp. 1740–1747 (2013)
25. Zeng, L., Benatallah, B., Dumas, M., Kalagnanam, J., Sheng, Q.Z.: Quality driven web services composition. In: Proceedings of the 12th International Conference on World Wide Web, pp. 411–421. ACM (2003)

A CP-Net Based Qualitative Composition Approach for an IaaS Provider

Sheik Mohammad Mostakim Fattah[✉], Athman Bouguettaya,
and Sajib Mistry

School of Information Technologies, University of Sydney, Sydney, Australia
sfat5243@uni.sydney.edu.au,
{athman.bouguettaya,sajib.mistry}@sydney.edu.au

Abstract. We propose a novel CP-Net based composition approach to
qualitatively select an optimal set of consumers for an IaaS provider.
The IaaS provider's and consumers' qualitative preferences are captured
using CP-Nets. We propose a CP-Net composability model using the
semantic congruence property of a qualitative composition. A greedy-
based and a heuristic-based consumer selection approaches are proposed
that effectively reduce the search space of candidate consumers in the
composition. Experimental results prove the feasibility of the proposed
composition approach.

Keywords: Cloud service composition
IaaS composition · Qualitative preference composition
CP-Net composability model · Monte Carlo simulation

1 Introduction

Infrastructure-as-a-Service (IaaS) model is a cloud service delivery model where
computational resources are usually delivered as Virtual Machines (VMs) to
cloud consumers [1,14]. The functional properties of an IaaS service are usually
CPU, storage, and memory [6]. The Non-functional properties (e.g., availability,
throughput, and price) are usually attached with VMs or IaaS services as Quality
of Services (QoS). IaaS services are generally configured based on the functional
and non-functional requirements of consumers. For example, Amazon EC2 IaaS
provider has different types of VMs (e.g., CPU-intensive, Memory-intensive, and
Network-intensive) that are targeted for different types of consumers (e.g., individuals, small enterprises, and large organizations).

The long-term IaaS composition is a topical research issue [12]. The composition from an IaaS provider's perspective is defined as the selection of a set of
optimal consumer requests [11]. An effective IaaS composition achieves the economic expectations, i.e., revenue and profit maximization of the provider. The
IaaS composition ensures the optimal utilization of available computing resources
for an IaaS provider. The selection of optimal consumer requests is essential to
achieve the IaaS composition. For example, selecting service requests from a

© Springer Nature Switzerland AG 2018
H. Hacid et al. (Eds.): WISE 2018, LNCS 11234, pp. 151–166, 2018.
https://doi.org/10.1007/978-3-030-02925-8_11

group of small enterprises may be more profitable than the single service request from a large organization due to the scale of economy.

We focus on the qualitative IaaS composition, i.e., selecting the optimal consumer requests according to the qualitative preferences of the provider. Qualitative preference models are effective tools for the selection where there exists uncertainties or incomplete information. The service requirements of future consumers are uncertain and probabilistic in nature [2,10,13]. A provider's preference may change with the requirements of the consumers. The dynamic business environment may also trigger a change in the provider's qualitative preferences. For example, the provider may observe a very high demand for Network-intensive services in the Christmas or holiday period. The provider may prefer to compose Network-intensive services than CPU-intensive services to increase its revenue.

IaaS consumers' requirements can be represented in a natural and intuitive manner using qualitative approaches [19]. Qualitative models provide the necessary tool to select appropriate providers where quantitative models are not applicable. A consumer requires to explicitly indicate the exact values of the functional and non-functional properties of a service in a quantitative model. It may not be possible to find providers that can meet the exact requirements of the consumers. For example, a consumer may require 10 units of CPU and 20 units of memory at 20 dollars/month at the level of 100% availability. Such requirements may not be exactly fulfilled by any service provider. In contrast, qualitative preferences can be expressed by comparison. For instance, a content provider (IaaS consumer) prefers a "high" network bandwidth to a "low" network bandwidth. The content provider may also specify conditional preferences. For example, if the price of network bandwidth is very low, a "high" network bandwidth is preferred over a "low" network bandwidth. These qualitative preferences are used to select suitable providers for consumers.

The IaaS composition problem is modeled in both quantitative and qualitative approaches [8,12,14,20,21]. The quantitative approaches do not consider the qualitative preferences of the provider. The composition of requests is transformed into an optimization problem in quantitative approaches. The proposed approaches (e.g., metaheuristic optimization and integer programming) are not applicable in the qualitative IaaS composition. A heuristic based sequential optimization approach is proposed in the qualitative IaaS composition [11]. This approach considers quantitative requirements of the consumers and matches them with the qualitative preferences of the provider. To the best of our knowledge, existing composition approaches are not applicable where both the provider and consumer have qualitative preferences.

We propose a Conditional Preference Network (CP-Net) based qualitative composition approach for an IaaS provider. We represent qualitative conditional preferences using CP-Nets. The CP-Net is a very effective tool to represent and reason with qualitative conditional preferences under ceteris paribus ("everything else being equal") semantics. A CP-Net creates a directed graph where each node is an attribute of a service preference. The edge between nodes defines the priority among service preferences. The rank of service preferences is generated by traversing the graph. We assume that the CP-Nets of the provider and the consumers are provided for simplicity. *Our target is to select the optimal composition of consumers' CP-Nets that has the highest similarity measure with the CP-Net of the provider.*

We propose a CP-Net based qualitative composition approach for an IaaS provider. First, we propose a novel CP-Net composability model to compose CP-Nets of multiple consumers using the semantic congruence property of a qualitative composition. Next, we propose a similarity checking mechanism between CP-Nets using the *coefficient of correlations*. It directs us to apply the brute-force approach where all possible composition of consumers' CP-Nets is considered to select the optimal composition. The brute-force approach is not a practical solution for composing a large set of consumer's CP-Nets due to its exponential runtime. we propose a heuristic-based algorithm and a greedy algorithm that reduces search space for compositions. The key contribution of our research is summarized below:

- A CP-Net composability model for the qualitative composition of IaaS consumers using the semantic congruence property.
- A qualitative similarity measure approach using the correlation coefficient.
- A heuristic-based and a greedy-based consumer selection approaches to reduce the search space using semantic similarities between the provider's and consumer's qualitative preferences.

2 Motivation Scenario

Let us assume an IaaS provider offers VM services based on a fixed set of computational resources for a specific period of time. Its resource capacity is up to 100 virtual CPU units and 100 memory units. For simplicity, we consider "price" as the only QoS in a VM and we omit "network bandwidth (NB)" functionality from a VM. We also assume that both the consumers' qualitative requirements and the providers' qualitative preferences on CPU, memory, and price are following the same semantic levels in Fig. 1(a). Three levels of semantics, i.e., high, moderate, and low for each attribute are specified in the semantic table in Fig. 1(a). The IaaS provider builds different types of strategies of service provisions. As there exist uncertainties on future consumers' requirements, it builds the strategies using qualitative models. We assume that the provider receives qualitative requests from three consumers. The target is to select an optimal set of consumer requests that matches with its preferred ways to service provisions.

We consider a "CPU-intensive" and a "memory-intensive" service provisions strategies for the provider. As the CP-Net provides an effective way to represent conditional qualitative preferences, we represent the "CPU-intensive" strategy as CP1 and the "memory-intensive" strategy as CP2 in Fig. 1(b). The "CPU-intensive" strategy is to offer CPU intensive services at relatively moderate prices to attract consumers with CPU intensive requests. Therefore, the CPU is the most important attribute in the dependency graph of CP1 followed by memory and price. The arc from "CPU" to "memory" implies that the provision of "memory" levels in a VM depends on the selection of "CPU" levels. The preference of CPU provisions is expressed as $c3 \succ c2$ in the CPT of CPU. It implies that the provider does not want to offer the "low" CPU services. The choice of CPU levels decides the choices of memory levels in CP1. The provider prefers the "moderate" memory to the "high" memory if the "high" CPU is chosen in the service provision according to CP1. If the choice of CPU is "moderate", the provider prefers the "high" memory to the "moderate" memory. Based on the selection of memory levels, the provider selects the price levels if the "high" memory is chosen according to the CPT of price in CP1. The provider prefers the "high" price to the "moderate" price. However, the provider prefers the "moderate" price to the "high" price for the "moderate" memory units. Similarly, CP2 represents the "memory-intensive" strategy where memory is the most important attribute in the dependency graph of CP2 followed by CPU and price.

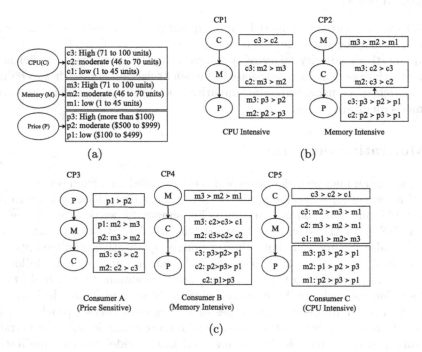

Fig. 1. (a) Semantic table for service attributes (b) CP-nets of an IaaS provider (c) CP-nets of consumers

The requirements of consumer A, B, and C are captured by CP3, CP4, and CP5 respectively in Fig. 1(c). Consumer A prefers "price-sensitive" services. The price is the most important attribute in the dependency graph of CP3 followed by memory and CPU. The Consumer A mentions its inability to consume "high" priced (P3) services due to budget constraints and prefers the "low" price to "moderate" priced services i.e., $p1 \succ p2$ according to the CPT of price in CP3.

Its memory requirements depend on the choice of price. In the CPT of memory in CP3, the "moderate" memory is preferred over the "high" memory in the required service if the "low" price is already chosen. If the "moderate" price is chosen, it sets a higher priority to "high" memory than the "moderate" memory. Preferences of CPU requirements are defined based on the choice of memory in a similar way. The consumer B has "memory-intensive" qualitative requirements. Therefore, memory is the root attribute in CP4. The choice CPU depends on the choice of memory and the choice of price depends on the choice of CPU in CP4. The consumer C defines "CPU-intensive" preferences in a similar way.

The IaaS provider can select the optimal composition of consumer's requests, i.e., CP-Nets from three consumers in $2^3 - 1 = 7$ possible ways. Multiple consumer's CP-Nets may not be composable for their semantic differences. For example, the "Price-sensitive" requirement (CP3) of consumer A prefers to receive the "low" memory level services if the price is at a "low" level. The "memory-sensitive" requirement (CP4) of consumer B prefers to receive the "high" memory level services even at a "high" level. Hence, CP3 and CP4 are not composable as they create semantic ambiguity in the composition. We propose a composability model to determine whether CP-Nets of two consumers are composable or not.

Finding all possible compositions of preferences in a brute force approach can be very inefficient. The number of all possible combination of N consumers is $(2^N - 1)$. We propose a heuristic algorithm to reduce the search space of consumers by finding the relative similarity between the consumers' CP-Net and the provider's CP-Net. For example, if the provider's strategy is "CPU-intensive", i.e., CP1; selecting the "price-sensitive" consumer A in a composition may not yield the optimal result as they are semantically very dissimilar. The search space reduces to 2 consumers (consumer B and C) and the optimal composition is selected from $2^2 - 1 = 3$ compositions out of the 7 possible ways.

3   CP-Nets for Qualitative IaaS Requests and Provisions

A CP-Net is a graphical model to formally represent and reason about qualitative preference relations. A CP-Net consists of a directed dependency graph and conditional preference tables (CPTs). The dependency graph is defined over a set of functional and non-functional attributes $V = \{X_1, ..., X_n\}$. A child node in a dependency graph depends on a set of direct parent nodes $Pa(X_i)$. The child node is connected by an arc from $Pa(X_i)$ to X_i in the dependency graph. Parent attributes affect the user's preferences over the value of X_i. Each node X_i in the dependency graph has $Pa(X_i)$ except for the root nodes.

The CPT of each variable X_i is defined over the finite, discrete domain $D(X_i)$ and semantic domains $S(X_i)$. Each value x_n in $D(X_i)$ is mapped into a semantic value in $S(X_i)$ using a semantic mapping table, $SemTable(X_n, x_n)$. Figure 1(a) is a semantic table that maps 71–100 units of CPU as a "high" CPU value. We only focus on the attributes that are compatible with additive operations. Hence, we define $s_i + s_j = s_k$ for $S(X_i)$. For example, $c1 + c1 = c2$ implies two "low" CPU can be added and generates a "moderate" CPU unit.

The preference between two values of an attribute X_i is specified by $\succ$ for a given value of the paraent attribute $Pa(X_i)$. A user explicitly defines its preferences over the semantic values of X_i for each complete outcome on $Pa(X_i)$. The preferences take the form of total or partial order over $S(X_i)$. For example, the attributes of CP1 are C, M, and P with semantic domains containing x_i if X is the name of the feature. The preferences statements are as follows: $c3 \succ c2$, $c3 : m2 \succ m3$, $c2 : m3 \wedge m2$, $m3 : p3 \succ p2$, $m2 : p3 \succ p2$. The statement $x_1 \succ x_2$ represents the unconditional preference for $X = x_1$ over $X = x_2$.

A preference outcome is a combination of values of all attributes of a CP-Net. For example, $\{c3, m2, p3\}$ and $\{c2, m2, p1\}$ are two preference outcomes for CP1 denoted by o_1 and o_2. According to the value of attribute P, it can be shown that $o_1 \succ o_2$ or o_1 dominates o_2. The dominance relationship of two preference outcomes is defined as a *pre-order* between them. Figure 2 depicts the induced graph [5] of a CP-Net with all preference outcomes.

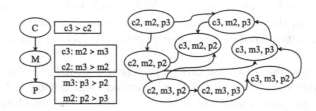

Fig. 2. Induced graph

4　Composability Model for Qualitative Preferences

Let us consider two CP-Nets CP_A and CP_B. We define the composability of two CP-Nets as $composable(CP_A, CP_B) = \{true, false\}$ to find CP_C where $CP_C = compose(CP_A, CP_B)$. Two CP-Nets are composable if their composition is semantically congruent. *A composition is called semantically congruent if the relative importance order of preference attributes for each consumer is preserved without any ambiguity.* The relative importance order is represented as (a, b) which means a is preferred over b. Let us consider a consumer who has the relative importance order of attributes as $(CPU, memory)$. Another consumer has the relative importance order of attributes as $(memory, price)$. If the importance order of their composition is $(CPU, memory, price)$, the composition is considered as semantically congruent.

Definition 1. *Semantic Congruence of a Qualitative Composition. A composition is called semantically congruent when the importance order of preference attributes for each consumer is preserved without any ambiguity.*

The semantic congruence of a composition can be efficiently represented using the directed acyclic graph of CP-Nets. We use semantic congruence property to define the composability of two CP-Nets. A CP-Net contains a directed dependency graph (DDG) and conditional preference table for each node in the graph. We define the composability of the DDG and the CPT to define the composability of CP-Nets. Two DDG are considered to be composable if their combined DDG does not contain any cycle. Figure 3(a) shows two DDGs from two different CP-Nets (CP1 and CP2). The CPU is the root of CP1. The memory depends on the CPU and the price depends on the memory in CP1. The DDG of CP2 has the order of memory, CPU, and price. To merge them in a single DDG (i.e., CP12) (Fig. 3(a)), we create a new DDG where all the attributes (i.e., CPU, memory, and price) are added from both CP-Nets. The next step is to create edges between the attributes. First, we take a pair of attributes (e.g., CPU and price) from the new DDG. If the same pair of attributes has an edge in either DDG (i.e., CP1 or CP2), we add an edge to the new DDG. We run this process for each pair of nodes until we cover all edges from both DDGs. If the resulted DDG contains any cycle, then DDGs are not composable. Two CP-Nets are composable if their dependency graphs and CPTs are composable.

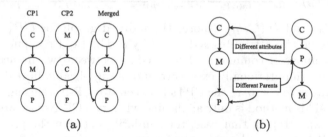

Fig. 3. (a) Dependency graph composability (b) CPT composability

Definition 2. *CP-Net Dependency Graph Composability. Two dependency graphs are composable if their combined dependency graph does not contain any cycle.*

Definition 3. *CPT Composability. Two CPTs from two different CP-Nets are composable if their preference attributes are same and their values depend on the same set of parent nodes.*

Two CPTs are composable if they have two properties. First, both CPTs' attribute nodes should be the same. In Fig. 3(b), "CPU" nodes from both DDGs can be valid candidates to be composable. A "CPU" node from one CP-Net and a "memory" node from another CP-Net are not composable. Second, the parent nodes of both nodes should have the same attribute. If two nodes have the same set of parent nodes, then preference statements of both nodes will depend on the same set of attributes. Therefore, the preference statements will be composable. For example, "CPU" nodes of both CP-Nets are unconditional nodes (i.e., no parent) in Fig. 3(b). However, "price" nodes have different parents (i.e., "memory" and "CPU"). The preference statements will have conditional attributes. Therefore, the CPTs of the "price" nodes are not composable.

5 Similarity Measurement with Qualitative Preferences

We assume that the IaaS provider expresses its qualitative preferences using the CP-Net, CP_A. The provider requires to find an optimal set of consumer CP-Nets from $\{CP3, CP4, CP5\}$ that matches with CP_A. We apply the composability model described in the above section and find the composed CP-Net CP_B for a set of composable consumers. If CP_A completely matches with CP_B, we say it is the optimal composition. To compare the composed CP-Net with provider's CP-Net we need to find a similarity measurement algorithm. The similarity measurement between two CP-Nets can be performed in two ways [17,18]. One way is to generate the induced graph of two CP-Nets and compute the number of common edges between two CP-Nets. This similarity can be computed by:

$$Sim(CP_A : CP_B) = \frac{|\{e : e \in In(CP_A) \wedge e \in In(CP_B)\}|}{|\{e : e \in In(CP_A) \vee e \in In(CP_B)\}| - |\{e : e \in In(CP_A) \wedge e \in In(CP_B)\}|} \quad (1)$$

where $In(CP_A)$ and $In(CP_B)$ denotes the induced graph for CP_A and CP_B. The edge between two attributes is denoted by e. The Eq. 1 computes the ratio between the number of common and total edges between two induced graphs. This method is computationally expensive and not applicable in real time [19].

Another way is to compare the CPTs between two CP-Nets using the dependency graphs. This method is only applicable when two CP-Nets share the same dependency graph [18]. In that case, the similarity between the provider's and consumers' CP-Net is calculated by the following equation:

$$Sim(CP_A : CP_B) = \frac{\sum_{X_i} \left(|CPT_A(X_i) \cap CPT_B(X_i)| \times \prod_{X_j \notin Pa(X_i)} |SemTable(X_i)| \right)}{\sum_{X_i} \left(|CPT_A(X_i) \cup CPT_B(X_i)| \times \prod_{X_j \notin Pa(X_i)} |SemTable(X_j)| \right)} \quad (2)$$

where CPT_A and CPT_B are the conditional preference table of CP_A and CP_B. $Pa(X_i)$ denotes the parent attributes of X_i and $SemTable(X_i)$ represents all values that can be assigned into X_i. We assume the composed CP-Net of consumers and the provider's CP-Net have the same dependency graph.

6 Heuristic-Based Consumer Selection Approaches

The number of composable consumers grows exponentially with the increase in the number of consumers (2^n). Finding all possible combinations of consumers is inapplicable as it may require a very large time depending on the number of consumers [12]. Our target is to reduce the search space for the consumer selection. We propose a greedy-based and a heuristic consumer selection algorithm using the similarity between a provider's and a consumer's CP-Nets.

6.1   Greedy-Based IaaS Consumer Selection Approach

We choose the first consumer who has the highest relative similarity with provider's CP-Net in the greedy selection approach. We iteratively choose the next consumers to achieve the maximum similarity with the provider's CP-Net. The following steps are performed in the greedy based approach:

1. Select a consumer CP-Net that is has the maximum coefficient of correlation with the provider's CP-Net.
2. Create a new CP-Net based on the difference between the provider's CP-Net and the selected consumer's CP-Net.
3. Find and select a consumer CP-Net who has the maximum correlation coefficient with the new CP-Net.
4. Create a new CP-Net based on the difference between the consumer CP-Net and the new CP-Net.
5. Perform steps 3 and 4 until the difference is zero or minimum.

6.2   IaaS Consumer Selection Based on Correlation Coefficients

The greedy approach may not always provide the accurate results as it considers only consumers with maximum correlation with the provider's CP-Net. We proposed a heuristic approach where we find those consumers who have relatively similar CP-Nets with the provider's CP-Net. Relatively similar CP-Nets are more likely to form a composition that will match the CP-Nets of the provider. Two CP-Nets are relatively similar if (1) they have the same dependency graph (2) Nodes with similar attributes have similar preferences statements in their CPTs. CP1 and CP5 have the same dependency graph in Fig. 1. The relative similarity between two preferences statements is measured based on their relative ordering. For example, consider two preferences statements $c1 \succ c3 \succ c2$ and $c8 \succ c10 \succ c9$. Although values of the attributes are different, patterns of both statements are same. We consider $c1 \succ c3 \succ c2$ and $c8 \succ c10 \succ c9$ are relatively similar. Let us consider two conditional preferences statements $c1 \land m2 : p1 \succ p2 \succ p3$ and $c8 \land m10 : p6 \succ p7 \succ p8$. The condition of the first statement $c1 \land m2$ can be fulfilled by the condition of the second statement $c8 \land m10$. A similar statement can be found in the CPT of the same attribute of the provider's CP-Net for each statement of the CPT of an attribute from the consumer's CP-Net. We perform the following steps to find the relative similarity between a consumer's and the provider's CP-Net:

1. Compare the dependency graph of the provider's and the consumer's CP-Net. If the dependency graphs are not the same, the CP-Nets are not similar.
2. If the dependency graphs are same, find an unconditional node from provider's CP-Net for each unconditional node of the consumer CP-Net.
3. Compute similarity between the unconditional nodes selected in step 2.
4. Store the similarity measurement in a global variable.
5. Find similar conditional nodes for each attribute from provider's and consumer's CP-Nets.

6. For each preference statement in a CPT of an attribute of the consumer's CP-Net, find a similar preference statement in the CPT of the same attribute of provider's CP-Net. The attributes and the conditions of both statements should be also relatively similar.
7. The similarity between the conditional nodes is computed. Update the total similarity measurement.

Algorithm 1. Similarity Checking between two CP-Nets

Input : CP_A, CP_B, $SemanticTable$
Output: Similarity $Sim(CP_A, CP_B)$

1 Integer $commonEdges \leftarrow 0$
2 Integer $allEdges \leftarrow 0$
3 $CPT_A \leftarrow$ find all CPT in CP_A
4 $CPT_B \leftarrow$ find all CPT in CP_B

5 **foreach** X_i attribute in CP_A **do**
6 $\quad$ $visitedPreferences \leftarrow \varnothing$
7 $\quad$ **foreach** P_A in $CPT_A[X_i]$ **do**
8 $\quad\quad$ boolean $flag \leftarrow false$
9 $\quad\quad$ **foreach** P_B in $CPT_B[X_i]$ **do**
10 $\quad\quad\quad$ **if** P_A has similar pattern P_B **then**
11 $\quad\quad\quad\quad$ $visitedPreferences \leftarrow P_B$
12 $\quad\quad\quad\quad$ $flag \leftarrow true$
13 $\quad\quad\quad\quad$ $commonEdges \leftarrow commonEdges + \prod_{X_j \notin P(X_i)}|SemTable(X_j)|$
14 $\quad\quad\quad\quad$ $allEdges \leftarrow allEdges + \prod_{X_j \notin P(X_i)}|SemTable(X_j)|$
15 $\quad\quad\quad$ **end**
16 $\quad\quad\quad$ **if** $!flag$ **then**
17 $\quad\quad\quad\quad$ $allEdges \leftarrow allEdges + \prod_{X_j \notin P(X_i)}|SemTable(X_j)|$
18 $\quad\quad\quad$ **end**
19 $\quad\quad$ **end**
20 $\quad$ **end**
21 $\quad$ **foreach** P_B in $CPT_B[X_i]$ **do**
22 $\quad\quad$ **if** $P_B \notin visitedPreferences$ **then**
23 $\quad\quad\quad$ $allEdges \leftarrow allEdges + \prod_{X_j \notin P(X_i)}|SemTable(X_j)|$
24 $\quad\quad$ **end**
25 $\quad$ **end**
26 **end**
27 **return** $Sim(CP_A, CP_B) \leftarrow commonEdges/allEdges$

We propose Algorithm 1 to find relatively similar consumers based on the provider's CP-Nets. Algorithm 1 calculates the coefficient of correlation between two CP-Nets using Eq. 2. The algorithm takes two CP-Nets with same dependency graphs. Two variables are defined to calculate the number of common edges and all edges between the CP-Nets (*commonEdges* and *allEdges*). According to our assumption, both CP-Nets have the same number of attributes.

For each attribute, we perform a check if the conditional preferences from both CP-Nets have a similar pattern. When a preference has a similar pattern in both CP-Nets, we update the number of common edges ($commonEdges$) and all edges ($allEdges$). The preference is added in $visitedPreferences$. However, if there is no preference from CP_A is found, the algorithm updates only the number of all edges ($allEdges$). $allEdges$ is updated with every iteration. The relative similarity is calculated by the ratio of the number of common edges ($commonEdges$) and the number of all edges ($allEdges$).

7   Experiments and Results

We have conducted a set of experiments to evaluate the efficiency and the feasibility of the proposed heuristic based composition approach. The heuristic and the greedy approaches are compared with the brute force approach in term of accuracy and time. We conducted the experiment on computers with Intel Core i7 (3.60 GHz and 8 GB RAM) using Java and Matlab.

7.1   Simulation Setup

It is difficult to find the real-world preferences of IaaS consumers. We have generated 20 CP-Nets to represent consumers' preferences. We have also generated a semantic table for consumers which is a subset of provider's semantic table. The provider has the entire view of its resource capacity. As the simulation has been performed based on randomly generated CP-Nets, the result varies depending on the type of the CP-Nets. We run the experiment based on Monte Carlo [3] simulation method for a conclusive result. We have run the simulation several times for each approach and taken the average accuracy and time for the different size of consumers. Table 1 shows the simulation variables and their corresponding values that we have used in the experiment to perform the performance analysis.

Table 1. Experiment variables

Variable names	Values
Simulation run	100
Number of consumers	2 to 23
Coefficient of correlation	0.15, 0.20, 0.25
Homogeneous domain size	20

7.2   Baseline: The Brute-Force Approach

We generate all combination of the consumers for the brute force approach. For each combination, we compose them using the composability model. The

set of composable CP-Nets are composed and compared with the provider's
CP-Net using Eq. 2. A composed CP-Net that has maximum similarity with
the provider's CP-Net is selected. As the brute force approach considers all
consumers, it achieves maximum similarity up to 90% with provider's CP-Net.

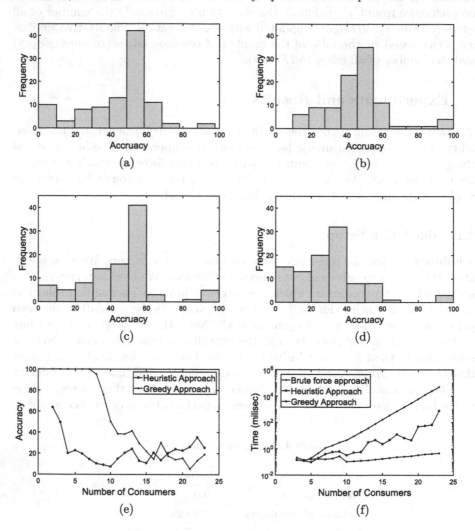

Fig. 4. Accuracy of the proposed heuristic approach with coeffecient (a) 0.15 (b) 0.20
(c) 0.25 (d) accuracy of the greedy approach (e) accuracy of the heuristic and greedy
approach (f) execution time in log scale

7.3   Accuracy Analysis

We have applied the proposed heuristic approach to select and compose relatively similar consumers according to the provider's CP-Net. The accuracy of

the heuristic approach is calculated with respect to the brute force approach. We also compose the consumers based on the greedy selection approach. The accuracy of the greedy approach is calculated in a similar manner.

The brute force approach normally provides more accurate result then the heuristic approach as the brute force approach considers all possible combination of consumers. For the heuristic approach, we select a consumer in the composition only if its CP-Net is more than 20% relatively similar to the provider's CP-Net. We have run the experiments several times to find the optimal threshold. For a specific provider and a service, this threshold should be set manually before composition. Figure 4(a), (b), and (c) show the histogram of the accuracy of the heuristic approach where the correlation coefficients are 0.15, 0.20, and 0.25. The brute force approach provides the optimal results. Compared to the brute force result, the heuristic approach generates almost 60% accurate result on average when the coefficient is 0.2. Figure 4(d) shows the accuracy result of the greedy based approach. The result shows the histogram of the outcomes. Here, outcomes have below 50% accuracy most of the time. Figure 4(e) shows the average accuracy of the proposed heuristic approach and the greedy approach. The greedy approach provides very good accuracy if the number of consumers is low. The accuracy of the greedy approach becomes low with the increase of consumers, as it starts to discard more consumers. The heuristic approach provides better accuracy with the increase in the number of consumers because it finds more similar consumers according to the provider's CP-Net.

7.4   Runtime Analysis

Figure 4(f) depicts the time comparison between the brute force, heuristic, and greedy approaches in log scale. The figure shows that with the increase in the number of consumers, time for composition in the brute force approach grows exponentially (i.e., linearly in log scale). The heuristic approach does not show exponential behavior. For a particular IaaS provider, the composition time increases very slowly compared to the brute force approach. The greedy approach shows a very interesting result. It composes consumers with a constant time for a particular service. The accuracy of the greedy approach is unreliable.

8   Related Work

Qualitative user-preferences are represented by graphical models, especially in multi-objective decision-making domain [7]. IaaS consumers can express their preferences more directly, intuitively using qualitative representations. A conditional preference network (CP-Net) provides a natural and an efficient way to represent consumers' preferences in qualitative manner [5]. CP-Nets are widely used to represent user's preferences to select and compose services [18]. A web service selection mechanism is proposed to incorporate incomplete or inconsistent user preferences from historical preferences [18]. A CP-Net based similarity measurement approach is proposed to find users with similar preferences [19].

Several variations of CP-Nets are also proposed to enhance the expressiveness of users. A UCP-Net is a graphical representation of utility functions that combines generalize additive models and CP-Nets [4]. The TCP-Net is another variation of CP-Net where relative importance between attributes can be captured through weighted edges [15]. The WCP-Net is another weighted CP-Net that is proposed to capture user's preference more precisely to select web services. A deterministic temporal CP-Net is used to express IaaS provider's long-term business strategies [11]. A probabilistic CP-Net is proposed to capture the IaaS provider's business strategies in a probabilistic manner [12].

Several existing studies propose different methods to compose multiple CP-Nets. A multi-agent CP-Net or mCP-Net is proposed as an extension of CP-Net [16]. Preferences from multiple users are aggregated into a single CP-Net. The proposed method aggregates preferences according to a voting analogy where preferences are selected based on the preference of majority user. A Majority-rule-based preference aggregation method is proposed based on a hypercube-wise composition to optimize the composition process. An aggregation method is proposed to capture multi-valued CP-Nets based on majoritarian aggregation rule [9]. The proposed method can aggregate CP-Nets even if they are cyclic. Most existing works on the aggregation of CP-Nets consider the composition problem as a multi-agent voting system. These approaches are not applicable in the case of resource allocation based on multi-agent preferences. We compose CP-Nets based on the composability model and the resource constraints in our proposed composition approach. The composed CP-Nets capture the preferences of multi-users instead of just considering the common preferences.

9   Conclusion

We propose a CP-Net based composition approach for an IaaS provider. The proposed approach allows the IaaS provider and consumers to express their qualitative conditional preferences using CP-Nets. We propose a composability model for IaaS consumers using the semantic congruence property of a qualitative composition. Finding the optimal composition may be difficult when the number of consumers is large. A greedy-based and a heuristic-based selection approaches are proposed to reduce the search space of candidate consumers. Both approaches utilize correlation coefficients between CP-Nets to find consumers who have similar preferences with the provider. Experimental results show that the proposed heuristic-based approach is applicable to the runtime and the performance is acceptable. One key limitation of the proposed approach is that it considers only the deterministic model of CP-Nets. In the future, we want to explore the composability model of probabilistic CP-Nets in the context of the long-term IaaS composition.

Acknowledgement. This research was made possible by NPRP 7-481-1-088 grant from the Qatar National Research Fund (a member of The Qatar Foundation). The statements made herein are solely the responsibility of the authors.

References

1. Armbrust, M., et al.: A view of cloud computing. Commun. ACM **53**(4), 50–58 (2010)
2. Balke, W.T., Wagner, M.: Towards personalized selection of web services. In: WWW (Alternate Paper Tracks), pp. 20–24 (2003)
3. Binder, K., Heermann, D., Roelofs, L., Mallinckrodt, A.J., McKay, S.: Monte Carlo simulation in statistical physics. Comput. Phys. **7**(2), 156–157 (1993)
4. Boutilier, C., Bacchus, F., Brafman, R.I.: UCP-networks: a directed graphical representation of conditional utilities. In: Proceedings of the Seventeenth Conference on Uncertainty in Artificial Intelligence, pp. 56–64. Morgan Kaufmann Publishers Inc. (2001)
5. Boutilier, C., Brafman, R.I., Domshlak, C., Hoos, H.H., Poole, D.: CP-nets: a tool for representing and reasoning with conditional ceteris paribus preference statements. J. Artif. Intell. Res. (JAIR) **21**, 135–191 (2004)
6. Chaisiri, S., Lee, B.S., Niyato, D.: Optimization of resource provisioning cost in cloud computing. IEEE Trans. Serv. Comput. **5**(2), 164–177 (2012)
7. Lang, J., Xia, L.: Sequential composition of voting rules in multi-issue domains. Math. Soc. Sci. **57**(3), 304–324 (2009)
8. Li, L., Liu, D., Bouguettaya, A.: Semantic based aspect-oriented programming for context-aware web service composition. Inf. Syst. **36**(3), 551–564 (2011)
9. Li, M., Vo, Q.B., Kowalczyk, R.: Aggregating multi-valued CP-nets: a CSP-based approach. J. Heuristics **21**(1), 107–140 (2015)
10. Liu, X., Bouguettaya, A., Wu, J., Zhou, L.: Ev-LCS: a system for the evolution of long-term composed services. IEEE Trans. Serv. Comput. **6**(1), 102–115 (2013)
11. Mistry, S., Bouguettaya, A., Dong, H., Erradi, A.: Qualitative economic model for long-term IaaS composition. In: Sheng, Q.Z., Stroulia, E., Tata, S., Bhiri, S. (eds.) ICSOC 2016. LNCS, vol. 9936, pp. 317–332. Springer, Cham (2016). https://doi.org/10.1007/978-3-319-46295-0_20
12. Mistry, S., Bouguettaya, A., Dong, H., Erradi, A.: Probabilistic qualitative preference matching in long-term IaaS composition. In: Maximilien, M., Vallecillo, A., Wang, J., Oriol, M. (eds.) ICSOC 2017. LNCS, vol. 10601, pp. 256–271. Springer, Cham (2017). https://doi.org/10.1007/978-3-319-69035-3_18
13. Mistry, S., Bouguettaya, A., Dong, H., Qin, A.K.: Predicting dynamic requests behavior in long-term IaaS service composition. In: ICWS, pp. 49–56. IEEE (2015)
14. Mistry, S., Bouguettaya, A., Dong, H., Qin, A.: Metaheuristic optimization for long-term IaaS service composition. IEEE Trans. Serv. Comput. **11**, 1 (2017)
15. Mukhtar, H., Belaïd, D., Bernard, G.: A quantitative model for user preferences based on qualitative specifications. In: ICPS, pp. 179–188. ACM (2009)
16. Rossi, F., Venable, K.B., Walsh, T.: mCP nets: representing and reasoning with preferences of multiple agents. AAAI **4**, 729–734 (2004)
17. Wang, H., Shao, S., Zhou, X., Wan, C., Bouguettaya, A.: Web service selection with incomplete or inconsistent user preferences. In: Baresi, L., Chi, C.-H., Suzuki, J. (eds.) ICSOC/ServiceWave -2009. LNCS, vol. 5900, pp. 83–98. Springer, Heidelberg (2009). https://doi.org/10.1007/978-3-642-10383-4_6
18. Wang, H., Tao, Y., Yu, Q., Lin, X., Hong, T.: Incorporating both qualitative and quantitative preferences for service recommendation. J. Parallel Distribut. Comput. **114**, 46–69 (2017)
19. Wang, H., Wang, H., Guo, G., Tang, Y., Zhang, J.: Measuring similarity of users with qualitative preferences for service selection. Knowl. Inf. Syst. **51**(2), 561–594 (2017)

20. Ye, Z., Bouguettaya, A., Zhou, X.: QoS-aware cloud service composition based on economic models. In: Liu, C., Ludwig, H., Toumani, F., Yu, Q. (eds.) ICSOC 2012. LNCS, vol. 7636, pp. 111–126. Springer, Heidelberg (2012). https://doi.org/10.1007/978-3-642-34321-6_8
21. Ye, Z., Mistry, S., Bouguettaya, A., Dong, H.: Long-term QoS-aware cloud service composition using multivariate time series analysis. IEEE Trans. Serv. Comput. 9(3), 382–393 (2016)

LIFE-MP: Online Virtual Machine Consolidation with Multiple Resource Usages in Cloud Environments

Deafallah Alsadie[1]($\boxtimes$), Zahir Tari[1], Eidah J. Alzahrani[1],
and Ahmed Alshammari[2]

[1] School of Science, RMIT University, Melbourne, Australia
{deafallah.alsadie,zahir.tari,eidah.alzahrani}@rmit.edu.au
[2] Northren Border University, Arar, Saudi Arabia
ab.alshammari2012@gmail.com

Abstract. Efficient management of computing resources in cloud data centers is critical to minimize the power consumption and subsequent operating costs of the data centers. However, most of the existing approaches have several limitations during VM consolidation, including a limited number of computing resources, a higher number of VM migrations, service-level agreement (SLA) violations, and performance degradation. This paper proposes the Multiple Resource based VM Selection (MRVMS) approach for VM selection, and the Lowest Interdependence Factor Exponent Multiple Resources Predictive (LIFE-MP) approach for VM placement, by considering multiple computing resources being used simultaneously. The MRVMS approach selects a VM with high CPU requirements and optimal memory requirement for reducing the workload of overloaded PMs with minimum migration cost. The LIFE-MP approach selects a PM at which to place the migrating VM, based on the PM with the lowest correlation coefficient value among the already-running VMs and the migrating VM to reduce performance degradation because of the VM migration. Comparative results show that the proposed approaches offer better performance with respect to a power-aware best-fit decreasing (PABFD) scheme, including reducing power consumption by 29.02%, SLA violations by 32.68%, and the number of VM migrations by 66.09%.

Keywords: Power consumption · VM placement · VM selection

1 Introduction

As per the 'Infrastructure as a Service' (IaaS) delivery model of cloud computing systems, it is assumed that a data center can provide the required computing resources to cloud customers. Such computing resources are provided as VMs running on PMs. To meet the changing demands of cloud customers and provide the service at an economical cost, data centers must offer sufficient computing

© Springer Nature Switzerland AG 2018
H. Hacid et al. (Eds.): WISE 2018, LNCS 11234, pp. 167–177, 2018.
https://doi.org/10.1007/978-3-030-02925-8_12

resources, as per the SLA, in the form of VMs, and simultaneously minimize the operating costs of the data center. It has been observed that the major operating cost component is the huge amount of power consumed by PMs. This power consumption can be reduced by optimizing resource utilization through balancing the workload among the minimum number of PMs. The idle machines can be powered off, thereby optimizing power consumption. Optimized utilization of computing resources can be achieved through a VM consolidation process consisting of migrating VMs from the overloaded PMs to appropriate under-loaded PMs, while maintaining QoS as per the SLA.

In this research, we specifically consider the VM consolidation problem of a cloud-based data center, divided into the following phases: (1) detection of the overloaded host; (2) detection of the under-loaded host; (3) selection of VMs for migrating from an overloaded host; (4) placement of selected VMs in an appropriate under-loaded host, while maintaining QoS as per the SLA, and minimizing the number of VM migrations.

This paper proposes two new approaches to address the issue-related phases listed above. The first approach, the MRVMS approach, selects VMs to migrate from overloaded PMs. The second approach, the LIFE-MP approach, implements VM placement while maintaining QoS as per the SLA, as well as minimizing the number of VM migrations. It has been observed that the VM memory size is a large contributor to VM migration costs [1]. Thus, MRVMS selects a VM for migration that has high CPU usage and low-enough memory usage, so that the workload of the overloaded PM can be reduced, thereby minimizing the VM migration costs. For the placement of a migrating VM, LIFE-MP selects a PM that has the lowest correlation coefficient of multiple resources (e.g., CPU, memory, and bandwidth) of the migrating VM with already-running VMs on the candidate PM. The lowest value of the correlation coefficient enables minimization of the influence of migrating the VMs.

2   Related Work

In the recent past, many research efforts have been devoted to developing approaches to optimize cloud data center power. VM consolidation remains a prime focus and emerging approach for optimizing the power of cloud data centers by moving VMs from overloaded PMs, and placing them into underused PMs. This enables minimization of the active number of PMs by turning excess PMs to low power states. The concept of VM consolidation involves dynamic migrations among PMs, and the process is repeated continuously [2]. Thus, the VM consolidation process involves identification of overloaded PMs, selection of appropriate VMs for migration to other PMs, identification of underutilized PMs, and placement of selected VMs in identified under-utilized PMs [3].

Many researchers have examined these phases separately and proposed efficient approaches for saving power. Heuristic-based resource allocation strategies enable the assignment of computing resources of PMs in a manner that ensures a target processing time [4]. As a result, the CPU resource is mainly considered for completing the task. However, these strategies focus on only one resource at a time and ignore the other computing resources, such as memory and network bandwidth, which have a major effect on the processing time of the tasks. These computing resources must be considered during the process of VM consolidation.

The authors of another study [5] suggested an approach for selecting a VM with a maximum positive correlation with the utilization of other VMs on PMs. The suggested approach selects the VM for migration, but negatively affects the other VM that has maximum correlation with the migrating VM. Beloglazov et al. [3] proposed the PABFD approach for placing VM in the identified PM. They empirically compared and demonstrated the superiority of the PABFD approach over the representative approaches in the field. The authors considered a number of VM migrations and a number of SLA violations during the placement of VMs. However, they did not consider the performance deterioration of the VMs during their migration.

3   The Propsed Algorithm

The proposed work contributes to the design of MRVMS and LIFE-MP approaches to identify VMs (from overloaded PMs) and to migrate them to under-loaded PMs to ensure QoS as per the SLA, as well as minimizing the number of VM migrations, thereby considering the usages of multiple computing resources, such as CPU, memory, and bandwidth. This work also considers the influence of migrating VMs to already-running VMs on the PM, which is mostly ignored by the representative research in the field. Let N denote the total count of heterogeneous PMs in a data center-that is, $P = \{p_1, p_2, ..., p_N\}$. The VMs in the system are referred as $VM = \{vm^1, vm^2, ..., vm^M\}$, where M indicates the total number of VMs of the data center. Each PM is uniquely identified in the form of $< p_i, vm, R^d >$, where $d \in \{1, ..., D\}$, denotes the existence of D types of multiple resources; p_i denotes a unique identifier of a PM; vm is a set of VMs that are assigned to PM p_i; and $R^d = \{R^1, R^2, ..., R^D\}$ describes the types of computing resources, where each dimension corresponds to the type of physical resource (e.g., CPU, memory, storage, and network bandwidth). Thus, a VM can be assumed to be similar to the resource dimensions of a PM. In the present work, a VM is represented as $< vm^j, R^d >$ for its unique identification.

3.1   The Proposed MRVMS Approach

The various steps of the proposed VM selection approach are described below.
Step 1: Initialize the PMs list as $P = \{p_1, p_2, ..., p_N\}$, with VMs running on each machine described as $VM = \{vm^1, vm^2, ..., vm^M\}$. Then, the CPU usage of each PM is computed at time t as $U_i^d(t)$, using Eq. 1.

$$U_i^d(t) = \frac{\alpha_i^d + \sum_{j=1}^{M} U_{vm_i^j}^d(t)}{R_i^d} \tag{1}$$

where $d = $ CPU and α_i^d is the initial load of the resource d of the ith PM, $\sum_{j=1}^{M} U_{vm_i^j}^d(t)$ is the CPU usage of all VMs allocated to PM p_i at time slot t, and R_i^d is the total resource capacity of the CPU of p_i.

Step 2: Traverse P list by observing their $U_i^{CPU}(t)$ to find an overloaded PM at time t.

    if $(U_i^{CPU}(t) > THR^{UP})$ **then**
      Go to Step 3
    else
      Go to Step 7
    end if

Step 3: Arrange the VMs, running on the overloaded machine p_i, in decreasing order of CPU usage as $VM = \{vm_i^1, vm_i^2, ..., vm_i^M\}$. The CPU usage of VM is $\sum_{j=1}^{M} U_{vm_i^j}^d(t)$. Then Compute the difference of CPU usage and upper limit of threshold value for each PM using Eq. 2.

$$Diff_i(t) = U_i^{CPU} - THR^{UP} \tag{2}$$

where $Diff_i(t)$ is the excess utilization of CPU than its upper threshold value for PM p_i

Step 4: Select the VM from the list of VMs at the overloaded PM p_i by computing the following (a)-(c) below.
(a) Compute the resource utility (RU) of each VM at time t for various resource types (e.g., CPU and memory) as per Eq. 3.

$$RU_{vm_i^j}(t) = U_{vm_i^j}^d - Diff_i(t), \ d = CPU \tag{3}$$

where $RU_{vm_i^j}(t)$ returns the Resource Utility of the jth VM of ith PM at time t.
(b) Compute RU of PM p_i at time t as per Eq. 4.

$$RU_i(t) = \sum_{j=1}^{M} RU vm_i^j(t) \tag{4}$$

where $RU_i(t)$ returns the RU of ith PM at time t.
(c) Compute memory utilization of PM p_i at time t as per Eq. 5.

$$U_i^{Mem}(t) = \frac{\alpha_i^d + \sum_{j=1}^{M} U_{vm_i^j}^d(t)}{R_i^d}, \ for \ d = Memory \tag{5}$$

where α_i^d is the initial load of resource d (=memory here) of ith PM, $\sum_{j=1}^{M} U_{vm_i^j}^d(t)$ is the memory utilization of all VMs allocated to PM p_i at time t.

Step 5: Create a list X of VMs that satisfies the condition mentioned in Eq. 6.

$$\frac{RU_{vm_i^j}(t)}{U_{vm_i^j}^{Mem}(t)} \geq \frac{RU_i(t)}{U_i^{Mem}(t)} \qquad (6)$$

where $U_{vm_i^j}^{Mem}(t)$ represents the amount of memory currently utilized by the jth VM of ith PM.

Step 6: Append the selected VM to the queue $ToMigrateList$ for further processing of migration.

   if (list X is Empty) **then**

      Select VM having max $\left\{ \dfrac{RU_{vm_i^j}(t)}{U_{vm_i^j}^{Mem}(t)} \right\}$

   else

      Select VM from List X having max $\left\{ \dfrac{RU_{vm_i^j}(t)}{U_{vm_i^j}^{Mem}(t)} \right\}$

   end if

Step 7: If a PM p_i is under utilized, i.e. its current utilization is less than the lower limit of threshold value (i.e., $U_i^{CPU}(t) \leq THR^{LW}$), then all the VMs running at p_i will be considered for migration to another under-loaded PM. These VMs are queued $ToMigrateList$ and p_i will be turned off to optimize the power consumption.

3.2   LIFE-MP - the Proposed VM Placement Approach

After selecting a VM that needs migration from an overloaded/underutilized PM, as per the proposed VM selection approach described above, there is a need to find an appropriate PM to place the migrating VM to minimize both the effect of VM migration and the number of VM migrations. The proposed LIFE-MP approach addresses this problem by considering the correlation coefficient and the predicted values of the future requirements of computing resources to find an appropriate PM for the VM to be migrated. Several techniques [6,7] have been proposed to predict the future requirements of computing resources. However, recently, the autoregressive integrated moving average (ARIMA) model has been advocated as useful because it is linear in predicting future values based on historical observations. Linear models are known to provide simplicity in predicting CPU usage, and this is the reason we used such a model. For the purpose of formally formulating the problem of VM placement, we assumed the existence of $P = \{p_1, p_2, ..., p_N\}$ as a set of all PMs in a data center. We focused on finding an optimal PM, say $p_i \in P$, for placement of migrating VM vm^j that satisfied the following conditions:

1. PM p_i satisfied the requirements of computing resources for VM vm^j's.
2. $P_{OPT} = \{p_i\} | \forall p_i \in \text{CP}, min(i)$ having

$$LIFE^{Mem}(p_i) < AvgLIFE^{Mem}(p)$$

The steps of the proposed approach to find an "optimal" PM for placement of a migrating VM are listed below.

Step 1: Initialize $P_{cands} \subseteq P = \{p_1, p_2, ..., p_N\}$ as a set of candidate PMs for placement of migrating VM, satisfying Conditions 1 and 2 above.

Step 2: Compute the historical CPU values and Memory utilization for each PM in P_{cands} for a slot time t as per the matrix below.

$$
U_i^{CPU}[M][t] = \begin{bmatrix} U_{11}^{CPU} & U_{12}^{CPU} & \cdot & \cdot & U_{1t}^{CPU} \\ & \cdot & \cdot & \cdot & \\ \cdot & & \cdot & U_{jk}^{CPU} & \cdot \\ & \cdot & \cdot & \cdot & \\ U_{M1}^{CPU} & U_{M2}^{CPU} & \cdot & \cdot & U_{Mt}^{CPU} \end{bmatrix}
$$

$$
U_i^{Mem}[M][t] = \begin{bmatrix} U_{11}^{Mem} & U_{12}^{Mem} & \cdot & \cdot & U_{1t}^{Mem} \\ & \cdot & \cdot & \cdot & \\ \cdot & & \cdot & U_{jk}^{Mem} & \cdot \\ & \cdot & \cdot & \cdot & \\ U_{M1}^{Mem} & U_{M2}^{Mem} & \cdot & \cdot & U_{Mt}^{Mem} \end{bmatrix}
$$

where M is the number of VMs on each $P_i \in P_{cands}$, U_{jk}^{CPU} and U_{jk}^{Mem} represent the CPU utilization and memory utilization of vm^j on PM p_i at time k, respectively.

Step 3: Compute usage of CPU and memory for PM p_i within a time slot k, as per Eqs. 7 and 8 respectively.

$$
U_i^{CPU}[k] = \sum_{j=1}^{M} U_i^{CPU}[j][k] \tag{7}
$$

$$
U_i^{Mem}[k] = \sum_{j=1}^{M} U_i^{Mem}[j][k] \tag{8}
$$

Step 4: Predict the future CPU requirements and memory usage using the ARIMA prediction model for each PM $P_i \in P_{cands}$ on the basis of historical CPU usage values at time slot k, as per Eqs. 9 and 10.

$$
FU_i^{CPU}[k] = ARIMA(U_i^{CPU}[k]) \tag{9}
$$

$$
FU_i^{Mem}[k] = ARIMA(U_i^{Mem}[k]) \tag{10}
$$

Step 5: Compute the values of $LIFE^{CPU}$ and $LIFE^{Mem}$ variables between migrating VM mvm and each PM $P_i \in P_{cands}$, as per Eqs. 11 and 12, respectively.

$$
LIFE_i^{CPU} = \frac{E[(U_{mvm}^d - \frac{1}{t}\sum_{k=1}^{t} FU_{mvm}^{CPU})(U_i^{CPU} - \frac{1}{t}\sum_{k=1}^{t} FU_i^{CPU}[k])]}{\sqrt{Var(U_{mvm}^{CPU})}\sqrt{Var(UU_i^{CPU})}} \tag{11}
$$

$$
LIFE_i^{Mem} = \frac{E[(U_{mvm}^{Mem} - \frac{1}{t}\sum_{k=1}^{t} FU_{mvm}^{Mem})(U_i^{Mem} - \frac{1}{t}\sum_{k=1}^{t} FU_i^{Mem}[k])]}{\sqrt{Var(U_{mvm}^{Mem})}\sqrt{Var(UU_i^{Mem})}} \tag{12}
$$

where U_{mvm}^{CPU} and U_{mvm}^{Mem} represent the ongoing rate of CPU and memory usage, respectively, of the VM that is to be migrated; the current CPU and memory usage of PM p_i is represented by U_i^{CPU} and U_i^{Mem}, respectively; $\frac{1}{t}\sum_{k=1}^{t} FU_{mvm}^{CPU}$, $\frac{1}{t}\sum_{k=1}^{t} FU_{mvm}^{Mem}$, $\frac{1}{t}\sum_{k=1}^{t} FU_i^{CPU}[k]$ and $\frac{1}{t}\sum_{k=1}^{t} FU_i^{Mem}[k]$ refer to the average of CPU and memory usage of the VM that is being migrated mvm and the corresponding PM, respectively, over time slots t in the future. The denomination factors of Eqs. 11 and 12 refer to variance of VM and PM, respectively, and are computed as per Eqs. 13 and 14.

$$Var(U_{mvm}^d) = E[(U_{mvm}^d - \frac{1}{t}\sum_{k=1}^{t} FU_{mvm}^d)^2] \tag{13}$$

$$Var(U_i^d) = E[(U_i^d - \frac{1}{t}\sum_{k=1}^{t} FU_i^d)^2] \tag{14}$$

where, d is a resource type, $d = \{CPU, memory\}$.

Step 6: Compute the average of $LIFE^{CPU}(P_{cands})$ and $LIFE^{Mem}(P_{cands})$, as per Eqs. 15 and 16.

$$AvgLIFE^{CPU}(P_{cands}) = \frac{1}{N}\sum_{i=1}^{N} LIFE_i^{CPU} \tag{15}$$

$$AvgLIFE^{Mem}(P_{cands}) = \frac{1}{N}\sum_{i=1}^{N} LIFE_i^{Mem} \tag{16}$$

where N is the number of candidate PMs.

Step 7: Find a sorted list all PMs $CPList$ on the basis of $LIFE^{CPU}$ first, and then $LIFE^{Mem}$ as described in Eq. 17.

$$CPList = Sort_{LIFE^{CPU,Mem}\{p_i\}} |\forall p_i havingLIFE_i^{CPU}$$
$$< AvgLIFE^{CPU}(P_{cands}) \tag{17}$$

Step 8: Select the first occurrence of PM as an optimal machine P_{OPT} from the $CPList$ that satisfies the condition $LIFE^{CPU} < AvgLIFE^{Mem}(P_{cands})$ as described by Eq. 18.

$$P_{OPT} = \{p_i\} \mid \forall p_i \in CP, min(i) having$$
$$LIFE^{Mem}(p_i) < AvgLIFE^{Mem}(p) \tag{18}$$

4   Experimental Setup

In this set of experiments, we developed a simulation environment similar to a real environment by using a total of 800 heterogeneous PMs as servers, of HP ProLiant ML110 G4 and HP ProLiant ML110 G5 make. Each server was equipped with two cores, 4 GB memory, and 1 GB/s bandwidth for network communication. The CPU frequency of the servers was mapped onto MIPS ratings of 1,860 MIPS each core of the HP ProLiant ML110 G5 server, and 2,660 MIPS each core of the HP ProLiant ML110 G5 server. Each server was modeled to have 1 GB/s bandwidth. The CPU MIPS rating and memory amount represented four VM instances used in CloudSim. They corresponded to Amazon EC2 1–that is, high CPU medium instance (2,500 MIPS, 0.87 GB), extra large instance (2,000 MIPS, 1.74 GB), small instance (1,000 MIPS, 1.74 GB), and micro instance (500 MIPS, 613 MB).

In our experiments, we initially allocated the VMs as per the requirements of computing resources as fixed for types of VMs. With the passage of time, VMs consume computing resources as per the actual cloud workload. Thus, this provides the scope to perform the dynamic consolidation of VMs. Therefore, we followed the approach of allocating a different set of resources to VMs, and then consolidating the computing resources dynamically as per the actual requirements of the VM workload.

We used Google cloud traces data as benchmark data traces to evaluate the proposed approach. The Google cloud traces data consist of the real data of a Google cluster over about a one-month period in May 2011, and are known as the second version of Google cloud traces. We extracted the duration of the cloud tasks based on their run time in addition to utilization of CPU and memory by the cloud tasks over the first 10 days. Each extracted job was uniquely identified by an identifier denoted as jobID. The extracted parameter for the jobs consisted of the CPU rate and canonical memory usage. The CPU rate determined the average use of CPU for a sample period of five minutes, while the canonical memory use provided the average consumption of memory for the specified sampling period.

5   Results and Discussion

In the recent past, many researchers have proposed numerous approaches for detecting overloaded PMs and selecting VM for migration to alternate PMs. The most representative approaches for detecting overloaded PMs include IQR, LR, MAD, and LRR [8]. The approaches for selecting VMs for migration to another PM include MMT, RS, MC, and MU [8]. In this set of experiments, we measured the performance of the proposed LIFE-MP and MRVMS approaches in combination with the above cited techniques to compare their results. The comparison of results was undertaken with the objective of determining the optimal combination of different approaches with LIFE-MP and MRVMS in terms of the identified metrics of energy consumption, count of SLA violations, and count of VM migrations. The results were recorded using a 10-day workload, with particular initialization parameters of overload detection approaches as LR = 1.2, LRR = 1.2, MAD = 2.5, and IQR = 1.5.

We ran 32 combinations of different approaches with the proposed approach of MRVMS and LIFE-MP in the CloudSim environment, and recorded their

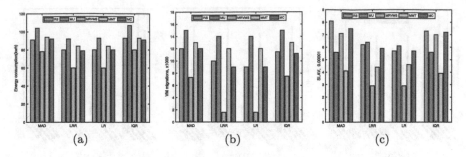

(a)   (b)   (c)

Fig. 1. Comparative results of the proposed MRVMS approach and other VM selection policies in combination with PABFD allocation approach in terms of (a) Energy consumed, (b) count for VM migrations, and (c) count for SLA violations.

results. The recorded results of the proposed approaches were compared with the baseline approaches to indicate the effectiveness of the proposed approaches. The current set of experiments can be broadly divided into four subsets of experiments for the sake of comprehensive analysis and discussion.

Experiment Subset 1: The first subset of experiments involved the use of representative approaches for VM selection in addition to the proposed MRVMS approach, in combination with PABFD as the allocation approach. The reported results are depicted in Fig. 1(a) to (c).

The comparative results depicted in Fig. 1(a) to (c) highlight that the proposed MRVMS approach was a better option than the other VM selection approaches in terms of the identified metrics of energy consumption, count of SLA violations, and count of VM migrations. The following observations can be made from the reporting results depicted in Fig. 1:

(a) The reported results for the IQR and MAD algorithms indicated that they behaved similarly in their performance. The similar behavior of local regression algorithms—namely, LRR and LR—was also observed. However, local regression-based algorithms exhibited lower consumption of energy, VM migrations count, and SLA violations count in comparison with the IQR and MAD approaches.

(b) The results of the proposed MRVMS approach indicated that it leads to low energy consumption and a lower count of VM migrations in comparison with the other VM selection approaches, irrespective of the use of any approach for detecting overloaded PM.

(c) It can be observed that the combination of the proposed MRVMS approach led to a higher number of SLA violations in combination with IQR and MAD approaches, in comparison with the combination with LR and LRR approaches.

Experiment Subset 2: This subset of experiments involved evaluation of the proposed LIFE-MP approach and PABFD approach as VM allocation

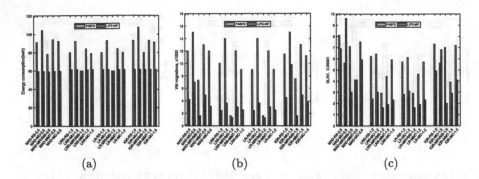

Fig. 2. Comparative results of the proposed LIFE-MP approach in comparison with the PABFD approach for VM allocation in terms of: (a) energy consumed, (b) count of VM migrations, and (c) count of SLA violations.

approaches. The comparative results are depicted in Fig. 2: The following observations can be made from the reporting results depicted in Fig. 2:

(a) The proposed LIFE-MP approach performed more successfully than the PABFD approach in terms of energy consumption, irrespective of the combination approaches for detecting overloaded PM and selecting VM.

(b) The LIFE-MP approach led to a reduction in the count of VM migrations by half, in comparison with the PABFD approach.

(c) The LIFE-MP approach provided better results than the PABFD approach in all combinations, except IQR-MU, in terms of the number of SLA violations.

6   Conclusion

In this paper, we addressed the limitations of the existing approaches for VM consolidation by proposing new approaches–MRVMS for selection of migrating VM and LIFE-MP for placing the migrating VM on an optimal PM by considering the multiple computing resources being used and the correlation between the migrating VM and already-running VMs on the PM–with the objective of minimizing the number of migrations, number of SLA violations, and power consumption of PMs. The proposed approaches were implemented as part of the VM consolidation process and evaluated using real-world Google cloud traces. The comparative results and analysis revealed the superiority of the proposed approaches for selection and placement of VMs to enable efficient reduction of power consumption, thereby minimizing the operating cost.

References

1. Xiao, Z., Song, W., Chen, Q.: Dynamic resource allocation using virtual machines for cloud computing environment. IEEE Trans. Parallel Distrib. Syst. **24**(6), 1107–1117 (2013)

2. Feller, E., Morin, C., Esnault, A.: A case for fully decentralized dynamic VM consolidation in clouds. In: 2012 IEEE 4th International Conference on Cloud Computing Technology and Science (CloudCom), pp. 26–33. IEEE (2012)
3. Beloglazov, A., Buyya, R.: Optimal online deterministic algorithms and adaptive heuristics for energy and performance efficient dynamic consolidation of virtual machines in cloud data centers. Concurr. Comput.: Pract. Exp. **24**(13), 1397–1420 (2012)
4. Gutierrez-Garcia, J.O., Sim, K.M.: A family of heuristics for agent-based elastic cloud bag-of-tasks concurrent scheduling. Futur. Gener. Comput. Syst. **29**(7), 1682–1699 (2013)
5. Cao, Z., Dong, S.: Dynamic VM consolidation for energy-aware and SLA violation reduction in cloud computing. In: 2012 13th International Conference on Parallel and Distributed Computing, Applications and Technologies (PDCAT), pp. 363–369. IEEE (2012)
6. Zhang, Q., Zhani, M.F., Zhang, S., Zhu, Q., Boutaba, R., Hellerstein, J.L.: Dynamic energy-aware capacity provisioning for cloud computing environments. In: Proceedings of the 9th International Conference on Autonomic Computing, pp. 145–154. ACM (2012)
7. Fang, W., Lu, Z., Wu, J., Cao, Z.: RPPS: a novel resource prediction and provisioning scheme in cloud data center. In: 2012 IEEE Ninth International Conference on Services Computing (SCC), pp. 609–616. IEEE (2012)
8. Beloglazov, A., Abawajy, J., Buyya, R.: Energy-aware resource allocation heuristics for efficient management of data centers for cloud computing. Futur. Gener. Comput. Syst. **28**(5), 755–768 (2012)

Stance and Credibility Based Trust in Social-Sensor Cloud Services

Tooba Aamir[1]([⊠]), Hai Dong[1], and Athman Bouguettaya[2]

[1] School of Science, RMIT University, Melbourne, Australia
{tooba.aamir,hai.dong}@rmit.edu.au
[2] School of Information Technologies, The University of Sydney, Sydney, Australia
athman.bouguettaya@sydney.edu.au

Abstract. We propose a users' stance and credibility based social-sensor cloud service trust model. We represent social media data streams, i.e., image meta-data and related posted information, as social-sensor cloud services. We use the textual features of the social-sensor cloud services, i.e., comments, and meta-data, e.g., spatio-temporal information, to gather the trust-rate of the services and the credibility of users' comments. The analytical results present the performance of the proposed model.

Keywords: Social-sensor · Social-sensor cloud service
Trust in social-sensor cloud service

1 Introduction

A social media platform can be viewed as a social-sensor cloud. The social sensor cloud stores the information sensed by its users through their smart-phones (also called social-sensors). These social-sensor clouds have a considerably different role than conventional information portals during special events [1,2]. Thousands even millions of posts are delivered from social-sensors to social-sensor clouds in a single second [2,3]. This social-sensor cloud data is mostly subjective and based on people's personal experience. For example, a significant amount of misinformation and disinformation in social media was observed during events like the 2010 earthquake in Chile [7], Hurricane Sandy in 2012 [6] and the Boston Marathon blasts in 2013. Such untrustworthy information or rumors spread quickly over the social media and adversely affect thousands of people because most people tend to trust the information regarding such events [6,7]. It is hard to identify the reliable information due to the free and unmonitored nature of available information. Therefore, it is essential to develop a trust model to present the trustworthy information along with the credibility of information provided.

Our proposed solution takes a service-centric approach for improved information flow and flexible organization of data. The integration of the social-sensor cloud data with service paradigm, i.e., social-sensor cloud services [3] presents

© Springer Nature Switzerland AG 2018
H. Hacid et al. (Eds.): WISE 2018, LNCS 11234, pp. 178–189, 2018.
https://doi.org/10.1007/978-3-030-02925-8_13

an open, flexible, and reconfigurable platform for numerous monitoring applications, e.g., urban management and scene analysis [4,19]. In contrast, we utilize the users' comments to gather the evidence in favor of or against the information. The proposed model considers various indicators such as the stance embedded in the services' comments, their meta-data, e.g., time, along with the users' credibility. We model the interactions between commenters and sub-comments in terms of their comments. The commenters' credibility is assessed based on the sub-commenters' confidence on the commenters' comments. This model helps to accumulate the collective stance (i.e., support or refute) regarding the service.

Untrustworthy information detection has been investigated in the past. The existing approaches mainly focus on natural language processing, data mining and information retrieval techniques [6,10,11,15,16]. Most of the current studies in the area of trust and credibility of social media focus on the topical news [10,12]. Some use social networks, social network users' credibility and users' interaction [6,13]. A major limitation of the existing research is the assumption that the information available over social media must be in the specific formats such as agreeing or disagreeing to a specific topic [8,10]. The studies utilizing the textual claims, i.e., news and views, require both evidence or counter-evidence of those claims must be easily accessible from the Web [13]. These, however, are difficult to search in social media considering the much faster-spreading of untrustworthy information [11]. Moreover, the methods to interpret a user's verdict regarding information's honesty depend on comparing the user's verdict against a selected set of data from researchers' predefined sources [5].

In our previous work, we measured the trustworthiness of a social-sensor cloud service based on users' stance only [9]. In this paper, we aim to define a stance and credibility based trust model for the evaluation of the trustworthiness of the social-sensor cloud services. We intend to study that how user comments can be used to build a trust system for the social-sensor cloud services. This trust-rate can help to identify the trustworthiness of social media information. Our main contribution is:

– We define a social-sensor cloud service trust problem, and propose the stance and credibility based trust model for the social-sensor cloud service. We introduce a stance based credibility model for the commenters, i.e., the service users.
– We study the stance of the service users to determine whether service is trustworthy or not. Our approach uses textual information, e.g., comments from the social media posts (i.e., the social-sensor cloud service) to gather the users' opinion or the *stance* regarding the service.
– We model the relationship between the users' comments and their sub-commenters to accumulate the collective stance (i.e., support or refute) regarding the service.

The paper focuses on proposing a novel approach to evaluate trustworthy and untrustworthy services and propagate trust in the social-sensor cloud services. The experiments are based on the real and synthetic datasets. The commenters' stance along with their (or social-sensor cloud service users) and the trustworthiness of the social media information (i.e., the social-sensor cloud service) is used to classify the services as trustworthy or untrustworthy. The results of the experiment demonstrate the effectiveness of the proposed model.

The rest of the paper is structured as follows: Sect. 2 describes the motivating scenario. Section 3 formally defines the trust model for the social-sensor cloud service. Section 4 describes the experiments and evaluation of the proposed model. Section 5 concludes the work.

2   Motivating Scenario

This paper aims to define a stance and credibility based trust model to evaluate the trustworthiness of the social-sensor cloud services and the reliability of the users' stance. We use the textual features (e.g., comments) and meta-data (e.g., spatio-temporal information) to gather the trust-rate of the service, the credibility of the users' belief (Fig. 2). For this purpose, we first identify the most common snippets (i.e., keywords and phrases) associated with the fake images. Next, we identify the most common snippets used by the commenters, i.e., the users of the service, to assess their belief regarding the untrustworthy services. We also utilize the spatio-temporal footprint of the services to authenticate their location. Subsequently, we classify the service into trustworthy or untrustworthy. The classification is based on the features like shared snippets and annotations. These features are commonly associated with the fake images and spatio-temporal attributes. For example, in Fig. 1, "Call you a lair" by commenter1(CM1) and "inaccurate Photo" by commenter2(CM2) show the users' opinion that the service is untrustworthy. Similarly, "verify origin" by sub-commenter3(SCM3) and "I agree" by sub-commenter4(SCM4) show the sub-commenters' opinion about the service user's comment, i.e., commenter3(CM3). The probability of the service being untrustworthy is high if the majority of the users deem the image (i.e., service) untrustworthy. Besides, the opinion of users regarding the service being trustworthy or untrustworthy is credible if the users are credible. The evaluation of the user credibility is achieved by assigning a score or rating depending upon the commenter's previous credibility and sub-commenter's confidence in his/her stance.

The stance of the user, i.e., commenter, is an aggregation of his/her stance score, i.e., the measure of their belief in the service and the credibility score, i.e., the measure of a user's stance being trustworthy. For example, in Fig. 2, $User_2$ and $User_3$'s stance is against and $User_1$'s stance is in the favor of $Service_1$. The confidence in $User_1$'s stance is less than $User_2$'s and $User_3$'s stance, based on the sub-commenters' stance and their credibility. The confidence in $User_1$'s stance is -0.05, $User_2$'s stance is 0.65 and $User_3$'s stance is 0.05 (Fig. 2). The credibility of the users for this specific service is -0.05 for $User_1$, 0.39 for $User_2$ and 0.035 for $User_3$. The credibility of the commenters, e.g., Cred $= 0.6$ for $User_2$, along with their stance (i.e., -1), and confidence in the stance, i.e., 0.65, is propagated for the trust calculation of the service, i.e.,

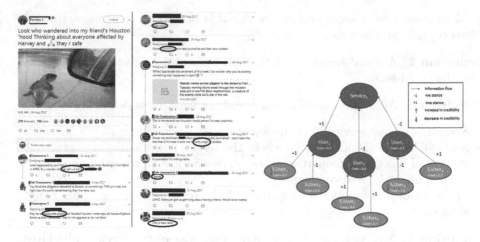

Fig. 1. Social media images to be abstracted as crowd sourced social-sensor cloud services

Fig. 2. Trust propagation between Service and Users

$Service_1$. The final trust-rate of the service is the aggregation of the opinion of all the credible commenters. Therefore, the trust-rate for a service can be calculated by aggregating the users' stance and credibility, i.e., trust-rate for $Service_1 = ((-0.05 * 0.5 * 1) + (0.65 * 0.6 * -1) + (.05 * 0.7 * -1))/3 = -0.15$. The opinions from the credible commenters, stating the service as untrustworthy results in decreases of the trust-rate. The credibility score of a commenter is increased in cases where the opinion of the commenter matches the opinion of the other credible commenters and the overall stance of the service.

It is essential that the commenters of the service have positive stance regarding the service to characterize and identify a service as trustworthy. For this purpose, we identify the following stakeholders: (1) stance of the commenter of the service, (2) a credibility model for a commenter or users of the service and (3) a trust model for the service.

3 Stance and Credibility Based Trust Model for Social-Sensor Cloud Service

In this section, we discuss several concepts related to the proposed stance and credibility based trust model. We offer a generic trust framework including the users' (or commenters') credibility model, a service trust model and their evolution models. The proposed model utilizes the stance in the comments of the social-sensor cloud service along with the users' credibility for trust management.

We define several concepts to determine the stance and credibility based trust-rate.

Definition 1. A crowd-sourced social-sensor $SocSen$ is a user of a social media. A sensor posts content on social media, i.e., social-sensor cloud. It is assumed, the data shared by a social-sensor has visual information, i.e., image, textual reference, time and location.

Definition 2. A social-sensor cloud *SocSenCl* is a social network hosting data from the social-sensors.

Definition 3. An atomic social-sensor cloud service *Serv* is defined by

- *Serv_id*, a unique service id of the service provider *SocSen*.
- *SocSenCl_id*, an ID of the cloud where the service is available.
- *F*, a set of functional properties of the service *Serv*, e.g., special mentions *M*, a set of users $\langle U \rangle$ and comments $\langle C \rangle$.
- *nF*, a set of non-functional properties of the service *Serv*, e.g., time *t* and location *l*.

Definition 4. A service user *U* is an individual user of the service, who gives an opinion on the service by *comments*.

Definition 5. A user's stance $U.\theta$ or the stance of the comment means that whether the user, i.e., commenter considers the service as trustworthy or not. The stance score is between -1.0 (least trustworthy) and 1.0 (most trustworthy). For example, a service might be a fake image, and users might reply to it by further supporting it to some degree or providing counter-evidence.

The aggregated stance of the users of the service is computed by computing the average of the stance of all the users of the service. It is calculated by:

$$Stance(U_{i=1}^n) = \frac{\sum_{i=1}^n (U_i.\theta)}{n} \qquad (1)$$

where $U_i.\theta$ is the stance of a specific user and n is the total number of users.

Definition 6. *Confidence* in a user's comment is the trust of the sub-commenters in the comment. The confidence is calculated as:

$$Confidence(U_i, SubComm_k) = SubComm_k.\theta \times SubComm_k.C \qquad (2)$$

Where, $SubComm_k.S$ is the stance of the sub-commenter and $SubComm_k.C$ is the credibility of the sub-commenter. The aggregated confidence in U_i's stance for $Serv_j$, is calculated as:

$$Confidence(U_i, Serv_j) = \frac{1}{l} \times \sum_{k=1}^{l} (SubComm_k.\theta \times SubComm_k.C) \qquad (3)$$

l is the total number of the sub-commenters of the user's comment. The value of the confidence is used to determine the credibility of the user U_i for the service $Serv_j$.

Definition 7. User's *credibility* $U.C$ is the quality of being trusted, i.e., user's stance is trustworthy or not. The credibility of the user $U.C$ depends upon the aggregated stance regarding the service, past credibility score of the user, confidence in the stance and the stance score for this service.

The generic formula for $Cred_S(U_i, Serv_j)$, i.e., credibility of U_i for a comment on the $Serv_j$ is:

$$Cred_S(U_i, Serv_j) = |U_i.S_\theta| \times Confidence(U_i, Serv_j) \qquad (4)$$

Where $|U_i.S_\theta|$ is the absolute stance score of U_i for a comment on $Serv_j$. The credibility score is between -1.0 (the lowest) and $+1.0$ (the highest). The positive and negative credibility of U_i is based on U_i's stance and the overall stance of all the users of $Serv_j$, i.e.,

$$Cred_S(U_i, Serv_j) = \begin{cases} +Cred_S(U_i, Serv_j) \, if \, U_i.S = Stance(U_{i=1}^n) \\ -Cred_S(U_i, Serv_j) \, if \, U_i.S \neq Stance(U_{i=1}^n) \end{cases} \qquad (5)$$

Definition 8. Trust-rate $S.T$ of the service measures the users' belief on trustworthiness based on the service's reviews. The trust in the service helps to ascertain whether the service description is an accurate description of the image. The value of the trust is within $[-1,1]$, where -1.0 is the least trustworthy (or untrustworthy), and 1.0 is the most credible.

The trustworthiness of a service at an given time t is computed as:

$$Trust(Serv_j) = \omega(t) \times \frac{\sum_{i=1}^n (U_i.\theta \times U_i.C)}{n} \qquad (6)$$

Where $\omega(t)$ is time decay function defined by:

$$\omega(t) = \frac{t_p - t_c}{\gamma} \qquad (7)$$

t_p refers to current time, t_c is the time when the comment was made, and γ refers to trust-decay ratio. Service users can define the value of γ according to the preference of the historical stance or the recent stance. For a significant value of γ, the decay of trust-rate over time is slow. Whereas, if γ is small, the decline of trust-rate is faster.

The implementation process of the proposed social-sensor cloud service trust model includes:

3.1   Stance Model for Social-Sensor Cloud Service

The first step is to compute the stance of all the users towards a service. We analyze conversations arisen from the service, i.e., all the comments of the service posted by the users. These comments (and sub-comments) result in a nested discussion triggered by a service.

First, we identify the service keywords and snippets associated with (1) disagreement to the service statement or stating it untrustworthy $\langle Serv_j.A_d \rangle$, (2) agreement to the service statement or stating it trustworthy $\langle Serv_j.A_a \rangle$. Second, we extract the snippets of crucial annotation(s), i.e., $\langle U_i.A \rangle$ from the user's comment and sub-comments of the user's feedback. Each user's stance is labeled based on their observation. User's stance is marked as trustworthy or untrustworthy. Next, if a sub-commenter: (1) agrees with the user's comment, we label it 'supporting user's stance,' or (2) disagrees with the user's comment, we label it 'denying user's stance'.

We use natural language processing to gather a better understanding of users' comments. Aylien Text Analysis[1] is used to gather the snippets and annotations related to the service's agreement and disagreement. The API helps to generate words or phrases that are semantically similar to the service keywords, i.e., agreement phrases or disagreement phrases. These phrases help to extract what concept is mentioned in a piece of text. The concept extraction endpoint performs accurate word sense disambiguation to find out what the user meant by each mentioned snippet. The concept helps in determining whether the commenter refutes or supports the claim of the social-sensor cloud service. We calculate a semantic distance between these annotations/snippets and the comments' annotations/snippets. A WordNet-based approach LIN [14] is used to calculate the semantic similarity as a number between [0, 1]. This approach measures the semantic relatedness of concepts based on the ratio of the amount of information needed to state the commonality of the information content of the services. The measure is determined by $related_{LIN}(Serv.A, U.A)$ [14]. It is determined by the information content of the lowest concept in the hierarchy that subsumes both $Serv$ and U. $related_{LIN}(Serv.A, U.A)$ measures the relatedness of the two descriptions.

A Java-based library, WS4J (WordNet Similarity for Java) is used to implement the similarity measure. The use of this library is defined in an online documentation[2]. We have used θ to define $related_{lin}(Serv.A, U.A)$. The higher value of θ shows higher similarity to the stance. For example, if θ of $\langle U_i.A \rangle$ has more similarity with $\langle Serv_j.A_d \rangle$; then the user considers the service untrustworthy. The stance score is between -1.0 (the lowest) to $+1.0$ (the highest). -1.0 shows a negative stance, i.e., disagreement with the service statement and higher similarity with $\langle Serv_j.A_d \rangle$, $+1.0$ shows a positive stance, i.e., agreement with the service statement and the higher similarity with $\langle Serv_j.A_a \rangle$. 0.0 shows a neutral stance.

The sub-commenters' stance is used to determine the sub-commenters' confidence in the user's stance of each comment. The sub commenter's stance is further used to gather the user's credibility (discussed in Sect. 3.2) and the final trust in the service (discussed in Sect. 3.3).

---

[1] AYLIEN Text Analysis - https://aylien.com/text-api/.
[2] JWNL - JavaWordNet Library - Dev Guide", http://jwordnet.sourceforge.net/handbook.html.

3.2   Credibility Model for Social-Sensor Cloud Service

The credibility of a social-sensor cloud service user is a decisive factor to measure the trustworthiness of a given piece of information. The general credibility of the user's comment depends upon:

- The past credibility score of the user's comments
- The confidence of the sub-commenters in the user's stance
- The current credit score depending on the user's confidence score and stance score

The overall credibility of U_i can be computed by:

$$U_i.C_{new} = U_i.C = \delta(Cred_S(U_i, Serv_j) + (1 - \delta)(U_i.C_{old})$$ (8)

Where, (δ) and $(1 - \delta)$ are the weight assigned to $Cred_S(U_i, Serv_j)$ and the previous credibility of U_i. Given a set of SU, i.e., all the users associated with $Serv.j$, $SU_i.C$, i.e., the credibility score of all the users is calculated. For each user of $Serv_j$ the following are calculated: (1) $Confidence(U_i, Serv_j)$ based on Eq. 3, 2 $Cred_S(U_i, Serv_j)$ based on Eqs. 4 and 3 $U_i.C_{new}$ based on Eq. 8.

3.3   Trust Model for Social-Sensor Cloud Service

The service trust helps to ascertain whether the service description is an accurate description of the image. The proposed trust model collects trust information from two sources: (1) the service users and (2) the similar services. The trust-rate is determined on the experience of the service users, i.e., the collective stance of the commenters, the time of the comment and trust in the services similar in description, location and time. The trust model combines the collective stance and the trust-rate of the same services with the proper weight. The value of the trust is within the range $[-1, 1]$, where -1 is the least trustworthy (or untrustworthy) and 1 is the most reliable. The trust-rate of the service might change over the period, depending upon the feedback and presence of the similar services. Trust-rate is recalculated each time when a service gets a new review, i.e., comment.

The trustworthiness of the service at the given time t is computed using Eq. 6. The value of trust changes continuously over the time based on the dynamics of trust and decay by time. The most recent commenters' stance and the current credibility of the users best reflect the confidence in a service. The earlier comments have less impact on the current trust rate. The trust of all the users in the service is computed as the weighted mean of the stance of the users, where the credibility of all the users is the weight.

4   Evaluation

In this section, we present a set of experiments to evaluate the performance of the stance and credibility based trust model. We focus on evaluating the proposed model using the real dataset. The set is a collection of 500 images uploaded to the social networks (e.g., flicker, twitter, google+) by multiple users. We have extracted their geo-tagged location, time, textual description and comments to create the services based on the images. Altogether, we collected 500 posts with the images in the period between November 2017 to December 2017. The number of comments per image varies from 27 to 300 with an average of 100 comments per image. The average number of likes and shares per image is 250. All the experiments are implemented in Java and MATLAB. All the experiments are conducted on a Windows 7 desktop with a 2.40 GHz Intel Core i5 processor and 8 GB RAM.

Human annotators manually labeled the candidate services as either trustworthy or not to create ground truth. The annotators made the decisions based on the service description, meta-data and facts through a Web search.

Table 1. Correctly classified social-sensor cloud services

	Trustworthy			unTrustworthy		
	Selected	True positive	True positive %age	Selected	True Positive	True positive % age
SVM	251	187	67.75362319	149	115	47.13114754
Stance	259	210	76.08695652	241	172	70.49180328
Stance+credibility	270	242	87.68115942	230	193	79.09836066

The conducted experiments evaluate the services' trustworthiness classification, the service users' credibility, and their accuracy. We study multiple quantitative metrics from the perspective of the precision and sensitivity, to analyze the behavior and effectiveness of the proposed model. We calculated the percentage of true positives to evaluate the accuracy of the trustworthiness of a service. We computed the root mean square error (RMSE) to measure how close the estimated trust-rate is to the ground truth. The lower RMSE shows, the more accurate trust-rate calculation is. Moreover, we analyzed and evaluated the sensitivity of trust-rate over the period using analysis of variance (ANOVA) matrix over time.

True Positive Classification. Accurate classification represents the ability to predict that a service can be trusted or not. The true positive percentage is used to compare the accuracy of the proposed model in predicting the trust-rates. We tested three different models to obtain a comprehensive understanding of the effectiveness of the proposed model. The first model is based on the classification of the trustworthy services using the traditional *SVM* classifier. *SVM* classifies the services based on their popularity, e.g., shares and likes, the number of comments, the number of negative and positive stance words using the trained SVM classifier. The other two models are (1) the stance and credibility based trust model and (2) the stance based trust model. The results (Table 1) depict the better performance and higher percentage of true positives for the stance and credibility based trust model.

Root Mean Square Error. We analyzed the trust estimation accuracy by comparing the RMSE of the stance and credibility based trust model and the stance based trust model. We calculated the average RMSE@N of the social-sensor cloud services with users between 0 to 200 based on the top N comments. We set four levels of N, i.e., $N \leq 50$, $N \leq 100$, $N \leq 150$ and $N \leq 200$ to identify the relationship between the services' trust-rate and the number of comments, i.e., users considered. As the number of users increases, the RMSE of the stance based trust gets higher (Graph 3). Noncredible users' stance is eliminated in the stance and credibility based trust model. Therefore, the stance and credibility based trust model gives overall lower and almost uniform RMSE, in comparison to the stance based trust model. Thus, the stance in conjunction with the credibility of the user helps to decrease the error rate in trust-rate calculation.

The SVM based trust model predicts trust-rate in binary values. Therefore, every squared error is 1, resulting in the mean squared error to be the miss-classification rate, and the RMSE is the square root of the miss-classification rate, i.e., 1 or 0. Therefore the SVM based trust model cannot be used for this comparison.

Analysis of Variance. Analysis of variance (ANOVA) matrix is used to analyze the sensitivity of trust-rate over the period. The change in variance is due to change in the stance of the users' over a specified period. The variance in trust-rate is based on the dynamics of trust over time. We analyzed the trust-rate variance by comparing the stance and credibility based trust model to the stance based trust model. The variance for the SVM based trust model fluctuates between absolute values of 0 and 1. Therefore, the SVM based trust model cannot be used for ANOVA. The users' stance and credibility over the specified period are considered to evaluate the variance in the trust-rate of trustworthy and untrustworthy services. We analyzed two sets of trustworthy and untrustworthy services. The variance in trust-rate of the both sets of services is calculated and updated over multiple time periods. We set the period lengths as 1, 2, ...20 units. We record and observe the changes in the variance over these periods. The ANOVA, i.e., the variance in the stance of the trustworthy and untrustworthy services for the both models, i.e., the stance and credibility based trust model and the stance based trust model is shown in Graph 4(a) and (b). The results depict the variance in trust-rate stabilizes and approaches zero more quickly for the stance and credibility based trust model. Moreover, the overall variance in trust-rate is also lower for the stance and credibility based model. The result shows that the proposed model is quicker to adapt to the changes in the users' trust in the service.

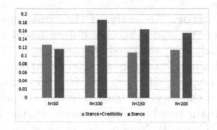

Graph 3. RMSE of correctly classified social-sensor cloud services

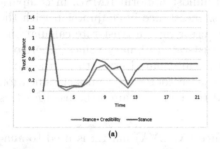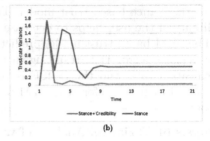

(a)   (b)

Graph 4. (a) Trust variance of trustworthy services (b) trust variance of untrustworthy services

5   Conclusion

We propose a stance and credibility based trust model to assess the trustworthiness of social sensor cloud services. The service users' stance towards another user and a social sensor cloud service is used to assess the credibility of the other user and the trustworthiness of the service. Our experiments demonstrate the effectiveness of the proposed model. We plan to focus on the trust enabled social-sensor clouds services composition.

Acknowledgement. This research was partly made possible by NPRP 9-224-1-049 grant from the Qatar National Research Fund (a member of The Qatar Foundation) and DP1501 00149 and LE180100158 grants from Australian Research Council. The statements made herein are solely the responsibility of the authors.

References

1. Rosi, A., Mamei, M., Zambonelli, et al.: Social sensors and pervasive services: approaches and perspectives. In: Proceedings of PERCOM (2011)
2. Aggarwal, C.C., Abdelzaher, T.: Social sensing. In: Aggarwal, C. (ed.) Managing and Mining Sensor Data, pp. 237–297. Springer, Heidelberg (2013). https://doi.org/10.1007/978-1-4614-6309-2_9
3. Aamir, T., Bouguettaya, A., Dong, H., et al.: Social-sensor cloud service selection. In: Proceedings of IEEE ICWS, pp. 508–515 (2017)

4. Aamir, T., Bouguettaya, A., Dong, H., Mistry, S., Erradi, A.: Social-sensor cloud service for scene reconstruction. In: Proceedings of ICSOC, pp. 37–52 (2017)
5. Popat, K., Kashyap, S., et al.: Where the truth lies: explaining the credibility of emerging claims on the web and social media. In: Proceedings of WWW, pp. 1003–1012 (2017)
6. Gupta, A., Lamba, H., Kumaraguru, P., Joshi, A.: Faking sandy: characterizing and identifying fake images on twitter during hurricane sandy. In: Proceedings of WWW, pp. 729–736 (2013)
7. Mendoza, M., Poblete, B., Castillo, C.: Twitter under crisis: can we trust what we RT?. In: Proceedings of the First Workshop on Social Media Analytics, pp. 71–79. ACM (2010)
8. Gupta, A., Kumaraguru, P., Castillo, C., Meier, P.: TweetCred: real-time credibility assessment of content on Twitter. In: Proceedings of SocInfo, pp. 228–243 (2014)
9. Aamir, T., Bouguettaya, A., Dong, H.: Trust in social-sensor cloud service. In: Proceedings of IEEE ICWS (2018)
10. Allcott, H., Gentzkow, M.: Social media and fake news in the 2016: election. J. Econ. Perspect. **31**, 211–236 (2016)
11. Schifferes, S., Newman, N., Thurman, N., Corney, D., Gker, A., Martin, C.: Identifying and verifying news through social media: developing a user-centered tool for professional journalists. Digit. Journalism **2**, 406–418 (2014)
12. Shu, K., Sliva, A., Wang, S., et al.: Fake news detection on social media: a data mining perspective. In: Proceedings of ACM SIGKDD, pp. 22–36 (2017)
13. Du, W., Lin, H., Sun, J., et al.: A new trust model for online social networks. In: Proceedings of IEEE ICCCI, pp. 300–304 (2016)
14. Lin, D.: An information-theoretic definition of similarity. In: Proceedings of ICML, pp. 296–304 (1998)
15. Malik, Z., Bouguettaya, A.: RATEWeb: reputation assessment for trust establishment among web services. VLDB J. **18**, 885–911 (2009)
16. Malik, Z., Bouguettaya, A.: Rater credibility assessment in web services interactions. In: Proceedings of WWW, pp. 3–25 (2009)
17. Liu, X., et al.: Ev-LCS: a system for the evolution of long-term composed services. IEEE Trans. Serv. Comput. **6**(1), 102–115 (2013)
18. Li, L., Liu, D., Bouguettaya, A.: Semantic based aspect-oriented programming for context-aware Web service composition. Inf. Syst. **36**(3), 551–564 (2011)
19. Aamir, T., Dong, H., Bouguettaya, A.: Social-sensor composition for scene analysis. In: Proceedings of ICSOC (2018)

Data Stream and Distributed Computing

StrDip: A Fast Data Stream Clustering Algorithm Using the Dip Test of Unimodality

Yonghong Luo, Ying Zhang$^{(\boxtimes)}$, Xiaoke Ding, Xiangrui Cai, Chunyao Song, and Xiaojie Yuan

College of Computer Science, Nankai University, Tianjin, China
{luoyh,zhangying,dingxk,caixr,songcy,yuanxj}@dbis.nankai.edu.cn

Abstract. Data stream clustering is an important problem of data mining. As the infinite growth of data stream's length, excessive data is making great troubles to the storage of data. A number of algorithms have been proposed for data stream clustering, such as CluStream, DenStream, DStream and StrAP. With the Big Data era's coming, the amount of data in one timestamp is growing at a great speed, so the time efficiency of data stream clustering algorithms is drawing huge attention from researchers while some state-of-the-art algorithms are excellent in cluster purity but intolerable in time efficiency. In this paper, we propose the **StrDip**, a fast data stream clustering algorithm which combines the Dip Test of Unimodality with the online/offline two-stage stream clustering framework. The **StrDip** also adapts a novel clustering feature vector and some *microcluster* pruning methods. Comparing to others algorithms, results of experiments on synthetic and real-world datasets show that, the **StrDip** gains a huge advantage in time efficiency and the clustering purity and quality are also good.

Keywords: Data stream · Stream clustering · Fast clustering · Dip

1 Introduction

Traditional data mining usually focuses on static datasets [16]. However, with the increasing amount of the generated data streams, data streams have attracted reseachers' huge attention [8]. A data stream is a continuous sequence of digital signals [3] and our world is filled with various data streams such as telephone records, network records, multimedia data and business logs. Data mining consists of a collection of technologies which can find hidden patterns from data. Data stream clustering has played a very important role in data mining technologies [23]. Data stream clustering algorithms can summarize information from data streams and transform it into an understandable structure for further using. In other words, clustering streams can track the evolution of various phenomena in medical, meteorological, astrophysical, and seismic studies.

© Springer Nature Switzerland AG 2018
H. Hacid et al. (Eds.): WISE 2018, LNCS 11234, pp. 193–208, 2018.
https://doi.org/10.1007/978-3-030-02925-8_14

When processing data streams, there are two features that we need to consider. The first one is the infinite length of data stream. As time goes by, the length of data stream is getting bigger and bigger. For this reason, we can't store all the data. We can only process this data for once while traditional data mining methods can process the data for many times [19]. The second one is the concept drift problem. The concept drift means the statistical property of the data stream randomly changes over time. So the model in last minute may not be applicable in next minute [21] and the old clusters may be transformed into noises as time goes by.

Many algorithms aiming at solving stream clustering problems have been proposed. Most of these algorithms first adopt traditional clustering approaches such as K-Means [27], DBSCAN [13], Affinity Propagation Clustering [14], Spectral Clustering [20] and Density Peaks based Clustering [25], then modify them to adapt to data stream clustering. Beyond that, many stream clustering algorithms also employ the online/offline two-stage stream clustering framework [5, 7]. The online part of the two-stage framework aims to summarize the statistical knowledge of recent points as time goes by. This summarization of the statistical knowledge is known as *microcluster*. The offline part of the two-stage framework takes advantage of the traditional clustering approaches and clusters the *microclusters* to get the final clusters.

With the Big Data era's coming, the number of points arrive in one timestamp increases drastically, preventing the state-of-the-art algorithms from rapidly processing of stream clustering. In order to achieve the goal of immediate processing of stream clustering, the **StrDip** is proposed in this paper. The **StrDip** is a fast data stream clustering algorithm which combines the SkinnyDip [18] with the two-stage stream clustering framework. The **StrDip**'s features are described in the following paragraph:

First, we adopt the classic online/offline two-stage stream clustering framework in **StrDip**. The online part introduces a novel Clustering Feature (CF) vector to maintain the statistical information of the past data points. By only storing the brief information of the past data points, we solve the infinite length problem of the data streams. The concepts of *normal microcluster* and *noise microcluster* are designed to distinguish normal data points and noise data points. Some pruning approaches are introduced to limit the number of *microclusters*. The pruning approaches make the effective handling of concept drift problem practical. Second, the proposed algorithm utilizes and modifies the SkinnyDip [18] which is based on the Dip Test of Unimodality [17]. It is a clustering algorithm that has a practically *linear* run-time growth in data size n. We find that this fast clustering algorithm can be adopted in the offline part of the two-stage stream clustering framework. The offline part uses the weighted SkinnyDip to cluster the *normal microclusters* and then generates the final results. By appropriately limiting the number of *microclusters*, we accelerate the online processing part of the continuous data stream. By employing the weighted SkinnyDip to cluster the *normal microclusters*, we also get an additional advantage in **StrDip**'s time efficiency.

The contributions of this paper are:

1. We propose a novel CF vector which maintains not only the statistical information of the points' attributes, the statistical information of the *microclusters*' timestamps, but also the weighted statistical information of the *microclusters*.
2. We effectively reduce the number of *microclusters* by employing some pruning approaches of *microclusters* and deleting the overdue *microclusters*. In this way, we speed up the processing of the data streams in the online stage and handle the problem of concept drift successfully. And by taking the advantage of the SkinnyDip in the offline stage, we get fast and precise stream clustering results finally.
3. Our performance evaluation on a number of real-world and synthetic datasets show that the **StrDip** gains a huge advantage in time efficiency while the clustering purity and quality are also good.

2   Related Work

In order to cluster the data streams with the difficulties of the infinite length of data streams and the concept drift problem, researchers have proposed many data stream clustering algorithms in recent years. These algorithms can be classified into **divide-and-conquer** based algorithms, **online/offline two-stage** stream clustering framework based algorithms, **grid** based approaches, **probabilistic** and **spectral** based methods.

The **divide-and-conquer** based methods. This type of method takes advantage of the divide-and-conquer strategy, and the very first work is STREAM [15]. This method first uses K-Means to find exemplars and then clusters these exemplars to get the final clusters by K-Means again. Later, Callaghan et al. present a K-Median based algorithm called LOCALSEARCH [22] that uses local search techniques. Then they give theoretical justification for its success, and present experimental results to show how it out-performs K-Means based algorithms.

The **online/offline two-stage** stream clustering framework based methods. This type of method is divided into an online component which periodically stores detailed summary statistics and an offline component which uses these summary statistics. The summary statistics which are also called CF vectors (or *microclusters*) contain each dimension's sum of the squares of the data values, sum of the data values, sum of the squares of the timestamps, sum of the timestamps and number of data points. CluStream [1] is an opening stream clustering algorithm that uses the CF vector structure and the two-stage stream clustering framework. This algorithm also makes use of a pyramidal time frame to store CF vectors. In order to solve the problem of arbitrary clusters and outliers, Cao et al. propose a DBSCAN [13] based stream clustering method: DenStream [5]. DenStream introduces the definitions of *core-microcluster*, *outlier-microcluster* and *density area* corresponding to DBSCAN's *core-object*, *outlier-object* and *density-reachable area* respectively. DenStream also applies a fading function which implicit decreases the weights of old points.

The **grid** based stream clustering algorithms. Such algorithms divide the spatial area into rectangle cells. DStream [7] is an example of the grid based stream clustering method. DStream also makes use of the online/offline two-stage clustering framework and a fading function. The online component maps each input data point into a grid and the offline component tries to cluster the dense grids and remove the sparse grids. Exclusive and Complete Clustering (ExCC) algorithm [4] is also a grid based stream clustering algorithm which emploies a synopsis structure to consolidate incoming data points.

The **probabilistic** and **spectral** based algorithms. PStream [10] proposes the concepts of "strong cluster", "transitional clusters" and "weak cluster" and uses probabilistic knowledge to find clusters. SSC [28] is a streaming spectral clustering algorithm which maintains an approximation of the normalized Laplacian of the data stream over time and efficiently updates the changing eigenvectors of this Laplacian in a streaming fashion. Besides, there are researchers who employ the Affinity Propagation (AP) [14] algorithm to find the best representative of one cluster. This algorithm is called StrAP [29]. Extensions of those all above works including applications [11,14,28,29]. Even though these algorithms have gained great achievements in clustering purity and quality, they run slowly. In other words, the data volume these algorithms process in a unit time is relatively less.

3   The Stream Clusering Algorithm Using the Dip Test of Unimodality (StrDip)

This section shows the processing procedures of the **StrDip**. We begin with the definition of the data stream clustering problem. Then we show the general process of the **StrDip**. We combine the online/offline two-stage stream clustering framework with the SkinnyDip. In Subsect. 3.1, we show the details of the online part. Subsection 3.2 demonstrates the offline part (i.e. the weighted SkinnyDip) of the framework. The data stream clustering problem is defined as follows:

Definition 3.1 *(Data stream clustering problem). A data stream is a continuous sequence of points* $X = \{X_0, X_1, X_2, ..., X_n, ...\}$. *Points' arriving timestamps* $T = \{T_0, T_1, T_2, ..., T_n, ...\}$, *where* $T_0 \leq T_1 \leq \cdots$. *The speed* v *means how many points arrive in a timestamp unit. In other words, points* $X_{v*i}, ..., X_{v*(i+1)-1}$ *are collected by stream clustering algorithm at the timestamp of* T_i. *The* **clients** *or* **users** *require to get stream clustering results at the timestamp of* $\{T_{k0}, T_{k1}, T_{k2}, ..., T_{ki}, ...\}$, *where* T_{ki} *is a member of* T.

The overall processing steps of the **StrDip** is shown in Fig. 1 and Algorithm 1. Consider lines 1–4 of Algorithm 1, we initialize the parameters. Then we use SkinnyDip to get the first batch of *microclusters*. Lines 6–18 are the processing procedures for the online stage. As long as the data stream does not end, we verify whether each incoming point belongs to a certain *microcluster* (line 8). If the answer is yes, then we add this point into the corresponding *microcluster* (lines 9–11); otherwise, we create a new *microcluster* (lines 12–14). We adjust

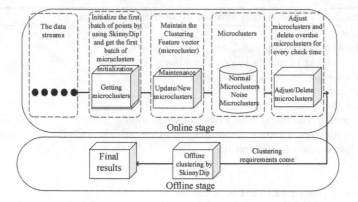

Fig. 1. The general processing steps of the StrDip

all *microclusters'* types for every certain time so that *noise microclusters* and *normal microclusters* can be transformed into each other. At the same time, we also delete the overdue *microclusters* permanently (lines 16 and 17). The above process is for the online period. When clustering requirement comes, we use the wiehgted SkinnyDip to cluster all *normal microclusters* (lines 19–21), this period is called the offline stage.

3.1   The Online Part of the StrDip

Consider Algorithm 1, we first initialize the timestamp to $t = 0$ (line 1 of Algorithm 1). After the number of *initialLength* points have been collected, we employ SkinnyDip (showed in Subsect. 3.2) to find the first batch of *microclusters*. Then we create CF vectors for these *microclusters* (lines 3 and 4). Before demonstrating the CF vector, we first show the *weight* of one point.

In this paper, we use the damped window [24] to record the *weight* of one point for the reason that the old point is getting less and less important as time goes by. The damped window means the *weight* of one point decreases in an exponential way.

Definition 3.2 *(Weight of point). The weight of one point is:* $\omega(t) = 2^{-\lambda t}$ *where $\lambda > 0$ and t means the time difference between current timestamp $t_{current}$ and the arriving time of the point $t_{arrival}$, i.e., $t = t_{current} - t_{arrival}$.*

We can know that when $t_{current}$ is equal to $t_{arrival}$, the *weight* of that point is 1. As we know, the stream *speed* v means how many points will arrive in one timestamp, then the total *weight* of the stream is:

$$W = v \sum\nolimits_{t=0}^{t_{current}} \omega(t) = v \sum\nolimits_{t=0}^{t_{current}} 2^{-\lambda t} = \frac{v}{1 - 2^{-\lambda}}, \text{where} \, t_{current} \to \infty.$$

Next, we demonstrate the CF vector. Based on CluStream's CF vector [2], our novel CF vector prefers to the weighted statistical information of the *microcluster* and the *weight* of the *microcluster*.

Algorithm 1 StrDip: A fast stream clustering algorithm

Input: A data stream;
        The first batch of points' size: $initialLength$;
        The number of points in one timestamp, v.
Output: final clusters of the data stream.
 1: timestamp $t=0$
 2: calculate $checktime$
 3: Use SkinnyDip to cluster first $initialLength$ points
 4: For every clusters, create CF vectors, and they are also called $microcluster$s
 5: **while** data stream is not end **do**
 6:    $t++$
 7:    $p_t \leftarrow \{X_{v*t}, ..., X_{v*(t+1)-1}\}$
 8:    **for** each point $nowPoint$ in p_t **do**
 9:        **if** $nowPoint$ belongs to one microcluster mc **then**
10:            add $nowPoint$ into mc, update mc's CF vector
11:        **end if**
12:        **if** $nowPoint$ doesn't belong to any $microcluster$ **then**
13:            create a new $microcluster$ and its CF vector
14:        **end if**
15:    **end for**
16:    **if** $t\%checktime == 0$ **then**
17:        adjust $microclusters$' types and adopt pruning strategies
18:    **end if**
19:    **if** clustering requirement comes **then**
20:        apply SkinnyDip to all $normal\ microclusters$
21:    **end if**
22: **end while**

Definition 3.3 *(Clustering Feature Vector). A microcluster contains n points with d dimensions, they are $\{x_0, x_1, ..., x_{n-1}\}$. Their arriving timestamps are $\{t_0, t_1, ..., t_{n-1}\}$. Clustering Feature vector is defined as $(\overrightarrow{CFx^2}, \overrightarrow{CFx^1}, \overline{CFt^2}, \overline{CFt^1}, t_l, weight, n, t_0)$, where $\overrightarrow{CFx^2}$ and $\overrightarrow{CFx^1}$ are d-dimensional vectors.*

- $\overrightarrow{CFx^2}$ *is a d-dimensional vector, and in every dimension, the value of $\overrightarrow{CFx^2}$ is the weighted sum of the squares of the data values in the corresponding dimension of the points in microcluster. The p-th dimension of $\overrightarrow{CFx^2}$ is:*
  $$\overrightarrow{CFx_p^2} = \sum_{i=0}^{n-1} w\,(t_{current} - t_i)\,x_{ip}^2$$
- $\overrightarrow{CFx^1}$ *is a d-dimensional vector too, and in every dimension, the value of $\overrightarrow{CFx^1}$ is the weighted sum of the values in the corresponding dimension of the points in microcluster. The p-th dimension of $\overrightarrow{CFx^1}$ is:*
  $$\overrightarrow{CFx_p^1} = \sum_{i=0}^{n-1} \omega\,(t_{current} - t_i)\,x_{ip}$$
- $\overline{CFt^2}$ *is the sum of the squares of the points' arriving timestamps.*
- $\overline{CFt^1}$ *is the sum of the points' arriving timestamps.*
- t_l *is the microcluster's last arrived point's timestamp.*

- *weight is the sum of the weights of the points in the microcluster.*
  $weight = \sum_{i=0}^{n-1} \omega\,(t_{current} - t_i)$
- *n is the number of points in the microcluster.*
- *t_0 is the creation time of this microcluster, i.e., the first point's arriving times-tamp of this microcluster.*

The CF vector records the statistical information of one *microcluster* by which we can get its radius and approximate distribution of the arriving time.

According to Definition 3.3, suppose there is no point be absorbed into one microcluster mc until t_c, and mc received the last point in t_l. Then the *weight* of mc will be updated in the following way:

$$\omega\,(mc, t_c) = 2^{-\lambda(t_c - t_1)}\omega\,(mc, t_l) + 1.$$

By this way, we can update the new *weight*, new $\overrightarrow{CFx^2}$ and new $\overrightarrow{CFx^1}$ of one *microcluster* quickly.

After getting the *weights* of *microclusters*, we can classify the *microclusters* into *noise microclusters* and *normal microclusters*. *Normal microcluster*'s *weight* must be equal to or larger than μ, and *noise microcluster*'weight should be smaller than μ.

It is worth noting that *noise microclusters* and *normal microclusters* can be transformed into each other frequently. After getting the statistical information of the *microcluster*, we define the *microcluster*'s center and radius as follows:

Definition 3.4 *(Microcluster center and radius). Microcluster's center and radius*

can be defined as follows: $c(center) = \frac{\overrightarrow{CFx^1}}{\omega}$, $r(radius) = \sqrt{\frac{|\overrightarrow{CFx^2}|}{\omega} - \left(\frac{|\overrightarrow{CFx^1}|}{\omega}\right)^2}$

Let's consider lines 7–15 of the Algorithm 1. After getting the first group of *micruclusters*' CF vectors, when a new point comes, we traverse all the *microclusters*. If the new point is close enough to one *microcluster*, then it will be absorbed into this *microcluster* (lines 9–11). The extent of "close enough" is measured by the boundary factor β. It means that if one point falls within the bound of βr ($\beta > 1$), then it will be merged into this *microcluster*, and r is this *microcluster*'s radius. We also update the CF vector of this *microcluster*. If the new point keeps distance from all *microclusters*, we create a new *noise microcluster* and its CF vector (lines 12–14). The radius of this new *noise microcluster* is the smallest distance between any two *microclusters*.

As time goes by, the distribution of the data stream may change. Some *normal microclusters* may accept no new point for a long time and be transformed into *noise microclusters*. This phenomenon is called by concept drift. So we should check the *weights* of *microclusters* every certain time. Suppose that one *normal microcluster*'s original *weight* is just precise μ and after T_g timestamps, this *microcluster* become a *noise microcluster*. Then, we have: $2^{-\lambda T_g}\mu + 1 < \mu$. The solution of above inequality is:

$$T_g > \lambda^{-1}\log\mu/(\mu - 1).$$

Therefore, we check the *microclusters' weights* and update *microclusters'* types for every T_g timestamps. This strategy ensures that we don't miss any conversion from *normal microcluster* to *noise microcluster*.

To save the **StrDip**'s running time, we have employed two pruning strategies and applied them every T_g time (lines 16–18 of Algorithm 1). The first pruning strategy is based on the idea that the *microcluster* with small enough *weight* should be deleted [5]. The second pruning strategy is based on the idea that the *microcluster* accepts no new points for a long time should be deleted too [2].

The method to measure whether *weight* is small enough [5] is showed as:

$$\theta\left(t_n, t_0\right) = \frac{2^{-\lambda(t_n - t_0 + T_g)} - 1}{2^{-\lambda T_g} - 1}$$

where t_n is the current timestamp, t_0 is the create time of this *microcluster*. $\theta\left(t_n, t_0\right)$ is the threshold that determines whether one *microcluster* should be deleted. If one *microcluster* mc_0's *weight* is smaller than θ, mc_0 will be deleted. We can see that θ is an increasing function as t_n increases, and the limit of θ is μ. This means that we won't delete any *normal microclusters*.

When using this pruning strategy, we will introduce some errors in *weight*. To be specific, the *weight* after the pruning will be smaller than the true *weight* of the *microcluster*. But we can ensure that the deleted *microclusters'* weight will be always smaller than μ. In other words, we only delete a part of *noise microclusters* and don't delete any *normal microcluster* [5].

The second pruning strategy is to delete the *microclusters* that accept no new point for a long time. The simplest way to detect which *microcluster* should be deleted is to calculate the average arriving time of one *microcluster*'s last 25% points and delete the earliest *microcluster*.

We use an estimation method to approximate the average arriving time of one *microcluster*'s last 25% points. Consider the CF vector, we can see that $\overline{CFt^2}$ stores the sum of the squares of the points' arriving timestamps, $\overline{CFt^1}$ stores the sum of the points' arriving timestamps and n stores the number of points of one *microcluster*. By using these parameters, we can calculate the mean value $\rho = \frac{\overline{CFt^1}}{n}$ and the standard deviation $\sigma = \sqrt{\frac{\overline{CFt^2}}{n} - \left(\frac{\overline{CFt^1}}{n}\right)^2}$ of one *microcluster*'s arriving timestamps.

Suppose that the distribution of the arriving time of one *microcluster* obeys normal distribution, $T \sim N(\mu, \sigma^2)$, then the average arriving time of last 25% points is $0.68 * \sigma + \rho$, according to the standard normal distribution table [12]. If the time difference between the present time and the last 25% arriving time is bigger than Δ, we will delete this *microcluster* for it may be an outdated *microcluster* in a significant probability.

3.2   The Offline Part of the StrDip

The offline part of the **StrDip** is based on the weighted SkinnyDip [18]. After the online stage of the two-stage stream clustering framework, we get a number of *microclusters*. We treat these *microclusters* as virtual points whose locations are the *microclusters'* centers and apply the weighted SkinnyDip algorithm to these virtual points to get the final clusters. When adopting the weighted SkinnyDip, we gennerate the same "n" virtual points where "n" is equal to the *weight* of the corresponding *microcluster*.

The Dip Test of Unimodality [17] is a statistical test for measuring the unimodality of a random variable and SkinnyDip applies The Dip Test of Unimodality to each dimension of a high dimensional vector, then combines every dimension's result to get the final cluster results.

Figure 2(a) is a two-dimensional example dataset, where there are five clusters with some noise points. In Fig. 2(b), the algorithm gets the cumulative distribution function (CDF) of the x-axis projection of the example dataset. It's easy to see that every cluster's gradient is a "concave-convex-concave" shape in CDF. For the segment of **B-E**, the gradient of segment **B-C** is relatively small, the gradient of segment **C-D** is relatively big and **D-E**'s gradient is also small, then the segment of **C-D** is a good cluster. In other words, the segment of **B-E** is unimodal because there is only one cluster in this segment.

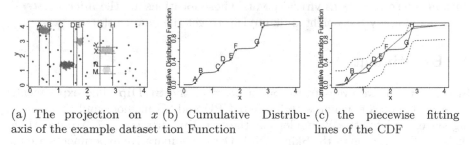

(a) The projection on x axis of the example dataset

(b) Cumulative Distribution Function

(c) the piecewise fitting lines of the CDF

Fig. 2. Steps of the dip test of unimodality (1. Get cumulative distribution function of the dataset. 2. Get piecewise fitting of the cumulative distribution function. 3. Use hypothesis test to evaluate the unimodality).

In order to evaluate the unimodality of the CDF, the algorithm uses a piecewise fitting function to fit the line into a "concave-convex-concave" shape and applies the hypothesis test to evaluate how the line deviates the "concave-convex-concave" model. In Fig. 2(c), there are two dotted lines which are copies of CDF. They are generated by moving CDF up d distance and moving CDF down d distance respectively. The d is the minimum shifting distance of CDF that makes the piecewise fitting line, i.e., the red line of **A-H**, not act against the "concave-convex-concave" model (if d decreases, the red line of **A-H** will get out of the area of the two copies of CDF). The d, in other words, is the "dip".

In hypothesis test, the algorithm sets $H_0 =$ "CDF is unimodal", and compares d with certain unimodal distribution [17], then get the *p-value*. The *p-value* means the probability of d's occurrence under the condition of "CDF is unimodal". If p is smaller than the threshold α (usually 0.01), we can say that CDF is not unimodal (or multimodal).

The Dip Test of Unimodality can not only judge whether a distribution is unimodal, but also get the max gradient area. In our example, it's the segment of **A-H**. By iteratively using the Dip Test of Unimodality, the algorithm can get all the clusters in x dimension, which are segment **A-B**, segment **C-D**, segment **E-F** and segment **G-H**.

After applying the Dip Test of Unimodality to find clusters in one dimension, now, the problem is how to process the multidimensional data. SkinnyDip [18] applies The Dip Test of Unimodality to each dimension of a high dimensional vector, then combines every dimension's result to get the final cluster results. For example, when the Dip Test of Unimodality find 4 clusters in x-axis direction, consider Fig. 2(a), the cluster **G-H** in x-axis direction actually contains two clusters, which are $[X, Y]$ and $[M, N]$, from which the SkinnyDip constructs two hyperclusters.

The Dip Test of Unimodality is fast. In the worst case, the time complexity of this algorithm is $O(n)$ [17]. For the purpose of immediate processing of stream clustering, the Dip Test of Unimodality with characteristic of rapid processing is a suitable solution.

In the offline part of the two-stage stream clustering framework, we treat *normal microclusters* as virtual points whose locations are the *microclusters'* centers and apply the weighted SkinnyDip to get the final clusters.

4   Experiments

In this section, we evaluate various aspects of the **StrDip**. The experiments are executed on a machine with 3.4 GHz CPU and 8G memory and the operating system is Ubuntu. We implement the **StrDip** with Java 1.7. We use Maurus's implementation of the SkinnyDip[1]. Other comparative experiments are all implemented by Java 1.7. The comparative algorithms are CluStream [1], DenStream [5] and StrAP [29]. **These comparative algorithms' parameters are exactly same as their papers**.

4.1   Datasets and Measurement Methods

We use one synthetic dataset D_s and one real-world dataset to evaluate the **StrDip**'s clustering quality, clustering purity, time performance and the ability to detect concept drift. The synthetic dataset D_s is made up of three small datasets with different distributions: D_1 (Fig. 6(a)), D_2 (Fig. 6(b)) and D_3 (Fig. 6(c)). Each of them contains 400 K points with 0.1% noise. We generate the synthetic dataset D_s by randomly taking 1 K points per timestamp from D_1 when the range of timestamp is 0–400, taking points from D_2 when timestamp is between 400 to 800 and taking points from D_3 when timestamp is between 800 to 1200. D_s's distribution is identical with D_1's, D_2's and D_3's at different stages. In other words, D_s's statistical property changes over time so that D_s meets with concept drift at the timestamp of 400 and 800.

---

[1] Available at the following website: github.com/samhelmholtz/skinny-dip.

The real-world dataset is KDD CUP'99 Network Intrusion Detection dataset [9] which has been used earlier [1,5–7,10,26] to evaluate the stream clustering quality and purity. The KDD CUP'99 dataset has 34 continuous attributes out of the total 42 attributes. KDD CUP'99 dataset is a series of TCP connection records from two weeks of LAN network traffic managed by MIT Lincoln Labs. Each record belongs to one of the five clusters, they are Normal connection, DOS attaction, R2L attaction, U2R attaction and PROBING attaction. Just like [5], we only use the 34 continuous attributes to cluster the data stream.

The clustering algorithm is evaluated by quality, i.e., sum of the squared distances (SSQ), purity and time efficiency. The SSQ in a pre-defined horizon (H) from current time is defined as:

$$SSQ = \sum d\,(p_i, c_i)$$

where p_i is the points in horizon H, c_i is the neareast cluster center of p_i and $d\,(p_i, c_i)$ is the Eulerian distance between p_i and c_i. For a clustering algorithm, the smaller the SSQ, the better the clustering quality is.

The clustering purity is the percent of the total number of data points that are classified correctly. It is defined as follows:

$$purity = \frac{1}{k} \sum_{i=1}^{k} \frac{p_i^{dom}}{p_i} \cdot 100\%$$

where k is the number of discovered clusters, p_i denotes the number of points in cluster i. And p_i^{dom} denotes the number of points with the dominant class label in cluster i. Our goal is to maximize the clustering purity as much as possible.

Since the data stream's distribution is in change forever, we only compute the purity and SSQ in a pre-defined horizon (H) from current time. Our parameters are: $H = 5$, $\beta = 3$, $\mu = 2$, $\lambda = 1$, $\Delta = 25$, $initialLength = 2000$, $v = 1000$. We empirically find that the algorithm's performance is robust to these parameters.

4.2   Clustering Quality and Purity

This subsection mainly evaluates the **StrDip**'s clustering quality and purity. The comparative algorithms are CluStream, DenStream and StrAP. The dataset is KDD CUP-99's dataset. The comparison results are shown in Fig. 3(a) and (b), Fig. 4(a) and (b). Consider Fig. 3(a) and (b), we can see that when processing the complete dataset of KDD CUP-99's data, the average SSQ of the **StrDip** is about four orders of magnitude and is obviously smaller than CluStream, DenStream and StrAP. When handling the incomplete dataset of KDD CUP-99' data (only using 10 dimensions), in all instances, the SSQs of the **StrDip** are always smaller than or equal to other algorithms. Let's consider Fig. 4(a) and (b). These two figures show the comparison of the clustering purity on KDD CUP-99's dataset. We can learn that the average purity of the **StrDip** is 1 when the dataset is complete while other algorithms' average purities are smaller than or equal to the **StrDip**, especially CluStream which is 0.9881. When the dataset is incomplete (10 dimensions), the average purity of the **StrDip** is also 1.

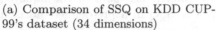

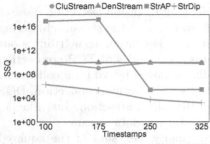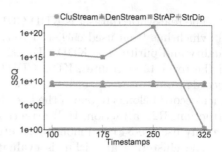

(a) Comparison of SSQ on KDD CUP-99's dataset (34 dimensions)

(b) Comparison of SSQ on KDD CUP-99's dataset (10 dimensions)

Fig. 3. Comparison of SSQ on KDD CUP-99's dataset, horizon = 5, speed = 1000, initialLength = 2000. The smaller the SSQ, the better the clustering quality is. The **StrDip**'s SSQs (the purple lines) are always smaller than the comparative algorithms in different dimensions. (Color figure online)

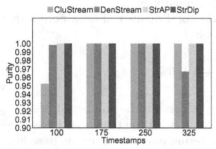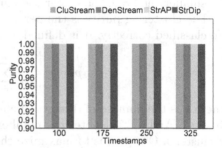

(a) Comparison of purity on KDD CUP-99's dataset (34 dimensions)

(b) Comparison of purity on KDD CUP-99's dataset (10 dimensions)

Fig. 4. Comparison of purity on KDD CUP-99's dataset, horizon = 5, speed = 1000, initialLength = 2000. Our goal is to maximize the clustering purity as much as possible. The **StrDip**'s purities (the dark blue, i.e., the last histograms) are always 1 and are bigger than others algorithms in different dimensions. (Color figure online)

4.3   Time Efficiency

We compare the **StrDip**'s time efficiency with CluStream, DenStream and StrAP. All the algorithms are tested on different data sizes of the KDD CUP-99's dataset with the same stream *speed*. Consider Fig. 5(a), the **StrDip** is about **4** times faster than DenStream, **10** times faster than CluStream and **14** times faster than StrAP. The StrAP's relative large execution time is caused by the AP (Affinity Propagation) algorithm's high time complexity.

Figure 5(b) is the experimental results excuted on different dimensions of the KDD CUP-99's dataset. We can see that the **StrDip**'s time efficiency is always superior to other comparison algorithms, especially CluStream and StrAP, whose

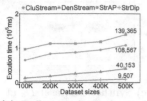
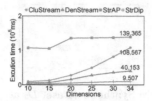
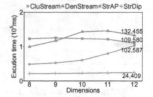

(a) Different data sizes of the KDD CUP-99's dataset   (b) Different dimensions of the KDD CUP-99's dataset   (c) Different dimensions of the synthetic dataset

Fig. 5. Comparison of excution time on different datasets, $speed = 1000$. All the experimental results including different datasets, different data sizes and different dimensions show that the **StrDip** (the purple lines) is about average **5** times faster than the DenStream, average **6** times faster than CluStream and average **11** times faster than StrAP. (Color figure online)

execution time is about **11** to **13** times of the **StrDip**'s when the dimension of the KDD CUP-99's dataset changes.

We also compare the **StrDip**'s execution time with the synthetic dataset. Consider Fig. 5(c), as the dimension of the dataset grows, the execution time of **StrDip** grows linearly too. And the results show that the **StrDip** is about **5** times faster than DenStream, **4** times faster than CluStream and **4** times faster than StrAP.

We can see that as the data size and dimension grows, the **StrDip**'s excution time grows linearly. This is because the *microcluster*'s pruning strategies can delete part of the *noise microclusters* and overdue *microclusters* permanently. In this way, the number of *microclusters* is limited so that the most time-consuming action of searching neareast *microcluster* won't cost so much time. And the **StrDip**'s linear growth of execution time can also be explained partly by the $O(n)$ time complexity of the Dip Test of Unimodality.

4.4 Detection of Concept Drift

In order to test the **StrDip**'s ability of handling concept drift, we generate the synthetic dataset D_s by composing D_1, D_2 and D_3 which are described in Fig. 6(a), (b) and (c) respectively. D_1 contains five clusters and three of them obey Gaussian distribution with standard deviation of 0.1 and the other clusters follow uniform distribution with range of 0.15. The distributions of D_2 and D_3 are similar to D_1. D_s's statistical property changes over time so that D_s will meet with concept drift at the timestamp of 400 and 800.

Figure 6(d)–(f) are the clustering results at the timestamp of 398, 798 and 1198. The circles in Fig. 6(a)–(c) represent raw points in the dataset, and the circles in Fig. 6(d)–(f) represent *normal microclusters*. We can see that the **StrDip** can successfully detect concept drift and get correct clusters with the disturbation of noise. The shapes and locations of final clusters are same as D_s's clusters. The only difference between the final clusters and the original

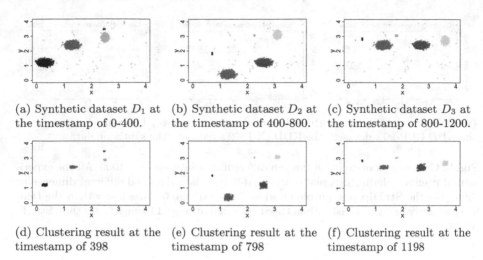

(a) Synthetic dataset D_1 at the timestamp of 0-400.

(b) Synthetic dataset D_2 at the timestamp of 400-800.

(c) Synthetic dataset D_3 at the timestamp of 800-1200.

(d) Clustering result at the timestamp of 398

(e) Clustering result at the timestamp of 798

(f) Clustering result at the timestamp of 1198

Fig. 6. The synthetic dataset D_s and the clustering results on D_s when timestamps are 398, 798 and 1190. Stream *speed* v is 1000. D_s consists of three different small synthetic datasets: D_1, D_2 and D_3. D_s's statistical property changes over time so that D_s will meet with concept drift at the timestamp of 400 and 800.

clusters is the clusters' sizes. The sizes of the final clusters are usually smaller than the original clusters' sizes, this is because one circle in results stands for one *microcluster* which may contains many raw points. And the results' circles' locations are corresponding to *microclusters*' centers' locations, this is why there are few circles in final results and the final clusters' sizes are smaller than the raw clusters' sizes.

5   Conclusion

In this paper, we have proposed the **StrDip**, a new fast stream clustering algorithm which combines the SkinnyDip with the online/offline two-stage stream clustering framework. The algorithm first gets some initial *microclusters* and stores these *microclusters*' feature vectors, then maps the new come points into *normal* and *noise microclusters*. After the pruning of *microlcusters*, when clustering requirements come, we apply the weighted SkinnyDip to *normal microclusters*, then we get the final clusters. By appropriately limiting the number of *microclusters*, the **StrDip** accelerates the online processing part of the continuous data stream. And by employing the wieghted SkinnyDip to cluster the *normal microclusters*, the **StrDip** also gets an additional advantage in time efficiency. The experiments show that the **StrDip** is about **5–11** times faster than the comparative algorithms and the clustering quality and purity are also the best one. This algorithm makes the immediate processing of stream clustering feasible without depressing the clustering quality and purity.

Acknowledgements. This research is supported by National Natural Science Foundation of China (No. 61772289), Natural Science Foundation of Tianjin (No. 17JCQNJC00200) and Fundamental Research Funds for the Central Universities.

References

1. Aggarwal, C.C., Han, J., Wang, J., Yu, P.S.: A framework for clustering evolving data streams. In: Proceedings of the 29th International Conference on Very Large Data Bases-Volume 29, pp. 81–92. VLDB Endowment (2003)
2. Aggarwal, C.C., Han, J., Wang, J., Yu, P.S.: A framework for projected clustering of high dimensional data streams. In: Proceedings of the Thirtieth International Conference on Very Large Data Bases-Volume 30, pp. 852–863. VLDB Endowment (2004)
3. Arasu, A., et al.: STREAM: the stanford data stream management system. Data Stream Management. DSA, pp. 317–336. Springer, Heidelberg (2016). https://doi.org/10.1007/978-3-540-28608-0_16
4. Bhatnagar, V., Kaur, S., Chakravarthy, S.: Clustering data streams using grid-based synopsis. Knowl. Inf. Syst. **41**(1), 127–152 (2014)
5. Cao, F., Estert, M., Qian, W., Zhou, A.: Density-based clustering over an evolving data stream with noise. In: Proceedings of the 2006 SIAM International Conference on Data Mining, pp. 328–339. SIAM (2006)
6. Chen, J.Y., He, H.H.: A fast density-based data stream clustering algorithm with cluster centers self-determined for mixed data. Inf. Sci. **345**, 271–293 (2016)
7. Chen, Y., Tu, L.: Density-based clustering for real-time stream data. In: Proceedings of the 13th ACM SIGKDD International Conference on Knowledge Discovery and Data Mining, pp. 133–142. ACM (2007)
8. Cugola, G., Margara, A.: Processing flows of information: from data stream to complex event processing. ACM Comput. Surv. (CSUR) **44**(3), 15 (2012)
9. Cup, K.: Dataset. available at the following website (1999). http://kdd.ics.uci.edu/databases/kddcup99/kddcup99.html
10. Dai, D.B., Zhao, G., Sun, S.L.: Effective clustering algorithm for probabilistic data stream. J. Softw. **20**(5), 1313–1328 (2009)
11. De Francisci Morales, G., Bifet, A., Khan, L., Gama, J., Fan, W.: IoT big data stream mining. In: Proceedings of the 22nd ACM SIGKDD International Conference on Knowledge Discovery and Data Mining, pp. 2119–2120. ACM (2016)
12. Dixon, W.J., Massey Frank, J.: Introduction To Statistical Analsis. McGraw-Hill Book Company Inc., New York (1950)
13. Ester, M., Kriegel, H.P., Sander, J., Xu, X., et al.: A density-based algorithm for discovering clusters in large spatial databases with noise. KDD **96**, 226–231 (1996)
14. Frey, B.J., Dueck, D.: Clustering by passing messages between data points. Science **315**(5814), 972–976 (2007)
15. Guha, S., Mishra, N., Motwani, R., O'Callaghan, L.: Clustering data streams (2000)
16. Han, J., Pei, J., Kamber, M.: Data Mining: Concepts and Techniques. Elsevier, Amsterdam (2011)
17. Hartigan, J.A., Hartigan, P.: The dip test of unimodality. Ann. Stat. **13**, 70–84 (1985)
18. Maurus, S., Plant, C.: Skinny-dip: clustering in a sea of noise. In: Proceedings of the 22nd ACM SIGKDD International Conference on Knowledge Discovery and Data Mining, pp. 1055–1064. ACM (2016)

19. Namiot, D.: On big data stream processing. Int. J. Open Inf. Technol. **3**(8), 48–51 (2015)
20. Ng, A.Y., Jordan, M.I., Weiss, Y.: On spectral clustering: analysis and an algorithm. In: Advances in Neural Information Processing Systems, pp. 849–856 (2002)
21. Nguyen, D.T., Jung, J.J.: Real-time event detection on social data stream. Mob. Netw. Appl. **20**(4), 475–486 (2015)
22. O'callaghan, L., Mishra, N., Meyerson, A., Guha, S., Motwani, R.: Streaming-data algorithms for high-quality clustering. In: Proceedings of 18th International Conference on Data Engineering, pp. 685–694. IEEE (2002)
23. Pietruczuk, L., Rutkowski, L., Jaworski, M., Duda, P.: How to adjust an ensemble size in stream data mining? Inf. Sci. **381**, 46–54 (2017)
24. Pramod, S., Vyas, O.: Data stream mining: a review on windowing approach. Glob. J. Comput. Sci. Technol. Softw. Data Eng. **12**(11), 26–30 (2012)
25. Rodriguez, A., Laio, A.: Clustering by fast search and find of density peaks. Science **344**(6191), 1492–1496 (2014)
26. Tavallaee, M., Bagheri, E., Lu, W., Ghorbani, A.A.: A detailed analysis of the KDD cup 99 data set. In: IEEE Symposium on Computational Intelligence for Security and Defense Applications, CISDA 2009, pp. 1–6. IEEE (2009)
27. Wagstaff, K., Cardie, C., Rogers, S., Schrödl, S., et al.: Constrained k-means clustering with background knowledge. ICML **1**, 577–584 (2001)
28. Yoo, S., Huang, H., Kasiviswanathan, S.P.: Streaming spectral clustering. In: 2016 IEEE 32nd International Conference on Data Engineering (ICDE), pp. 637–648. IEEE (2016)
29. Zhang, X., Furtlehner, C., Germain-Renaud, C., Sebag, M.: Data stream clustering with affinity propagation. IEEE Trans. Knowl. Data Eng. **26**(7), 1644–1656 (2014)

Classification and Annotation of Open Internet of Things Datastreams

Federico Montori[1]([✉]), Kewen Liao[2], Prem Prakash Jayaraman[2],
Luciano Bononi[1], Timos Sellis[2], and Dimitrios Georgakopoulos[2]

[1] University of Bologna, Bologna, Italy
{federico.montori2,luciano.bononi}@unibo.it
[2] Swinburne University of Technology, Melbourne, Australia
{kliao,pjayaraman,tsellis,dgeorgakopoulos}@swin.edu.au

Abstract. The Internet of Things (IoT) is springboarding novel applications and has led to the generation of massive amounts of data that can offer valuable insights across multiple domains: Smart Cities, environmental monitoring, healthcare etc. In particular, the availability of open IoT data streaming from heterogeneous sources constitute a novel powerful knowledge base. However, due to the inherent distributed, heterogeneous and open nature of such data, metadata that describe the data is generally lacking. This happens especially in contexts where IoT data is contributed by users via cloud-based open data platforms, in which even the information about the type of data measured is often missing. Since metadata is of paramount importance for data reuse, there is a need to develop intelligent techniques that can perform automatic annotation of heterogeneous IoT datastreams. In this paper, we propose two novel IoT datastream classification algorithms: CBOS and TKSE for the task of metadata annotation. We validate our proposed techniques through extensive experiments using public IoT datasets and comparing the outcomes with state-of-the-art classification methods. Results show that our techniques bring significant improvements to classification accuracy.

Keywords: Internet of Things · Classification · Open data
Metadata · Sensors

1 Introduction

The connected future is set to be dominated by a significant growth of the heterogeneous Internet of Things (IoT) devices that are estimated to surpass the total number of mobile phones by 2022 [1]. The IoT (underpinned by the principles of the Internet) has led to a phenomenal increase in the generation of IoT data that are contributed by users across the globe and can be publicly accessed via the Internet. In fact, a lot of data for such domains is available in open access forms from public platforms [14] either official, such as the Environmental Protection Agency (EPA) (https://www.epa.gov/), or user-produced, such as ThingSpeak

© Springer Nature Switzerland AG 2018
H. Hacid et al. (Eds.): WISE 2018, LNCS 11234, pp. 209–224, 2018.
https://doi.org/10.1007/978-3-030-02925-8_15

(https://thingspeak.com/) and, until few years ago, Xively (https://xively.com) and SparkFun (https://www.sparkfun.com/). Such "open data" is a powerful source of information for developing novel IoT applications in domains such as smart cities, defense, healthcare and environmental monitoring, to name a few. A fundamental requirement in successfully re-purposing such open IoT data, in order to enable interoperability as envisioned by Semantic Web 3.0, is to be able to automatically characterize its metadata i.e. information such as observation type (e.g. temperature, humidity), unit of observation (e.g. Celsius, Fahrenheit), location etc. However, as validated by a recent study in the literature [20], most publicly available IoT data lack availability of such accurate metadata and, in most cases, even the observation type is unclear (i.e. what is actually being measured).

Open IoT datastream[1] classification is a novel problem and has not been addressed well in the literature. Publicly available open IoT data is very diverse with varying degree of availability and accuracy of metadata (partial to none). In order to support the vision of the IoT-driven Web 3.0 – i.e. moving from streaming data to smart data (data that is well described, allowing interoperability and re-purposing across several domains) – in this paper, we propose algorithms to tackle the challenge of annotating open IoT datastreams produced by *heterogeneous* IoT environments. Such heterogeneity, attributed to the nature of the Internet that allows everyone to contribute, is due to diversities in the way data is captured (e.g. location, the accuracy and range of the sensor producing the IoT datastream, non-alignment in time, etc.) and imposes additional challenges in solving the IoT datastream classification and annotation problem. In particular, our focus is on classifying the observation type of an IoT datastream under the following two cases (1) lack of metadata and (2) partial, incomplete or inaccurate textual metadata e.g., an IoT datastream that produces temperature may be described by the user using non machine interpretable names such as "temp", or "t1", or "T (°C)". To this end, this paper makes the following contributions:

- A novel Class-wise Bag of Summaries (CBOS) approach, based solely on the numerical characteristics of the sensor readings.
- A novel Top-k Sequential Ensemble (TKSE) approach, which uses a combination of any available textual metadata describing the datastream and numerical summaries.
- Extensive experimental evaluations to validate the accuracy and the performances of the proposed approaches against the current state-of-the-art approaches in time series classification.

Our proposed sequential ensemble approach is a pioneering effort in the classification of IoT datastream that considers a combination of the numerical characteristics and partially available metadata of the IoT datastream. The rest of

---

[1] We use the term "datastream" to refer to an individual stream of data, together with its metadata, produced by a sensor on an IoT device. In contrast, we refer to "IoT data" as the collection of all the data produced by several IoT devices.

the paper is organized as follows: Sect. 2 provides the state-of-the-art in similar data classification and annotation tasks, Sect. 3 introduces our IoT data classification problem and describes the proposed CBOS and TKSE approaches, Sect. 4 describes our experimental design and the IoT datasets used in our evaluations, Sect. 5 presents the results of our experimental evaluations and, finally, Sect. 6 concludes the paper with recommendations for future work.

2   Related Work

IoT datastreams are ordered sequences of sensor readings which can naturally be seen as attributes that form what in the literature is called a "time series". Time Series Classification (TSC) problems, indeed, differ from ordinary classification problems in that features are ordered (not necessarily in the dimension of time). Within the last years, several TSC approaches have been proposed [4] as an alternative to the one-nearest-neighbor (1NN) approach using the simple pointwise Euclidean Distance as a similarity measure between series. The common agreement accepted among researchers as a "hard to beat" standard distance measure between series has been Dynamic Time Warping (DTW) [5], for which several alternatives have been proposed in order to contrast its high time complexity [10]. The above mentioned methods consider the whole series, since they extract series similarities by pointwise comparison. Other recent TSC approaches aim to find a subsequence, called "shapelet", yielding the highest information gain that can discriminate among classes and using a tree-based classification algorithm [22]. Such *shapelet-based approaches* have been improved over time, especially due to their high time complexity [16]. Finally, a third type of well-performing methods, namely *dictionary-based approaches*, split the time series in time windows and extract patterns out of each window as new features. Such methods tend to be faster than the aforementioned ones due to feature numerosity reduction. Examples of such methods are Bag-of-Patterns (BOP) [13], which uses piecewise aggregate approximation (PAA) through Symbolic Aggregation approXimation (SAX) words [12] and *Bag-of-SFA-Symbols* (BOSS) [19], which encodes subsequences through Discrete Fourier Transform (DFT).

The classification of IoT data stemming from heterogeneous IoT devices has been already considered in literature. In [7], the authors imply a PAA-based approach which treats the sensor data classification as a dictionary-based TSC problem and uses interval slopes as features. A different approach has been taken in [6], which aims to assess open IoT data validity by comparing it with a certified ground truth. In [14] The data streams are classified only on top of the metadata provided, i.e. the user-assigned stream name. However, no approach currently employs ensemble methodologies in order to consider varying degree of metadata quality and availability for classification; i.e. if no metadata is available it relies only on the numerical characteristics of data, whereas if limited and inaccurate metadata in the form of text is available, it uses a combination of numerical data and textual metadata.

3   CBOS and TKSE: Approaches for Classification and Annotation of IoT Datastreams

In this section we first formulate the problem of IoT datastream classification and annotation and then present our proposed algorithms: Classwise Bag of Summaries (CBOS) and Top-k Sequential Ensemble (TKSE). Our algorithms are based on generic data mining and machine learning approaches that are extensible to other problem domains.

3.1   Problem Formulation

Formally, we are given n IoT sensor datastreams[2] $\mathbf{NS} = \{\mathbf{S_1}, \mathbf{S_2}, \ldots, \mathbf{S_n}\}$, each of which can be represented as an ordered tuple that resembles a time series with metadata. That is, $\forall i \in [n] : \mathbf{S_i} = \langle \mathbf{D}_i, r_{i,1}, r_{i,2}, \ldots, r_{i,m} \rangle$ where $[n]$ represents the interval from 1 to n, $\mathbf{D}_i$ denotes a dictionary of metadata on sensor stream i with or without annotations, and each $r_{i,j}$ is a numerical sensor reading (typically a double value) from i-th stream at time τ_j. For instance, consider a temperature stream $\mathbf{S_i}$ with annotated metadata 'name' and 'description' and the annotation for metadata 'type' missing. Then $\mathbf{D}_i = \{(name : \text{"}outdoorTemp\text{"}), (description : \text{"ESP8266 with DHT11"}), (type : \text{""}), \ldots\}$. Without loss of generality and for simplicity, we assume datastreams being of the same length m (thus, $\mathbf{NS}$ can be viewed as a column-ordered matrix of size $n \times m$) and time intervals $\{\tau_{j+1} - \tau_j\}$ between two consecutive readings in each datastream being near-uniform. As, in our scenario, the textual metadata 'type' values that indicate the classes of datastreams are missing, our *goal* is to recover the datastream class (the 'type' value in $\{\mathbf{D}_i\}$) from the information of sensor readings and/or other aiding metadata. To achieve this, types or classes are mapped to numerical labels $\{y_i\}$, i.e. datastreams become of the form $\{\mathbf{S_i}, y_i\}$. Specifically, streams with known classes in $\mathbf{NS}$ form the training set of size t, otherwise the testing set of size $n - t$. From the training set, existing classes are mapped to c distinct numerical labels $\mathbf{L} = \{l_1, l_2, \ldots, l_c\}$[3] and, normally, $c \ll n$. In training phase classifiers are built for the later testing phase to inference, from $\mathbf{L}$, which class each missing y_i in the test set belongs to. Throughout the paper, we use bold symbols to denote multi-dimensional data structures such as vectors, matrices and dictionaries.

For the above formulated classification and annotation problem, we first propose CBOS, a preliminary data mining solution based on statistical summaries of numerical readings $\{r_{i,j}\}$ – what we call as bag of summaries (BOS). Then, we introduce the best performer TKSE that sequentially and incrementally ensembles multiple classifiers trained from the textual metadata and numerical summaries respectively.

---

[2] We use the term IoT datastream and IoT sensor datastream interchangeably.

[3] For convenience, we can treat 'class' and 'label' equivalently.

3.2   Classwise Bag of Summaries (CBOS)

Our first approach took place in investigating TSC algorithms for inferencing the type of data measured by each datastream. Given the noisy fashion of each datastream, we attempted to tackle the problem using Bag-of-Patterns-Features (BOPF) [11], a recent phase-invariant dictionary-based approach based on SAX words to encode local patterns. In essence, such approach splits the z-normalized time series through a sliding window in time-ordered chunks and extracts a SAX word by aggregating symbolic mappings of the chunks' mean values. Features are then represented by the distribution of SAX words across sliding windows. In the training set, features are first ranked by their global ANOVA-F value (the mean variance of a feature value across the whole dataset divided by the sum of within-class variances of that feature) and then by means of a leave-one-out cross validation together with the computed ANOVA-F ranking, a subset of features yielding the highest cross-validation accuracy is selected.

Although TSC approaches are widely known to work well on homogeneous datasets, such algorithm, as well as other TSC approaches we attempted, did not perform the way we expected on IoT datasets (as shown later in Sect. 5) due to heterogeneity in IoT data. To cope with this phenomenon, after some experimentation, we observed that a set of global statistical summaries of each non-normalized datastream can be highly discriminative for certain classes. For example, pressure values tend to have a mean around 10,000 hPa, thus the mean has a higher information gain since is far from any other; likewise, values like RSSI (the received signal strength by a wireless appliance) are often negative, thus the minimum value of the datastream is likely to be indicative. Hence, we adopt a set of meaningful statistical summary features $\{F_1, F_2, \dots F_f\}$ – defined as *bag of summaries* (BOS) – i.e. mean, median, minimum value, maximum value, root mean squared error (RMS) and standard deviation. Moreover, we propose the algorithmic approach CBOS, built on BOS features, that differentiates classwise features, i.e. each class of instances is trained to have its own/different discriminative subset of BOS features. This is in contrast with the aforementioned BOPF approach, where identical global features are shared by all classes. The proposed CBOS algorithm is presented in Algorithm 1.

In Algorithm 1, instead of ranking global features by ANOVA-F values, we compute *classwise* ANOVA (CANOVA) distributions **AF** on the training set (line 4), where $AF_{i,j}$ is the global variance of feature F_j divided by the local variance of F_j among instances of class i ($i \in [c]$ and $j \in [f]$). Moreover, values in $\mathbf{AF_i}$ are normalized into *weights* (for our later weighted cross validation and testing) by forcing them to sum up to 1. Similar to BOPF, both our cross validation and testing phases rely on a simple 1NN feature distance classification against the training class centroids $\mathbf{Cen_i}$ (line 5), obtained by averaging out the BOS feature values within the same class. The main difference is our leave-one-out cross validation (line 7), which performs classwise discriminative feature selection on all classes, i.e. it incrementally tries the $h \in \{1, 2, \dots, f\}$ highest CANOVA-ranked features of each class and eventually finds the best h^* that yields the maximum cross-validation accuracy. For each class with its respective

---

Algorithm 1. CBOS Algorithm

Require: Training set $\textbf{TRAIN} = \{(\textbf{S}_1, y_1), \ldots, (\textbf{S}_t, y_t)\}$, test example $(\textbf{T}, y)$
Ensure: $y \in \{1, 2, \ldots, c\}$
 1: {TRAINING PHASE}
 2: $\{(\textbf{F}_1, y_1), \ldots, (\textbf{F}_t, y_t)\} := ExtractSummaryFeatures(\textbf{TRAIN})$
 3: **for** $i := 1$ **to** c **do**
 4:     $\textbf{AF}_i := NormalizedCANOVA(i, \{(\textbf{F}_1, y_1), \ldots, (\textbf{F}_t, y_t)\})$
 5:     $\textbf{Cen}_i := CalculateCentroids(\{(\textbf{F}_1, y_1), \ldots, (\textbf{F}_t, y_t)\}_i)$
 6: **end for**
 7: $h^* := LeaveOneOutCV(\{(\textbf{F}_1, y_1), \ldots, (\textbf{F}_t, y_t)\}, \textbf{AF}, \textbf{Cen})$
 8:
 9: {TESTING PHASE - 1NN}
10: **for** $i := 1$ **to** c **do**
11:     $d_i := 0$
12:     **for** $j := 1$ **to** h^* **do**
13:         $d_i := d_i + \|(\textbf{F}[j] - \textbf{Cen}_i[j]) \cdot \textbf{AF}_i[j]\|$ # $\textbf{F}$ is the feature value vector of $\textbf{T}$
14:     **end for**
15: **end for**
16: **return** $y := \arg\min_i d_i$

---

centroids and CANOVA values, the distance between a new or cross-validated example and the centroids of such class is calculated as their sum of feature distances *weighted by* the respective feature CANOVA values (lines 10 to 16).

3.3 Top-k Sequential Ensemble (TKSE)

Within the scope of classification on data streams and time series, ensemble algorithms have been shown to be effective and able to capture different facets of the type of data [18]. However, most of the existing ensemble algorithms are designed in a parallel fashion, in that a number of classifiers are built and trained on the original data and the class of an unseen example is typically guessed via majority vote. In our case, as stated in Sect. 3.1, IoT datastreams may come with partial and inaccurate metadata (e.g. the ThingSpeak dataset in Sect. 4.1), which can provide a powerful source of information from a different dimension, for instance the dimension of the natural language. For such reason, we propose a novel *sequential ensemble* of classifiers that sequentially combines the text-based Natural Language Processing (NLP) and numerical value-based classification techniques (on the metadata and sensor readings respectively), so that they both contribute in classifying a datastream enriched with annotated metadata. In particular, our proposed *Top-k Sequential Ensemble* (TKSE) algorithm aims to independently train two or more classifiers of different nature and then classify a new example in a pipeline, rather than doing it in parallel. Our choice of a sequential ensemble relies on the fact that data is *noisy* and presents features in *several dimensions*. Hence, we think that a parallel ensemble of classifiers, each of them trying to guess one class, would be hardly sufficient to get rid of the noise and does not have a way to assign weights to classifiers operating on different dimensions. Conversely, sequential classifiers iteratively get rid of sets of classes that are highly unlikely to be the correct one.

Suppose that two classifiers Γ_1 and Γ_2 can be independently trained on the same training set $\mathbf{TRAIN} = \{(\mathbf{S_1}, y_1), \ldots, (\mathbf{S_t}, y_t)\}$ with training/ground truth classes $|\mathbf{L}| = c$ and $\forall i \in [t] : \mathbf{S_i} = \langle \mathbf{D}_i, r_{i,1}, r_{i,2}, \ldots, r_{i,m} \rangle$. Then, for an unseen example $(\mathbf{T}, y)$, these classifiers' TOP-1[4] predicted classes are $y_1 = \Gamma_1(\mathbf{T}, 1, \mathbf{L})$ and $y_2 = \Gamma_2(\mathbf{T}, 1, \mathbf{L})$ respectively. During testing, TKSE instead first applies the classifier $\Gamma_1(\mathbf{T}, k, \mathbf{L})$ that outputs TOP-k ranked classes $\subseteq \mathbf{L}$ (this can also be viewed as a filtering classifier) based on learning from annotated textual metadata. Then, these output classes from Γ_1 are fed as the input class labels of TOP-1 Γ_2 trained from sensor readings. Therefore, the final ensembled prediction result becomes $y = \Gamma_2(\mathbf{T}, 1, \Gamma_1(\mathbf{T}, k, \mathbf{L}))$. Note that if $k = 1$ then TKSE reduces to $\Gamma_1(\mathbf{T}, 1, \mathbf{L})$, and if $k = c$ it reduces to $\Gamma_2(\mathbf{T}, 1, \mathbf{L})$.

In all our experiments we use as Γ_1 a simple supervised dictionary-based NLP classifier, which we will refer to as a Dictionary Damerau-Levenshtein NLP (DDL-NLP) classifier, first introduced in [14]. We chose to use it as Γ_1, because of its TOP-k accuracy (i.e. the probability that the correct class falls into its TOP-k guessed classes) is significantly higher than one of the other classifiers we have considered. The algorithm focuses on the similarity of the metadata 'name' attributed to data streams as a classifying parameter. Algorithm 2 outlines the proposed TKSE algorithm: in training phase, a "dictionary" for each class L_j is constructed in the form of Bag-of-Words (BOW_j) including all stream names in metadata attributed to data streams within the same class (line 2); in testing phase, for each class L_j, the classwise minimum edit distance $d_j = min\{ed(w, s) \mid s \in BOW_j\}$ of a testing example w is computed (lines 4–5). For the distance function ed, we leverage the Damerau-Levenshtein edit distance [8] normalized by the maximum length between the two words. The algorithm then picks the closest classes through the TOP-k smallest distances among $\{d_1, \ldots, d_c\}$ (line 7), and subsequently inputs these to a second vanilla machine learning classifier Γ_2 (line 8) such as decision tree, random forest and SVM (as experimented in Sect. 5.1) trained on all the BOS features.

In order to properly combine two classifiers and achieve the best accuracy, the optimal value of k has to be determined, since, an inappropriate k would impact negatively on the performance by either missing many classes or introducing much noise. The straightforward approach would try all *discrete* values of $k \in [c]$ and choose the value k^* which yields the highest accuracy in k cross-validation rounds of Γ_2 on the training set, however, such method could be slow when c is large. This is also unlike applying stochastic gradient descent on approximating continuous non-convex functions. Instead, we perform a faster logarithmic search heuristic as a simple approximation: first let ACC_k denote the cross-validation accuracy for the chosen k, then from 1 to k we incrementally try values $\{2^0, 2^1, 2^2, \ldots, k\}$ and pick $k' = 2^p$ with the highest $ACC_{k'}$. As intervals between $[2^{p-1}, 2^p]$ and $[2^p, 2^{p+1}]$ might get larger, we can then recursively

---

[4] In many cases, classifiers produce a (probabilistic) rank of classes based on classification accuracy and the best ranked class is selected as the final predication.

perform the above logarithmic guesses in these intervals until they are small enough and find the overall best guess $k^* = \arg\max_{k'} ACC_{k'}$. Note that if we have more useful annotated metadata, TKSE can then be extended to a chain of more than two classifiers.

4 Experimental Design

In this section we describe in detail the datasets we used to validate the proposed approaches, our experimental objectives and design consideration.

4.1 Public Open IoT Datasets

We chose a combination of unannotated and partially annotated open source IoT datasets, namely: *The Swiss Experiment* dataset from [2] and a dataset we extracted from the public channels on the *ThingSpeak* cloud platform [3] in order to validate the performance and accuracy of the proposed approaches.

The Swiss Experiment is a platform that allows publishing environmental sensor data located within the Swiss Alps mountain range on the web in real-time. Data is highly noisy, comes from different microscopic locations and it is taken within different time spans. The sampling rate is also different among sensors making the phase shift of data series very significant. Neither semantic annotation nor timestamps were originally provided, thus data is unusable as it is, since the only information provided is the numerical datastream. To the best of our knowledge, the Swiss Experiment dataset (for which a manually annotated version is available at http://lsirpeople.epfl.ch/qvhnguye/benchmark/) is one of the few heterogeneous datasets used in research for our type of problem [7]. The dataset consists in datastreams measuring 11 different environmental parameters: CO_2, humidity, lysimeter, moisture, pressure, radiation, snow height, temperature, voltage, wind speed and wind direction. Time series are of different length, however, some of the algorithms we have tested require the time series to have the same length, therefore we cut each time series to the length of the shortest stream in the dataset. The dataset is composed by 346 datastreams without metadata and 445 data points each.

Algorithm 2. TKSE Algorithm with Γ_1 as DDL-NLP

Require: Training set $\textbf{TRAIN} = \{(\textbf{S}_1, y_1), \ldots, (\textbf{S}_t, y_t)\}$, test example $(\textbf{T}, y)$, k
Ensure: $y \in \{1, 2, \ldots, c\}$
1: **for all** $(\textbf{S}_i, y_i) \in \textbf{TRAIN}$ **do**
2:    $BOW_{y_i} \leftarrow \textbf{D}_i('name')$ # $\textbf{S}_i = \langle \textbf{D}_i, r_{i,1}, r_{i,2}, \ldots, r_{i,m} \rangle$
3: **end for**
4: **for** $j := 1$ **to** c **do**
5:    $d_j = min\{ed(\textbf{D}('name'), s) \mid s \in BOW_j\}$ # $\textbf{T} = \langle \textbf{D}, r_{i,1}, r_{i,2}, \ldots, r_{i,m} \rangle$
6: **end for**
7: $\textbf{C} = \{L_j \mid d_j \in kmins\{d_1, \ldots, d_c\}\}$ # $kmins$ picks the k lowest values
8: $y = \Gamma_2(\textbf{T}, 1, \textbf{C})$

ThingSpeak is an online cloud platform to which users can subscribe and push sensor data produced by their personal device onto their personal "channel" through dedicated APIs. Channels are optionally public and are composed by a set of time series (one for each sensor) and partially and mostly inaccurate user-annotated metadata: each channel has a name, a description, a name for each datastream and, optionally, a geolocation in GPS coordinates. Each name (as well as the other metadata) is user-assigned, thus it can be informative as well as useless. Each channel can be updated at any time rate above 10 s and the cloud keeps permanently in memory the last 8000 measurements. We built our dataset by scraping the first 500,000 channels through a dedicated HTTP call[5] returning a JSON object. With such call, the metadata and the last 8000 readings in year 2018 from all the datastreams belonging to the queried channel are retrieved and, subsequently, all the datastreams were made independent from their channel, thus the initial dataset is composed by more than 10,000 datastreams, each including a time series of sensor readings, with associated timestamps, and a set of metadata. We further filtered out private streams, non geolocated streams and streams with less than 5,000 readings in 2018; clustered the rest both by geo-location and time, selecting the most "populated" area and the 24-h time period for which we have the highest number of measurements. We obtained 1,091 streams having a consistent number of data points on a defined day in an area in central Europe, which includes parts of Germany, Poland, Czech Republic, Slovakia, Hungary and Austria. We clustered, for each series, the data points in 15-min time chunks and interpolated the missing points by means of cubic splines. Datastreams have been manually annotated in 16 different classes: non-air temperature, humidity, pressure, wind speed, wind direction, UV, light, sound, air quality, electrical parameter, RSSI, indoor air temperature, outdoor air temperature, heath index, rain index, and dew point. In summary, the dataset is composed by 1,091 datastreams with metadata and 96 data points each. We made the dataset available for download at https://github.com/stradivarius/TSopendatastreams.

4.2   Experimental Objectives

In order to evaluate the effectiveness and efficiency of our proposed IoT datastream classification algorithms, we performed extensive experiments against other state-of-the-art approaches on the above open IoT datasets.

For the purpose of experimental evaluation, we use the following three metrics: *Accuracy, Macro averaged F1-Score* – defined in [21] as the harmonic mean of macro averaged precision and recall over all classes – and the *average runtime performance* over the number of folds in seconds. We chose to report the macro averaged F1-Score as the number of instances per class in each of the datasets is unbalanced and, while a high accuracy indicates the overall success, a high F1-Score implies that all classes have been equally considered. Through the usage of such metrics, we performed the following experiments:

---

[5] https://thingspeak.com/channels/{ch}/feed.json?results=8000&start=2018
-01-0100:00:00.

Evaluation of Accuracy and F1-Score. The objective of this experiment is to validate the proposed algorithms on both our IoT datasets against time-series and BOS-based algorithms from the literature. Each algorithm has been validated through a k-fold cross validation. On Swiss Experiment we used 5-fold cross validation (same as [7]), whereas on ThingSpeak, we used 10-fold, as the number of instances is much higher. The algorithms are tested both on the original and z-normalized time series, as the conventional homogeneous time series analysis often requires normalized data.

Impact of K on TKSE. This experiment has the goal of validating the behavior of TKSE for different values of k. It is designed such that our BOS-based algorithms are tested as a second step in TKSE together with the DDL-NLP method (as described in Sect. 3.3) in a 10-fold cross validation on the ThingSpeak dataset (with annotated names) on all values of k.

Evaluation of Runtime Performance. This experiment has the purpose of validating the efficiency of the considered algorithms in time. They are tested on both datasets and the average runtime over the number of folds is reported.

4.3   Experimental Design

Our approaches CBOS and TKSE together with our adopted vanilla classifiers: decision trees (C4.5), random forest (RF)[6] (which have been shown to perform well on remote sensing scenarios [15]) and Support Vector Machines (SVM) are compared with the following time series classification (TSC) algorithms and the results obtained with the slope-based algorithm in [7] for sensor data:

The Golden Standard. As sensor reading streams can be easily interpreted as time series, we took into account the most widely used TSC algorithm: *One Nearest Neighbor with Dynamic Time Warping* (1NN-DTW) [17] – considered as the golden standard for TSC [5].

Dictionary Approaches. Within the scope of TSC, we also included two recent dictionary-based approaches: *Bag-of-Pattern-Features* (BOPF) [11], which our CBOS is based on as outlined in Sect. 3.2 and it adopts SAX sequence encoding, and *Bag-of-SFA-Symbols* (BOSS) [19], which encodes subsequences through Discrete Fourier Transform.

All experiments are performed on a computer with an Intel Core i7-7700HQ CPU @ 2.80 GHz × 4 and 8GB RAM while running Linux Mint 18.2 64-bit. All algorithms are implemented in Python 3.5.2 except from BOPF [11] and BOSS [19], for which we used the original C++ code provided by the authors.

---

[6] Although [9] has considered a more sophisticated RF, this is not the focus in dealing with heterogeneous IoT data.

5   Experimental Results and Discussions

In this section we provide results and insights about the experiments outlined in Sect. 4.2.

5.1   Evaluation of Accuracy and F1-Score

We evaluated ours proposed algorithms with the ones outlined in Sect. 4.2 using the Swiss Experiment and ThingSpeak datasets. Both accuracy and F1-Score are reported for the datasets (Swiss Experiment in Fig. 1 and ThingSpeak in Fig. 2) in their original and z-normalized form. Observing the outcomes of such analysis, it is immediately clear how data normalization causes the loss of important features and hence impacts the accuracy of classification making this a much harder problem. In fact, only BOSS and BOPF achieve similar results for both normalized and unnormalized datasets, in that such algorithms were designed specifically to operate on z-normalized series and some parameters are hard coded to cope with the underlying data. Nevertheless, they still achieve a similar accuracy on the original series. Our first summary-based approach CBOS performed poorly on z-normalized data. This is expected, as z-normalization loses most of the information based on statistical summaries. But, on the non z-normalized data CBOS improves on BOPF, which it is based on, and is later shown to be faster than BOSS. In summary, the above mentioned TSC approaches still do not achieve the desirable IoT data classification accuracy. On the other hand, the golden standard DTW tends to perform better on non z-normalized data, sometimes achieving good results on ThingSpeak. However, this further validates our findings i.e. the trend of the time series is not as indicative as the absolute value ranges in heterogeneous IoT data. These absolute value ranges are better captured when the data is not z-normalized.

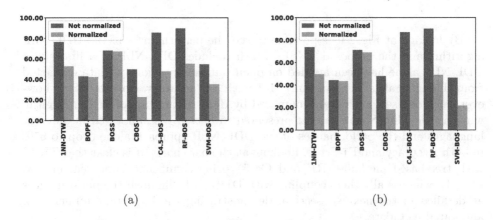

(a)   (b)

Fig. 1. (a) Accuracy and (b) F1-score for swiss experiment dataset

Based on this inference, we further found through our experiments that the vanilla machine learning methods performed using our BOS features achieve significant accuracy and F1-Score gains. Our justification for such phenomenon lies behind the fact that, due to heterogeneity in IoT data, different classes exhibit different behavior (and, therefore, bias) in terms of such features. For such reason, tree-based classifiers (C4.5 and RF), built in a way in which the attribution of an example to a class is driven by a sequence of decisions based on the threshold of the feature itself, seem intuitively suitable for open sensor data. This is not the way in which SVM are designed, in fact, they do not seem to work as well. It is also interesting to notice that SVM tend to assign more likely the most populated classes, resulting in a fairly poor F1-Score. Furthermore, on the Swiss Experiment dataset, the authors in [7] have reported on normalized data in the form of a pattern-based time series approach, achieving an accuracy of 67.7%, whereas our RF approach on BOS easily achieves above 80%.

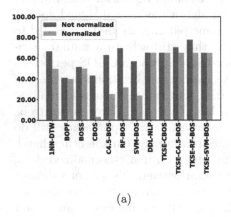
(a)

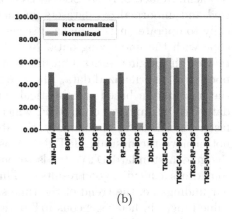
(b)

Fig. 2. (a) Accuracy and (b) F1-score for ThingSpeak dataset

By looking at Fig. 2 it is possible to see the performances of the DDL-NLP algorithm and the proposed TKSE (which includes DDL-NLP as a first step). DDL-NLP alone has been applied on open datasets previously (data extracted from ThingSpeak and SparkFun) [14] with an accuracy close to 88%, however, such datasets are purely composed by data streams annotated in English, whereas the ThingSpeak dataset presented in Sect. 4.1 is annotated in different languages and the performances of the DDL-NLP approach alone drop to 65% in both Accuracy and F1-Score. Looking at the bar charts, it is clear that TKSE with tree-based methods (RF and C4.5) bring significant improvements (at $k^* = 4$), whereas all others coupling with DDL-NLP diminish the performances as detailed in the next subsection, demonstrating the important influence of annotated metadata.

5.2   Impact of k on TKSE

The performance of the proposed TKSE approach has been evaluated via coupling the DDL-NLP classifier with other feature-based classifiers included in our previous test. For the purpose of this experiment we only used original data due to their preserved distinguishing power. For completeness in the illustration we tried each value of k rather than using a recursive logarithmic search. As stated before and shown in Fig. 3(a), only tree-based approaches display a performance increase with TKSE, whereas the others tend to be pejorative. It is also interesting to notice how the curves are similar in their trend, in particular they all display a local maximum for $k \simeq 4$ while performance starts dropping for greater values (as more noises are introduced). On the other hand, F1-Score, as shown in Fig. 3(b), is not positively affected by TKSE on ThingSpeak, since DDL-NLP is already the highest among the algorithms – this is mostly due to the reason that the proposed TKSE searching/tuning process is designed with the objective of optimizing the classification accuracy instead of the F1-Score. This phenomenon would require our future investigation, but nevertheless TKSE-RF keeps the F1-Score to approximately the same as DDL-NLP alone. Therefore, along with its consistent superior accuracy, TKSE-RF on BOS is by far the best choice among all the algorithms.

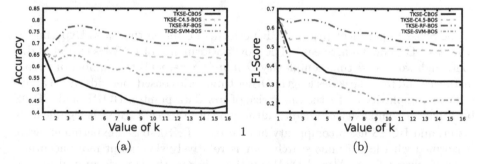

(a)   1   (b)

Fig. 3. Accuracy (a) and F1-score (b) of TKSE approaches evaluated on different algorithms. Tree-based algorithms display an evident increase in accuracy.

5.3   Evaluation of Runtime Performance

We have tested extensively the runtime performances of all the evaluated methods. The results are reported in Table 1 and time is measured in seconds over the cross-validation process divided by the number of the folds (which gives the actual training and testing time over a training and a test set). In particular, the Swiss Experiments has a training set size of 273 and a test set size of 73, while the ThingSpeak dataset has a training set size of 974 and a test set size of 117. It is interesting to notice how algorithms perform differently on such datasets due to the relatively larger number of instances in ThingSpeak and longer series

Table 1. Runtime performances in training and testing of all the algorithms on both Swiss Experiment and ThingSpeak datasets.

	Time in seconds (Swiss experiment)	Time in seconds (ThingSpeak)
1NN-DTW [17]	5989.694	1528.244
BOPF[9] [11]	5.87	36.589
BOSS[9] [19]	79.028	15.628
CBOS	0.068	0.265
C4.5-BOS	0.045	0.138
RF-BOS	0.589	1.178
SVM-BOS	24.194	845.437
DDL-NLP	–	1.286
TKSE-CBOS	–	11.670
TKSE-C4.5-BOS	–	7.195
TKSE-RF-BOS	–	21.510
TKSE-SVM-BOS	–	5082.260

[9]This algorithm is written in C++ since the original code provided by the authors has been used – implying stronger runtime baselines to compare with

in the Swiss Experiment (e.g. BOPF performs better on the Swiss Experiment, whereas BOSS on ThingSpeak). The slow performance of DTW is expected due to its high worst-case time complexity $\mathcal{O}(n^2m^2)$, where n is the number of stream instances and m is the series length. Other time series based methods mostly have a linear complexity (w.r.t the input size of nm data points): BOPF and CBOS both have a complexity of $\mathcal{O}(nm)$, although, in practice, CBOS is significantly faster, and BOSS has a complexity of $\mathcal{O}(nm^{\frac{3}{2}})$. TKSE-based experiments were performed with logarithmic search that is relatively slower but more accurate. Overall, except for SVM and DTW all the other methods perform well in runtime, and, in particular standalone RF and TKSE-RF flexibly trade off some runtime for classification quality and exhibit better performances than others on both datasets.

6   Conclusion

In this paper, we proposed novel algorithms to tackle the challenge of annotation and classification of open IoT datastreams produced from *heterogeneous* IoT environments. In particular, we first proposed CBOS, a bag of summary-based approach to classify IoT datastreams based on numerical characteristic of the underlying IoT datastream. Through experimental evaluations and validation we conclude, although IoT datastreams are reminiscent of time series datasets, due to the heterogeneity in the sensor readings produced by IoT devices, classic

TSC approaches perform poorly while vanilla classifiers such as decision trees and random forest based on bag-of-summaries (BOS) perform significantly better when only considering the numerical characteristics of the IoT datastream. Our second proposed algorithm namely TKSE uses a novel sequential ensemble approach to take advantage of both (1) partially available textual metadata that describes the IoT datastream and (2) the numerical characteristics of the IoT datastream. Through extensive experimental evaluations and comparisons with state-of-the-art approaches in the literature, we validated the significant gain in accuracy of the proposed TKSE algorithm while imposing minimal impact on runtime performance. Future work can be devoted into further improving annotation quality with more sophisticated features and ensembles that leverage all useful metadata to some extent, as well as studying the behavior of sequential ensembles in the presence of more than two classifiers.

References

1. Internet of Things Forecast. https://www.ericsson.com/en/mobility-report/internet-of-things-forecast
2. The Swiss Experiment. http://www.swiss-experiment.ch/
3. ThingSpeak Channels. https://thingspeak.com/channels/public
4. Bagnall, A., Lines, J., Bostrom, A., Large, J., Keogh, E.: The great time series classification bake off: a review and experimental evaluation of recent algorithmic advances. Data Min. Knowl. Discov. **31**(3), 606–660 (2017)
5. Batista, G.E., Wang, X., Keogh, E.J.: A complexity-invariant distance measure for time series. In: Proceedings of the 2011 SIAM International Conference on Data Mining, pp. 699–710. SIAM (2011)
6. Borges Neto, J.B., Silva, T.H., Assunção, R.M., Mini, R.A., Loureiro, A.A.: Sensing in the collaborative Internet of Things. Sensors **15**(3), 6607–6632 (2015)
7. Calbimonte, J.P., Corcho, O., Yan, Z., Jeung, H., Aberer, K.: Deriving semantic sensor metadata from raw measurements (2012)
8. Damerau, F.J.: A technique for computer detection and correction of spelling errors. Commun. ACM **7**(3), 171–176 (1964)
9. Deng, H., Runger, G., Tuv, E., Vladimir, M.: A time series forest for classification and feature extraction. Inf. Sci. **239**, 142–153 (2013)
10. Jeong, Y.S., Jeong, M.K., Omitaomu, O.A.: Weighted dynamic time warping for time series classification. Pattern Recognit. **44**(9), 2231–2240 (2011)
11. Li, X., Lin, J.: Linear time complexity time series classification with bag-of-pattern-features. In: 2017 IEEE International Conference on Data Mining (ICDM), pp. 277–286. IEEE (2017)
12. Lin, J., Keogh, E., Wei, L., Lonardi, S.: Experiencing SAX: a novel symbolic representation of time series. Data Min. Knowl. Discov. **15**(2), 107–144 (2007)
13. Lin, J., Khade, R., Li, Y.: Rotation-invariant similarity in time series using bag-of-patterns representation. Intell. Inf. Syst. **39**(2), 287–315 (2012)
14. Montori, F., Bedogni, L., Bononi, L.: A collaborative Internet of Things architecture for smart cities and environmental monitoring. IEEE Internet Things J. **5**(2), 592–605 (2018)
15. Pal, M.: Random forest classifier for remote sensing classification. Int. J. Remote. Sens. **26**(1), 217–222 (2005)

16. Rakthanmanon, T., Keogh, E.: Fast shapelets: a scalable algorithm for discovering time series shapelets. In: SIAM International Conference on Data Mining, pp. 668–676 (2013)
17. Ratanamahatana, C.A., Keogh, E.: Three myths about dynamic time warping data mining. In: SIAM International Conference on Data Mining, pp. 506–510 (2005)
18. Rokach, L.: Ensemble-based classifiers. Artif. Intell. Rev. **33**(1–2), 1–39 (2010)
19. Schäfer, P.: The BOSS is concerned with time series classification in the presence of noise. Data Min. Knowl. Discov. **29**(6), 1505–1530 (2015)
20. Siow, E., Tiropanis, T., Wang, X., Hall, W.: TritanDB: time-series rapid Internet of Things analytics. arXiv preprint arXiv:1801.07947 (2018)
21. Sokolova, M., Lapalme, G.: A systematic analysis of performance measures for classification tasks. Inf. Process. & Manag. **45**(4), 427–437 (2009)
22. Ye, L., Keogh, E.: Time series shapelets: a new primitive for data mining. In: Proceedings of the 15th ACM SIGKDD International Conference on Knowledge Discovery and Data Mining, pp. 947–956. ACM (2009)

Efficient Auto-Increment Keys Generation for Distributed Log-Structured Storage Systems

Jianwei Huang[1], Jinwei Guo[1], Zhao Zhang[1(✉)], Weining Qian[2],
and Aoying Zhou[2]

[1] School of Computer Science and Software Engineering,
East China Normal University, Shanghai 200062, China
{jwhuang,guojinwei}@stu.ecnu.edu.cn
[2] School of Data Science and Engineering, East China Normal University,
Shanghai 200062, China
{zhzhang,wnqian,ayzhou}@dase.ecnu.edu.cn

Abstract. Recent years, writing-intensive workloads on big data make log-structured style storage popular in distributed data storage systems, which provides both large-volume storage capacity and high-performance data updates. Rapidly generating valid keys for append records can significantly improve the data write performance of log-structured storage systems. In distributed and high concurrency environment, however, both the huge disk IO and the interaction overhead of a traditional lock manager limit the transactional throughput for generating auto-increment keys. In this paper, we design an efficient *auto-increment keys generation manager* (AKGM), a memory management structure that cannot only avoid disk IO but also eliminate the interaction overhead of traditional lock manager for transactions of generating auto-increment keys. We also propose a protocol called *adaptive batch processing* (ABP), which enables systems implementing AKGM to achieve high transactional throughput even under high contention workloads. We implement these protocols in an open-source database based on log-structured storage, and our experimental results show the superior performance of AKGM and ABP.

Keywords: Keys generation · Log-structured
Lock manager · Auto-increment key generation manager
Adaptive batch processing

1 Introduction

To efficiently support write-intensive mission critical on big data (e.g. second-kill activities of E-commerce, social user-generated data streams, etc.), increasingly many transaction processing applications keep the bulk of their active datasets in main memory [9,17]. Due to the huge amount of data and the high price

© Springer Nature Switzerland AG 2018
H. Hacid et al. (Eds.): WISE 2018, LNCS 11234, pp. 225–239, 2018.
https://doi.org/10.1007/978-3-030-02925-8_16

of memory, however, only in-memory storage cannot be entirely applicable to many commercial applications. Therefore, the log-structured storage is widely used in distributed data storage systems (represented by Google's BigTable [7], Apache's HBase [1] and Cassandra [13]) and commercial database systems (e.g. SQLite [3], OceanBase [2], etc.) to provide both large-volume storage capacity and high-performance data updates.

Generally, the log-structured storage systems only support append operations. For many real-world scenarios, a large number of user-generated append records lack unique identifiers. Therefore, rapidly generating a valid key for a new record is very important to increase the transactional throughput of records append operations. The auto-increment in keys is a simple and effective way to guarantee that the generated keys are unique and meaningful.

To generate a new auto-increment key, the key tasks are to efficiently get the current maximal key of the data table and to resolve conflicts captured the current maximum key under high contention workloads. For distributed log-structured storage systems, data are partitioned into the in-memory storage or disk storage of different nodes. This makes accessing the data on the disk a bottleneck for executing auto-increment keys transactions. Furthermore, as the increase in concurrency (and therefore lock contention), the traditional lock manager implemented by hash table will be another bottleneck of such transactions. One study reported that more than 30% of transaction time is spent interacting with the lock manager in the main memory DBMS. Moreover, there is an even larger lock manager overhead when the transaction running on multiple cores competing for access lock manager [11, 12, 15, 16].

On the other hand, to avoid the impact of distributed transaction processing, many online web applications prefer physically separated log-structured storage architecture. The recently updated data are stored in a single memory transaction node while a large amount of static immutable data are stored in multiple disk storage nodes. However, the single transaction processing node may become the performance bottleneck in those systems, especially under the auto-increment-intensive workload. Therefore, it is challenging to efficiently generate auto-incremental keys in distributed and high concurrency environment.

In this paper, we explore two major techniques for generating auto-increment keys for the append operations in distributed log-structured storage systems. First, we maintain an *auto-increment cache table* that caches the current maximum keys of data tables in the in-memory component of the log-structured storage, instead of accessing the disk to capture the current maximum key. Second, we also propose a towards *auto-increment keys generation concurrency control mechanism* (AKGCC) to further improve the performance of keys generation. Instead of the traditional pessimistic concurrency control represented by two-phase locking (2PL) protocol, AKGCC eliminates the interaction overhead of the traditional lock manager by the lightweight atomic operations and multi-threaded schedule.

We call the combination of these two techniques as *auto-increment keys generation manager* (AKGM), which incurs far significantly increasing the transactional throughput for generating auto-increment keys. For high contention workloads on the same auto-increment table, however, AKGM occupies excessive worker threads on transaction processing node. This can result in performance degradation and poor CPU utilization. To ameliorate this problem, we also propose an optimization called *adaptive batch processing* (ABP), which allows the clients' requests be batched as needed.

Specifically, we summarize the following contributions proposed in the paper:

- We design an auto-increment keys generation manager (AKGM), a memory management structure that can significantly increase the transactional throughput for generating auto-increment keys in distributed log-structured storage systems.
- We propose an adaptive batch processing (ABP) protocol, which allows AKGM to achieve good performance even under workloads with high contention.
- We integrate these algorithms and optimizations into OceanBase [2], a commercial database that employs log-structured storage. Extensive experiments show the superior performance of the proposed methods.

The rest of paper is organized as follows: First, Sect. 2 introduces the background knowledge and related work. Then, we describe the detailed implementation of AKGM in Sect. 3. Next, Sect. 4 explains the ABP optimization. Section 5 evaluates the performance of AKGM and ABP. Finally, we conclude the paper in Sect. 6.

2   Background and Related Work

Firstly, we review in this section the log-structured storage and introduce the architecture of the distributed log-structured database system OceanBase [2]. Secondly, we show the implementation of the traditional lock manager in the database system.

2.1   Log-Structured Storage

The log-structured merge tree (LSM-tree) was proposed by O'Neil et al. [14], a data structure used to optimize write performance of database systems. It maintains two or more tree-like separate structures in the database system, one of them is in-memory store and others are disk stores. The recently updated data are stored in memory structure while a large amount of static immutable data are stored in disk structures, and data are synchronized among these structures in batch efficiently.

Compared to the traditional B-tree structure [8], LSM-tree dramatically improves write performance of the storage systems because it reduces the overhead of random reads from disk. Reads of the LSM-tree system is not performing well, however, owing to its need to merge in-memory store with the disk storage for getting consistent data.

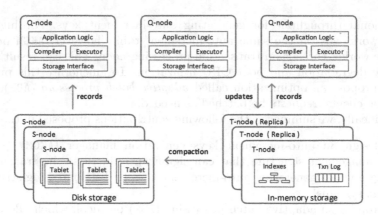

Fig. 1. Architecture of OceanBase.

We implemented the efficient auto-increment key generation method in OceanBase [2], a commercial database based on log-structured storage. Figure 1 provides an overview of the architecture of OceanBase, which consists of an in-memory transaction engine, a distributed storage engine, and a query processing engine. T-node is an in-memory transaction processing node, which stores all newly submitted updates and supports transaction management. A number of S-nodes form a distributed storage engine. The in-memory store in T-node and the disk storage in all S-nodes constitute the two-layer of the LSM-tree structure, and the recent committed updates are merged from T-node back into S-nodes. Both T-node and S-nodes have data replication to guarantee the system's high availability. Multiple Q-node constitute a query processing engine, which is responsible for handling client's query requests and business logic execution.

2.2   The Traditional Lock Manager

The most common way to implement a lock manager is as a hash table that maps lockable records primary key to a linked list of lock requests for that record [5,6,10,18]. These linked lists usually have a lock head that tracks the current lock state for the records. Each lock release needs to traverse the linked list for the sake of determining which lock request will inherit the lock next. The size of the lock request linked lists will increase, under workloads with high contention, which will also increase the cost of traversing the linked lists each lock release. Figure 2 shows the working principle of the traditional lock manager.

3   Auto-Increment Keys Generation Manager

In this section, we describe the working mechanism of AKGM in a distributed log-structured storage system. Compared to the traditional approach, AKGM first

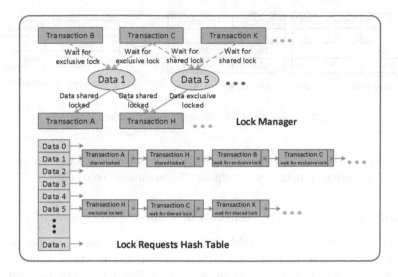

Fig. 2. Principle of traditional lock manager.

avoids huge disk IO overhead by caching the current maximum keys of all data tables in an *auto-increment cache table*; Second, AKGM eliminates the interaction overhead of the traditional lock manager with an *auto-increment key genera-tion concurrency control* (AKGCC) mechanism based on the lightweight atomic operations and multi-threaded schedule.

3.1 The Auto-Increment Cache Table

In distributed log-structured storage systems, the data are stored in multiple tree-like separate structures. A large amount of static immutable data are parti-tioned by the primary keys into many disk storage, and the recently updated data are in-memory storage. Therefore, transactions of generating auto-increment keys have to access disk storage to determine the current maximum key of a data table, thereby producing a huge disk IO.

Fortunately, it is not necessary to access the disk storage to determine whether the new keys generated is unique since we adopt an auto-increment way for keys. Moreover, there is only one current maximum key for a data table, so it does not increase too much storage cost even if we cache all the current maximum keys in the in-memory storage. As a result, it is feasible to cache the current maximum keys of data tables in system's in-memory storage (the in-memory component of the log-structured storage).

We maintain an auto-increment cache table in system's memory that man-ages the current maximum keys of all data tables. Figure 3 depicts an exam-ple of an auto-increment cache table. We organize the auto-increment cache table as the ordinary data table for two reasons. First, the modifications of the data table and the auto-increment cache table are on the same node, which is

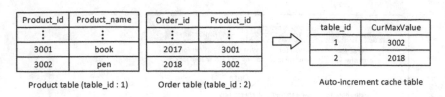

Fig. 3. Example of an auto-increment cache table.

important to a distributed write-intensive system. Second, it is well coupled with the LSM-tree index structure to facilitate the maintenance of data.

3.2 Auto-Increment Key Generation Concurrency Control

In a database system, the hash table is the most common way to implement the lock manager (see Sect. 2.2). If transactions of generating auto-increment keys use a lock manager based on two-phase locking (2PL) protocol to implement concurrency control, these hash table lookups, lock acquisitions, and the related linked list operations are all memory operations. The overhead of these operations is negligible if transactions need to access disk storage. When we cached the current maximum keys of data tables in in-memory storage, however, these operations are not negligible. As the number of concurrent transactions increases, the lock contention for the current maximum key becomes more intense—along with the associated an increase in the cost of traversing the linked lists each lock release.

On the one hand, all the transactions of generating auto-increment keys only compete for the current maximum key in the local in-memory storage, and other transactions do not access that data; this therefore guarantees that there will be no deadlock between transactions. On the other hand, transactions of generating auto-increment keys also do not need to coordinate the partition's lock requests because all the read-write sets of the transaction are in the same local partition.

We argue that it is unnecessary to implement transactional concurrency control for generating auto-increment keys with a traditional lock manager. Therefore, instead of a traditional lock manager, we design a towards *auto-increment keys generation concurrency control* (AKGCC) mechanism. The biggest difference between auto-increment keys generation concurrency control and traditional lock manager implementations is that AKGCC is not the complex lock auto-increment operation, but rather implemented by the lightweight atomic operation.

In addition, as the increase in cores and processors per machine, transactions of generating auto-increment keys will concurrently access the same current maximum key in a multi-threaded environment. The throughput of atomic auto-increment operations under workloads with high contention is very low, however, it even performs be not as good as the traditional lock manager. To solve this problem, we also design a set of queues (called TempKeyQueue) that store the temporary auto-increment keys for the keys generation concurrency control mechanism.

```
//Generate a batch of temporary auto-increment keys in single-threaded mode.
//The CurMaxValue x is from the auto-increment cache table.
function  AtomicAutoIncrementing( CurMaxValue x)
     TempMaxValue Tx = x;
     //Add  increment values to the TempKeyQueue
     while ( TempKeyQueue.size < MaxSize )
        TempMaxValue Tx = ++Tx(atomically);
        TempKeyQueue.Enqueue(Tx);

//An ordinary append transaction accesses the auto-increment key generation manager.
function  AKGCCOrdinaryAppendTxn( OrdinaryAppendTxn T)
     // Update the CurMaxValue and generate a new batch of temporary auto-increment keys.
     if (T is successful)
        AutoIncrementTable.CurMaxValue = T.key;
        TempKeyQueue.clear();
        AtomicAutoIncrementing( T.key );

//Auto-increment key generation transactions access the auto-increment key generation manager.
function  AKGCCAutoIncrementTxn( AutoIncrementTxn T )
     T.Type == blocked;
     // the size of the TempKeyQueue is less than a threshold
     if ( TempKeyQueue.size <= threshold)
        AtomicAutoIncrementing( CurMaxValue x );
     //Txn is executed after getting a key from the TempKeyQueue.
     while (T.Type == blocked && TempKeyQueue is not empty)
        T.key = TempKeyQueue.pop();
        T.Type = Free;
        Execute(T);
        //Transaction is successful and update the CurMaxValue in memory.
        if (T is successful)
           AutoIncrementTable.CurMaxValue = T.key;
```

Fig. 4. Pseudocode for the AKGCC algorithm.

Before transactions of generating auto-increment keys initiating requests to get the current maximum keys, the transaction node executes atomic auto-increment operations based on the value in the local auto-increment cache table, which quickly generate a number of temporary auto-increment keys and add them to TempKeyQueue in order. Multi-threaded transactions of generating auto-increment keys can quickly get the keys from the front of the TempKeyQueue at the same time. Transactions that acquire auto-increment keys for append records are termed free. Those which fail to acquire auto-increment keys for append records are termed blocked. The free transaction can continue to execute the append operation of the new record (with an auto-increment key), and synchronously update the current maximum key in the auto-increment cache table if that transaction is successful. The blocked transactions are not allowed to begin executing until they acquire an auto-increment key from the TempKeyQueue.

Although atomic auto-increment operations of current maximum keys are single-threaded, AKGCC can use a thread to serve a number of auto-increment data tables, which conserves the transaction node valuable processing resources. This

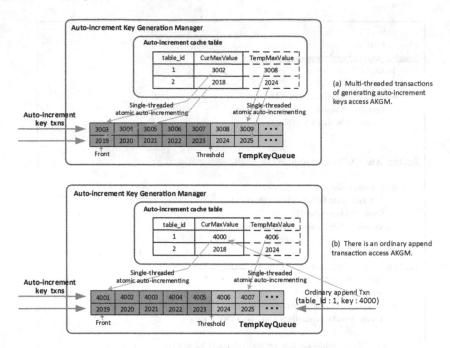

Fig. 5. Two examples of transactions access the AKGM.

specific parameter can be adjusted according to the actual application scenarios of the database system. Furthermore, we can also artificially control the time to generate temporary auto-increment keys. When the size of the TempKeyQueue is less than a threshold, AKGCC will call the particular thread to generate a new batch of temporary auto-increment keys, and then shifts its processing resources to serve other tasks. In practice we have found that this threshold should be tuned depending on the number of blocked transactions on a transaction node. AKGCC should run a larger threshold when there are more blocked transactions to ensure the TempKeyQueue can provide more auto-increment keys for transactions of generating auto-increment keys. A smaller threshold is also available for less blocked transactions.

One problem that AKGCC sometimes faces is that an ordinary append transaction (with a specified key) and many transactions of generating auto-increment keys access concurrently the transaction node. After the append operation of the ordinary append transaction is completed, it should update the current maximum key of the corresponding data table in time based on the specified key in the ordinary append transaction, and then clear out the TempKeyQueue. Finally, the atomic auto-increment thread is invoked to generate a new batch of temporary auto-increment keys and add them to TempKeyQueue. The purpose of this process is to guarantee that the current maximum keys in the auto-increment cache table are correct and effective.

These the above-mentioned processes and optimizations are termed a towards auto-increment keys generation concurrency control (AKGCC) mechanism, which incurs far less overhead than maintaining a traditional lock manager.

Figure 4 shows the pseudocode for the basic AKGCC algorithm. Transactions of generating auto-increment keys access the AKGM through the AKGCCAutoIncre-mentTxn function. Those blocked transactions become free after getting the auto-increment keys from the TempKeyQueue. Free transactions must update the current maximum key in the auto-increment cache table when the auto-increment records are successfully appended. When the size of the TempKeyQueue is less than a threshold, the AtomicAutoIncrement function is run in a single thread to generate a new batch of temporary auto-increment keys and adds them into TempKeyQueue. Ordinary append transactions access the AKGM through the AKGCCOrdinaryAppendTxn function. There are separate worker threads to execute three functions in the algorithm, and the AKGM becomes efficient under the coordination of a number of worker threads.

Figure 5 depicts two examples of transactions access the AKGM. (a) shows the multi-threaded transactions of generating auto-increment keys access the AKGM. We use the example of the auto-increment cache table in Sect. 3.1 to explain the AKGM, and the current maximum key of the *Product* table and the *Order* table are *3002* and *2018*, respectively. (b) shows an example of an ordinary append transaction and many transactions of generating auto-increment keys concurrency access the AKGM, and the specified key for that ordinary append transaction is *4000*.

4   Adaptive Batch Processing (ABP)

The main disadvantage of AKGM is a performance degradation under workloads with high contention. Consider, a large number of transactions of generating auto-increment key concurrent access the AKGM, and they are appended operations for the same data table. To get an auto-increment key, these transactions have to compete to access the front of the same TempKeyQueue. At this point, there will be many transactions that do not get an auto-increment key in time are blocked at the transaction node.

For high contention workloads on the same auto-increment table, AKGM occupies a growing percentage of worker threads on transaction node, where result in overall performance degradation. In order to maximize resource utilization of transaction node, we introduce the idea of *adaptive batch processing* (ABP).

In a distributed log-structured storage system, ABP greatly relieves the pressure of transaction node under high-contention workloads. It does this by batching the clients' auto-increment keys generation requests based on the system workload (i.e., excessive blocked transactions for the same auto-increment table). ABP therefore reduces the amount of competing for access AKGM and thereby increases transactional throughput.

In order to maximize the performance of our implementation of ABP, we include the adaptive module in ABP that controls the trigger time of ABP and

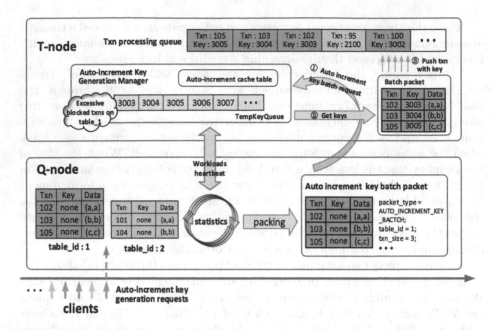

Fig. 6. Example execution of a sequence of auto-increment key generation requests using ABP.

the number of batch transactions. In our system model, the transaction node and query processing nodes interact with each other through heartbeat packets (auto-increment table information, the number of blocked transactions, etc.), and the query processing node collects these statistics to determine the timing and quantity of batch processing.

As with ordinary auto-increment key transactions, the transaction node uses a worker thread to handle the auto-increment key batch request. These batch requests can get increment keys only by accessing AKGM once, and then add these transactions into the transaction processing queue to complete the append operation of the new record (with an auto-increment key).

Figure 6 depicts an example execution of a sequence of auto-increment key generation requests using ABP. At this point, the transaction node (T-node) has blocked excessive auto-increment key generation transactions of *table_1*. The query processing node (Q-node) triggers batch processing after collecting relevant statistics. It packs the three keyless auto-increment transactions {*102*, *103*, *105*} of *table_1* into an auto-increment key batch packet and sends it to the T-node to access AKGM. The auto-increment key batch request then simultaneously acquires three auto-increment keys {*3003*, *3004*, *3005*} from the TempKeyQueue. Finally, ABP adds these three transactions into the transaction processing queue to complete the append operation of records.

5   Experimental Evaluation

We implement auto-increment keys generation manager (AKGM) and adaptive batch processing (ABP) in OceanBase [2], a commercial database based on distributed log-structured storage (see Sect. 2.1). To evaluate AKGM and ABP, we ran several experiments comparing AKGM (with and without ABP) against alternative schemes in a number of contexts. We separate our experiments into three groups: overall performance experiments for generating auto-increment keys, efficiency experiments of AKGM, and scalability experiments.

Cluster Platform: We ran all the experiments on a Linux cluster with 5 machines. Each machine has two 2.4 GHz 8-Core E5-2630 processors, 64GB DRAM and 3TB Raid5 while running CentOS version 6.5. All machines are connected by a gigabit Ethernet switch.

Database Deployment: OceanBase deploys two T-nodes (in-memory transaction nodes, Master-Slave) on separate servers. It deploys both an S-node (disk storage node) and a Q-node (query processing node) on each of the remaining servers.

Benchmark: We used Sysbench [4] to evaluate all performance experiments, a popular open-source benchmarking tool for evaluating the database system. We extend Sysbench by adding an *Order table* in which each tuple has a unique *order_id* as the primary key column, in addition, it has a *product_id* column and a *product_price* column. We test the performance of generating auto-increment keys by appending records without the *order_id* predicate attribute. The SQL instance is: *REPLACE INTO Order (product_id, product_price) VALUES ("?","?")*. Performance that we compare in the experiments are evaluated by transaction processed per second (TPS). We tuned the workload of the system by manually increasing the number of clients until throughput decreased due to too high workloads.

5.1   Overall Performance Experiments

This section compares the performance of AKGM against traditional lock manager to generate auto-increment keys. For AKGM, we analyze performance with and without the ABP optimization. As comparison points, we also implemented two versions of the traditional lock manager (with and without cache).

In our traditional lock manager without cache experiments, we did not implement the technology for the auto-increment cache table (exactly as described in Sect. 3.1). In contrast, in lock manager with cache experiments, we cached the current maximum keys of the data tables in the system's in-memory storage. Note, however, that both of these two experiments use a traditional lock manager to implement transactional concurrency control for generating auto-increment keys.

Figure 7 shows the throughput and latency of the transaction for generating auto-increment keys under four alternative schemes.

If the current maximum keys are not cached in in-memory storage, obviously, the performance of generating auto-increment keys is poor. As previously

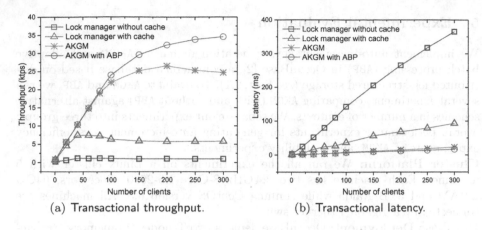

(a) Transactional throughput.   (b) Transactional latency.

Fig. 7. Overall performance under three different implementations.

analyzed, it is inevitable to access disk storage nodes to get a new auto-increment key in this case. The huge disk IO overhead in this process is the major reason for the high latency of the transaction for generating auto-increment keys—along with the associated transactional throughput is poor. When we add the auto-increment cache table into the system's in-memory storage, and the bottleneck of the disk IO is eliminated, which leads to a great improvement in transactional throughput and latency. In fact, we can see that disk IO has a greater impact on performance than the traditional lock manager.

In this experiment, AKGM improves the performance of generating auto-increment keys to traditional lock manager by 4 to 5 times. When the workload is low (below 100 clients), AKGM (with and without the ABP) yields near-optimal throughput; while the interaction overhead of the traditional lock manager increases rapidly, so its transactional throughput will soon reach the upper limit. As the workload increases (and contention is high), excessively auto-increment keys transactions are blocked at the transaction node, and the ABP optimization is able to improve performance relative to the basic AKGM scheme by adaptive batch processing. In this experiment, ABP boosts AKGM's performance by up to 30% under the highest contention levels.

5.2 Efficiency Experiments of AKGM

In this section, we tested the performance of AKGM itself with two experiments. First, we compare the auto-increment performance of AKGM against the atomic operations. Second, we collected the overhead ratio of AKGM and the traditional lock manager in the transaction for generating auto-increment keys.

To evaluate the performance of AKGM itself, we abandoned the other processes and overhead of auto-increment key transaction in a distributed database system. Instead of the actual transaction for generating auto-increment keys to access the AKGM, we imitate worker threads to access the AKGM at the transaction node.

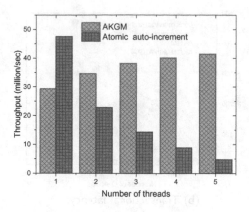

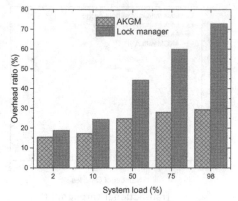

Fig. 8. Auto-incrementing throughput (AKGM vs. Atomic operation).

Fig. 9. Transactional overhead ratio (AKGM vs. Lock manager).

These worker threads are not intended to execute record appending transactions, but only to get the auto-increment keys from the AKGM. As a comparison point, we tested the performance of atomic auto-increment operations under the same condition.

Figure 8 shows the auto-increment throughput of AKGM and atomic operations as we vary the number of worker threads. Single-threaded atomic auto-increment operations can achieve high throughput. As the worker threads increase, however, its throughput has fallen sharply. On the contrary, although the performance of the single-threaded AKGM is not as good as that of the atomic operations, its performance is obviously better than the atomic operations as the number of threads increases.

Figure 9 shows the overhead ratio of AKGM and the lock manager in the transaction for generating auto-increment keys, respectively. When the system load is low (and contention is low), AKGM and the lock manager take a similar transaction time. As the system load increases (and contention is high), the interaction overhead of the traditional lock manager costs a great deal of transaction time for generating auto-increment keys, and the overhead of AKGM has been less all the way.

After the above analysis, we demonstrated AKGM is efficient.

5.3 Scalability Experiments

In our first set of experiments, we evaluated the overall performance of AKGM and ABP. For these experiments, we vary the number of clients connected to query processing nodes (Q-nodes) to get the best throughput when there are three Q-nodes in the database. From the experiment data, ABP boosts AKGM's performance by up to 30% under the highest contention levels.

In this section, we tested the scalability of the different schemes at high contention. We scale from 1 to 8 Q-nodes in the cluster and connect more than 50 clients on each Q-node.

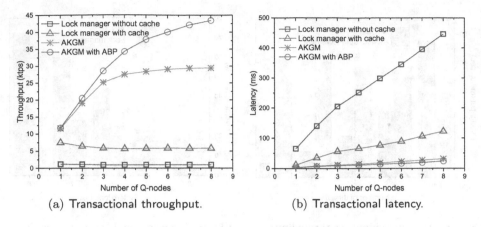

(a) Transactional throughput.   (b) Transactional latency.

Fig. 10. Scalability.

Figure 10 shows the results of the scalability experiments in which we tested the transactional throughput and latency. This figure shows that the throughputs of AKGM (with and without ABP) increase when more Q-nodes are deployed. In contrast, the performance of traditional lock manager (with and without cache) actually deteriorates, for the following reason. Lots of transaction time is spent interacting with the lock manager. These transactions block all the working threads of the transaction node. With more Q-nodes being used, it becomes more expensive for such request to be processed. As a result, lock manager achieves its best performance when a single Q-node is used. Meanwhile, AKGM with ABP significantly more scalable than AKGM. ABP optimization improving performance relative to regular AKGM by 30% to 55% when more Q-nodes are deployed. The primary reason is that ABP relieves the processing pressure of a single transaction node under high-contention workloads by batching the clients' requests.

6   Conclusion

In order to rapidly generate valid keys for the append records in distributed log-structured storage systems, we have presented an efficient auto-increment keys generation manager (AKGM). AKGM is a memory structure cannot only avoid the costs of disk IO but also eliminate the interaction overhead of the traditional lock managers, and therefore yields higher transactional throughput than traditional implementations. AKGM caches the current maximum keys of the data tables in the system's in-memory storage. AKGM eliminates the interaction overhead of the traditional lock manager with an *auto-increment key generation concurrency control* (AKGCC) mechanism based on the lightweight atomic operations and multi-threaded schedule. Although the high contention workloads on the same auto-increment table can result in overall performance degradation, adaptive batch processing (ABP), allows the clients' requests be batched as needed. This optimization allows our AKGM to achieve good performance even under workloads with high contention.

We implemented the AKGM and ABP in OceanBase, a commercial database based on log-structured storage. The experiments we presented demonstrate that AKGM can outperform the implementations of the traditional approach.

Acknowledgements. The project is partially supported by National Key R&D Plan Project under grant numbers 2016YFB1000905 and 2018YFB1003400, National Science Foundation of China under grant numbers U1401256, 61432006 and 61332006, and Shanghai Agriculture Applied Technology Development Program, China (T20170303).

References

1. Hbase website. http://hbase.apache.org/
2. OceanBase website. https://github.com/alibaba/oceanbase/
3. SQLite website. https://sqlite.org/
4. Sysbench website. http://dev.mysql.com/downloads/benchmarks.html/
5. Agrawal, R., Carey, M.J., Livny, M.: Concurrency control performance modeling: Alternatives and implications. In: Performance of Concurrency Control Mechanisms in Centralized Database Systems, pp. 58–105 (1996)
6. Bernstein, P.A., Goodman, N.: Concurrency control in distributed database systems. ACM Comput. Surv. **13**(2), 185–221 (1981)
7. Chang, F., Dean, J., et al.: Bigtable: a distributed storage system for structured data. In: Proceedings of the 7th USENIX Symposium on Operating Systems Design and Implementation, vol. 7, p. 15 (2006)
8. Comer, D.: The ubiquitous b-tree. ACM Comput. Surv. **11**(2), 121–137 (1979)
9. Diaconu, C., et al.: Hekaton: SQL server's memory-optimized OLTP engine. In: Proceedings of the ACM SIGMOD International Conference on Management of Data, SIGMOD 2013, New York, NY, USA, 22–27 June 2013, pp. 1243–1254 (2013)
10. Gray, J., Reuter, A.: Transaction Processing: Concepts and Techniques. Morgan Kaufmann, Burlington (1993)
11. Harizopoulos, S., Abadi, D.J., Madden, S., Stonebraker, M.: OLTP through the looking glass, and what we found there. In: SIGMOD Conference, pp. 981–992. ACM (2008)
12. Johnson, R., Pandis, I., Ailamaki, A.: Improving OLTP scalability using speculative lock inheritance. PVLDB **2**(1), 479–489 (2009)
13. Lakshman, A., Malik, P.: Cassandra: a decentralized structured storage system. Operating Syst. Rev. **44**(2), 35–40 (2010)
14. O'Neil, P.E., Cheng, E., Gawlick, D., O'Neil, E.J.: The log-structured merge-tree (LSM-tree). Acta Inf. **33**(4), 351–385 (1996)
15. Pandis, I., Johnson, R., Hardavellas, N., Ailamaki, A.: Data-oriented transaction execution. PVLDB **3**(1), 928–939 (2010)
16. Ren, K., Thomson, A., Abadi, D.J.: Lightweight locking for main memory database systems. PVLDB **6**(2), 145–156 (2012)
17. Stonebraker, M., Madden, S., Abadi, D.J., Harizopoulos, S., Hachem, N., Helland, P.: The end of an architectural era (it's time for a complete rewrite). In: Proceedings of the 33rd International Conference on Very Large Data Bases, University of Vienna, Austria, 23–27 September 2007, pp. 1150–1160 (2007)
18. Thomasian, A.: Two-phase locking performance and its thrashing behavior. In: Performance of Concurrency Control Mechanisms in Centralized Database Systems, pp. 166–214 (1996)

A Parallel Joinless Algorithm for Co-location Pattern Mining Based on Group-Dependent Shard

Peizhong Yang, Lizhen Wang$^{(\boxtimes)}$, Xiaoxuan Wang, and Yuan Fang

School of Information Science and Engineering, Yunnan University,
Kunming 650091, China
pzyang0924@163.com, wangxiaoxuan1037@163.com,
{lzhwang, fangyuan}@ynu.edu.cn

Abstract. Spatial co-location patterns, whose instances are frequently located together in geography, are particularly valuable for discovering spatial dependencies. Since its inception, lots of co-location pattern mining algorithms have been developed, but the computational cost remains prohibitively expensive with large data size. In this work, we propose to parallelize joinless algorithm on MapReduce framework. Our approach partitions computation in such a way that each machine independently executes joinless algorithm to finish a group of mining tasks. Such partitioning eliminates computational dependencies and reduces communication cost between machines. Moreover, a novel pruning technique is suggested to improve mining performance. The experimental results on synthetic and real-world data sets show that the parallel joinless algorithm is efficient and scalable.

Keywords: Spatial data mining · Co-location patterns
Group-dependent shards · Parallel algorithm · MapReduce

1 Introduction

The spatial co-location pattern mining is a part of spatial knowledge discovery techniques, and it is a useful tool for discovering spatial dependencies. A co-location pattern is a subset of spatial features whose instances are frequently located together in geography. Co-location patterns may yield important insights in many applications [1–5]. Various co-location pattern mining techniques have been developed, but most of them are serial processing and have no ability to process massive spatial data. With the development of spatial database technologies, the spatial data presents explosive growth trend which brings a huge challenge for co-location pattern mining. Therefore, developing parallel algorithms for co-location pattern mining is becoming necessary. In this work, we propose a parallel joinless (PJoinless) algorithm to discover co-location patterns from massive spatial data. The algorithm is implemented on Apache Spark and extensive experiments are conducted to evaluate the efficiency. Experimental results demonstrate that PJoinless algorithm is efficient and scalable.

© Springer Nature Switzerland AG 2018
H. Hacid et al. (Eds.): WISE 2018, LNCS 11234, pp. 240–250, 2018.
https://doi.org/10.1007/978-3-030-02925-8_17

The main contributions of this work are summarized as follows: (1) A partitioning scheme is proposed to divide database of star neighborhoods into some group-dependent shards so that joinless algorithm can be executed in each shard independently and correctly. (2) We propose a novel pruning strategy to improve mining efficiency. (3) A MapReduce-based parallel joinless algorithm is presented.

The rest of the paper is organized as follows: Sect. 2 reviews related work. Section 3 gives relevant basic concepts. In Sect. 4, we discuss the parallel joinless algorithm. Experimental evaluations are reported in Sect. 5. Conclusion and future work are presented in Sect. 6.

2   Related Work

A large number of approaches have been developed to discover co-location patterns from spatial data. Huang et al. [2] defined the participation index to measure the prevalence of a co-location pattern and an *Apriori-like* approach called join-based algorithm has been proposed by the authors. On the basis of the participation index, a variety of co-location pattern mining techniques are proposed to improve efficiency [7–10], such as join-less algorithm [7], partial-join algorithm [9], and CPI-tree algorithm [8], etc. However, above methods are ineffective for massive spatial data because of the limitation for a single machine in the computation and storage ability. With the help of MapReduce framework, Yoo et al. [6] proposed a parallel co-location pattern mining algorithm, called PCPM_SN for short. PCPM_SN algorithm has the capability to process massive spatial data, but it is inefficient. In this work, we put forward a parallel joinless algorithm to improve mining efficiency.

3   Preliminary

This section presents the basic concept of co-location pattern mining and joinless algorithm. Moreover, MapReduce framework will be introduced in this section.

3.1   Co-location Pattern Mining

In a spatial database, let $F = \{f_1, ..., f_m\}$ be a set of different spatial features and a spatial instance set $S = S_1 \cup ... \cup S_m$, where S_i is a set of spatial instances of the feature f_i, $1 \leq i \leq m$. Figure 1 shows an example of spatial data, where each spatial instance is described by its feature type and instance id, e.g., A.1. Two instances have the spatial neighbor relationship R if the distance between two instances is less than a threshold d. In Fig. 1, if two instances are neighboring, they are connected by a black line. A co-location pattern $Cl = \{f_1, ..., f_k\}$, is a subset of spatial features, whose instances are frequently observed in a nearby area. A set of spatial instances I is a row instance (or co-location instance) of Cl, if (1) I contains all features of Cl and no proper subset of I does so, and (2) all instances of I form a clique. In Fig. 1, {B.1, C.1, D.1} is a row instance of the pattern {B, C, D}. All row instances of Cl constitute the table instance of Cl, denoted as *table_instance(Cl)*.

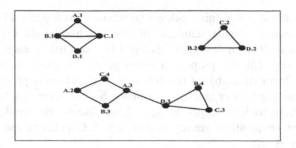

Fig. 1. An example spatial data set

The prevalence strength of a co-location pattern is often measured by the participation index [2]. The participation index of a size-k co-location pattern Cl $PI(Cl)$ is the minimum $PR(Cl, f_i)$, $1 \leq i \leq k$, where $PR(Cl, f_i)$ is the participation ratio of the feature f_i in Cl. $PR(Cl, f_i)$ can be calculated as:

$$PR(Cl, f_i) = \frac{\pi_{f_i}(|table\_instance(Cl)|)}{|table\_instance(\{f_i\})|} \tag{1}$$

where π is the relational projection operation with duplication elimination. If the participation index of Cl is not less than a user-specified minimum prevalence threshold min_prev, Cl is a prevalent co-location pattern. The participation index is anti-monotone [2], which means the participation index of a co-location pattern is not bigger than the participation index of its sub-patterns.

3.2 The Joinless Approach

The joinless algorithm is an excellent co-location pattern mining approach proposed by Yoo et al. [7]. It is based on a neighborhood materialization model, the *star neighborhood*. The star neighborhood and star instance are defined as:

Definition 1 (Star Neighborhood). Given a spatial instance $o_i \in S$, the feature type of o_i is $f_i \in F$, the *star neighborhood* of o_i is a set of instances $SN = \{o_j \in S \mid o_j = o_i \vee (R (o_j, o_i) \wedge f_j > f_i)\}$, where $f_j \in F$ is the feature type of o_j and R is a neighbor relationship.

Definition 2 (Star Instance). Let $I = \{o_1, ..., o_k\}$ be a set of spatial instances whose feature types $\{f_1, ..., f_k\}$ are different. If all instances in I are neighbors to the first instance o_1, I is called a *star instance* of the co-location pattern $\{f_1, ..., f_k\}$.

The star neighborhood of an instance is a set of the center instance and some instances which are located in the center's neighborhood and their feature types are greater than the feature type of the center instance in a lexical order. In joinless algorithm, an *instance-lookup* scheme is used to filter co-location instances from a spatial data set by using star neighborhoods. For a co-location pattern $\{f_1, ..., f_k\}$, its star instances can be generated from star neighborhoods which the feature type of center instances is f_1. However, star instance is not guaranteed a clique. Then, the clique relationship checking can be implemented by searching other star instances or by looking up the co-location instances of pattern $\{f_2, ..., f_k\}$. The expensive spatial or instance join operation for identifying co-location instances would be replaced by the instance-lookup scheme in joinless algorithm. The joinless algorithm has three steps. The first step converts spatial data into a set of star neighborhoods. The second step generates star instances of candidate co-location patterns from star neighborhoods and coarsely filtering the candidate co-location patterns by the prevalence value of their star instances. The third step filters clique instances from star instances and selects prevalent co-location patterns. The second and third steps are repeated with the increasing of co-location pattern size until having no prevalent pattern be discovered.

3.3 MapReduce Framework

MapReduce is a programming model which provides a highly scalable and flexible framework for data-oriented parallel computing. A MapReduce job is executed in two defined data transformation functions, namely, *map* and *reduce*. In map phase, the *<key,value>* pairs are processed by Mapper instances and a series of new intermediate *<key',value'>* pairs are output. The process of moving intermediate *<key',value'>* pairs from map tasks to assigned Reducers is called *shuffle*. At the completion of the shuffle phase, all values associated with *key'* will be gathered into a set (say *Set(key')*) that located in a Reducer. Then, reduce tasks are performed to process pairs *<key', Set (key')>* , and each Reducer generates a series of *<key",value">* pairs as output.

4 Parallel Joinless Approach

In this section, we present the parallel joinless algorithm based on MapReduce, called PJoinless algorithm for short.

4.1 Partitioning Scheme

Definition 3 (Prefix Feature). Given a spatial instance o_i whose feature type is $f_i \in F$, $\exists o_j \in S_j$, $R(o_i, o_j)$, if f_j is smaller than f_i in alphabetical order, f_j is a prefix feature of o_i, where S_j is the spatial instances set of the feature f_j.

In the example spatial data set, A is a prefix feature of instance B.1 because instance A.1 is adjacent with B.1, and A is smaller than B in alphabetical order.

Definition 4 (Group-Dependent Star Neighborhood Shard). Let all star neighborhood records represents as T. A group-dependent star neighborhood shard is a part of T and the partitioning is at the level of spatial features. The identification (group-id) for each shard is a unique spatial feature, for a spatial feature $f \in F$, $T(f)$ presents the group-dependent star neighborhood shard whose group-id is f. $T(f)$ includes some star neighborhood records:

(1) The star neighborhood of center o_i, $SN(o_i)$, if the feature type of o_i is f.
(2) The star neighborhood of center o_j, $SN(o_j)$, if f is a prefix feature of o_j.

Figure 2 shows the partitioning process for the example spatial data set in Fig. 1. All star neighborhood records constitute a database, presented in Fig. 2(a). Then, the database of star neighborhoods is divided into three group-dependent shards ($T(A)$, $T(B)$, $T(C)$) according to Definition 4.

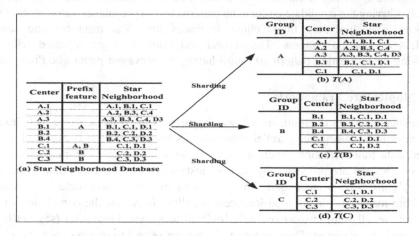

Fig. 2. The partitioning process on the example spatial data

Given a group-dependent shard $T(f)$, joinless algorithm can be executed in $T(f)$ independently to discover all prevalent co-location patterns whose first feature type is f. The partitioning scheme eliminates the computational dependencies because each shard contains all neighbor relationships that may be required in the mining process.

4.2 Pruning Strategy

Given a group-dependant shard $T(f)$, a spatial feature $x \in F$, the number of center instances whose feature type is x in $T(f)$ presents as $num(x, T(f))$, the total number of instances of the feature x presents as $total(x)$. The probability that spatial feature x is included in $T(f)$ is defined as:

$$p(x, T(f)) = \frac{num(x, T(f))}{total(x)} \tag{2}$$

In Fig. 2(b), just one instance (B.1) of the spatial feature B is included in the shard $T(A)$, and the number of global instances of feature B is 4. So, $p(B, T(A)) = 1/4$.

Lemma 1. Given a size-k ($k > 2$) co-location pattern $Cl = \{f_1, ..., f_{k-1}, f_k\}, f_i < f_j$ holds for every $1 \leq i < j \leq k$. If $p(f_i, T(f_1)) < min\_prev, 1 < i < k$, Cl must be not prevalent.

Proof. Given an instance o, whose feature type is f_i ($1 < i < k$) and the star neighborhood is $SN(o)$. If the shard $T(f_1)$ includes the star neighborhood of o, f_1 must be a prefix feature of o by Definition 4. Therefore, o is adjacent with one or more instances of f_1 according to Definition 3. Given an instance o' whose feature type is f_k, if $R(o, o')$, o' must be an element of $SN(o)$ by the definition of star neighborhood. As a consequence, the number of instance which the feature type is f_i and has adjacent simultaneously with the instances of f_1 and the instances of f_k must not exceed $num(f_i, T(f_1))$. That is, $PI(Cl) \leq PR(Cl, f_i) \leq p(f_i, T(f_1))$. Therefore, if $p(f_i, T(f_1)) < min\_prev$, $PI(Cl)$ must be less than min_prev and Cl is not prevalent.

With the help of Lemma 1, some candidate co-location patterns can be filtered so that narrowing the searching space.

4.3   MapReduce Algorithm

The parallel joinless (PJoinless) algorithm uses three MapReduce phases to complete the mining task. The algorithm is presented in Fig. 3.

```
1:procedure Mapper(null, value=o)              5:procedure Reducer(key=f, value=[1])
2:    f(o)←the feature type of o               6:   count=sum(value)
3:    emit(f(o),1)                             7:   save(f, count)
4:end procedure                               8:end procedure
```
(a) Job1: Parallel Counting

```
1:procedure Mapper(key=oᵢ, value=oⱼ)           8:      append x into SN
2:   emit(oᵢ, oⱼ), emit(oⱼ, oᵢ)                9:   else
3:end procedure                               10:       append f(x) into PFS
4:procedure Reducer(key=o, value=S(o))        11:   end if
5:   SN=[o], PFS=[]                            12:  end for
6:   foreach x in S(o) do                      13:  emit(o, (PFS, SN))
7:   if f(x)>f(o) then                         14: end procedure
```
(b) Job2: Generating Star Neighborhoods

```
1: procedureMapper(key=o,                      7:procedure Reducer(key=f, value=T(f))
       value=PFS(o), SN(o))                    8:   PatternSet←
2:   emit( f(o), SN(o) )                              call joinless algorithm in T(f)
3:   foreach f in PFS(o)  do                    9:   foreach pattern in PatternSet do
4:     emit(f, SN(o))                          10:       emit(pattern, PI)
5:   end                                       11:  end for
6: end procedure                               12: end procedure
```
(c) Job3: Parallel Joinless

Fig. 3. The parallel joinless algorithm

Firstly, as shown in Fig. 3(a), the parallel counting MapReduce job is performed to count the number of instances per spatial feature. It is preparation for future participation index calculation.

The job that generating star neighborhood for each instance is presented in Fig. 3 (b). The instance pair $<o_i, o_j>$ which o_i and o_j are neighboring is fed for Mapper. Spatial neighbor relationships can be obtained in advance using the parallel neighbor searching approach [6]. Then, Mapper outputs the pair $<o_i, o_j>$ and $<o_j, o_i>$ so that all instances which is adjacent with same instance o can be collected into a set $S(o)$ after the shuffle process. In reduce phase, building and emitting the star neighborhood set SN (o) and the prefix feature set $PFS(o)$ for instance o.

The third job presented in Fig. 3(c) is mainly to divide star neighborhood records into some group-dependent shards and perform joinless algorithm in each shard. In map phase, an instance, its star neighborhood and prefix feature set which generated in Job2 are input for Mapper. Then, for each star neighborhood record, we get the group-dependent shards which the star neighborhood record belongs to by Definition 4. The pair <group-id, star neighborhood record> is the output of Mapper. After all Mapper instances finished, star neighborhood records corresponding to same shard will be gathered into a set in the shuffle process. In reduce phase, joinless algorithm is executed in each shard concurrently to discover a group of co-location patterns whose first feature type is same with the group-id of the shard. In the phase of generating candidate co-location patterns in joinless algorithm, the pruning strategy introduced in Sect. 4.2 can be used to filter candidate patterns.

4.4   Completeness and Correctness

Completeness means PJoinless algorithm discovers all prevalent co-location patterns, and correctness means all co-location patterns discovered by PJoinless algorithm have participation index values above the minimum threshold.

Lemma 2. Given a group-dependent shard $T(f)$, any co-location instance whose first feature type is f can be generated from $T(f)$.

Proof. All star neighborhood records that the feature type of center instance is f are included in $T(f)$ by Definition 4, so all star instances whose first feature type is f can be generated from $T(f)$. Given a star instance $\{o_1, o_2, ..., o_k\}$, the feature type of o_1 is f. If all instances in a star instance except the first instance form a clique, the star instance is a co-location instance. The clique checking of the subset $\{o_2, ..., o_k\}$ can be implemented by searching other star instances from star neighborhood records, e.g., the subset $\{o_2, ..., o_k\}$ can be checked by searching the star neighborhood of o_2, the subset $\{o_3, ..., o_k\}$ can be checked by searching the star neighborhood of o_3, and so on. The star neighborhood records of $o_2, o_3, ..., o_k$ must be included in $T(f)$ by Definition 4, because all of them are neighbor with o_1 and the feature type f is a prefix feature of them by Definition 3. Thus, the clique checking operation of the star instance $\{o_1, o_2, ..., o_k\}$ can be performed independently in $T(f)$.

Theorem 1. The parallel joinless algorithm is complete.

Proof. Given any candidate co-location pattern $\{f_1, ..., f_k\}$, it must be searched by joinless algorithm in the group-dependent shard $T(f_1)$. Pruning strategies just filter the co-location patterns whose participation index is less than the threshold.

Theorem 2. The parallel joinless algorithm is correct.

Proof. The correctness of PJoinless algorithm can be guaranteed by the correctness of joinless algorithm. Given a group-dependent shard $T(f)$, executing joinless algorithm in $T(f)$ to discover all prevalent co-location patterns whose first feature type is f is correct according to Lemma 2, because all co-location instances whose first feature type is f can be generated completely from $T(f)$, which means the calculation of participation index is correct.

5 Experimental Evaluation

We implement PJoinless algorithm in Spark library function. The performance evaluation is conducted on the distributed cluster which deployed Hadoop and Spark. One master node and six worker nodes make up the cluster and each node consists of the following characteristics (1) CPU: Intel Core i7-6700, @3.4 GHz, 8 cores; (2) Memory: 8 GB. Experimental evaluations are conducted on real and synthetic data sets. We use two real data sets, the plant dataset of "Three Parallel Rivers of Yunnan Protected Areas" which contains 25 spatial features and 13,348 spatial instances, the POI of Beijing city which contains 63 spatial features and 303,895 spatial instances. The synthetic data sets are generated by the spatial data generator described in [2].

5.1 PJoinless Versus Joinless

In the first experiment, we use a small real data set, the plant dataset, because joinless algorithm easily overflows on large data set. In Fig. 4(a), we set the participation index threshold *min_prev* to 0.1. The running time of two algorithms increases with the increasing of distance threshold *d*. PJoinless algorithm is more efficient than joinless algorithm especially when *d* becomes larger. A larger value of *d* means more spatial neighbor relationships and more spatial instances could form cliques. The efficiency of joinless algorithm is slightly better than PJoinless algorithm when *d* is smaller, because PJoinless algorithm requires extra cost. In Fig. 4(b), the distance threshold is 3,000 m. The running time of two algorithms decreases as *min_prev* becomes higher, because more patterns dissatisfy the condition of prevalent co-location patterns when *min_prev* is larger. Similarly, PJoinless algorithm outperforms joinless algorithm obviously when *min_prev* is smaller, and this advantage will exist no longer when *min_prev* is larger.

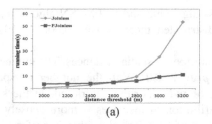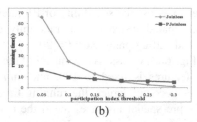

(a)   (b)

Fig. 4. Running time over the plant data set: (a) by distance thresholds, (b) by participation index thresholds

5.2 PJoinless Versus PCPM_SN

In the second experiment, we evaluate the efficiency of PJoinless algorithm compared with the parallel algorithm PCPM_SN [6]. The POI of Beijing city is used. First, the effect of the distance threshold is evaluated. In Fig. 5(a), *min_prev* is 0.4. The running time of PCPM_SN algorithm increases dramatically as the distance threshold becomes bigger, but PJoinless algorithm is not violent relatively. The performance of PJoinless algorithm is better than PCPM_SN algorithm, and this advantage will be more obvious when *d* is larger. Next, we assess the impact of the participation index threshold. In Fig. 5(b), we set *d* to 300 m. Similarly, the performance of PJoinless algorithm is preferable to PCPM_SN algorithm especially when *min_prev* is smaller.

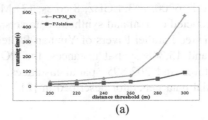 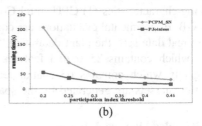

(a)   (b)

Fig. 5. Running time over the POI of Beijing city: (a) by distance thresholds, (b) by participation index thresholds

Both algorithms are built on the star neighborhood materialization model. However, PJoinless algorithm can narrow the searching space by pruning techniques in joinless algorithm and the novel pruning strategy introduced in Sect. 4.2. Furthermore, the communication cost of PJoinless algorithm is lower than PCPM_SN algorithm. Because the partitioning scheme eliminates computational dependencies so that joinless algorithm can be executed in local independently and there is no clique instances need to shuffle in network. Therefore, PJoinless algorithm is more efficient than PCPM_SN algorithm.

5.3   Scalability Evaluation

In order to assess the scalability of PJoinless algorithm and PCPM_SN algorithm, we generate spatial instances and randomly distributed them into a 10,000 × 10,000 space, and setting parameter *d* = 20.

In Fig. 6(a), we assess the impact of the number of spatial instances. The number of spatial features is fixed to 100, and *min_prev* is 0.2. With the number of spatial instances becomes larger, the running time of two algorithms is raised. In the same space, more spatial instances mean the distribution of instances is more densely and more instances form cliques. The performance of PJoinless algorithm is preferable to PCPM_SN algorithm especially when the number of spatial instances is larger.

In Fig. 6(b), the effect of the number of worker nodes is evaluated. The number of spatial features is 100, the number of total spatial instances is 1,000,000, and *min_prev* is 0.1. Increasing the number of worker nodes, the running time of two algorithms is reduced obviously, because identical computing tasks are allocated to more worker nodes to execute. In the condition of the same number of worker nodes, the performance of PJoinless algorithm is better than PCPM_SN algorithm.

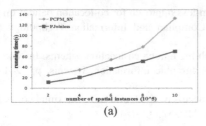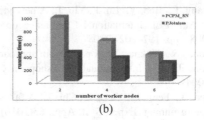

(a)   (b)

Fig. 6. Running time over synthetic data sets: (a) by the number of spatial instances, (b) by the number of worker nodes

6   Conclusion

In this work, we presented a parallel joinless algorithm to discover co-location patterns from massive spatial data. The proposed algorithm is based on a novel star neighborhood records partitioning scheme which guarantees joinless algorithm can be executed independently in each group-dependent shard. Extensive experiments show that PJoinless algorithm has a significant improvement in efficiency and has a better scalability. However, the issue of load balance in PJoinless algorithm is universal because the partitioning scheme does not ensure the computational cost is equal in each shard. Therefore, the problem of load balance in PJoinless algorithm must be considered in future work.

Acknowledgement. This work is supported by the National Natural Science Foundation of China (61472346, 61662086, 61762090), the Natural Science Foundation of Yunnan Province (2015FB114, 2016FA026), the Project of Innovative Research Team of Yunnan Province (2018HC019), and the Project of Yunnan University Graduate Student Scientific Research (YDY17110).

References

1. Shekhar, S., Huang, Y.: Discovering spatial co-location patterns: a summary of results. In: Jensen, C.S., Schneider, M., Seeger, B., Tsotras, V.J. (eds.) SSTD 2001. LNCS, vol. 2121, pp. 236–256. Springer, Heidelberg (2001). https://doi.org/10.1007/3-540-47724-1_13
2. Huang, Y., Shekhar, S., Xiong, H.: Discovering colocation patterns from spatial data sets: a general approach. IEEE Trans. Knowl. Data Eng. **16**(12), 1472–1485 (2004)
3. Yu, W.: Spatial co-location pattern mining for location-based services in road networks. Expert Syst. Appl. **46**(2016), 324–335 (2016)
4. Wang, L., Bao, X., Zhou, L.: Redundancy reduction for prevalent co-location patterns. IEEE Trans. Knowl. Data Eng. **30**(1), 142–155 (2018)
5. Wang, L., Bao, X., Chen, H., Cao, L.: Effective lossless condensed representation and discovery of spatial co-location patterns. Inf. Sci. **436–437**(2018), 197–213 (2018)
6. Yoo, J.S., Boulware, D., Kimmey, D.: A parallel spatial co-location mining algorithm based on MapReduce. In: IEEE International Congress on Big Data, pp. 25–31 (2014)
7. Yoo, J.S., Shekhar, S.: A joinless approach for mining spatial colocation patterns. IEEE Trans. Knowl. Data Eng. **18**(10), 1323–1337 (2006)

8. Wang, L., Bao, Y., Lu, J., Yip, J.: A new join-less approach for co-location pattern mining. In: 8th IEEE International Conference on Computer and Information Technology (CIT 2008), pp. 197–202 (2008)

9. Yoo, J.S., Shekhar, S.: A partial join approach for mining co-location patterns. In: The 12th Annual ACM International Workshop on Geographic Information Systems, pp. 241–249 (2004)

10. Yao, X., Peng, L., Yang, L., Chi, T.: A fast space-saving algorithm for maximal co-location pattern mining. Expert Syst. Appl. **63**(2016), 310–323 (2016)

Data Mining Techniques

Improving Maximum Classifier Discrepancy by Considering Joint Distribution for Domain Adaptation

Zehang Lin[1], Zhenguo Yang[1(✉)], Runwei Situ[1], Feitao Huang[1],
Jianming Lv[2], Qing Li[3], and Wenyin Liu[1(✉)]

[1] School of Computer Science and Technology,
Guangdong University of Technology, Guangzhou, China
gdutlin@outlook.com, yzgcityu@gmail.com,
siturunwei@163.com, hfttao@gmail.com,
liuwenyin@gmail.com
[2] School of Computer Science and Engineering,
South China University of Technology, Guangzhou, China
jmlv@scut.edu.cn
[3] Department of Computer Science, City University of Hong Kong,
Hong Kong, China
itqli@cityu.edu.hk

Abstract. Recently, domain adaptation has gained great popularity, while most researchers are focusing on domains in homogenous modalities, e.g., image domains. In reality, heterogeneous domains are pretty common and more challenging. In this paper, we present MCD-JD—a Maximum Classifier Discrepancy model which considers the joint distribution of the source and target domain data for heterogeneous domain adaption. MCD-JD derives from Generative Adversarial Networks (GAN) consisting of two parts, i.e., minimizing the discrepancy of joint distribution, and maximizing classifier discrepancy. Specifically, the first part uses the Maximum Mean Discrepancy (MMD) regularization to adapt the data distributions between source and target domains. The second part utilizes two different classifiers to maximize their discrepancy of making predictions on the target domain data, which further minimizes the discrepancy of data distributions between source and target domains. We collect a dataset depicting real-world events (e.g., protests, explosions, etc.) from multiple heterogeneous data domains, including news media textual articles, social media (Flickr) images, and YouTube videos. Extensive experiments conducted on the real-world dataset manifest the effectiveness of MCD-JD, which outperforms state-of-the-art benchmark models.

Keywords: Domain adaptation · Generative Adversarial Networks
Transfer learning · Multimedia · Social media

© Springer Nature Switzerland AG 2018
H. Hacid et al. (Eds.): WISE 2018, LNCS 11234, pp. 253–268, 2018.
https://doi.org/10.1007/978-3-030-02925-8_18

1 Introduction

In the past, people know what happened around them mainly through reading newspapers and/or watching TVs, but nowadays the Internet has become the most dominant information channel. Indeed, it is quite convenient for users to share information about human activities through such Internet platforms like Flickr, Twitter, and YouTube. Many researchers also try to make use of such data to develop a variety of applications, e.g., content-based image retrieval [1, 2], event detection [3, 4], etc. However, the data from the Internet platforms have a lot of noise due to the easy-and-lazy way of sharing data, making the data difficult to use, particularly for event detection.

In terms of event detection, existing works can be divided into three categories according to data modalities, i.e., text [5, 6], image [7, 8], and video [9, 10]. However, users may also be interested in the multi-domain data which describes the same event in reality. For instance, online news written by journalists are often more accurate and detailed reports, while social media images and videos shared by different users are noisy but more timely. The data from different Internet platforms may also describe the same real world events from different perspectives with different distributions due to the heterogeneity of data. Therefore, domain adaption techniques are highly expected to adjust the different data distributions for further pattern analysis and knowledge discovery.

Recently, Saito et al. [11] proposed a Maximum Classifier Discrepancy (MCD) model, which utilized two classifiers to maximize the discrepancy to align the distributions of source and target domain data. MCD assumes that the features generated by the Generator for source and target domain should have connections if the data are in the same class. MCD achieves impressive performance on image classification and semantic segmentation tasks. However, such an assumption may not be reasonable for heterogeneous domain adaption, where the aforementioned connections may not be obvious, making it falling short in classifying heterogeneous data in the target domain.

In this paper, we propose to extend the Maximum Classifier Discrepancy model by considering the joint distribution of the source and target domain data for heterogeneous domain adaption. We term the proposed approach as MCD-JD. More specifically, MCD-JD incorporates Maximum Mean Discrepancy (MMD) [12] to reduce the mismatch of both marginal and conditional distributions (i.e., joint distribution) between the source and target domains. MMD is a non-parametric distance measure on probability distribution, which is used to minimize the distance of the features generated by the generator from source and target domains in each class. Furthermore, MCD is utilized to maximize the discrepancy of the two different classifiers to make the data distributions of the source and target domains as close as possible. Consequently, MCD-JD is able to obtain domain-invariant features from heterogeneous domains, and transfer event knowledge across heterogeneous data, such as online news media and social media (Flickr, and YouTube).

The main contributions of our work in this paper are summarized as follows:

- We propose the MCD-JD model to reduce the mismatch of joint distribution of the heterogeneous data from source and target domains. In particular, MCD-JD obtains domain-invariant features for heterogeneous data from different domains.
- We collect a real-world dataset from three heterogeneous data domains, i.e., textual articles from online news media, images from Flickr, and videos from YouTube. The dataset is open for peer researchers to download and use.
- We conduct extensive experiments on the real-world dataset. The experimental results manifest the effectiveness of our proposed MCD-JD which outperforms other four categories of methods on domain adaption.

The rest of the paper is organized as follows. In Sect. 2, related work is reviewed. In Sect. 3, the problem and the motivation are introduced. In Sect. 4, the proposed MCD-JD model is presented. In Sect. 5, extensive experiments are conducted and analyzed, Finally, Sect. 6 offers a few concluding remarks and future work.

2   Related Work

In this section, we review some existing research works on event detection from the Internet platforms and unsupervised domain adaptation, respectively.

2.1   Event Detection from the Internet Platforms

The research work on event detection from the Internet platforms can be divided into three categories according to the main data modalities of the platforms, i.e., text, image, and video. Event detection from the text-oriented Internet platforms (e.g., online news media) is derived from the topic detection and tracking task [13], which aims to monitor a stream of broadcast news stories. For instance, Wei et al. [14] combined information extraction and text categorization techniques to detect the event from online news. In recent years, more and more researchers are paying attention to event detection from the image-oriented Internet platforms (e.g., Flickr and Twitter) and video-oriented Internet platforms (e.g., YouTube), which are more challenging. Petrovic et al. [15] utilized cosine similarity between documents to detect events in Twitter, and Sakaki et al. [16] detected earthquakes in Twitter. Chen et al. [17] detected events from Flickr photos by exploiting the tags. Petkos et al. [18] published a series of event detection challenges on Flickr, Instagram, and YouTube. However, the data in different domains can complement with each other for describing events, while these methods are focusing on single domain data alone.

2.2   Unsupervised Domain Adaptation

Domain adaptation refers to the scenarios where labeled data from the target domain are insufficient, while the source domain has much more labeled data. It aims to solve the problem of different data distributions between the source and target domains [19]. In particular, unsupervised domain adaptation focuses on the unknown target domain, which is meaningful and challenging. In recent years, many researchers are utilizing deep neural networks in domain adaptation (known as deep domain adaptation). Deep neural networks are powerful methods to learn feature representations, and can be combined with the traditional transfer learning methods [20–22]. The work on deep domain adaptation can be divided into two categories, i.e., statistics-based methods [23–25], and adversarial methods [11, 26, 27]. Statistics-based methods are developed from traditional transfer learning methods, which aim to minimize domain shift by aligning the higher-order statistics of the source and target distributions. For instance, Long et al. [23] explored multiple kernels for adapting deep representations. Sun et al. [24] extended a simple yet effective unsupervised domain adaptation method called CORAL [22] by using deep neural networks. In addition, the adversarial methods exploit the idea of Generative Adversarial Networks (GAN) [28] to train the system to become unable to distinguish whether the data representation resulting from the source domain or the target domain. Ganin et al. [26] designed a domain-adversarial neural network, which integrated a gradient reversal layer to ensure that the feature distributions of the two domains are similar. Saito et al. [11] used two classifiers to maximize the discrepancy between the source and target domains, and trained the feature generator to minimize such discrepancy. However, these approaches deal with homogeneous domain adaption, e.g., transferring an image domain to another image domain, which cannot be used to deal with the domain adaption between heterogeneous data domains, such as online news media and social media (Flickr and YouTube).

3   Problem Statement and Motivation

Without loss of generality, we assume there are source domain data X_S with labels Y_S drawn from a distribution $P_S(x, y)$, and target domain data X_T drawn from a distribution $P_T(x, y)$. The two distributions are not consistent, i.e.,

$$P_S(x, y) \neq P_T(x, y) \tag{1}$$

Unsupervised domain adaptation aims to obtain data representations for the two-domain data that obey the same distribution, so that a classifier trained on the source domain can be used to make predictions on the target domain data directly.

Quite a few researchers utilize GAN to learn a common feature space between the source and target domains by training a feature generator G, and fool a discriminator D to make it unable to distinguish whether the feature representation is learned from the source domain or the target domain, i.e., making $P_S(G(x))$ and $P_T(G(x))$ close. In particular, Saito et al. [11] proposed a Maximum Classifier Discrepancy (MCD) model using two different classifiers in the target domain to align the distribution of each class between the source and target domains. More specifically, MCD trains two classifiers in the source domain, which will be used to make predictions on the features generated by the Generator for the target domain data. MCD maximizes the discrepancy between the distributions of labels predicted by the classifiers, and further trains the Generator to minimize the discrepancy between such distributions. Consequently, the data of each class between the source and target domains can be aligned.

However, MCD suffers from the initial performance of the classifiers, while poor classification performance at the beginning could make the adversarial training to fail. Especially for the heterogeneous domain adaptions, it is hard to obtain high initial classification performance due to the heterogeneity problem. Consequently, we introduce the Maximum Mean Discrepancy (MMD) regularization by considering the joint distribution of the source and target domain data, so as to bridge the gap between the class features generated from the source domain and target domains. The details are presented in the next section.

4   MCD-JD: The Model and Algorithm

As shown in Fig. 1, we divide MCD-JD into two parts: minimizing the discrepancy of joint distribution and maximizing classifier discrepancy. The first part utilizes the Maximum Mean Discrepancy (MMD) regularization to adapt the data distribution between source and target domains. The second part uses two classifiers to maximize their discrepancy of making predictions on target domain data, which further minimizes the discrepancy of data distribution generated by the generator between source and target domains.

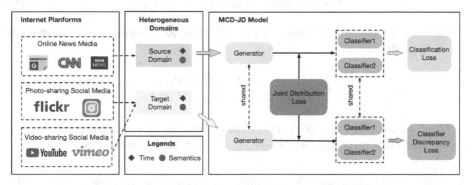

Fig. 1. Overview of the MCD-JD framework (assume the source domain is news media and the target domain is Flickr or YouTube). One stream operates on the source data and the other on the target data.

(1) **Minimizing the Discrepancy of Joint Distribution.** Domain adaptation aims to reduce the discrepancy of joint distribution between $P_S(G(x), y)$ and $P_T(G(x), y)$, which cannot be estimated directly from the observed data. Therefore, we decompose them according to the following Bayes formula:

$$\begin{cases} P_S(G(x), y) = P_S(G(x))P_S(y|G(x)) \\ P_T(G(x), y) = P_T(G(x))P_T(y|G(x)) \end{cases} \tag{2}$$

which transforms the joint distribution into marginal distributions ($P_S(G(x))$ and $P_T(G(x))$) and conditional distributions ($P_S(y|G(x))$ and $P_T(y|G(x))$).

Furthermore, we introduce the maximum mean discrepancy (MMD) [29] to measure the distances between two distributions. Without loss of generality, let $\mathcal{H}_k$ denote the reproducing kernel Hilbert space (RKHS), and we assume that source/target domain data X_S/X_T obey distributions p/q, respectively, and the mean embedding of distribution p in $\mathcal{H}_k$ is $\mu_k(p)$. We have the relation $\mathbf{E}_{x\sim p}f(x) = \langle f(x), \mu_k(p)\rangle_{\mathcal{H}_k}$ for all $f \in \mathcal{H}_k$. The MMD regularization term in RKHS is defined as follows:

$$
\begin{aligned}
MMD^2(p,q) &= \left\| \mathbf{E}_p[k(x_S)] - \mathbf{E}_q[k(x_T)] \right\|_{\mathcal{H}_k}^2 \\
&= \mathbf{E}_{x_S,x_s'}\left[k\left(x_S, x_s'\right)\right] - 2\mathbf{E}_{x_S,x_T}[k(x_S, x_T)] + \mathbf{E}_{x_T,x_T'}\left[k\left(x_T, x_T'\right)\right]
\end{aligned}
\tag{3}
$$

where $k(\cdot)$ is a kernel function, which can be the Gaussian kernel $k(a,b) = \exp\left(-\frac{\|a-b^2\|}{\gamma}\right)$ with width γ; and x_S, x_s' and x_T, x_T' are samples from X_S and X_T, respectively. Specially, we have $MMD^2(p,q) = 0$ if $p = q$ according to [21].

For the marginal distribution, we minimize the MMD loss to align the source and target domain data as follows:

$$
\ell_{Margin} = MMD^2(P_S(G(x)), P_T(G(x)))
\tag{4}
$$

For the conditional distribution, as $P_T(y|G(x))$ is not observable, we transform it according to Bayes' theorem as follows:

$$
\begin{cases}
P_S(y|G(x)) \propto P_S(G(x)|y)P_S(y) \\
P_T(y|G(x)) \propto P_T(G(x)|y)P_T(y)
\end{cases}
\tag{5}
$$

Here, we assume $P_S(y) \approx P_T(y)$, therefore we only need to minimize the discrepancy between $P_S(G(x)|y)$ and $P_T(G(x)|y)$, which can be defined as follows:

$$
\ell_{Condition} = \frac{1}{|C|}\sum_{c\in C} MMD^2(P_S(G(x)|c), P_T(G(x)|c))
\tag{6}
$$

where C is a set of classes, $P_S(G(x)|c)$ and $P_T(G(x)|c)$ are the distributions of features for source and target samples of class c, respectively. In particular, $P_T(G(x)|c)$ cannot be estimated as the labels of target data are not available. Therefore, we use the pseudo label generated by the source classifiers after a few iterations of training, which facilitates us to minimize intra-class distance between the source and target domain data.

Finally, we train the feature generator G and two classifiers (F_1 and F_2) by using the data from the source and target domains to minimize the aforementioned three losses, i.e., classification loss, marginal loss, and conditional loss, as specified below:

$$\overset{min}{G, F_1, F_2} \; \ell_C + \alpha\left(\ell_{Margin} + \ell_{Condition}\right) \tag{7}$$

$$\ell_C = \mathbf{E}_{(x_S, y_S) \sim (G(X_S), Y_S)} J(F_1(x_S), y_S) + \mathbf{E}_{(x_S, y_S) \sim (G(X_S), Y_S)} J(F_2(x_S), y_S) \tag{8}$$

where J denotes the cross-entropy loss function, and α control the importance of the regularization terms.

(2) **Maximize Classifier Discrepancy.** MCD aims to maximize the discrepancy between the distributions of predictions made by the two classifiers on the features generated by G. Simultaneously, MCD minimizes such discrepancy loss of two classifiers and optimizes G to generate similar distributions of predictions between the source and target domain data.

Algorithm 1: The MCD-JD Algorithm

Input: source data X_S; target data X_T; randomly initialized generator G; randomly initialized classifiers F_1 and F_2; the number of steps for joint distribution training n; the number of steps for conditional distribution training k.

Output: generator G; classifiers F_1 and F_2

```
1:    step ← 0
2:    repeat
3:        Sample {x_i^s, y_i^s}_{i=1}^m, {x_i^t}_{i=1}^m from X_S and X_T
4:        for t = 1, ..., n do
5:            if step < k then
6:                update G, F_1 and F_2 by minimizing Equation (4) and (8)
7:            else
8:                update G, F_1 and F_2 by minimizing Equation (7)
9:            step ← step + 1
10:           end if
11:       end for
12:       update G, F_1 and F_2 by minimizing Equation (9)
13:       update G, F_1 and F_2 by minimizing Equation (11)
14:   until G, F_1 and F_2 converge
```

More specifically, in order to maximize the discrepancy loss to obtain the data samples resulting in large discrepancy loss, we use the target data to train F_1 and F_2 with fixing the parameters of G as follows:

$$\overset{min}{F_1, F_2} \; \ell_C - \ell_{adv} \tag{9}$$

$$\ell_{adv} = \mathbf{E}_{x_T \sim G(X_T)} [d(p_1(y|x_T), p_2(y|x_T))] \tag{10}$$

where d denotes the distance between two distributions, which can be measured by using L_1-distance; $p_1(y|x_T)$ and $p_2(y|x_T)$ are the distributions generated by F_1 and F_2. In particular, we use ℓ_C to prevent both classifiers from decreasing accuracy of predictions.

In order to minimize the discrepancy loss to make G generate data representation obeying the same distribution, we use the target data to train G by fixing the parameters of F_1 and F_2 as follows:

$$\underset{G}{min} \; \ell_{adv} \tag{11}$$

For completeness, the proposed MCD-JD is summarized in Algorithm 1.The convergence of Algorithm 1 is shown in Sect. 5.7.

5   Experiments

In this section, we introduce the dataset we have collected for multiple domain event detection tasks, and evaluate the performance of the approaches.

5.1   Dataset

The dataset denoted as MDED consists of data samples from three data domains in three data modalities, i.e., text (online news articles), image (Flickr) and video (You-Tube). The dataset currently contains 68 real-world events, and will be further enlarged. MDED includes 10,678 online news articles collected from different web-sites, such as CNN News, Google News, BBC News, the Star, etc., 23,874 images shared by Flickr users, and 1,337 videos uploaded by YouTube users. Table 1 sum-marizes the statistics of the MDED dataset. Figure 2 shows the distributions of each domain corresponding to each event. Figure 3 shows an example of data samples from the MDED dataset.

Table 1. Statistics of the MDED dataset.

Domain	#events	#items	Time (%)	Title (%)	Username (%)	Description (%)
Online news media	68	10,678	100	100	N/A	100
Flickr	68	23,874	100	99.53	100	99.17
YouTube	68	1,337	100	100	N/A	94.76

We investigate six cross-domain scenarios, i.e., Online News Media → Flickr, Online News Media → YouTube, Flickr → Online News Media, Flickr → YouTube, YouTube → Flickr and YouTube → Online News Media.

5.2   Evaluation Metrics

The evaluation metrics include F1-measure and Normalized Mutual Information [3, 18, 30]. Both metrics range from 0 to 1, and the higher value indicates better performance.

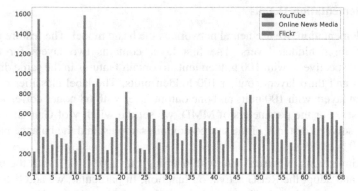

Fig. 2. The distributions of the number of items in each event class.

Fig. 3. An example of data samples from the MDED dataset.

(1) **F1-measure (F1).** It combines both precision and recall for the event clusters.
(2) **Normalized Mutual Information (NMI).** It quantifies the "amount of information" between source domain labels and the target domain labels.

5.3 Data Preparation

An event can be characterized according to its time and semantics (e.g., Who, What, How, Why, etc.). In our experiment, we extract the time and semantic information from the time, title, and textual descriptions as features, which are critical aspects for describing events. For the time features, we use the year, month and day directly. For semantics, we utilize the Skip-Thoughts model [31] trained on the BookCorpus dataset [32] to obtain a 4800-dimensional vector as the semantic feature. In particular, we utilize zero-mean normalization to normalize the time feature and semantic feature, respectively, as the input data.

5.4 Implementation Details

In our implementation, we use neural network as the basic model. The feature generator consists of three hidden layers. The first layer contains two layers for time and semantics, respectively, with 100 hidden units to obtain features in the same dimension. The second and third layers contain 100 hidden units. The label classifier consists of three hidden layers with 100 units and one output layer with 68 nodes corresponding to the 68 classes. For the parameters of MMD, we set the width γ of Gaussian kernel as the median squared distance of the source-domain samples, and the balance parameters α is set to 0.3.

For optimizations, we use Adam with learning rate 1.0×10^{-3} with the batch size set as 180, and adopt the ReLU activation function. In particular, we add batch normal layers [33] before each activation function and dropout layer [34] before each layer to prevent over-fitting. Since the number of samples in each class is unbalanced, we randomly sample the same number of data in each class during each training epoch. We set the number of step for joint distribution training n as 20 and the number of steps for training conditional distribution k as 10.

5.5 Baselines

We compare our proposed MCD-JD with a number of baselines in four categories:

(1) **Classification models**: these include Random Forests (RF) [35], and neural network (NN) [36].
(2) **Common space based domain adaptation**: this category includes Geodesic Flow Kernel (GFK) [37]. Specifically, GFK aims to bridge the source and target domains by interpolating across an infinite number of intermediate subspaces. In our experiment, we set the dimension of the geodesic flow kernel as 100.
(3) **Statistical regularized deep learning models**: these include Deep Correlation Alignment (CORAL) [24], Deep Adaptation Network (DAN) [23], and Joint Adaptation Network (JAN) [25]. Specifically, CORAL minimizes different domain discrepancy by aligning the second-order statistics of the source and target distributions. DAN embeds features of multiple task-specific layers into reproducing kernel Hilbert spaces (RKHSs) and uses multi-kernel MMD to minimize different distributions from the source and target domains. JAN learns transferable features by aligning the joint distributions of multiple domain-specific layers based on a joint maximum mean discrepancy (JMMD) criterion. For a fair comparison, we use the same network structures as our MCD-JD for these approaches, and the other parameters are set by following their papers.
(4) **Adversarial models for domain adaptation**: this category includes Domain-Adversarial Neural Networks (DANN) [26], Adversarial Representation learning approach (ARDA) [27], and Maximum Classifier Discrepancy (MCD) [11]. DANN uses a gradient-reversal trick to make the source and target domains indistinguishable. ARDA incorporates Wasserstein distance of Wasserstein GAN to measure distribution divergence. MCD attempts to align distributions of the source and target domains by utilizing the task-specific decision boundaries.

5.6 Performance on the MCED Dataset

The performance of the various approaches on the MCED dataset is shown in Tables 2 and 3. MCD-JD outperforms JAN and MCD, which only minimizes the distance of distribution or maximizes the classifier discrepancy alone, demonstrating the significance of integrating them jointly by MCD-JD. For the scenarios of F → Y and Y → F, GFK achieves better performance. The reason as we think is that the performance of the classifier on the source domain has an impact on the adversarial training our MCD-JD, while the number of data samples collected from YouTube is limited for training classifiers. We also note that statistical regularized deep learning models (i.e., CORAL, DAN, and JAN) mostly outperform the adversarial methods (i.e., DANN, ARDA, and MCD), demonstrating the obvious divergence between the source and target domains. Overall, our proposed MCD-JD outperforms the baselines for most of the transfer scenarios.

Table 2. F1 performance of the approaches on MDED dataset (O: Online news media, F: Flickr, Y: YouTube)

Method	O → F	O → Y	F → O	F → Y	Y → O	Y → F	Mean
RF	0.1740	0.1203	0.0499	0.1095	0.0463	0.1790	0.1132
NN	0.3756	0.4061	0.1764	0.2269	0.1546	0.2887	0.2714
GFK	0.3932	0.2814	0.1943	**0.2688**	0.1740	**0.3735**	0.2809
CORAL	0.3805	0.3991	0.1705	0.2156	0.1409	0.2757	0.2637
DAN	0.4019	0.3657	0.1640	0.2254	0.1704	0.3423	0.2783
JAN	0.3940	0.3844	0.1987	0.2153	0.1779	0.3414	0.2853
DANN	0.3651	0.3090	0.1863	0.1973	0.1395	0.3153	0.2521
ARDA	0.3316	0.2564	0.1603	0.1482	0.1622	0.2662	0.2208
MCD	0.4204	0.3194	0.2118	0.2240	0.1640	0.3118	0.2752
MCD-JD (ours)	**0.4822**	**0.4267**	**0.2202**	0.2360	**0.1873**	0.3319	**0.3140**

5.7 Convergence of MCD-JD

The training loss and performance of MCD-JD along with the iterations are shown in Fig. 4. From Fig. 4(a), we can see that all the loss are converged after training for a few epochs. In Fig. 4(b), we report the performance on the target domain. From the two subfigures, we can see that both performance metrics increase while the discrepancy loss decrease with training epochs. The experimental results demonstrate the effectiveness of the training and testing procedures.

5.8 Feature Visualization

We take the data samples related to two event classes (i.e., January 2016 United States blizzard and 2010 Pakistan floods) as an example, by using t-SNE [38] to show the features obtained by the model as shown in Fig. 5. From the figure, we have the

Table 3. NMI performance of the approaches on MDED dataset (O: Online news media, F: Flickr, Y: YouTube)

Method	O → F	O → Y	F → O	F → Y	Y → O	Y → F	Mean
RF	0.3392	0.3832	0.1134	0.3548	0.1202	0.3531	0.2773
NN	0.5931	0.6682	0.3699	0.4931	0.3260	0.5124	0.4938
GFK	0.6330	0.5723	0.4056	**0.5467**	0.3745	**0.6158**	0.5246
CORAL	0.6148	0.6711	0.3749	0.5263	0.3088	0.5138	0.5016
DAN	0.6356	0.6530	0.3724	0.5419	0.3623	0.5793	0.5241
JAN	0.6243	0.6641	0.4140	0.5347	0.3694	0.5813	0.5313
DANN	0.6064	0.6075	0.3953	0.5076	0.3117	0.5255	0.4923
ARDA	0.5676	0.5666	0.3658	0.4653	0.2865	0.5045	0.4594
MCD	0.6362	0.6163	0.4248	0.5272	0.3487	0.5306	0.5140
MCD-JD (ours)	**0.6801**	**0.6916**	**0.4383**	0.5441	**0.3776**	0.5585	**0.5484**

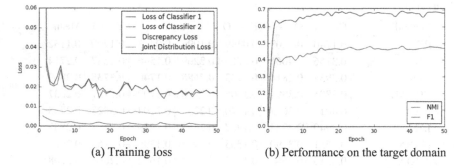

(a) Training loss   (b) Performance on the target domain

Fig. 4. Convergence of MCD-JD (online news media → Flickr)

following observations: (1) From Fig. 5(a) and (b), we can observe that it is hard for classifiers to distinguish the two classes in the target domain, as the regions of the two classes are upside down between source domain and target domain. (2) From Fig. 5(a) and (c), we can observe that it becomes possible to distinguish the classes in the target domain, as their regions are becoming close to the ones in the source domain after using JD. (3) From Fig. 5(b) and (d), we can find that the MCD obtain similar data distribution for the two-class data in the target domain, while the regions of the two classes are still upside down between source domain (refer to Fig. 5(a)) and target domain. (4) From Fig. 5(a) and (e), we can see that MCD-JD both obtains similar data distribution for the two-class data in the target domain, and the regions of the two classes are consistent with the ones in the source domain. Therefore, MCD-JD benefits to the classifications on the data in the target domain.

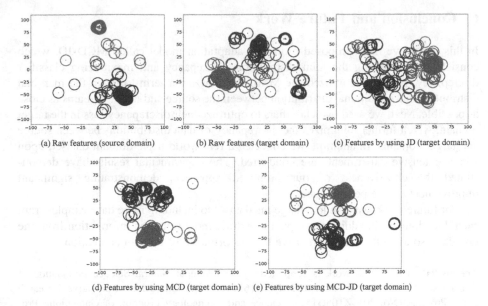

(a) Raw features (source domain)   (b) Raw features (target domain)   (c) Features by using JD (target domain)

(d) Features by using MCD (target domain)   (e) Features by using MCD-JD (target domain)

Fig. 5. Visualization of the features. Red and Green circles indicate the event samples from January 2016 United States Blizzard and 2010 Pakistan floods, respectively. (Color figure online)

5.9   Evaluation of Parameter α

The proposed MCD-JD has a parameter α controlling the impact of the regularization term for joint distribution of source and target domain data. We evaluate the impact of α in Fig. 6, from which we can observe that introducing the regularization term (i.e., $\alpha > 0$) achieves improvement on the performance. The experimental results manifest the effectiveness of taking into account the joint distribution for domain adaption. In particular, our MCD-JD obtains the best performance when α is set as 0.3.

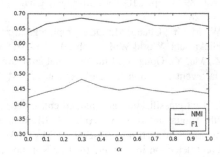

Fig. 6. Impact of α.(Online News Media → Flickr)

6 Conclusion and Future Work

In this paper, we have proposed a domain adaptation model called MCD-JD, which consists of minimizing the joint distribution discrepancy and maximizing classifier discrepancy. Firstly, we utilized the MMD regularization term to make the marginal distribution and conditional distribution between the source and target domains as close as possible. Next, we used two classifiers to optimize their discrepancy loss in the target domain, which further minimizes the discrepancy of data distributions between source and target domains. In addition, we collected a real-world triple-domain dataset, upon which extensive experiments are conducted. The experimental results have demonstrated the effectiveness of our proposed approach, demonstrating significant improvement on the performance.

For future work, we plan to enlarge the dataset to including more data samples from each data domain. In addition, we plan to extract more semantic information from the raw data, so as to the further improve the performance of domain adaptation.

Acknowledgments. This work is supported by the National Natural Science Foundation of China (No. 61703109, No. 91748107, No.U1611461), the Guangdong Innovative Research Team Program (No. 2014ZT05G157), Science and Technology Program of Guangdong Province, China (No. 2016A010101012), and CAS Key Lab of Network Data Science and Technology, Institute of Computing Technology, Chinese Academy of Sciences, 100190, Beijing, China. (No. CASNDST201703), and an internal grant from City University of Hong Kong (project no. 9610367).

References

1. Hsieh, L.C., Hsu, W.H.: Search-based automatic image annotation via flickr photos using tag expansion. In: ICASSP, pp. 2398–2401 (2010)
2. Ginsca, A.L., Popescu, A., Le Borgne, H., Ballas, N., Vo, P., Kanellos, I.: Large-scale image mining with flickr groups. In: He, X., Luo, S., Tao, D., Xu, C., Yang, J., Hasan, M.A. (eds.) MMM 2015. LNCS, vol. 8935, pp. 318–334. Springer, Cham (2015). https://doi.org/10.1007/978-3-319-14445-0_28
3. Yang, Z., Li, Q., Liu, W., Ma, Y., Cheng, M.: Dual graph regularized NMF model for social event detection from Flickr data. World Wide Web **20**, 995–1015 (2017)
4. Yang, Z., Li, Q., Lu, Z., Ma, Y., Gong, Z., Liu, W.: Dual structure constrained multimodal feature coding for social event detection from Flickr data. ACM Trans. Internet Technol. (TOIT) **17**, 19 (2017)
5. Kumaran, G., Allan, J.: Text classification and named entities for new event detection. In: Proceedings of the 27th Annual International ACM SIGIR Conference on Research and Development in Information Retrieval, pp. 297–304 ACM (2004)
6. Weng, J., Lee, B.S.: Event detection in Twitter. In: ICWSM, pp. 401–408 (2011)
7. Zaharieva, M., Zeppelzauer, M., Breiteneder, C.: Automated social event detection in large photo collections. In: Proceedings of the 3rd ACM Conference on International Conference on Multimedia Retrieval, pp. 167–174 ACM (2013)

8. Firan, C.S., Georgescu, M., Nejdl, W., Paiu, R.: Bringing order to your photos: event-driven classification of flickr images based on social knowledge. In: Proceedings of the 19th ACM International Conference on Information and Knowledge Management, pp. 189–198 ACM (2010)
9. Ye, G., Li, Y., Xu, H., Liu, D., Chang, S.F.: Eventnet: a large scale structured concept library for complex event detection in video. In: Proceedings of the 23rd ACM International Conference on Multimedia, pp. 471–480 ACM (2015)
10. Abhik, D., Toshniwal, D.: Sub-event detection during natural hazards using features of social media data. In: Proceedings of the 22nd International Conference on World Wide Web, pp. 783–788 ACM (2013)
11. Saito, K., Watanabe, K., Ushiku, Y., Harada, T.: Maximum classifier discrepancy for unsupervised domain adaptation. In: Computer Vision and Pattern Recognition (CVPR) (2018)
12. Sejdinovic, D., Sriperumbudur, B., Gretton, A., Fukumizu, K.: Equivalence of distance-based and RKHS-based statistics in hypothesis testing. Ann. Stat. **41**, 2263–2291 (2013)
13. Allan, J., Papka, R., Lavrenko, V.: On-line new event detection and tracking. In: Proceedings of the 21st Annual International ACM SIGIR Conference on Research and Development in Information Retrieval, pp. 37–45. ACM (1998)
14. Wei, C., Lee, Y.: Event detection from online news documents for supporting environmental scanning. Decis. Support Syst. **36**, 385–401 (2004)
15. Petrovic, S., Osborne, M., and Lavrenko, V.: Streaming first story detection with application to Twitter. In: The 2010 Annual Conference of the North American Chapter of the Association for Computational Linguistics, HLT 2010, pp. 181–189 (2010)
16. Sakaki, T., Okazaki, M., Matsuo, Y.: Earthquake shakes Twitter users: real-time event detection by social sensors. In: Proceedings of the 19th International Conference on World Wide Web, pp. 851–860. ACM (2010)
17. Chen, L., Roy, A.: Event detection from flickr data through wavelet-based spatial analysis. In: Proceedings of the 18th ACM Conference on Information and Knowledge Management, pp. 523–532 ACM (2009)
18. Petkos, G., et al.: Social event detection at MediaEval: a three-year retrospect of tasks and results. In: Proceedings ACM ICMR 2014 Workshop on Social Events in Web Multimedia (2014)
19. Pan, S.J., Yang, Q.: A survey on transfer learning. IEEE Trans. Knowl. Data Eng. **22**, 1345–1359 (2010)
20. Long, M., Wang, J., Ding, G., Sun, J., Philip, S.Y.: Transfer feature learning with joint distribution adaptation. In: 2013 IEEE International Conference on Computer Vision (ICCV), pp. 2200–2207. IEEE (2013)
21. Gretton, A., Borgwardt, K.M., Rasch, M.J., Schölkopf, B., Smola, A.: A kernel two-sample test. J. Mach. Learn. Res. **13**, 723–773 (2012)
22. Sun, B., Feng, J., Saenko, K.: Return of frustratingly easy domain adaptation. In: AAAI, p. 8 (2016)
23. Long, M., Cao, Y., Wang, J., Jordan, M.: Learning transferable features with deep adaptation networks. In: International Conference on Machine Learning, pp. 97–105 (2015)
24. Sun, B., Saenko, K.: Deep CORAL: correlation alignment for deep domain adaptation. In: Hua, G., Jégou, H. (eds.) ECCV 2016. LNCS, vol. 9915, pp. 443–450. Springer, Cham (2016). https://doi.org/10.1007/978-3-319-49409-8_35
25. Long, M., Zhu, H., Wang, J., Jordan, M.I.: Deep transfer learning with joint adaptation networks. In: International Conference on Machine Learning, pp. 2208–2217 (2017)
26. Ganin, Y., et al.: Domain-adversarial training of neural networks. J. Mach. Learn. Res. **17**, 2030–2096 (2016)

27. Shen, J., Qu, Y., Zhang, W., Yu, Y.: Adversarial representation learning for domain adaptation. arXiv preprint arXiv:1707.01217 (2017)
28. Goodfellow, I., et al.: Generative adversarial nets. In: Advances in Neural Information Processing Systems, pp. 2672–2680 (2014)
29. Borgwardt, K.M., Gretton, A., Rasch, M.J., Kriegel, H.P., Schölkopf, B., Smola, A.J.: Integrating structured biological data by kernel maximum mean discrepancy. Bioinformatics 22, e49–e57 (2006)
30. Daras, P., Manolopoulou, S., Axenopoulos, A.: Search and retrieval of rich media objects supporting multiple multimodal queries. IEEE Trans. Multimed. 14, 734–746 (2012)
31. Kiros, R., et al.: Skip-thought vectors. In: Advances in Neural Information Processing Systems, pp. 3294–330 (2015)
32. Zhu, Y., et al.:Aligning books and movies: towards story-like visual explanations by watching movies and reading books. In: Proceedings of the IEEE International Conference on Computer Vision, pp. 19–27 (2015)
33. Ioffe, S., Szegedy, C.: Batch normalization: accelerating deep network training by reducing internal covariate shift. In: International Conference on Machine Learning, pp. 448–456 (2015)
34. Srivastava, N., Hinton, G., Krizhevsky, A., Sutskever, I., Salakhutdinov, R.: Dropout: a simple way to prevent neural networks from overfitting. J. Mach. Learn. Res. 15, 1929–1958 (2014)
35. Breiman, L.: Random forests. Mach. Learn. 45, 5–32 (2001)
36. Bengio, Y., Lamblin, P., Popovici, D., Larochelle, H.: Greedy layer-wise training of deep networks. In: Advances in Neural Information Processing Systems, pp. 153–160 (2007)
37. Gong, B., Shi, Y., Sha, F., Grauman, K.: Geodesic flow kernel for unsupervised domain adaptation. In: 2012 IEEE Conference Computer Vision and Pattern Recognition (CVPR), pp. 2066–2073. IEEE (2012)
38. Maaten, L.V.D., Hinton, G.: Visualizing data using t-SNE. J. Mach. Learn. Res. 9, 2579–2605 (2008)

Density Biased Sampling with Locality Sensitive Hashing for Outlier Detection

Xuyun Zhang[1]([✉]), Mahsa Salehi[2], Christopher Leckie[3], Yun Luo[4], Qiang He[5],
Rui Zhou[5], and Rao Kotagiri[3]

[1] Department of Electrical and Computer Engineering, University of Auckland,
Auckland, New Zealand
xuyun.zhang@auckland.ac.nz
[2] Faculty of Information Technology, Monash University, Melbourne, Australia
[3] Department of Computing and Information Systems, University of Melbourne,
Melbourne, Australia
[4] Faculty of Computer Science and Technology, Guizhou University, Guiyang, China
[5] School of Software and Electrical Engineering, Swinburne University of Technology,
Melbourne, Australia

Abstract. Outlier or anomaly detection is one of the major challenges in
big data analytics since unusual but insightful patterns are often hidden
in massive data sets such as sensing data and social networks. Sampling
techniques have been a focus for outlier detection to address scalability on big data. The recent study has shown uniform random sampling
with ensemble can boost outlier detection performance. However, uniform sampling assumes that all points are of equal importance, which
usually fails to hold for outlier detection because some points are more
sensitive to sampling than others. Thus, it is necessary and promising
to utilise the density information of points to reflect their importance
for sampling based detection. In this paper, we formally investigate density biased sampling for outlier detection, and propose a novel density
biased sampling approach. To attain scalable density estimation, we use
Locality Sensitive Hashing (LSH) for counting the nearest neighbours of
a point. Extensive experiments on both synthetic and real-world data
sets show that our approach significantly outperforms existing outlier
detection methods based on uniform sampling.

Keywords: Outlier/anomaly detection · Locality-Sensitive Hashing
Density biased sampling · Big data · Unsupervised learning

1 Introduction

Huge amounts of data are being generated by deployed sensors, social media,
search engine logs and transactions, to name a few. Big data analytics play a
key role to extract insightful information from enormous volumes of data. The
outlier or anomaly detection problems are of paramount importance in big data

© Springer Nature Switzerland AG 2018
H. Hacid et al. (Eds.): WISE 2018, LNCS 11234, pp. 269–284, 2018.
https://doi.org/10.1007/978-3-030-02925-8_19

analytics since they are concerned with detecting unusual but insightful patterns from data sets. A variety of outlier detection methods have been proposed [6]. Traditional statistical methods for outlier detection have been studied by researchers where they assume an underlying distribution of normal data points as a priori [26]. Distance based and density based approaches form another category of outlier detection techniques where the focus is on the distances between data points and their k nearest neighbour (k-NN) distances for evaluating the outlierness scores of data points without assuming any underlying distributions [5,12]. There is another category of methods that exploits the variances of angles between data points to detect outliers [14]. These techniques have been adopted across a wide range of areas like fraud detection, traffic accident detection and network intrusion detection [6].

Sampling techniques have been extensively studied in the statistics community and widely adopted in diverse applications. An attractive feature of sampling is that problems can be analysed on a much smaller scale. This salient feature is particularly appealing in the era of big data where one critical challenge is to process large data volumes efficiently. Thus, the sampling techniques are widely adopted in many data analytics tasks like clustering analysis. In particular with outlier detection, recent studies surprisingly discovered that uniform random sampling can enhance the detection performance [1,23,28,29].

However, the underlying assumption in uniform sampling that all points are of equal importance fails to hold for outlier detection, because some points (e.g., weak outliers or weak inliers) are more sensitive to sampling than others (e.g., strong inliers). The sensitivity can be reflected by the consequent outlier rank inversion. Note that sampling also incurs outlier rank inversion while enhancing the contrast of outlierness. Mostly, strong inliers can still be detected as inliers after sampling, while weak inliers and weak outliers are at risk of being labeled incorrectly. As the outlierness used in distance based detection methods is often highly sensitive to density of points, it is helpful to exploit the density information in the sampling process to produce a sample where outliers can be detected more easily.

In this paper, we formally investigate density biased sampling for outlier detection, and propose a novel density biased sampling approach. The basic idea is to split points into two parts by a density boundary, and use different parameters of density biased sampling for each part. The major contributions of our work are fourfold. First, we study the sampling effects of density on outlierness formally and empirically. Then, we propose a piecewise density biased sampling approach. Moreover, we leverage Locality Sensitive Hashing (LSH) [10] to improve the scalability of density estimation. A parameter tuning rule specific to outlier detection is proposed for LSH. Finally, we conduct extensive experiments on both synthetic and real-world data sets over a variety of detection methods, showing the effectiveness and efficiency of our approach compared to uniform sampling.

The rest of the paper is organised as follows. The next section briefly surveys related work. The sampling effects of density on outlierness are formally investigated in Sect. 3. Our novel density biased sampling approach and the density estimation using LSH are formulated in Sect. 4. Experimental settings and results are presented in Sect. 5. We conclude our work in Sect. 6.

2 Related Work

In this section, we briefly survey the existing work on outlier detection approaches based on sampling or LSH.

The work [25] proposed to draw a random sample of a fixed size for each point and derive the outlierness scores (the k-NN distance) of a point from this sample. Furthermore, only one sample drawn from the original data was used to compute the outlierness scores for all points in [23]. As pointed out in [2], the detection performance gain from sampling is parameter- and data-dependent due to the unpredictable sampling effects on bias. But the introduced variance is more likely to degrade the performance. Thus, the ensemble technique has been exploited to overcome these problems by reducing the bias or variance. The work [1,2] has initially formulated the algorithmic framework of ensemble outlier analysis, and developed the theoretical foundation from the perspective of the bias-variance trade-off theory. Following the framework, two sequential ensemble approaches for outlier detection have been proposed in [19,21], where a set of dependent based anomaly detectors are produced. Besides, a series of isolation forest based anomaly detection methods were proposed in [15,16,27], which can derive anomaly scores from an ensemble (forest) of isolation trees in a very fast way. The work [22,29] formally showed that uniform sampling enlarges outlierness contrast. All the methods above adopted uniform sampling without considering the effects of density.

The work [13] proposed a density biased sampling approach for clustering and outlier detection. The probability that a point is included in the sample is determined by the point's local density. For outlier detection tasks, sparse regions are oversampled to avoid missing outliers. Very similarly, the work in [9] proposed a density biased sampling for local outlier detection. In addition to sampling sparser regions at higher sampling rates, our approach can also sample at lower sampling rates to strengthen outlierness contrast. The work in [29] has formally validated that outlierness contrast can be increased by subsampling. Since biased sampling is mainly used for clustering tasks [13], its usage for outlier detection is often treated as a byproduct. Thus, the benefits of biased sampling for outlier detection have not been well recognised. Even the effects of uniform sampling on outlier detection have only been recognised and studied recently as discussed above.

LSH has been recently incorporated in outlier detection methods due to its salient features [10]. The work [24] proposed a ranking based outlier detection method using LSH to reorder points by their likelihood of being an outlier. The counts of points falling into the same buckets after hashing are used to estimate the outlierness. A recently developed anomaly detection method ACE [17] also followed the same basic idea and provided more formal analysis on the variance of the estimation. In our approach, LSH is exploited for density estimation which is further used in the biased sampling process, and a parameter tuning rule specific to outlier detection is proposed. The work [20] used LSH to prune strong inliers based on the redundancy of a point in hash tables so that fewer points are used to find true outliers. The space cost of their approach is a problem for large data sets, as it is required to construct a number of hash tables for redundancy computation.

3   Sampling Effects of Density on Outlierness

Recently developed sampling based outlier detection methods have intensively exploited uniform sampling [16,23,29], a simple but effective technique. Uniform sampling can amplify the outlierness contrast, making it easier to distinguish between outliers and inliers. Another benefit is that it offers diversity to the ensemble methods which are often combined with sampling. These properties of uniform sampling are formally analysed in [22,29]. However, they failed to examine the sampling effects of density on outlierness, where density is an important factor for many outlier detection methods.

3.1   Expectation of Outlierness

Throughout this paper, we employ the commonly-used k-NN (Nearest Neighbour) distance to represent the distance based outlierness, while the analysis is also applicable to other distance based outlierness like the summed k-NN distance [4]. Let a sphere of radius r in a d-dimensional Euclidean space contain n uniformly distributed data points. Let s denote the uniform sampling rate. The expectation of the difference between the k-NN distance of a point before and after sampling can be derived from the analysis in [22,29]:

$$\Delta = r \left(\frac{k}{n+1} \right)^{\frac{1}{d}} \left(\frac{1 - s^{\frac{1}{d}}}{s^{\frac{1}{d}}} \right). \tag{1}$$

This is because the kth order statistic of a random sample of size n from the uniform distribution $U(0,1)$ follows the Beta distribution [11], i.e., $\mathcal{U}_{(k)} \sim$ Beta$(k, n+1-k)$, and the expectation of this distribution is $\frac{k}{n+1}$.

To understand how sampling benefits outlier detection, we formally define outlierness contrast as follows:

Definition 1 (Outlierness Contrast). Let o_1 and o_2 be the outlierness scores of two points p_1 and p_2, respectively. The outlierness contrast (denoted as δ) between p_1 and p_2 is defined as: $\delta \triangleq |o_1 - o_2|$.

Usually, a high outlierness contrast is preferred by outlier detection methods as outliers and inliers become more distinguishable with higher δ. To study the relationship between δ and density of a point, we change (1) into another form.

To facilitate the discussion, we replace $n + 1$ in (1) with n, which makes little difference if n is large according to [22]. Let $\rho = \frac{n}{V} = \frac{n}{cr^d}$ be density, where c is a positive constant. Then, Δ is further represented by:

$$\Delta = \left(\frac{k}{c\rho}\right)^{\frac{1}{d}} \left(\frac{1 - s^{\frac{1}{d}}}{s^{\frac{1}{d}}}\right). \tag{2}$$

From (2), it can be derived that $\Delta \propto \rho^{-\frac{1}{d}}$. As the first order and second order derivatives of function $f(x) = x^{-\frac{1}{d}}, x > 0$ are constantly less than and greater than 0 respectively, Δ will grow increasingly fast as the density decreases. As a result, the increase of distance in a sparse region is larger than that in a dense region. Formally, the increase of outlierness contrast after sampling between two points p_1 and p_2 with density ρ_1 and ρ_2 respectively, denoted as Δ_δ, can be calculated by:

$$\Delta_\delta = \left(\frac{k}{c}\right)^{\frac{1}{d}} \left(\frac{1 - s^{\frac{1}{d}}}{s^{\frac{1}{d}}}\right) \left|\frac{1}{\rho_1^{\frac{1}{d}}} - \frac{1}{\rho_2^{\frac{1}{d}}}\right|. \tag{3}$$

It can be seen from (3) that the density difference between the two points contributes to the increase of outlierness contrast Δ_δ. However, the contribution is fixed in uniform sampling as shown below. The density of a point after uniformly sampling is $\rho = s\rho'$, where ρ', the density before sampling, is a constant. Then, Δ_δ is reformulated as:

$$\Delta_\delta = \left(\frac{k}{c}\right)^{\frac{1}{d}} \left(\frac{1 - s^{\frac{1}{d}}}{s^{\frac{2}{d}}}\right) \left|\frac{1}{\rho_1'^{\frac{1}{d}}} - \frac{1}{\rho_2'^{\frac{1}{d}}}\right|. \tag{4}$$

The term $|1/\rho_1'^{\frac{1}{d}} - 1/\rho_2'^{\frac{1}{d}}|$ in (4) is a constant determined by the original data set. As a result, the increase of outlierness contrast is only influenced by the sampling rate s.

It can be seen from (3) and (4) that in addition to the sampling rate, the density difference can also contribute to the increase of outlierness contrast. This demonstrates the feasibility of using density biased sampling for outlier detection. In terms of the bias-variance analysis in [2], the sampling rate influences the effects of bias on the performance of an base detection method in an unpredictable manner. The density difference will have the same influence because it can change Δ_δ in the same way as s according to (3). Note that density biased sampling provides another source of diversity for outlierness ensemble. We further explore the change of distance variance below with respect to the density.

3.2 Variance of Outlierness

Following the Beta distribution, the variances before and after uniform sampling with sampling rate s of the 1-dimensional k-NN distance within a unit sphere ($r = 1$) are $\frac{k(n-k)}{n^2(n+1)}$ and $\frac{k(sn+1-k)}{(sn+1)^2(sn+2)}$, respectively [22]. Thus, the variance change rate, denoted as θ, can be approximated by

$$
\begin{aligned}
\theta &= \frac{k(sn+1-k)}{(sn+1)^2(sn+2)} \Big/ \frac{k(n-k)}{n^2(n+1)} \\
&\approx \frac{k(sn+1-k)}{(sn)^3} \Big/ \frac{k(n-k)}{n^3} \\
&= \frac{1}{s^3}\Big(\frac{sn+1-k}{n-k}\Big).
\end{aligned}
\tag{5}
$$

The approximation step is reasonable as sn is usually much greater than 2, e.g., the minimum recommended value in [2] is 50. Complying with the analysis in [22], the variance change rate θ increases polynomially when the sampling rate s declines. This formally explains why a lower sampling rate can cause a greater outlier rank inversion. The increase of outlierness contrast obtained by decreasing the sampling rate comes at the cost of enlarging the variance change rate. As a consequence, sampling often degrades the performance of a base detector. To tackle this problem, ensemble methods are often employed to improve the detection performance by exploiting the diversity translated from the variances [16,22,29].

Note that unlike the distance expectation case where the distance always increases after sampling, (5) fails to guarantee $\theta \geq 1$, i.e., it is still possible that the variance decreases after sampling, which is interestingly counterintuitive. To study the relationship between θ and 1, a function is derived from (5), $f(s) = (n - k)s^3 - ns + (k - 1), s \in (0, 1]$. Note that we work on the approximated result. But the same conclusion can be drawn if we use the first equation in (5), which is more complicated, though. If $f(s) \leq 0$, then we have $\theta \geq 1$. Otherwise, $\theta < 1$. Analytically, we can obtain the zero points of the cubic equation $f(s) = 0$ in a closed form and determine which intervals correspond to $\theta \geq 1$ and $\theta < 1$, respectively. Due to the complicated presentation of the formulae, they are omitted herein. Instead, we can estimate an interval for s that guarantees the increase in variance. Let $(n - k)s^3 \leq \frac{n}{2}s$ and $k - 1 \leq \frac{n}{2}s$. It can then be derived that $\frac{2(k-1)}{n} \leq s \leq \sqrt{\frac{n}{2(n-k)}}$, which ensures $f(s) \leq 0$. Since $\sqrt{\frac{n}{2(n-k)}} \geq \frac{\sqrt{2}}{2} \approx 0.707$, there is still a wide range of values to select for s in the 1-dimensional case. However, a narrower range is allowed to make $\theta \geq 1$ for multi-dimensional data, as shown empirically later.

Now we investigate the difference of the variance change rates in regions of different densities. By replacing n in (5) with $c\rho$, where c is a constant, θ will be $\frac{1}{s^3}\big(\frac{sc\rho+1-k}{c\rho-k}\big)$. Similar to (3), the difference of the variance change rate after sampling between two points p_1 and p_2 with density ρ_1 and ρ_2 respectively, denoted as Δ_θ, is calculated by:

$$
\begin{aligned}
\Delta_\theta &= \frac{1}{s^3}\left(\frac{cs\rho_1 + 1 - k}{c\rho_1 - k}\right) - \frac{1}{s^3}\left(\frac{cs\rho_2 + 1 - k}{c\rho_2 - k}\right) \\
&= \frac{1 - (1 - s)k}{cs^3}(\rho_2 - \rho_1).
\end{aligned}
\tag{6}
$$

The term $1-(1-s)k$ will be less than 0, as s is usually small. If $\rho_1 > \rho_2$, then $\Delta_\theta > 0$ in terms of (6), i.e., the variance change rate in a denser region will be larger than that in a lower density region. This interesting property can benefit sampling based outlier detection methods by introducing less severe outlier rank inversion to sparse regions where outliers are often located. As it is hard to extend (6) to the multi-dimensional case explicitly, we empirically show that the same trend applies.

3.3   Empirical Study

To empirically verify the above analysis, Fig. 1 illustrates the effects of sampling rates and density on outlierness contrast and variance change rate after uniform sampling. The data set contains 5000 points in a 10-dimensional space. More details of the data set will be described in Sect. 5.1. Specifically, Fig. 1(a) shows the effects on outlierness contrast, where the 5-NN distance is employed to signify the outlierness. The distances are sorted to approximately reflect the change in density. Three sampling rates, $s = 0.1, 0.05$ and 0.01, are used here as representatives. The distance of the last point of every 10 points is used in the figure for a concise presentation. Figure 1(b) shows the effects on the variance change rate θ. Three more sampling rates $s = 0.005, 0.2$ and 0.5 are examined in this experiment. The points are grouped into 10 regions of different densities. The groups are sorted in ascending order by their average density, and indexed from 1 to 10 accordingly. Note that we use a different y axis for $s = 0.005$ and 0.01 due to the large θ values they produce. It can be observed from the figure that the empirical results comply with the analysis above. Most interestingly, we observe in Fig. 1(b) that almost all the variance change rates decrease when the region becomes sparser, and some variance change rates are lower than 1 for large sampling rates 0.2 and 0.5.

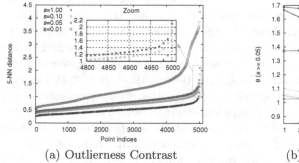

(a) Outlierness Contrast

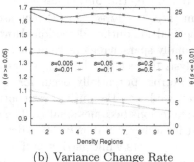

(b) Variance Change Rate

Fig. 1. Effects of the sampling rate θ and density on k-NN distance after uniform sampling.

As theoretically and empirically analysed above, the outlierness contrast gained from uniform sampling can be amplified by decreasing the sampling rate, at the cost of reversing outlier ranks though. An interesting question arises: Can we further improve the outlierness contrast when a fixed sampling rate is given? This question motivates the research herein, and the answer is yes in terms of our analysis of the density effects on outlierness. Our piecewise density biased sampling approach for outlier detection is formulated in Sect. 4.

4   Density Biased Sampling Using LSH

4.1   Piecewise Density Biased Sampling

To answer the question raised in Sect. 3.3, we propose to change the density contrast in a sample to further improve the performance of outlier detection when a fixed sampling rate is given. Therefore, we need to produce a density-aware sample. Let ρ_p be the density of a point p. Then, the probability of including p in a sample should be a function of ρ_p, denoted as $\Phi(\rho_p)$. In [13], $\Phi(\rho_p)$ is defined to be proportional to ρ_p^v. Specifically,

$$\Phi(\rho_p) = \frac{s|D|\rho_p^v}{\sum_{p \in D} \rho_p^v}, \tag{7}$$

where s is the sampling rate, and D is the original data set. It can be easily verified that $\sum_{p \in D} \Phi(\rho_p) = s|D|$, i.e., the sample size is $s|D|$. The tuning parameter v controls the density in a sample. Concretely, if $v = 0$, it corresponds to the uniform sampling. Denser regions will be sampled with a higher sampling rate if $v > 0$, and a lower one if $v < 0$. The work in [13] suggested $v > 0$ for clustering analysis, and $v < 0$ for outlier detection. However, $v < 0$ potentially leads to a detection performance degradation compared with uniform sampling because it weakens the outlierness contrast between points according to our analysis in Sect. 3.1. The effect of v on outlierness is qualitatively analysed as follows.

If $v > 0$, the outlierness contrast increases according to (3) because the term $|1/\rho_1^{\frac{1}{d}} - 1/\rho_2^{\frac{1}{d}}|$ in (3) increases due to a larger density contrast. This is beneficial for outlier detection. However, the variance change rate is larger in originally sparser regions than that in originally denser regions according to the analysis in Sect. 3.2, because sparser regions have lower sampling rates. Thus, more errors potentially occur for a base detection method. If $v < 0$, the case is reversed, i.e., the outlierness contrast declines as the variance change rate in sparse regions becomes smaller. The empirical study in Sect. 5 also demonstrates that for either $v > 0$ or $v < 0$, a detection method obtains performance gains on some data sets while getting performance reductions on other data sets. Since v is a user-specified parameter, an inappropriate value of v, especially the sign of v, can result in severely bad performance. To circumvent such a situation and make use of the benefits in the cases $v > 0$ and $v < 0$, we propose a piecewise density biased sampling technique as follows.

Let B denote a density boundary that divides the points in D into two parts according to their density, the left part D_l and the right part D_r. There are $D_l \cup D_r = D$ and $D_l \cap D_r = \emptyset$. If $\rho_p \geq B$, then $p \in D_l$. Otherwise, $p \in D_r$. We use $v_l < 0$ for D_l, and $v_r > 0$ for D_r. Then, the probability of including a point p in a sample, $\Phi_P(\rho_p)$, is computed by:

$$\Phi_P(\rho_p) = \begin{cases} \frac{s|D_l|\rho_p^{v_l}}{\sum_{p \in D_l} \rho_p^{v_l}}, & \rho_p \geq B, \\ \frac{s|D_r|\rho_p^{v_r}}{\sum_{p \in D_r} \rho_p^{v_r}}, & \text{o.w.} \end{cases} \tag{8}$$

It can be easily verified that $\sum_{p \in D} \Phi_P(\rho_p) = s|D|$ by summing the two parts in (8). As B represents the density value which is data specific, it is hard for a user to specify such a parameter. Instead, a user-friendly way is to just specify the quantile corresponding to B. Let ζ_B denote the quantile. We have $\zeta_B = |D_l|/|D|$. A user can set either B or ζ_B mannually in term of their domain knowledge and the Cantelli's inequality, which has been extensively leveraged in [19,21].

The rationale behind the proposed sampling technique for outlier detection is elaborated as follows. Unlike other data mining tasks, the outlier detection task needs to handle highly unbalanced classes, i.e., a small fraction of outliers versus a large number of inliers. Hence, the points in a data set are usually of different importance for outlier detection. Generally, we can classify the points into three types, strong inliers, strong outliers and critical points that include weak inliers and weak outliers. What challenges an outlier detection method is to correctly discriminate critical points. Although sampling will affect all types of points, what matters most is the sampling effects on the outlierness of critical points. Intuitively, the outlier rank inversion issues of strong inliers or strong outliers have much less impact on the detection results than within critical points. Thus, the piecewise density biased sampling can benefit the critical points by drawing fewer points within strong inliers and strong outliers but more within critical points, if the boundary B is placed between strong inliers and critical points.

Note that the benefits from the sampling processes in the two parts are different. In the left part where $v_l < 0$, points in sparse regions (including critical points and strong outliers) take a higher proportion than strong inliers. Thus, the outlierness variances of these points will become smaller due to the higher local sampling rate in sparse regions compared with uniform sampling, which has been quantitatively analysed by (5). The outlier rank inversion issue of critical points is mitigated because of lower outlierness variance. To engender an appropriate proportion of original strong inliers in a sample, v_l can be set as -1. Then, the density of original strong inliers in the sample will be roughly equal. Although $v_l < -1$ can make a smaller proportion of original strong inliers in a sample, some of these inliers are potentially at risk of being treated as outliers due to their low density in the sample. In the right part where $v_r > 0$, both critical points and strong outliers can still benefit from the increase of outlierness contrast as analysed in Sect. 3.1. As such, our approach shares the merits of both cases of traditional density biased sampling.

4.2   Density Estimation Using LSH

To make density biased sampling practical, we need to estimate the density in a highly efficient and scalable manner. Kernel density estimation methods are adopted in [13] for density approximation, which presents a significant time overhead as pointed out in [18]. We require the estimation is of linear complexity and parallelisable, making density biased sampling comparable to uniform sampling in terms of computational cost. Usually, linear complexity and parallelisability are necessary for an algorithm to handle very big data sets.

LSH Preliminary. To estimate the density of a point p, one way is to count the number of points within a ball centred at p with a given radius r. This direct method needs pair-wise computation and results in $O(n^2)$ time complexity, failing to satisfy the requirements above. Hence, we leverage Locality Sensitive Hashing (LSH) [10] for efficient and scalable density estimation. The core idea of LSH is to hash similar data points into the same buckets with high probability [10]. Given a family of $\alpha \cdot \gamma$ hash functions, this technique splits the hash values into γ bands of α rows, where $\alpha, \gamma > 0$ are two integers selected by the user. With this technique, two points will be hashed into the same bucket if all their α hash values in at least one out of the γ bands are equal. Let sim be the similarity between two points. The probability that they are hashed into the same bucket is $1 - (1 - sim^\alpha)^\gamma$. Since the Euclidean distance is involved in our work, we leverage the l_2 LSH family herein. Let v be a d-dimensional vector, the hash function that maps $\mathbf{v}$ into an integer hash value follows in the form [3]

$$h_{\mathbf{a},b}(\mathbf{v}) = \left\lfloor \frac{\mathbf{a} \cdot \mathbf{v} + b}{W} \right\rfloor, \tag{9}$$

where $\mathbf{a}$ is a d-dimensional random vector whose components are drawn from Gaussian distribution independently, b is a real number uniformly chosen from the interval $[0, W]$, and W is a user-specified parameter. The parameters α, γ and W controls the performance of l_2 LSH.

Density Estimation. Since LSH can produce the same hash values for similar objects (herein approximate nearest neighbours), we can count the number of objects having the same hash value as point p, i.e., those that have been hashed into the same bucket, to estimate its density. Let the hash key for p be denoted as $K(p) = \langle h_1(p), h_2(p), \cdots, h_\alpha(p) \rangle$, where $h_i(p), 1 \le i \le \alpha$, is computed using (9). Let random variable N^p denote the number of points hashed into the same bucket as p. The expectation of N^p, $E[N^p]$, is estimated by [24]:

$$E[N^p] = \frac{\sum_{i=1}^{\gamma} n_i^q}{\gamma}, \tag{10}$$

where n_i^q is the number of points hashed into the same bucket as p in the ith round. Note that we count only the number of neighbours without storing them. The number of hash tables γ here is only used to control the rounds of trials. Thus, the larger γ is, the more accurate estimate of $E[N^p]$ we can obtain. Then, we can estimate the density of a point p by $\rho_p = \frac{E[N^p]}{c}$, where c is a constant. We assume $c = 1$ to facilitate discussion, i.e., $\rho_p = E[N^p]$.

Intuitively, $E[N^p]$ is sensitive to parameters α and W. Previous work [7] has suggested how to tune these parameter for k nearest neighbour search. However, these methods cannot be applied directly to the density estimation case. We need to make the estimated density discriminative. One extreme case is that if α is too large or W is too small, the bucket size $E[N^p]$ will be very small accordingly, e.g., 1 or 2. The consequence is that most of the points have the same (and small) density, lacking of discrimination. Another extreme case is that if α is too small or W is too large, the bucket size $E[N^p]$ is very large because too many points are hashed into the same buckets. This also results in lacking of density distinguishability.

To address the issues above, we propose a rule for tuning α and W. Given a small fraction ϵ, the tuned α and W will produce such a density estimation that the portion of points whose densities are less than k is ϵ. Let D be the set of points, and $E[N^p]_{\alpha,W}$ denote the value of $E[N^p]$ produced under the parameters α, W. Formally, α and W need to satisfy the following equation:

$$\frac{\sum_{p \in D} \mathbb{1}(E[N^p]_{\alpha,W} < k)}{|D|} = \epsilon, \tag{11}$$

where $\mathbb{1}(\cdot)$ is an indicator function, i.e., $\mathbb{1}(A) = 1$ if A is true, while $\mathbb{1}(A) = 0$ otherwise.

The value of ϵ can be a small number like 0.01 or 0.05, or set by the rate of outliers in D if the information is available. So, most of the bucket sizes will be larger than or equal to k, making the density estimation meaningful and discriminative. Note that we still allow a small fraction (ϵ) of points to go into the buckets of size less than k. As these points will be outliers with high probability, we deliberately fail them in finding their k^{th} nearest neighbours, so that their k-NN distances will increase if other points are chosen as their neighbours. This can help to promote the outlierness contrast for potential outliers.

Given a parameter ϵ, we use the numerical method to solve α and W from equation (11). As integer $\alpha \geq 1$, we can initialise α as 1. For real number $W > 0$, we can initialise it as 1.0 or other values if certain priori knowledge is available. As the probability of putting two objects into the same bucket decreases exponentially as α increases linearly. Accordingly, the bucket size $E[N^p]$ will also decrease exponentially. Thus, if we increase α exponentially (with base 2), the upper bound of α can be identified quickly. Then, we can use binary search to find the desired α. The same search strategy can be applied to finding W. Note that the bucket size will increase $E[N^p]$ as W becomes larger. Thus, by increasing α and W, we can always achieve the fraction ϵ. Let $\epsilon(\alpha, W)$ be the fraction in each trial with parameters α and W. In each round of search, we compare ϵ and $\epsilon(\alpha, W)$. If their difference is less than a pre-specified threshold denoted as η, a solution is identified. The threshold η can be as small 0.001 or 0.0001.

5 Empirical Evaluation

5.1 Experimental Settings

In this section, we study the performance of our approach empirically by conducting extensive experiments on both synthetic and real-world data sets. Specifically, we compare our sampling approach with uniform sampling and traditional density biased sampling [13] over various state-of-the-art sampling based outlier detection methods [16,23,25] and [29]. We measure the AUC of the ROC curve [8] and the execution time to for evaluation. All approaches are implemented in Java and run on a Mac node with a 4-core Intel i7 CPU 2.2 GHz. The source code is available here.

In our experiments, we use 4 data sets including a synthetic one and 3 real-world ones from the UCI Machine Learning Repository[1]. The basic information of the 4 data sets are summarised in Table 1. As we mainly aim to compare the performance of our sampling methods with uniform sampling and traditional density biased sampling, the sampling rate is at 5% throughout all experiments. The k-NN parameter k is 5. The parameter settings for LSH tuning are: $\gamma = 100$, $\epsilon = 0.01$ and $\eta = 0.005$. The number of ensemblers is 10. Each experiment is repeated 10 times, and the mean and standard errors of the results are reported.

Table 1. Basic description of data sets.

Name	$|D|$	d	Outlier vs. inlier labels	Outlier rate
Synthetic	5,000	10	$distance \geq 0.975$ quantile	2.5%
Covertype	5,700	10	class 4 vs. class 2	0.96%
Gamma telescope	6,300	10	"h" vs. "g"	2.87%
Robot navigation	5,456	4	"Slight-left-turn" vs. others	6.01%

5.2 Experimental Process and Results

Comparison with the State-of-the-Arts. The outlier detection methods are listed in Table 2, with sampling types U, D+1, D−1 and P, for Uniform, Density biased ($v = +1$), Density biased ($v = -1$) and Piecewise density biased sampling, respectively. The parameters for the P sampling type are as follows: $\zeta_B = 0.7$, $v_l = -1$, $v_r = 1$.

The execution time results are reported in Fig. 2, and the performance results in Fig. 3. As per Fig. 3, the detectors with density biased sampling outperform those with uniform sampling in most cases. As to the traditional density biased sampling, the first three data sets prefer $v < 0$, while the last one get very good results with $v > 0$. However, the performance of piecewise density biased

---

[1] https://archive.ics.uci.edu/ml/datasets.html.

Table 2. Basic description of outlier detection methods.

Detection methods	Sampling types
1Sample	U [23], D + 1, D − 1, P
Iterative	U [25], D + 1, D − 1 [13], P
Iterative+Ensemble	U [29], D + 1, D − 1, P
Isolation forest	U [16], D + 1, D − 1, P

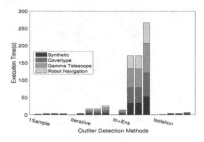

Fig. 2. Execution time w.r.t. different sampling.

sampling is mostly comparable to or even better than the cases where $v < 0$ or $v > 0$. The ensemble iterative methods mostly outperform others. However, the performance gains are at time cost as shown in Fig. 2.

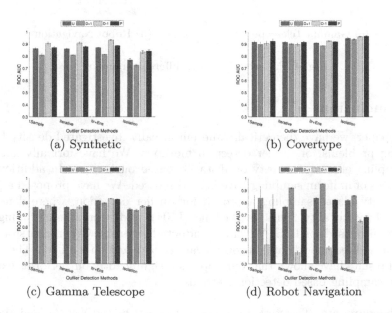

(a) Synthetic

(b) Covertype

(c) Gamma Telescope

(d) Robot Navigation

Fig. 3. Performance w.r.t. different sampling.

Density Boundary. The performance of our proposed sampling technique with respect to the density boundary B is studied herein by varying ζ_B in the interval $[0.5, 0.95]$ in steps of 0.05. The parameters are set as $v_l = -1$, $v_r = 1$. The experimental results are presented in Fig. 4. It is observed that when the boundary exceeds the third quartile (0.75), the change in performance becomes noticeable. Thus, it is empirically suggested that the critical points are probably located around regions corresponding to $\zeta_B = 0.75$. It is safe in most if not all cases to choose ζ_B from the interval $[0.6, 0.75]$ to avoid poor performance.

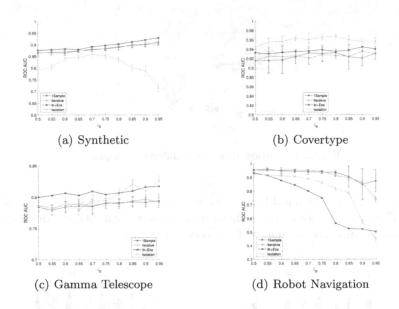

(a) Synthetic   (b) Covertype

(c) Gamma Telescope   (d) Robot Navigation

Fig. 4. Performance w.r.t. different boundary.

6   Conclusions

In this paper, we have theoretically and empirically investigated density biased sampling problems for outlier detection methods. We have formally analysed the sampling effects of density on distance based outlierness. In addition, the limitations of uniform sampling have been discussed. We have proposed a piecewise density biased sampling approach for outlier detection, where density is estimated by Locality Sensitive Hashing (LSH), and a parameter tuning rule has been proposed for LSH. We have conducted extensive experiments on both synthetic and real-world data sets over a variety of detection methods. The experimental results have shown that our approach significantly outperforms existing uniform sampling based detection methods.

Acknowledgments. This work was supported in part by the New Zealand Marsden Fund under Grant No. 17-UOA-248, the UoA FRDF under Grant No. 3714668, and the NJU Overseas Open fund under Grant No. KFKT2018A12.

References

1. Aggarwal, C.C.: Outlier ensembles: position paper. ACM SIGKDD Explor. Newsl. **14**(2), 49–58 (2013)
2. Aggarwal, C.C., Sathe, S.: Theoretical foundations and algorithms for outlier ensembles. ACM SIGKDD Explor. Newsl. **17**(1), 24–47 (2015)
3. Andoni, A., Indyk, P.: Near-optimal hashing algorithms for approximate nearest neighbor in high dimensions. In: FOCS, pp. 459–468 (2006)
4. Angiulli, F., Pizzuti, C.: Fast outlier detection in high dimensional spaces. In: Elomaa, T., Mannila, H., Toivonen, H. (eds.) PKDD 2002. LNCS, vol. 2431, pp. 15–27. Springer, Heidelberg (2002). https://doi.org/10.1007/3-540-45681-3_2
5. Breunig, M.M., Kriegel, H.P., Ng, R.T., Sander, J.: LOF: identifying density-based local outliers. ACM SIGMOD Rec **29**(2), 93–104 (2000)
6. Chandola, V., Banerjee, A., Kumar, V.: Anomaly detection: a survey. ACM Comput. Surv. (CSUR) **41**(3), 15 (2009)
7. Dong, W., Wang, Z., Josephson, W., Charikar, M., Li, K.: Modeling LSH for performance tuning. In: CIKM, pp. 669–678 (2008)
8. Fawcett, T.: An introduction to ROC analysis. Pattern Recogn. Lett. **27**(8), 861–874 (2006)
9. Fu, P., Hu, X.: Biased-sampling of density-based local outlier detection algorithm. In: ICNC-FSKD, pp. 1246–1253 (2016)
10. Indyk, P., Motwani, R.: Approximate nearest neighbors: towards removing the curse of dimensionality. In: STOC, pp. 604–613 (1998)
11. Jones, M.: Kumaraswamy's distribution: a beta-type distribution with some tractability advantages. Stat. Methodol. **6**(1), 70–81 (2009)
12. Knox, E.M., Ng, R.T.: Algorithms for mining distance-based outliers in large datasets. In: VLDB, pp. 392–403 (1998)
13. Kollios, G., Gunopulos, D., Koudas, N., Berchtold, S.: Efficient biased sampling for approximate clustering and outlier detection in large data sets. IEEE Trans. Knowl. Data Eng. **15**(5), 1170–1187 (2003)
14. Kriegel, H.P., Zimek, A., et al.: Angle-based outlier detection in high-dimensional data. In: ACM SIGKDD, pp. 444–452 (2008)
15. Liu, F.T., Ting, K.M., Zhou, Z.-H.: On detecting clustered anomalies using SCi-Forest. In: Balcázar, J.L., Bonchi, F., Gionis, A., Sebag, M. (eds.) ECML PKDD 2010. LNCS (LNAI), vol. 6322, pp. 274–290. Springer, Heidelberg (2010). https://doi.org/10.1007/978-3-642-15883-4_18
16. Liu, F.T., Ting, K.M., Zhou, Z.H.: Isolation-based anomaly detection. ACM Trans. Knowl. Discov. Data **6**(1), 3 (2012)
17. Luo, C., Shrivastava, A.: Arrays of (locality-sensitive) count estimators (ACE): anomaly detection on the edge. In: WWW, pp. 1439–1448 (2018)
18. Nanopoulos, A., Manolopoulos, Y., Theodoridis, Y.: An efficient and effective algorithm for density biased sampling. In: CIKM, pp. 398–404 (2002)
19. Pang, G., Cao, L., Chen, L., Lian, D., Liu, H.: Sparse modeling-based sequential ensemble learning for effective outlier detection in high-dimensional numeric data. In: AAAI (2018)
20. Pillutla, M.R., Raval, N., Bansal, P., Srinathan, K., Jawahar, C.: LSH based outlier detection and its application in distributed setting. In: CIKM, pp. 2289–2292 (2011)
21. Rayana, S., Zhong, W., Akoglu, L.: Sequential ensemble learning for outlier detection: a bias-variance perspective. In: ICDM, pp. 1167–1172 (2016)

22. Schubert, E.: Generalized and efficient outlier detection for spatial, temporal, and high-dimensional data mining. Ph.D. thesis (2013)
23. Sugiyama, M., Borgwardt, K.: Rapid distance-based outlier detection via sampling. In: NIPS, pp. 467–475 (2013)
24. Wang, Y., Parthasarathy, S., Tatikonda, S.: Locality sensitive outlier detection: a ranking driven approach. In: ICDE, pp. 410–421 (2011)
25. Wu, M., Jermaine, C.: Outlier detection by sampling with accuracy guarantees. In: ACM SIGKDD, pp. 767–772 (2006)
26. Yang, X., Latecki, L.J., Pokrajac, D.: Outlier detection with globally optimal exemplar-based GMM. In: SDM, pp. 145–154 (2009)
27. Zhang, X., et al.: LSHiForest: a generic framework for fast tree isolation based ensemble anomaly analysis. In: ICDE, pp. 983–994 (2017)
28. Zimek, A., Campello, R.J., Sander, J.: Ensembles for unsupervised outlier detection: challenges and research questions a position paper. ACM SIGKDD Explor. Newsl. 15(1), 11–22 (2014)
29. Zimek, A., Gaudet, M., Campello, R.J., Sander, J.: Subsampling for efficient and effective unsupervised outlier detection ensembles. In: ACM SIGKDD, pp. 428–436 (2013)

A Novel Technique of Using Coupled Matrix and Greedy Coordinate Descent for Multi-view Data Representation

Khanh Luong$^{(\boxtimes)}$, Thirunavukarasu Balasubramaniam, and Richi Nayak

Queensland University of Technology, Brisbane, QLD 4000, Australia
{khanh.luong,t.balasubramaniam}@hdr.qut.edu.au, r.nayak@qut.edu.au

Abstract. The challenge of clustering multi-view data is to learn all latent features embedded in multiple views accurately and efficiently. Existing Non-negative matrix factorization based multi-view methods learn the latent features embedded in each view independently before building the consensus matrix. Hence, they become computationally expensive and suffer from poor accuracy. We propose to formulate and solve the multi-view data representation by using Coupled Matrix Factorization (CMF) where the latent structure of data will be learned directly from multiple views. The similarity information of data samples, computed from all views, is included into the CMF process leading to a unified framework that is able to exploit all available information and return an accurate and meaningful clustering solution. We present a variable selection based Greedy Coordinate Descent algorithm to solve the formulated CMF to improve the computational efficiency. Experiments with several datasets and several state-of-the-art benchmarks show the effectiveness of the proposed model.

Keywords: Multi-view learning/clustering
Coupled matrix factorization · Greedy coordinate descent

1 Introduction

Multi-view clustering has attracted significant attention in recent years due to its effectiveness and applicability to many fields such as web mining and computer vision, where multi-faceted data is naturally generated [1–3]. An example of multi-view data is a web search system where web-pages can be represented with content or links or search queries or viewers. Another example is a multilingual website where content can be represented in multiple different languages. Multi-view data, where the data can be represented with multiple views, provides complementary and compatible information to a clustering algorithm and assists it to learn accurate and meaningful outcome [4]. Early approaches to cluster multi-view data concatenate the data from all views to obtain a single view and apply traditional clustering methods. This approach includes mutual

© Springer Nature Switzerland AG 2018
H. Hacid et al. (Eds.): WISE 2018, LNCS 11234, pp. 285–300, 2018.
https://doi.org/10.1007/978-3-030-02925-8_20

information from multiple views in the process, however, it ignores the specific characteristics and the structural information inherent in individual views [5,6]. In recent years, several methods have been proposed that exploit multi-view representation explicitly and showed improved clustering quality [6]. Amongst them, the most popular and effective methods are matrix factorization based that learn the embedded space by projecting high-dimensional data to a lower-dimensional space and seek the cluster structures in this new space.

The multi-view clustering methods based on Non-negative Matrix Factorization (NMF) [7] framework present each view data as a matrix, factorize each matrix independently and find the consensus factor matrix among all views (as shown in Fig. 1a). A multi-view clustering algorithm can only perform effectively when it is able to learn the latent features and typical information from multiple views in a consensus manner. Most of the NMF-based multi-view methods [3,8–10] first use a collection of matrices to formulate the multi-view problem then learn the latent features embedded in each view. In the later step called integration or fusion, the consensus latent feature matrix is achieved by linearly combining each factorized view matrix or using regularization to guide the consensus factor matrix learning process. The well-known NMF-based method, MultiNMF [8] incorporates the NMF framework with Probabilistic Latent Semantic Analysis (PLSA) constraint [11] to seek the common latent feature matrix from the low representations of all views. The multi-view representation setting in MultiNMF has been extended in many later works. The Adaptive Multi-View Semi-Supervised Non-negative Matrix Factorization (AMVNMF) method [9] uses label information as prior knowledge and uses it as a hard constraint to enhance the discriminating power in MultiNMF. MMNMF [1] incorporates a multi-manifold regularization into the multi-view NMF frame-work which can preserve the manifold learned from multiple views. Multi-component nonnegative matrix factorization (MC-NMF) [10] and Diver NMF (DiNMF) [3] aim to exploit the diverse information from different data views. This approach shows effectiveness as it understands the underlying structures in complex datasets by considering all compatible and complementary information from all views for the learning process.

Since the multi-view data contains the data where samples are represented via different features, we conjecture that all the views data should be decomposed jointly and simultaneously to discover the shared features directly from the views, instead of separately decomposing each view and then looking for the common latent representations in the fusion step as in most of the NMF-based multi-view clustering methods. We propose a novel solution by using a Coupled Matrix Factorization (CMF) model [12,13] to effectively formulate and solve the multi-view representation problem based on the NMF framework [7]. The CMF representation allows us to make a shared mode amongst all views. The goal is to find a distinct base matrix corresponding to each view and the consensus coefficient matrix generated from the shared mode which captures underlying structure of data samples as well as commonality among multiple views.

The NMF-based clustering methods are known to produce a quality solution at the cost of computation time. Majority of NMF methods use Multiplicative Update Rule (MUR) [14] for optimization. MUR has the advantage of good compromise between speed and the ease of implementation [7], however, it has been criticized for failing to guarantee the convergence of component matrices to a stationary point [15]. Recently, many Coordinate Descent (CD) based updating methods, that focus on updating one element at a time until convergence, have been proposed such as HALs or FHALs [16] where the algorithm will sequentially update each element of every updating matrix (HALs) or cyclic update all elements in each matrix before starting to update the others in the same manner (FHALs). Greedy Coordinate Descent (GCD) [17] is proposed to carefully and greedily select elements relying on their importance to update at each step rather than treating every element equally as in HALs or FHALs. GCD outperforms other updating methods in terms of convergence time as well as it can result in a better lower-order dimensional representation due to its ability to emphasize updating important elements at each iteration. We propose to use the Greedy Coordinate Descent (GCD) method which has been proved effective in traditional matrix completion problem as compared to MUR [17]. In this paper, we propose to extend GCD to solve the multi-view clustering problem.

While projecting the search space from high to lower-dimension, the NMF based methods do not guarantee that the geometric data structure will not change after the lower space mapping. Traditionally, NMF based methods rely on manifold learning to discover and preserve the local structures of data, i.e., to learn and maintain the similarities between data points within local areas [18,19]. Yet due to multi-view data represented through many views, it is not a trivial task to learn the consensus manifold for multi-view data [1,20], i.e., learning and preserving the consensus similarity information of all views. In this paper, taking the advantage of CMF, we propose to embed a similarity matrix into the factorization process to be decomposed as a new sample-feature matrix. The similarity matrix is the linear combination of similarity information derived from all views. This novel solution by factorizing the similarity matrix while higher to lower dimension mapping will ensure the consensus matrix to embed all similarity information during optimizing.

The proposed Multi-view Coupled Matrix Factorization (MVCMF) method has been tested with various diverse datasets and benchmarked with state-of-the-art multi-view clustering methods. The experimental results show significant improvement in terms of both accuracy and time complexity. It ascertains that (1) the proposed MVCMF representation is able to exploit the similarity information of data samples from multiple views more effectively, and (2) the proposed optimization method is able to produce a better representation with important elements selection at each optimizing iteration.

More specifically, the contributions of this paper are: (1) Formulate the multi-view representation problem effectively with CMF to cluster multi-view data; (2) Embed the similarities between samples on each view in CMF which can improve the quality of clustering while keeping the framework simple but comprehensive;

(3) Present a single variable update rule of GCD optimization for the proposed unified framework. To our best of knowledge, this is the first work presenting CMF and combining samples similarity into a unified framework to cluster multi-view data. This framework respects data characteristics from multiple views and at the same time respecting similarities between data points within a cluster. This is also a first work where GCD is applied in NMF optimization in multi-view setting.

2   The Proposed Multi-view Coupled Matrix Factorization (MVCMF) Clustering Technique

2.1   Problem Definition - Traditional NMF-based Multi-view Clustering

Suppose $X = \{X_1, X_2, \ldots, X_{n_v}\}$ is a multi-view dataset with n_v views in total. Data in view v is represented in data matrix $X_v \in R_+^{n \times m_v}$ where n is the number of samples and m_v is the number of features of the vth view.

The NMF-framework to factorize multiple views of data can be written as [8,10],

$$J_1 = \min \sum_{v=1}^{n_v} \|X_v - H_v W_v\|_F^2, \text{ s.t. } H_v \geq 0, W_v \geq 0 \tag{1}$$

where $H_v \in R_+^{n \times k}$ is the new low-rank representation of data corresponding to the basis $W_v \in R_+^{k \times m_v}$ under the vth view and k denotes the number of new rank.

In the fusion step, the consensus latent feature matrix, denoted as H_*, is calculated by taking the average [3],

$$H_* = \sum_{v=1}^{n_v} H_v / n_v \tag{2}$$

or linearly combining [10]

$$H_* = [H_1 \ldots H_{n_v}] \tag{3}$$

The objective function in Eq. (1) is able to learn different data representations from different data views. In the later step, the consensus data matrix that embeds the cluster structure of data samples will be learned based on the new learned data representations from all views. Different H_v may contribute different impact on computing H_*. However, since H_* will be computed by taking the average or linear combination of all H_v as in Eqs. (2) or (3), this will balance the importance of different data views in the latent feature matrix and will fail to ensure the correctness and meaningfulness of the final clustering solution. Moreover, data in multi-view has been well-known of complementary and compatible characteristics, independently learning embedded structures from different views before computing the consensus data will ignore the associative nature among different views during the learning process. Therefore it may not learn good latent features of the multi-view data.

Alternatively, some methods learn the coefficient matrix of each view and simultaneously seek the consensus coefficient matrix by (1) minimizing the disagreement between the consensus latent matrix and the new learned latent feature matrices of all views as in Eq. (4) [8], or (2) looking for the consensus manifold of data as in Eq. (5) [1].

$$J_2 = \min \sum_{v=1}^{n_v} \|X_v - H_v W_v\|_F^2 + \sum_{v=1}^{n_v} \|H_v - H_*\|_F^2, \text{ s.t. } H_v \geq 0, H_* \geq 0, W_v \geq 0 \quad (4)$$

$$J_3 = \min \sum_{v=1}^{n_v} \left(\|X_v - H_v W_v\|_F^2 + \|H_v - H_*\|_F^2 + \|L_v - L_*\|_F^2 \right) + Tr(H_*^T L_* H_*),$$

$$\text{s.t. } H_v \geq 0, H_* \geq 0, W_v \geq 0 \quad (5)$$

where the intrinsic manifold L_* is the linear approximation of all views manifolds, $L_* = \sum_{v=1}^{n_v} \lambda_v L_v$ and $L_v = D_v - S_v$ is the Laplacian matrix on each view, D_v is the diagonal matrix, $(D_v)_{ii} = \sum_j (S_v)_{ij}$ and S_v is the weighted adjacency matrix [18,21].

Additional terms in Eqs. (4), (5) help finding the optimal coefficient matrix by constraining the optimal matrix to be as close to different views as possible or by constraining the final space that lies on a convex hull of the data manifold, thus can bring the more realistic common latent matrix H_* as compared to learning H_* as in Eqs. (2) or (3). However, the step of constraining H_v to find the optimal latent feature matrix H_* dramatically increases computation as well as it introduces the computation approximation errors that affect the effectiveness of the model. For instance, objective function in Eq. (4) results in $\epsilon = \epsilon_1 + \ldots + \epsilon_{n_v}$, the sum of errors ϵ_v of computing H_* from each H_v.

To overcome these inherent problems, we propose to reformulate the NMF-based multi-view representation by using the CMF model as well as embed the intra-similarities among data objects in each view in the factorization process.

2.2   Proposed Multi-view Coupled Matrix Factorization (MVCMF)

We propose to formulate the multi-view representation using CMF as follows,

$$J = \min \sum_{v=1}^{n_v} \|X_v - H_* W_v\|_F^2, \text{ s.t. } H_* \geq 0, W_v \geq 0 \quad (6)$$

where $W_v \in R_+^{k \times m_v}$ is the basis representation of data under the vth view and H_* is the consensus coefficient matrix learned from all views. In this objective function, factor matrices to be updated include H_* and $\{W_v\}_{v=1\ldots n_v}$. Since data represented by multiple views share the same mode, the optimal coefficient matrix H_* should be shared among the low-dimensional spaces learned from data of different views. Figure 1b shows how CMF works as compared to traditional NMF-based framework (Fig. 1a). The latent low-dimensional space of data can be learned by integrating all information from multiple views and by considering the association relationship between different data views. Figure 2 shows an example of learning H_* by CMF model that can capture the truly latent features of multi-view data.

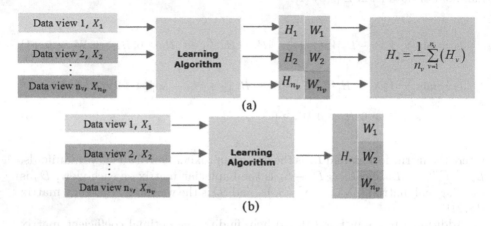

(a)

(b)

Fig. 1. Traditional NMF-based versus CMF for Multi-view Learning. H_v encodes the latent features learned from data view vth. W_v is the corresponding basis matrix and H_* is the common consensus matrix.

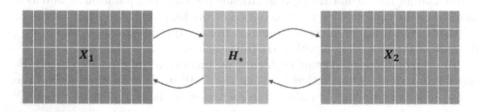

Fig. 2. The consensus latent matrix H_* is learned by a CMF model in a two-view dataset. H_* is first learned from data view 1, i.e., X_1 and then passed onto data view 2, i.e., X_2 to be updated. H_* is iteratively updated by simultaneously using data matrices from all views until converged.

By using CMF, we bring the following benefits. (1) Firstly, the number of components to be updated has been decreased by half in CMF as compared to traditional multi-view NMF framework. There are $n_v + 1$ components to be updated in the CMF model, i.e., $\{W_v\}_{v=1...n_v}$ and H_*, as compared to $2n_v + 1$ components in the multi-view NMF model, i.e., $\{W_v\}_{v=1...n_v}$, $\{H_v\}_{v=1...n_v}$ and

H_*. This significantly reduces computational time. (2) Secondly, the objective function in Eq. (6) ties factors of different views together to capture data as close as to the fine-grained clusters. This results in a unique and stable solution as possible. This is unlike traditional multi-view NMF formulation that is known to produce multiple non-unique solutions because each view is learned independently before seeking the consensus matrix that leads to different solutions of H_*. (3) Lastly, the latent consensus feature matrix is learned naturally since it has been incorporated during the learning process. This reduces approximation errors in contrast to the traditional NMF where they are accumulated as in Eqs. (4) or (5).

We further improve the objective function in Eq. (6) by adding the intra-similarities of data samples represented on all views. Consider view v, the similarity matrix S_v of data samples on this data view X_v is constructed using k nearest-neighbour graph (kNN) as in [18,21,22]. Specifically,

$$S_v(x_i, x_j) = \begin{cases} s_v(x_i, x_j) \text{ if } x_i \in \mathcal{N}_k(x_j) \text{ or } x_j \in \mathcal{N}_k(x_i) \\ 0, \text{ otherwise} \end{cases} \quad (7)$$

where $\mathcal{N}_k(x_i)$ denotes the kNN points of x_i and $s_v(i,j)$ is the weight of the affinity between x_i and x_j on view v. The affinity value can be represented as Binary, Heat Kernel, Cosine or Dot-Product [18]. We have different similarity matrices S_v of size $n \times n$ computed from different views showing how data samples are similar to each other on each view. We use a linear combination to combine the affinity information from these matrices into a single matrix as follows,

$$S = \sum_{v=1}^{n_v} \lambda_v S_v \quad (8)$$

We conjecture that S is able to encode different view semantics by including how close each sample is to others to an extent. λ_v is the parameter that can be tuned to set different levels of contributions of different views to the learning process. In this work, we aim to treat all views equally hence the parameters will be set to $1/n_v$ in our experiment. Regarding S as a new data view and embedding in the objective function J will ensure the learning process to find clusters where closeness between objects is high within each cluster.

The objective function J from Eq. (6) can be written as follows,

$$J = \min \sum_{v=1}^{n_v} \|X_v - H_* W_v\|_F^2 + \|S - H_* B\|_F^2, \text{ s.t. } H_* \geq 0, B \geq 0, W_v \geq 0 \quad (9)$$

B in Eq. (9) is the corresponding basis matrix when factorizing the similarity matrix S. We also note that this objective function can be modified to add various constraints such as $l1$ norm, $l2$ norm or orthogonal constraint depending on the property of dataset or the objective of learning algorithm, i.e., achieving sparseness or even distributions or uniqueness for the factor matrices. There exists several NMF methods considering these constraints [21,23]. We omit them and aim at introducing a fundamental and novel coupled-matrix framework for clustering multi-view data where all available information has been exploited.

2.3   The Proposed Optimization Solution

Existing NMF-based methods apply MUR [7] to solve optimization where every element in the matrix is equally updated in each iteration. Distinct from these methods, we conjecture that more important elements should be updated more frequently to learn a better low-order representation.

We propose to solve the objective function in Eq. (9) using GCD so that the important elements can be selected and important features for each factor matrix can be learned. The factor matrices to be updated from Eq. (9) are H_*, B and $\{W_v\}_{v=1\ldots n_v}$. The proposed updating framework will switch between the factor matrices similar to other CD based methods [17,24]. Specifically, we have the following outer updating,

$$(W_1^0, W_2^0, \ldots, W_v^0, B^0, H_*^0) \rightarrow (W_1^1, W_2^0, \ldots, W_v^0, B^0, H_*^0)$$
$$\rightarrow \cdots \rightarrow (W_1^n, W_2^n, \ldots, W_v^n, B^n, H_*^n)$$

and the following inner updating,

$$(W_1^i, W_2^i, \ldots, W_v^i, B^i, H_*^i) \rightarrow (W_1^{i,2}, W_2^i, \ldots, W_v^i, B^i, H_*^i)$$

GCD is a row-based updating scheme where each inner iteration will iteratively update variables in the ith row until the inner stopping condition is met. In order to solve objective function J utilizing GCD, we need to solve H_*, B and $\{W_v\}_{v=1\ldots n_v}$ via the below updating solutions.

Proposed Rigorous Single Variable Update Rule. We propose the rigorous one-variable sub problem for the CD update rule to solve the MVCMF objective function in Eq. (9). We explain the update rule for H_* while it can be derived for other factor matrices similarly. To update a single variable in the shared factor matrix H_*, the primary goal is to fix all the other factor matrices and modify the shared factor matrix by adding a scalar value s to the selected single variable. This scalar value implies how much the single variable value should be tuned to achieve minimal objective difference. It can be expressed as,

$$(H_*, W_v, B) = (H_* + sZ, W_v, B) \tag{10}$$

where Z is a zero matrix except the element, that is being updated, is set to 1.

In order to identify the value of s, the rigorous one-variable sub problem is defined,

$$\min_{s: H_* + s > 0} g^{H_*}(s) \equiv f(H_* + sZ, W_v, B) \tag{11}$$

Above equation can be rewritten as,

$$g^{H_*}(s) = g^{H_*}(0) + g'^{H_*}(0)s + \frac{1}{2}g''^{H_*}(0)s^2 \tag{12}$$

where g' and g'' are the first and second order derivatives of H_* representing the gradient and Hadamard product respectively. For simplicity we have

$$g' = \partial J/\partial H_* = G^{H_*} \tag{13}$$
$$g'' = \partial^2 J/\partial H_* = H^{H_*} \tag{14}$$

The update rule to identify the value of s that is added to H_* is

$$s = \frac{g'}{g''} = \frac{G^{H_*}}{H^{H_*}} \tag{15}$$

Using the one variable sub problem and respective gradients, the variable importance (Magnitude Matrix D^{H_*}) as the difference in objective function can be calculated as follows,

$$D^{H_*} = g^{H_*}(0) - g^{H_*}(s) \tag{16}$$

By substituting Eq. (12) in Eq. (16), the variable importance equation is as,

$$D^{H_*} = -(H_* * G^{H_*}) - \frac{1}{2} * (H^{H_*} * s^2) \tag{17}$$

Updating Solution for H_*. Taking the first and the second derivative of J corresponding to H_* we have:

$$\partial J/\partial H_* = -2\sum_v X_v W_v^T + 2H_* \sum_v W_v W_v^T - 2SB^T + 2H_* BB^T \tag{18}$$

$$\partial J^2/\partial H_* = 2\sum_v W_v W_v^T + 2BB^T \tag{19}$$

According to the CD method [16], H_* will be updated through the update rule,

$$(H_*)_{ir} = [(H_*)_{ir} + (\hat{H}_*)_{ir}]_+ \tag{20}$$

where the $[]_+$ notation is for non-negative constraint and the additional value $(\hat{H}_*)_{ir}$ to be added in element $(H_*)_{ir}$ at each update step is computed as,

$$(\hat{H}_*)_{ir} = (H_*)_{ir} - \frac{\partial J/\partial H_*}{\partial^2 J/\partial H_*} \tag{21}$$

setting,

$$G^{H_*} = \partial J/\partial H_* = -2\sum_v X_v W_v^T + 2H_* \sum_v W_v W_v^T - 2SB^T + 2H_* BB^T \tag{22}$$

For simplicity, let us represent

$$Ha^{H_*} = \frac{\partial^2 J}{\partial H_*} \tag{23}$$

Algorithm 1: The Proposed MVCMF Algorithm

Input   : Multi-view data matrices $\{X_v\}$; Randomly initiated factor matrices;
Rank k; number of samples n,
$init, inner stopping condition, tol, best value.$

Output: Factor matrices: H_*, $\{W_v\}_{v=1..n_v}$ and B.

Compute G^{H_*}, Ha^{H_*} using Eqs. (22), (23).

for *each i, r, $1 \leq i \leq n$, $1 \leq r \leq k$* **do**
   | $(\hat{H}_*)_{ir} \leftarrow G_{ir}^{H_*}/Ha_{ir}^{H_*}$
   | $(\hat{H}_*)_{ir} \leftarrow (H_*)_{ir} - (\hat{H}_*)_{ir}$
   | $(\hat{H}_*)_{ir} = \max((\hat{H}_*)_{ir}, 0)$
   | Update $(\hat{H}_*)_{ir}$ and $(D^{H_*})_{ir}$ as in Eqs. (21), (24)
   | $init = \max((D^{H_*})_{ir}, init)$
end

for *each p, $1 \leq p \leq n$* **do**
   | **while** *inner stopping condition* **do**
   | | $q = -1, best value = 0$
   | | **for** *each r, $1 \leq r \leq k$* **do**
   | | | $(\hat{H}_*)_{pr} \leftarrow G_{pr}^{H_*}/Ha_{pr}^{H_*}$
   | | | $(\hat{H}_*)_{pr} \leftarrow (H_*)_{pr} - (\hat{H}_*)_{pr}$
   | | | $(\hat{H}_*)_{pr} = \max((\hat{H}_*)_{pr}, 0)$
   | | | Update $(\hat{H}_*)_{pr}$ and $(D^{H_*})_{pr}$ as in Eqs. (21), (24)
   | | | **if** $(D^{H_*})_{pr} > best value$ **then**
   | | | | $best value = (D^{H_*})_{pr}, q = r$
   | | | **end**
   | | **end**
   | | *break if $q = -1$*
   | | $(H_*)_{pr} \leftarrow (H_*)_{pr} + (\hat{H}_*)_{pr}$
   | | $G_{pr}^{H_*} = G_{pr}^{H_*} + ((\hat{H}_*)_{pr} * Ha_{qr}^{H_*})$ **for all** $r = 1..k$
   | | *break if $(best value \leq init * tol)$*
   | **end**
end

For updates to W_v and B, repeats analogues these above steps.

With the pre-calculated gradient and Hadamard product, the magnitude matrix $D^{H_*} = \{D^{H_*}\}_{ir}^{n \times k}$, is calculated where $(D^{H_*})_{ir}$ encodes how much objective function value will be decreased when each corresponding element $(H_*)_{ir}$ is updated.

$$(D^{H_*})_{ir} = -G_{ir}^{H_*}(H_*)_{ir} - 0.5 Ha^{H_*}(H_*)_{ir}^2 \qquad (24)$$

Based on this magnitude matrix, for each ith row, the most important element is updated. By updating the gradient of all the elements in i^{th} row, we can identify the next most important element to be updated by recalculating the ith row of magnitude matrix. This process is repeated until the inner stopping condition or maximum inner iteration is reached. The working process is explained in Algorithm 1.

Updating Solution for W_v. Taking the first and the second derivative of J in Eq. (9) corresponding to W_v we have,

$$\partial J/\partial W_v = -2X_v H_*^T + 2H_* H_*^T W_v \tag{25}$$
$$\partial J^2/\partial W_v = 2H_* H_*^T \tag{26}$$

W_v will be updated through the update rule

$$(W_v)_{ir} = [(W_v)_{ir} + (\hat{W}_v)_{ir}]_+ \tag{27}$$

where the additional value $(\hat{W}_v)_{ir}$ to be added into element $(W_v)_{ir}$ at each update step is computed as,

$$(\hat{W}_v)_{ir} = (W_v)_{ir} - \frac{\partial J/\partial W_v}{\partial^2 J/\partial W_v} \tag{28}$$

We define the magnitude matrix $D^{W_v} = \{D^{W_v}\}_{ir}^{m_v \times k}$, where $(D^{W_v})_{ir}$ encodes how much objective function value will be decreased when each corresponding element $(W_v)_{ir}$ is updated as below,

$$(D^{W_v})_{ir} = -G_{ir}^{W_v}(W_v)_{ir} - 0.5\frac{\partial^2 J}{\partial W_v}(W_v)_{ir}^2 \tag{29}$$

Updating Solution for B. Similar to updating schemes for H_* and W_v, B will be updated through the update rule,

$$(B)_{ir} = [(B)_{ir} + (\hat{B})_{ir}]_+ \tag{30}$$

The additional value $(\hat{B})_{ir}$ to be added into element $(B)_{ir}$ at each update step is computed as,

$$(\hat{B})_{ir} = (B)_{ir} - \frac{\partial J/\partial B}{\partial^2 J/\partial B} \tag{31}$$

The magnitude matrix $D^B = \{D^B\}_{ir}^{n \times k}$ is calculated as below,

$$(D^B)_{ir} = -G_{ir}^B(B)_{ir} - 0.5\frac{\partial^2 J}{\partial B}(B)_{ir}^2 \tag{32}$$

where $G^B = \partial J/\partial B = -2SH_*^T + 2H_* H_*^T B$.

It can be noted that the steps of iteratively updating all important elements on each row is equivalent to selecting important features to be updated instead of equally updating all elements as in MUR or FHALs [14,16]. By these single variable update rules, the factor matrices can be updated as per Fig. 1b that avoids the frequent update of the shared matrix H_* and reduces the learning complexity.

3 Experiments and Results

The proposed MVCMF method is evaluated using several multi-view datasets exhibiting different sizes and different numbers of views (as detailed in Table 1). Movie (D1) is extracted from the IMDB Website[1] that contains a set of movies represented by two different views that are actors and keywords. Cora[2] (D2) is a scientific publication dataset with documents represented via two views, content and cites. Two multi-view benchmark datasets R-MinMax (D3) and R-Top (D4) are selected from the Reuters-21578[3], a well-known text dataset. R-MinMax (D3) includes 25 classes of the dataset with at least 20 and at most 200 document per class and R-Top (D4) contains 10 largest classes in the dataset. Apart from the first view where documents are represented via terms, we used external knowledge, i.e., Wikipedia, to represent the second view of components (following steps as in [21]). Two subsets MLR-36k4k (D5) and MLR-84k5k (D6) were extracted from the most well-known multi-view dataset of Reuters RCV1/ RCV2 Multilingual with English, French, Italian and Germany terms representing each view.

We utilise two popular measurement criteria, accuracy which measures the percentage of correctly obtained labels and Normalized Mutual Information (NMI) [25] which measures the purity against the number of clusters. We also investigate the runtime performance. Average results are reported after 5-fold runs of each experiment.

We empirically evaluate the performance of MVCMF against the following state-of-art methods.

1. ConcatK-Means: In this method, we concatenate the features of all views to create a single view data and apply K-means to achieve the clustering result.
2. ConcatNMF: In this method, we concatenate the features of all views to create a single view data and apply the traditional NMF [14].
3. ConcatGNMF: Data features from all views of dataset are concatenated before executing GNMF [18].
4. MultiNMF [8]: The method simultaneously learns the low-dimensional representations from all views before seeking the consensus low-dimensional matrix.
5. MVCMF-FHALs: This is variation of our proposed method MVCMF with the FHALs updating scheme [16], to check the effectiveness of using GCD in clustering problem.

3.1 Performance Analysis

As illustrated in Tables 2 and 3, MVCMF outperforms other state-of-the-art benchmark- ing methods on all datasets. This performance is exhibited due to

---

[1] http://www.imdb.org.

[2] http://www.cs.umd.edu/%7Esen/lbc-proj/data/cora.tgz.

[3] http://www.daviddlewis.com/resources/testcollections/reuters21578/.

Table 1. Characteristic of the datasets

Properties	D1	D2	D3	D4	D5	D6
≠ Classes	17	7	25	10	6	6
≠ views	2	2	2	2	4	4
≠ Samples	617	2,708	1,413	4,000	3,600	8,400
≠ features of view 1	1,398	1,433	2,921	6,000	4,000	5,000
≠ features of view 2	1,878	5,429	2,437	6,000	4,000	5,000
≠ features of view 3	-	-	-	-	4,000	5,000
≠ features of view 4	-	-	-	-	4,000	5,000

Table 2. Accuracy of each dataset and method

Methods	D1	D2	D3	D4	D5	D6	Average
ConcatK-Means	23.18	54.91	53.86	52.20	37.61	49.11	45.15
ConcatNMF	28.20	53.77	56.33	37.78	44.81	49.18	45.01
ConcatGNMF	23.99	23.23	42.18	52.70	23.86	31.49	32.91
MultiNMF	23.01	46.16	52.80	47.35	42.00	39.95	41.88
MVCMF-FHALs	27.23	39.07	56.55	53.20	45.75	**53.07**	45.81
MVCMF-GCD	**33.71**	**58.68**	**67.02**	**57.67**	**48.42**	49.61	**52.52**

Table 3. NMI of each dataset and method

Methods	D1	D2	D3	D4	D5	D6	Average
ConcatK-Means	20.86	35.59	68.92	**50.35**	21.23	31.92	38.15
ConcatNMF	28.56	34.48	64.51	37.96	30.18	**32.15**	37.97
ConcatGNMF	24.22	05.11	55.07	41.49	09.88	08.22	24.00
MultiNMF	20.87	30.49	66.24	43.07	23.39	21.49	34.26
MVCMF-FHALs	28.46	15.83	69.40	40.38	25.76	30.28	35.02
MVCMF-GCD	**33.96**	**37.96**	**72.22**	46.77	**34.89**	31.68	**42.92**

Table 4. Running time (in seconds) of each dataset and method

Methods	D1	D2	D3	D4	D5	D6	Average
ConcatK-Means	**0.49**	12.16	6.28	68.77	51.77	187.89	54.54
ConcatNMF	1.73	7.06	6.56	29.38	36.26	103.59	30.76
ConcatGNMF	1.77	7.12	6.63	29.69	34.43	103.92	30.59
MultiNMF	31.01	157.47	120.5	724.29	790.81	2377	701.18
MVCMF-FHALs	1.03	**0.86**	**0.86**	**7.57**	**0.58**	13.25	**4.02**
MVCMF-GCD	8.03	1.39	1.06	7.91	3.34	**8.7**	**5.98**

the fact that the optimal latent features matrix of multi-view data has been learnt jointly, simultaneously and directly from all views data. In addition, the similarity matrix embedded in the factorization helps MVCMF to exploit the similarity information of data samples from multiple views, therefore, resulting in the meaningful solution. More importantly, on datasets D1–3 and D5, MVCMF obtains significantly better clustering results in terms of both accuracy and NMI as compared to other benchmark methods. Though ConcatK-Means and ConcatNMF outperform MVCMF-GCD on D4 and D6 in terms of NMI, the difference is negligible. This validates the effectiveness of our approach of using coupled matrix for multi-view representation utilising GCD framework.

While exploiting the same amount of information, i.e., data from all views, Mult- iNMF achieves better performance as compared to ConcatNMF and ConcatGNMF. It is because MultiNMF simultaneously learns factor matrices from all views and seeks the consensus latent features instead of concatenating all features and applying single view method as the two methods ConcatNMF and ConcatGNMF do. Similarly, K-means fails to bring a good result on multi-view data, though it is the most well- known method for clustering one view data.

It can also be observed that ConcatGNMF gives a very low performance on high dimensional datasets D2, D5, D6. This method relies on manifold learning to discover the latent structures for data. Unfortunately, it is a non-trivial task of learning the embedded manifold of data based on the kNN graph. Since the kNN will become meaningless as the difference between distances from a point to its nearest and farthest points becomes non-existent in high-dimensional data [26], a poor performance is produced.

Comparing MVCMF-FHALs and MVCMF-GCD with only difference of the proposed optimization method, it can be ascertained by the improved clustering results on all datasets that the proposed optimization method is able to learn a better lower-order representation.

With regard to time complexity, it can be seen from Table 4, the proposed MVCMF easily outperforms other benchmark methods. MVCMF-GCD is 11 to 272 times faster than the existing methods for bigger datasets. The reason behind the better runtime performance is that the coupled matrix factorization avoids updating the same factor matrices multiple times and equally updates all the factor matrices. We also note that for small dataset D1, concatenation based methods are faster, but the runtime increases exponentially with increase in the data size, unlike MVCMF. Due to the frequent gradient updates that helps to learn the factor matrices more accurately, the runtime of GCD is higher than FHALs. In some cases like D4 and D6, it can be noted that GCD outperforms or performs equally with FHALs due to the fast convergence achieved through the single variable selection strategy.

4   Conclusion

In this paper, we have presented the novel unified coupled matrix factorization framework that incorporates similarity information of data samples calculated from all views to efficiently generate clusters. We also present the single variable update rule of GCD for the proposed framework which exploits variable importance strategy to accurately learn the factor matrices that improves clusters quality. Experiment results on many datasets with different sizes and different number of views show the effectiveness of the proposed framework MVCMF and learning algorithm MVCMF-GCD. In future work, we will further investigate to extend the proposed framework in tensor factorization to handle higher order multi-view clustering.

References

1. Zong, L., Zhang, X., Zhao, L., Yu, H., Zhao, Q.: Multi-view clustering via multi-manifold regularized non-negative matrix factorization. Neural Netw. **88**(Suppl. C), 74–89 (2017)
2. Gao, H., Nie, F., Li, X., Huang, H.: Multi-view subspace clustering. In: IEEE International Conference on Computer Vision (ICCV), pp. 4238–4246 (2015)
3. Wang, J., Tian, F., Yu, H., Liu, C.H., Zhan, K., Wang, X.: Diverse non-negative matrix factorization for multiview data representation. IEEE Trans. Cybern. (2017)
4. Zhang, X., Li, H., Liang, W., Luo, J.: Multi-type co-clustering of general heterogeneous information networks via nonnegative matrix tri-factorization. In: Proceedings of the International Conference on Data Mining (ICDM) (2016)
5. Hidru, D., Goldenberg, A.: EquiNMF: graph regularized multiview nonnegative matrix factorization. CoRR, arXiv:abs/1409.4018 (2014)
6. Zhao, J., Xie, X., Xu, X., Sun, S.: Multi-view learning overview. Inf. Fusion **38**(C), 43–54 (2017)
7. Lee, D.: Algorithms for non-negative matrix factorization. In: Advances in Neural Information Processing Systems 13, vol. 401, pp. 788–791
8. Jing, G., Jiawei, H., Jialu, L., Chi, W.: Multi-view clustering via joint nonnegative matrix factorization. In: SDM, pp. 252–260. SIAM (2013)
9. Wang, J., Wang, X., Tian, F., Liu, C.H., Yu, H., Liu, Y.: Adaptive multi-view semi-supervised nonnegative matrix factorization. In: Hirose, A., Ozawa, S., Doya, K., Ikeda, K., Lee, M., Liu, D. (eds.) ICONIP 2016. LNCS, vol. 9948, pp. 435–444. Springer, Cham (2016). https://doi.org/10.1007/978-3-319-46672-9_49
10. Wang, J., Tian, F., Wang, X., Yu, H., Liu, CH., Yang, L.: Multi-component nonnegative matrix factorization. In: IJCAI, pp. 2922–2928(2017)
11. Hofmann, T.: Probabilistic latent semantic indexing. In: ACM SIGIR (1999)
12. Acar, E., Gurdeniz, G., Rasmussen, M.A., Rago, D., Dragsted, L.O., Bro, R.: Coupled matrix factorization with sparse factors to identify potential biomarkers in metabolomics. Int. J. Knowl. Discov. Bioinform. **3**, 22–43 (2012)
13. Acar, E., Bro, R., Smilde, A.K.: Data fusion in metabolomics using coupled matrix and tensor factorizations. Proc. IEEE **103**, 1602–1620 (2015)
14. Dhillon, I.S.: Co-clustering documents and words using bipartite spectral graph partitioning. In: ACM SIGKDD, pp. 269–274 (2001)
15. Ding, C.H.Q., Li, T., Jordan, M.I.: Convex and semi-nonnegative matrix factorizations. IEEE Trans. Pattern Anal. Mach. Intell. **32**(1), 45–55 (2010)
16. Cichocki, A., Phan, A.H: Fast local algorithms for large scale nonnegative matrix and tensor factorizations (2008)

17. Hsieh, C.-J., Dhillon, I.S.: Fast coordinate descent methods with variable selection for non-negative matrix factorization. In: ACM SIGKDD, pp. 1064–1072 (2011)
18. Cai, D., He, X., Han, J., Huang, T.S.: Graph regularized nonnegative matrix factorization for data representation. IEEE Trans. Pattern Anal. **33**, 1548–1560 (2011)
19. Belkin, M., Niyogi, P., Sindhwani, V.: Manifold regularization: a geometric framework for learning from labeled and unlabeled examples. J. Mach. Learn. Res. **7**, 2399–2434 (2006)
20. Li, P., Bu, J., Chen, C., He, Z.: Relational co-clustering via manifold ensemble learning. In: CIKM, pp. 1687–1691 (2012)
21. Hou, J., Nayak, R.: Robust clustering of multi-type relational data via a heterogeneous manifold ensemble. In: Proceedings of the International Conference on Data Engineering (ICDE) (2015)
22. Luong, K., Nayak, R.: Learning association relationship and accurate geometric structure for multi-type relational data. In: Proceedings of the International Conference on Data Engineering (ICDE) (2018)
23. Ding, C., Li, T., Peng, W., Park, H.: Orthogonal nonnegative matrix t-factorizations for clustering. In: KDD 2006, pp. 126–135. ACM (2006)
24. Cichocki, A., Phan, A.-H.: Fast local algorithms for large scale nonnegative matrix and tensor factorizations. IEICE **92**, 708–721 (2009)
25. Zhong, S., Ghosh, J.: Generative model-based document clustering: a comparative study. Knowl. Inf. Syst. **8**(3), 374–384 (2005)
26. Beyer, K., Goldstein, J., Ramakrishnan, R., Shaft, U.: When is "Nearest Neighbor" meaningful? In: Beeri, C., Buneman, P. (eds.) ICDT 1999. LNCS, vol. 1540, pp. 217–235. Springer, Heidelberg (1999). https://doi.org/10.1007/3-540-49257-7_15

Data-Augmented Regression with Generative Convolutional Network

Xiaodong Ning[1(✉)], Lina Yao[1], Xianzhi Wang[2], Boualem Benatallah[1], Shuai Zhang[1], and Xiang Zhang[1]

[1] Universit of New South Wales, Kensington, Australia
xiaodong.ning@student.unsw.edu.au, lina.yao@unsw.edu.au
[2] University of Technology Sydney, Ultimo, Australia

Abstract. Generative adversarial networks (GAN)-based approaches have been extensively investigated whereas GAN-inspired regression (i.e., numeric prediction) has rarely been studied in image and video processing domains. The lack of sufficient labeled data in many real-world cases poses great challenges to regression methods, which generally require sufficient labeled samples for their training. In this regard, we propose a unified framework that combines a robust autoencoder and a generative convolutional neural network (GCNN)-based regression model to address the regression problem. Our model is able to generate high-quality artificial samples via augmenting the size of a small number of training samples for better training effects. Extensive experiments are conducted on two real-world datasets and the results show that our proposed model consistently outperforms a set of advanced techniques under various evaluation metrics.

1 Introduction

Classification and regression are two main types of machine learning applications. While previous research mostly focuses on classification tasks, regression has received less attention. Although a regression task can generally be converted into a multi-classification task by approximating continuous variables using discrete classes, regression can provide more meaningful and accurate insights in many real-world problems such as house price prediction [13], stock price forecasting [12], crime rate inference [10], and movie box prediction [9]. Some commonly used regression methods include parametric and semi-parametric spatial hedonic models, state frequency memory (SFM) recurrent network [3], (kernel-based) regression models [2], and Hodrick Prescott filter and regression hybrid models.

Different from the above work, we aim to propose a unified framework to solve general regression problems instead of targeting a specific topic. Our framework integrates three parts together: feature preprocessing, feature transformation, and numeric prediction model. Feature preprocessing is responsible for converting the raw meta-data into a unified feature format. The feature transformation

© Springer Nature Switzerland AG 2018
H. Hacid et al. (Eds.): WISE 2018, LNCS 11234, pp. 301–311, 2018.
https://doi.org/10.1007/978-3-030-02925-8_21

part applies a robust auto-encoder to eliminate outliers and noise and to extract high quality, non-linear latent features from the original feature. In addition, we alleviate the problem of small training samples by generating artificial samples based on a generative convolutional neural network (GCNN), a new variant of generative adversarial networks (GAN) [8] that we propose specially for regression tasks.

Recently, generative adversarial networks (GAN) has attracted a lot of attention for its capability of generating photorealistic images or videos that help visualize new interior/industrial designs, daily commodities or items for scenes in computer games. The basic GAN is implemented by a system of two neural networks contesting with each other in a zero-sum game framework. Many variants of GAN have been proposed since its invention. For example, Radford et al. [14] propose a deep convolutional generative adversarial network (DCGANs) that regards parts of the generator and discriminator networks as feature extractors in CNN. Liu et al. [6] propose a coupled generative adversarial network (CoGAN) for learning a joint distribution of multi-domain images. While most of the previous investigations about GAN focus on solving the classification or image generation problems, our model aims to utilize the generator for data-augmentation in regression tasks with a small number of training samples.

We make the following contributions in this paper:

- We propose a unified framework that combines feature transformation and numeric prediction for general regression tasks. To the best of our knowledge, our work is the first to address the regression problem using the idea of GAN in a data-augmented manner.
- We design a generative convolutional neural network based regression model to solve the continuous numeric prediction problem with small labeled samples. Our model is able to effectively augment the available number of training data by generating high-quality artificial samples.
- We conduct comprehensive experiments on two real-world datasets and demonstrate that our proposed method consistently outperforms several non-deep learning and deep learning methods.

2   A Unified Framework

We propose a unified framework that consists of three main parts: feature extraction, feature transformation and a generative convolutional neural network. The architecture of our model is shown in Fig. 1, where the meta-data is fed into our model first, then the original features are extracted and transformed via robust autoencoder. After that, the artificial samples are produced from the artificial feature generator (G1) and artificial label generator (G2). And finally, both the artificial samples and real samples are fed into discriminator for training. We will describe the details in the following subsections.

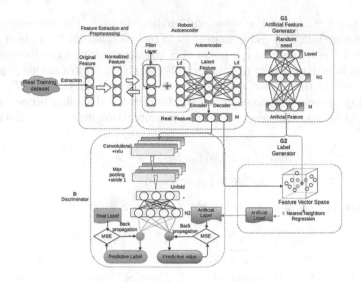

Fig. 1. Overall framework

2.1   Feature Preprocessing

We use two datasets to evaluate our model: IMDB dataset and American Community Crime dataset. We extract two types of features from meta-data from the datasets: the intrinsic attributes of each movie and the community properties such as population distribution and law enforcement distribution. To reduce the influence of missing values and noises in the extracted features on the final performance of the model, we first process it before feeding it into the prediction model. Our first step is to fill up the missing feature values by replacing the 'Null' values with the median value of all samples. As the value of different features can differ in wide range, we convert them into a unique range 0–1 by feature normalization. Now, we get a complete feature vector of all the samples where each feature value falls within 0–1.

2.2   Feature Transformation

We use a robust autoencoder network, which is a variant of the normal autoencoder proposed by Zhou et al. [7], to transform features and to discover high quality, non-linear features while eliminating outliers and noise to clean the training data. The robust autoencoder differs from normal autoencoders in adding a filter layer into its network. The filter layer is used to cull out the anomalous parts of the data that are difficult to reconstruct so as to represent the remaining portion of the data by the low-dimensional hidden layer with a small reconstruction error. First, it splits the input data X into two parts $X = L_D + S$, where L_D represents the part of the input data that is well represented by the hidden layer of the auto-encoder and S contains the noise and outliers which are difficult to reconstruct. By removing the noise and outliers from X, the auto-encoder can recover the remaining L_D more accurately. The loss function for a given input X can be either the l_1 norm or $l_{2,1}$ norm of S balanced against the reconstruction error of L_D. In a data-driven manner, we choose the $l_{2,1}$ norm in this paper. The overall process of the robust auto-encoder is shown in Eq. 1, where E_θ denotes the encoding process of auto-encoder, D_θ denotes the decoding process of auto-encoder, and λ tunes the weight between reconstruction error of L_D and sparsity of S.

$$\min_\theta ||L_D - D_\theta(E_\theta(L_D))||_2 + \lambda ||S||_{2,1}$$
$$s.t.\ X - L_D - S = 0 \tag{1}$$

In particular, $l_{2,1}$ norm is defined in Eq. 2, where $S \in R^{m \times n}$.

$$||S||_{2,1} = \sum_{j=1}^{n} ||S_j||_2 = \sum_{j=1}^{n} (\sum_{i=1}^{m} |S_{ij}|^2)^{1/2} \tag{2}$$

The objective of Eq. 1 is optimized by L_D and S, which are trained independently through iterations in a similar procedure as that described in [7]. The optimization of L_D is similar to the optimization of a traditional autoencoder while the minimization of the $l_{2,1}$ norm of S is a complicated proximal problem [7]. We present a modified minimization process of $||S||_{2,1}$ in Algorithm 1. This algorithm differs from [7] in that we set the value of $S[i,j]$ as a random value within the range $[-e_j, e_j]$ instead of setting it zero when the e_j is less than λ. By doing so, we can prevent the over-optimizing problem of $||S||_{2,1}$ and its influence on the autoencoder performance. Processed by the robust autoencoder, the original feature is transformed into an M-dimensional feature vector.

Algorithm 1. Random Proximal Method

Input: $S \in R^{m \times n}, \lambda$
  for j in 1 to n do
    $e_j = (\sum_{i=1}^{m} |S[i,j]|^2)^{1/2}$
    if $e_j > \lambda$ then
      for i in 1 to m do
        $S[i,j] = S[i,j] - \lambda \frac{S[i,j]}{e_j}$
      end for
    else
      for i in 1 to m do
        $S[i,j] = Random[-e_j, e_j]$
      end for
    end if
  end for
Output: S

2.3 Generative Convolutional Neural Network Based Regression Model

In recent years, deep learning has been proven to achieve promising performance in various domains such as twitter classification [5], rating prediction

[1] and brain disease diagnosis [4], etc. However, the requirement for a large amount of training data prevents deep learning networks from being applied into many real problems. In this regard, we propose a generative convolutional neural networks based regression model inspired by the generative adversarial network (GAN) [8]. Instead of contesting between generator and discriminator in GAN, our model utilizes a generator to generate artificial samples and combine them with real samples to co-train the entire network to solve the limited training samples problem. As shown in Fig. 1, our regression model consists of three components: artificial feature generator (G1), artificial label generator (G2) and discriminator (D).

Artificial Feature Generator G1. The artificial feature generator G1 works as follows. Firstly, we initialize $Nseed$ seed vectors with $Lseed$ length where each vector point is generated randomly from the range of $(0, 1)$. Then, the seed vectors are fed into the two fully-connection layers (the neurons in each layer are $N1$ and M). We apply Leaky Relu activation function on the dot product results generated from each layer.

Via the two fully connected layers, the random seed vector will be transformed into an M-dimensional artificial feature vector (with the same length as the real transformed feature vector from the robust auto-encoder).

Artificial Label Generator G2. We design an artificial label generator G2 to assign labels for the artificial features generated from G1. As the artificial feature vector and real transformed feature vector share the same dimension, we apply the weighted K nearest neighbors regression to label the artificial samples using the real training samples. The distance between artificial samples and real samples are calculated as their Euclidean distance, as described in Eq. 3, where D_{ij} denotes the distance between i_{th} artificial sample and j_{th} real sample, M denotes the length of feature, F_{ik} denotes k_{th} feature value of i_{th} artificial sample, and F_{jk} denotes k_{th} feature value of j_{th} real sample.

$$D_{ij} = \sum_{k=1}^{M} \frac{(F_{ik} - F_{jk})^2}{M} \tag{3}$$

Then, we apply weighted knn regression to calculate the labels for artificial samples as follows:

$$A_{ti} = \frac{\sum_{j=1}^{K} D_{ij} * L_j}{\sum_{j=1}^{K} D_{ij}} \tag{4}$$

where A_{ti} denotes the generated label of the i_{th} artificial feature; K denotes the top K nearest neighbors to the i_{th} artificial feature; D_{ij} denotes the distance between i_{th} artificial sample and j_{th} nearest real sample; L_j denotes the label value of j_{th} nearest real sample

Discriminator. The discriminator is a one-dimensional convolutional neural network (1DCNN) based regression network where the artificial samples feed in. It consists of four layers: one convolutional layer, one max pooling layer, and two fully-connected layers. The convolutional layer in discriminator takes a set of Fn independent filters and slides them over the whole feature vector. Along the way, the dot product is taken between the filters and chunks of the input features. Filters are used to generate the feature vectors in each filter length. The same padding is chosen for the convolutional layer in order to keep the same size of output as input. In this way, the feature vector is projected into a stack of feature maps (vector maps in our work). Followed by the convolutional layer, we add one max-pooling layer. The convolved and max-pooled feature vectors will be unfolded and fed into two fully-connected layers (the neurons in each layer are $N2$ and 1) applied for the high-level reasoning. Finally, the initial feature vector is transformed and projected into one-dimensional value via the second fully connected layer. The one-dimensional value generated from the final layer is the predictive label. The objective of our model is to predict the label as a continuous value instead of a class, so the commonly used cross entropy loss function should be modified. As there are two types of samples (artificial samples and real samples) in our model, we define two loss functions used in our model $\mathcal{L}_1$ and $\mathcal{L}_2$, where $\mathcal{L}_1$ represents the mean square rrror of artificial generated samples, and $\mathcal{L}_2$ represents the mean square error of real samples. Additionally, a multiplying factor μ(0-1) is used in $\mathcal{L}_1$ to tune the weight between two loss functions.

Network Training. In our model, only feature generator (G1) and discriminator should be trained. The feature generator G1 and discriminator are trained via back-propagation based on the Adam optimizer. As we have two loss functions ($\mathcal{L}_1$ and $\mathcal{L}_2$) from real samples and artificial samples, the discriminator is trained iteratively by them to achieve a satisfactory prediction performance. In comparison, the feature generator G1 is trained only with artificial samples ($\mathcal{L}_1$). The G1 generator evolves subsequently following the discriminator using $\mathcal{L}_1$ and updates its parameters via back-propagation to produce highly realistic artificial features. It is notable that the evolution of generator G1 will also improve the training effect of the discriminator.

By optimizing the two loss functions, feature generator G1 learns how to generate high-quality artificial features similar to the real samples, and the discriminator learns to predict the final labels accurately.

3 Experiment

3.1 Dataset

Dataset. We choose two datasets to evaluate our model. One is the IMDB dataset[1], where movie attributes and plot information are obtained from the

---

[1] https://www.kaggle.com/tmdb/tmdb-movie-metadata.

IMDB dataset in Kaggle and the IMDB website, respectively. We keep among all movies 1,471 US movies released after 2003 to ensure each has some plot descriptions and the history ratings and box office of directors and actors. Each movie record contains attributes including facebook likes of directors and cast, genres, country, MPAA rating[2], release date, budgets of the movie, box office records of directors and actors and plot latent topics. All the historical features of samples in our dataset are only extracted from the previous movies whose release dates are earlier than the samples. By doing so, we can avoid using any 'future information' in our dataset. The movies are then divided into two part: training samples (1219 movies released between 2003–2013) and testing samples (252 movies released after 2013).

The second dataset is the American community crime rate dataset[3]. This dataset contains various attributes of each community such as the percent of the population considered urban, the median family income, per capita number of police officers, etc. These attributes are used to predict the per capita violent crime for each community. Among the total 1994 samples, we use 1000 as training samples and keep the remaining 994 as testing samples.

Table 1. Mean absolute error of each approach on IMDB and Community Crime datasets under different percentage of training samples; MAE = Mean Absolute Error

Datasets	IMDB			Community crime		
Experimental methods	MAE (50%)	MAE (75%)	MAE (100%)	MAE (50%)	MAE (75%)	MAE (100%)
RF Regressor	0.914	0.88	0.87	0.104	0.099	0.095
Gradientboosting	0.89	0.872	0.87	0.109	0.103	0.10
Adaboosting	0.921	0.91	0.9	0.1369	0.133	0.121
Xgboosting	0.945	0.937	0.935	0.109	0.105	0.103
SVR	0.882	0.865	0.86	0.126	0.125	0.123
Kernel-1 [9]	0.86	0.841	0.83	0.121	0.117	0.111
Hypergrah Regression [11]	0.86	0.83	0.822	0.110	0.101	0.098
Deep Belief Network	0.78	0.77	0.76	0.112	0.105	0.103
MLP	0.801	0.765	0.751	0.101	0.097	0.095
1DCNN	0.812	0.763	0.744	0.096	0.095	0.091
1DCNN-SVR [2]	0.788	0.754	0.738	0.101	0.096	0.093
Ours	**0.712**	**0.701**	**0.68**	**0.093**	**0.090**	**0.089**

3.2   Comparison Methods

We set the default settings of our model as following: $M = 55$, $Nseed = 500$, $Lseed = 200$, $K = 8$, $\mu = 1.0$ for IMDB; $M = 90$, $Nseed = 200$, $Lseed = 60$, $K = 5$, $\mu = 0.9$ for Crime. We also tune the optimal parameters of each comparison method respectively for the two datasets for a fair comparison.

---

[2] https://en.wikipedia.org/wiki/Motion_Picture_Association_of_America_film_rating_system\#MPAA_film_ratings.

[3] https://archive.ics.uci.edu/ml/datasets/Communities+and+Crime.

Table 2. Hit ratio of each approach on IMDB and Community Crime datasets under different threshold θ (3%, 5%, 8%, 10%, 15%);

Datasets	IMDB					Community crime				
Experimental methods	3%	5%	8%	10%	15%	3%	5%	8%	10%	15%
RF Regressor	0.18	0.27	0.416	0.535	0.738	0.034	0.062	0.095	0.129	0.181
Gradientboosting	0.19	0.29	0.452	0.523	0.742	0.04	0.063	0.090	0.115	0.183
Adaboosting	0.11	0.23	0.388	0.46	0.66	0.031	0.043	0.071	0.089	0.135
Xgboosting	0.16	0.277	0.42	0.53	0.789	0.026	0.054	0.088	0.111	0.176
SVR	0.18	0.29	0.468	0.567	0.785	0.033	0.056	0.075	0.09	0.141
Kernel-1 [9]	0.16	0.26	0.44	0.531	0.747	0.025	0.041	0.070	0.078	0.138
Hypergrah Regression [11]	0.179	0.288	0.463	0.555	0.779	0.032	0.049	0.084	0.092	0.151
Deep Belief Network	0.177	0.30	0.471	0.588	0.788	0.031	0.057	0.098	0.130	0.166
MLP	0.183	0.285	0.489	0.544	0.781	0.044	0.058	0.101	0.141	0.183
1DCNN	0.202	0.293	0.468	0.561	0.787	0.050	0.07	0.111	0.148	0.20
1DCNN-SVR	0.202	0.30	0.471	0.565	0.791	0.048	0.065	0.107	0.144	0.194
Ours	**0.22**	**0.332**	**0.508**	**0.611**	**0.821**	**0.068**	**0.084**	**0.141**	**0.177**	**0.235**

- *Random Forest Regressor (RF)* with 200 estimators (IMDB); 20 estimators (Crime).
- *Gradientboosting (Gra)* with 30 estimators (IMDB); 25 estimators (Crime).
- *Adaboosting (Ada)* with base estimator as the decision tree regressor, 50 estimators (IMDB); 50 estimators (Crime)
- *Xgboosting (Xg)* with 100 estimators (IMDB); 30 estimators (Crime).
- *Support Vector Regressor (SVR)* with penalty parameter as 0.8 (IMDB); penalty parameter as 1.0 (Crime).
- *Kernel-1 Regression method* is a kernel-based approach with an improved version of KNN regression [9]. In our paper, we utilize the recommended parameter settings in the paper for both two datasets.
- *Hypergraph Regression* is a regression version of the hypergraph classification [11]. We define a hyperedge by each sample and its K nearest neighbors to form a hypergraph. The weight of each hyperedge is calculated using the mean similarities of pairs in this hyperedge. The predicted label of i_{th} sample is calculated using the weighted average ratings of the samples belonging to the same hyperedge with i_{th} sample. We set K as 10 and 8 for IMDB and Crime;
- *Multiple Layer Perceptron (MLP)* with three layers (100, 300, 1 neurons in each layer) (IMDB); with three layers (200, 350, 1 neurons in each layer) (Crime).
- *Deep Belief Network Regression* A deep belief regression network is proposed where deep belief network is placed at the bottom for unsupervised feature learning with a linear regression layer at the top of supervised prediction. We set three hidden layers (110, 200, 330 neurons in each layer) (IMDB); with three hidden layers (110, 300, 200 neurons in each layer) (Crime)
- *One Dimensional Convolutional Neural Network (1DCNN)* with one convolutional layer (5 filters), one max pooling layer and two fully connected layer

(300 and 1) for IMDB; one convolutional layer (4 filters), one max pooling layer and two fully connected layer (250, 1) for Crime.

- *One Dimensional Convolutional Neural Network with Support Vector Regressor (1DCNN-SVR)* is the regression version of 1DCNN which utilizes the support vector regressor to replace the output layer of 1DCNN. The parameters are set same as the 1DCNN with support vector regressor.

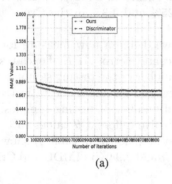
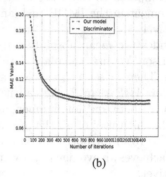

(a)   (b)

Fig. 2. Mean absolute value (MAE) comparison between our model and discriminator (excluding the artificial generator G1, G2) in training process for IMDB (a) and crime (b)

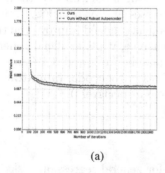
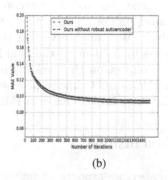

(a)   (b)

Fig. 3. Mean absolute value (MAE) comparison between our original model and our model (excluding the robust autoencoder) in training process for IMDB (a) and crime (b)

3.3   Comparison Results

Since the aim of our model is to accurately predict labels of new samples in the regression problem, we apply different percentages (50%, 75%, 100%) of training data to train all the models, predict the rating values of testing samples for both the two datasets and evaluate the performance of different approaches by the mean absolute error (MAE). The holdout results (shown in Table 1) show our model achieves the best performance for both the two datasets with MAE (0.712, 0.701, 0.68) and MAE (0.093, 0.090, 0.089), under 50%, 75% and 100% training samples. 1DCNN-SVR performs the best among the all the compared methods in the IMDB task while the 1DCNN achieves the best performance in Crime task. Compared to the other models, our model achieves a 9.6%, 7.02%, 7.85% improvement in the IMDB task and 6.9%, 6.25% and 4.3% in the Crime task, respectively.

Besides the mean absolute error (MAE), we define a new measure for the comparison. We first calculate the absolute percentage error APE of each testing sample. Then we set a threshold θ and count number of the testing samples with a smaller APE than θ. We call it 'hit number' and calculate the ratio (called 'hit ratio') between 'hit number' and the total number of testing samples. We show the 'hit ratio' of all the approaches under five different θ values (3%, 5%, 8%, 10%, 15%) for both two datasets. The results (shown in Table 2) show our model achieves an average improvement of 3.26% and 2.94% in IMDB and Crime, respectively.

4   Ablation Study

In addition, we carry out the ablation study to examine the effectiveness of the components in our model. We conduct comparative experiments by excluding artificial generator and robust autoencoder from our model on two tasks during training process. The MAE of training processes is shown in Figs. 2 and 3, respectively. We can observe that the two components indeed improve the training process by influencing the final MAE performance, while artificial generator has more impact on the performance (approximately 7.6% and 3.4% improvement in IMDB and Crime respectively) than robust autoencoder (approximately 4.3% and 3.1% improvement in IMDB and Crime respectively).

5   Conclusions

In this paper, we propose an integrated model for general regression tasks. Our model leverages a robust auto-encoder in combination with a generative convolutional neural network for feature transformation and numeric prediction. Owing to artificial sample generator (G1, G2), our model is capable of handling the numeric prediction tasks with a small size of training samples, which are originally insufficient for traditional deep learning methods. Extensive experiments on two real-world datasets have shown the superior performances of our model over a series of existing advanced techniques.

References

1. Ning, X., et al.: Rating prediction via generative convolutional neural networks based regression. Pattern Recognit. Lett. (2018)
2. Andreas, C., Steinwart, I.: Consistency and robustness of kernel-based regression in convex risk minimization. Bernoulli **13**(3), 799–819 (2007)
3. Hu, H., Qi, G.-J.: state-frequency memory recurrent neural networks. In: ICML (2017)
4. Suk, H.I., Lee, S.W., Shen, D.: Deep ensemble learning of sparse regression models for brain disease diagnosis. Med. Image Anal. **37**, 101–113 (2017)
5. Ning, X., Yao, L., Wang, X., Benatallah, B.: Calling for response: automatically distinguishing situation-aware tweets during crises. In: Cong, G., Peng, W.-C., Zhang, W.E., Li, C., Sun, A. (eds.) ADMA 2017. LNCS (LNAI), vol. 10604, pp. 195–208. Springer, Cham (2017). https://doi.org/10.1007/978-3-319-69179-4_14
6. Liu, M.-Y., Tuzel, O.: Coupled generative adversarial networks. In: Advances in Neural Information Processing Systems (2016)
7. Chong, Z., Paffenroth, R.C.: Anomaly detection with robust deep autoencoders. In: Proceedings of the 23rd ACM SIGKDD (2017)
8. Goodfellow, I., et al.: Generative adversarial nets. In: NIPS (2014)
9. Eliashberg, J., Hui, S.K., John, Z.J.: Assessing box office performance using movie scripts: a kernel-based approach. TKDE **26**(11), 2639–2648 (2014)
10. Gerber, M.S.: Predicting crime using Twitter and kernel density estimation. Decis. Support Syst. **61**, 115–125 (2014)
11. Huang, S., et al.: Regression-based hypergraph learning for image clustering and classification. arXiv preprint arXiv:1603.04150 (2016)
12. Zhang, L., Aggarwal, C., Qi, G.-J.: Stock price prediction via discovering multi-frequency trading patterns. In: Proceedings of the 23rd ACM SIGKDD (2017)
13. Montero, J.-M., Mínguez, R., Fernández-Avilés, G.: Housing price prediction: parametric versus semi-parametric spatial hedonic models. J. Geogr. Syst. **20**(1), 27–55 (2018)
14. Radford, A., Metz, L., Chintala, S.: Unsupervised representation learning with deep convolutional generative adversarial networks. arXiv preprint arXiv:1511.06434 (2015)

Towards Automatic Complex Feature Engineering

Jianyu Zhang[1]([envelope]) [ORCID], Françoise Fogelman-Soulié[1] [ORCID],
and Christine Largeron[2,3,4]

[1] Tianjin University, Tianjin 300350, China
{edzhang, soulie}@tju.edu.cn
[2] Université de Lyon, 42023 Saint-Etienne, France
christine.largeron@univ-st-etienne.fr
[3] CNRS, UMR 5516, Laboratoire Hubert Curien, 42000 Saint-Etienne, France
[4] Université de Saint-Etienne, Jean-Monnet, 42000 Saint-Etienne, France

Abstract. Feature engineering is one of the most difficult and time-consuming tasks in data mining projects, and requires strong expert knowledge. Existing feature engineering techniques tend to use limited numbers of simple feature transformation methods and validate on simple datasets (small volume, simple structure), obviously limiting the benefits of feature engineering. In this paper, we propose a general *Automatic Feature Engineering Machine* framework (*AFEM* for short), which defines families of complex features and introduces them one family at a time (block bottom-up). We show that this framework covers most of the existing features used in the literature and allows us to efficiently generate complex feature families: in particular, local time, social network and representation-based families for relational and graph datasets, as well as composition of features. We validate our approach on two large realistic competitions datasets and a recommendation system task with social network. In the first two tasks, *AFEM* automatically reached ranks 15 and 12 compared to human teams; in the last task, it achieved 1.5% regression error reduction, compared to best results in the literature. Furthermore, in the context of big data and web applications, by balancing computation time and number of features/performance, in one case, we could reduce 2/3 computation time with only 0.2% AUC performance loss. Our code is publicly available on GitHub (https://github.com/TjuJianyu/AFEM).

Keywords: Feature engineering · Machine learning for the web
Social network computing · Big data · Web application

1 Introduction

In recent years, it has been recognized in machine learning that *feature engineering* is the most critical factor for performance [1]: features are more important than the machine learning algorithm used. A *derived feature* is produced from the existing features set (originally the raw data) by some transformation: feature engineering (*FE*

© Springer Nature Switzerland AG 2018
H. Hacid et al. (Eds.): WISE 2018, LNCS 11234, pp. 312–322, 2018.
https://doi.org/10.1007/978-3-030-02925-8_22

in the rest of this paper) aims at producing "good" derived features, i.e. features improving performances, through a model simpler and easier to train (i.e. requiring less examples) than with the original features set. The standard approach to *FE* is to carefully hand-craft meaningful features, a very time-consuming process, more art than science, requiring domain knowledge and trials and errors. It is not rare to spend 90% of a machine learning project on *FE*, so automating it would save significant time and effort. As a consequence, research groups and companies, such as IBM for example [4–6, 8] have started to produce methods for (semi-) automatic *FE*.

Since the number of possible derived features is potentially infinite, it is necessary to be able to control the generation of features. There exist basically two ways: the *top-down* approach (expansion-reduction [5]) consists in generating all the features for a full set of transformations, and then do feature selection and performance evaluation of a model trained on the enriched feature set; this is of course very costly in terms of computation, if the sets of transformations and original features are of large sizes, which is why most publications limit these sizes. In the *bottom-up* approach (evolution-centric [5]), features are progressively entered and the most interesting ones are selected, either using meta-features to characterize their quality [3] or through learning [4, 5, 8]; this approach is more scalable than the previous one, but it still is compute-intensive, since it evaluates performances of derived features. Our contributions are as follows:

- We develop a modular, general and flexible framework, *AFEM*, for generating new features in relational datasets. The framework can handle many common types of features, both simple and more complex, structured in *families*, including e.g. timestamp, social network and representation related features.
- We introduce a *block-bottom-up* approach, intermediate between top-down and bottom-up, for progressively inserting and evaluating derived features, by introducing blocks of features and evaluating the most promising ones in each family.
- We evaluate *AFEM* on three realistic and relatively large datasets: in KDD cup 2014[1], *AFEM* got rank 15/472 on private leaderboard; in WSDM cup 2018[2] *AFEM* got 6th rank out of 1081 teams [10]; in Last.fm[3] rating prediction task, *AFEM* got 1.5% regression error reduction compared to literature.

The paper is organized as follows: in Sect. 2, we present definitions and notations, related work in Sect. 3 and our approach in Sect. 4. Experimental results are presented and analyzed in Sect. 5. Section 6 concludes with a discussion of the potential of our approach and perspectives for further research.

---

[1] https://www.kaggle.com/c/kdd-cup-2014-predicting-excitement-at-donors-choose.

[2] https://wsdm-cup-2018.kkbox.events/.

[3] http://files.grouplens.org/datasets/hetrec2011/hetrec2011-delicious-2k.zip.

2 Definitions and Notations

In this paper, we assume that we have (at least) one *entity* E and a predictive task associated to it; we want to derive new features for E. For example, in Last.fm[4] (Fig. 1-c), User and Artist are two entities containing features of users (e.g. age, city) and artists and other tables allow to derive features for these two entities, e.g. how many times a user listened to an artist's music.

Let be given an entity E, an individual i with entity index Id_E and a features set F, we will call *neighborhood* N of i, the result of a mapping *Neigh* from Id_E and p attributes of i in features set F: a neighborhood is a set of instances that are close with each other by some distance metrics. It could be a set of users, a time interval, a geographic region, etc. So we call S the space in which neighborhoods are and, for simplicity, do not further specify S. For example, in Fig. 1-c, the neighborhood of a user could be the set of his friends with the same age living in the same city:

$$Neigh : (Id_E, a_1, a_2, \ldots, a_p) \rightarrow N \qquad (1)$$

A *function* of arity n is a mapping for deriving a feature from n features (this is a *global function*) or from n features and a neighborhood (a *local function*). For example, we can define the count of songs by a specific artist which users in N listened to. A derived feature is of *order* k if it depends upon k attributes. k is thus the number of distinct elements in the set of attributes $\{b_1, b_2, \ldots, b_n; a_1, a_2, \ldots, a_p\}$ the feature depends upon (through f and N in Eq. (2)). Obviously, derived features of large order will require a lot more computing time than features of order 1:

$$f : (b_1, b_2, \ldots, b_n) \rightarrow b \qquad f : (b_1, b_2, \ldots, b_n; N) \rightarrow b \qquad (2)$$

A *transformation* is defined as $T = (T_{Func})$ (*global transformation*) or $T = (T_{Func}, T_{Neigh})$ (*local transformation*), where T_{Func} is a set of functions (global or local) and T_{Neigh} is a set of neighborhoods.

The *feature engineering process* consists in starting from an initial collection of features F_0, collected from the raw data, and progressively deriving enriched feature sets, $F_1, F_2, \ldots, F_t, \ldots$ through transformations $T_0, T_1, \ldots, T_t, \ldots$

$$F_t \rightarrow F_{t+1} = F_t \cup DF_t \qquad (3)$$

where DF_t is the *set of derived features* φ obtained as:

$$\varphi \in DF_t : \varphi = f(b_1, b_2, \ldots, b_n) \qquad \text{if } T_t = (T_{Func}^t) \text{ is global} \qquad f \in T_t$$
$$\varphi \in DF_t : \varphi = f(b_1, b_2, \ldots, b_n; N) \quad \text{if } T_t = (T_{Func}^t, T_{Neigh}^t) \text{ is local}, \quad (f, N) \in T_t$$

$$(4)$$

---

[4] http://files.grouplens.org/datasets/hetrec2011/hetrec2011-delicious-2k.zip.

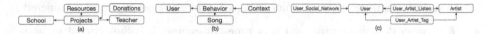

Fig. 1. Dataset structures: KDD'14 (a), WSDM'18 (b) and Last.fm (c). An arrow from one entity to another signifies that the first entity references the second in the dataset.

where $f \in T_{Func}^t$, $b_1, b_2, \ldots, b_n$; $a_1, a_2, \ldots, a_p \in F_t$, $N = Neigh(Id_E, a_1, a_2, \ldots, a_p)$ $\in T_{Neigh}^t$. By changing, at each time step t, the set of transformations T_t we can define various groups of features, we will describe some in Sect. 4.

3 Related Work

It has long been known in Machine Learning that original data might not be best for building a simple model with good performance. Various researchers have indeed demonstrated the benefit of using additional features: for example, the final winner in the Netflix prize [9] designed 16 *Global Effects* and 24 *Global Time Effects* features. Deep learning automatically learns features from raw data, but training requires careful design of the network structure, as well as enough examples to learn the large number of the network parameters (weights). In many common tasks with relational datasets, pure deep learning may not work well, because of the complex structure and semantic meaning of the relations. Moreover features learnt by deep learning are opaque, while derived features have explicit meaning. The literature on *FE* is not very abundant. It contains two main lines of approaches. In the *bottom-up approach* (or evolution-centric), features are progressively added and evaluated, either using meta-features (e.g. information gain for LFE [8]) or using the performance of a true prediction model (Cognito [4], or reinforcement learning [5]). In the *top-down approach* (or expansion-reduction), all features are generated and used to build a true prediction model; then, based on that model, most important features are selected. This approach is not feasible in practice with large datasets and/or large number of features. To the best of our knowledge, the first automatic *FE* work on "flat" data is ExploreKit [3]; it uses a predefined structured set of operations to generate all possible candidate features, and learns to rank and select them through a machine learning model with a series of meta-features. It is very compute-intensive due to the generation of all possible features and the evaluation of each feature through the performance of a true prediction model. Data Science Machine *DSM* [2] and One Button Machine *OneBM* [6] can do automatic *FE* from relational data through predefined aggregation operators, but cannot handle complex data structure and features. Until now, most research on automatic *FE* proposes very simple features, but, to the best of our knowledge, nobody uses more complex features, such as social network-based or representation-based features. Some works like *DSM* and *OneBM* extract features from timestamp, but do not get complex time-based features, which can actually be very significant in many time-related tasks.

To summarize, there are three major issues of feature engineering to be considered:

- *FE* is rather domain-specific, and very time-consuming. It is thus critical to automate this process as much as possible (a *holy grail of machine learning* according to [1]), with a *generic, domain-independent* method.
- Adding more features increases the dimension of the attributes space. Evaluating a large number of derived features may thus become a problem: the top-down approach might not be feasible at all.
- Incremental *evaluation of derived features* (bottom-up) is certainly dependent upon the order of inclusion of derived features. *AFEM* proposes a heuristic approach to this problem, but a full solution for this is not in the scope of this paper.

4   Method

We want to solve a prediction task P for an entity E and measure the performance through some indicator *Perf*. In any prediction task we want to solve, we assume that there is a unique index which labels each observation of entity E in the dataset.

4.1   Families of Features

A *family* of features is defined through a set of transformations of a given type (global or local transformations). Families should be generic (domain-independent) and computable for most datasets. We will use the following families:

- Ψ_{stat} is the family of *statistical features*. They are global, order-1 features such as: {max, min, sum, mean, var, std …} on numerical features or {count, count-distinct, most frequent …} on categorical features; or order 2 such as {ratio, mean-difference …}. This family is the one most commonly used in the literature.
- Ψ_{time} is the family of *time-based features*, for attributes with time-stamps. They include global features such as, for order-1 for example: {day-of-week, day-of-month, …, time-to-last, time-since-last, …}; and local features, obtained from local Ψ_{stat} defined on time-based neighborhoods, such as for example: {max, min, sum, mean, var, std …} or {count, count-distinct, most frequent …} in time windows {last-hour, last-week, …}. These features are very common in the literature.
- Ψ_{SN} is the family of *social graph-based features*, when observations of entity E are nodes in a social graph. They include global features such as for example: {degree, clustering coefficient, community index …}; and local features, obtained from local Ψ_{stat}, defined on graph-based neighborhoods, such as for example: {max, min, sum, mean, var, std …} in {first-circle, second-circle, community …}.
- Ψ_{Rep} is the family of *representation-based features*. These are mostly global features obtained through embedding of the original data, e.g. *SVD, PCA, AE* (deep learning auto-encoder) features.

Obviously, many more families could be defined: for example, when data on (longitude, latitude) are available, the Ψ_{Geo} family includes features in neighborhoods such as {street, district, city, region, country …}. (We will not use this family here).

4.2   Process for Deriving Features

Let us now describe our Automatic Feature Engineering Machine (*AFEM*) process to derive features. *AFEM* is an extension of *DSM* [2] and *OneBM* [6]. It can handle categorical and numeric features in different tables, but also history features with timestamp and social network features. It is a *block bottom-up approach* (like [2, 6]), but with blocks of features of a same family entered at successive time steps. We start from an initial features set F_0, collected from the raw data on entity E and progressively derive enriched feature sets, $F_1, F_2, \ldots, F_t, \ldots$ through transformations $T_0, T_1, \ldots,$ $T_t, \ldots$, by Eqs. (3–4). At each stage, transformations T_t are from one family of features only. To limit computation time, we will always generate global order-1 features first, then local order-1, global order-2, local order-2... We will choose the order in which we put the families in: usually, in this paper, we use the order Ψ_{Stat}, Ψ_{time}, Ψ_{SN}, Ψ_{Rep}. Of course any order is possible. Let us denote $\Psi^0, \Psi^1, \ldots, \Psi^t, \ldots$ the succession of family blocks successively applied. Once we have introduced, at time t, a block of features from one family (for example global order-1 Ψ_{Stat}), we build a model on $F_{t+1} = F_t \cup DF_t$, evaluate performance *Perf* and select the top-k_t most important features: we retain the k_t features with importance larger than some ratio α of total importance (for example, $\alpha = 1\%$). The process is described in Table 1. Obviously, through this process, features can be easily composed.

5   Experimental Results

5.1   Datasets

In this section, we evaluate our *AFEM* framework on three datasets, which we have chosen because they are large and complex enough to allow for significant evaluation, and comparison to challenges or the literature [2, 6]. The datasets are as follows and their characteristics and structures are shown in Table 2 and Fig. 1.

- KDD'14 (see footnote 1): participants are asked to predict whether a crowd-funded project will be "exciting" based on project descriptions and donation information.
- WSDM'18 (see footnote 2): participants are asked to predict whether a user will listen to a song again after the first observable listening event based on user, song and context.
- Last.fm (see footnote 3): in this task, we need to predict the rating that a user gives to an artist based on the user social network and user-artist tagging behaviors.

5.2   Evaluation Methodology

In our experiments, we use LightGBM, Random Forest or XGBoost[5] to build our models and select most important features. For all the experiments, we split the dataset in three parts: training, validation and test. We train the model in successive epochs on the training set and validate the results on the validation set, until there is no more

---

[5] http://lightgbm.readthedocs.io, http://scikit-learn.org, http://xgboost.readthedocs.io.

Table 1. *AFEM*, automatic feature engineering machine process for deriving features

---

For entity E, prediction task P, importance Imp, ratio α, performance evaluation $Perf$

Input: $\Psi^0, \Psi^1, ..., \Psi^T$, α, initial feature set F_0　　　**Output**: Feature set F_{T+1}

For t=0 to T do

- $\Psi^t = (T^t_{Func})$ or $\Psi^t = (T^t_{Func}, T^t_{Neigh})$
- $F_t \rightarrow F_{t+1} = F_t \cup DF_t$ with
  - $DF_t = \{\varphi = f(b_1, b_2, ..., b_n)/f \in T^t_{Func}\}$,　　　　　　　if Ψ^t is global
  - $DF_t = \{\varphi = f(b_1, b_2, ..., b_n; N)/f \in T^t_{Func}, N \in T^t_{Neigh}\}$　if Ψ^t is local
- Build model on F_{t+1}, evaluate its performance $Perf_t$, select Top-k_t features with importance Imp larger than α
- $F_{t+1} := \{$Top k_t most important features$\}$; t :=t+1

End

Table 2. Datasets used for evaluation

Dataset	# Rows	# Features	With timestamp	With graph
KDD'14	664,098	51	Yes	No
WSDM'18	9,934,208	19	Yes	No
Last.fm	92,834	7	Yes	Yes

improvement. We then retrain the model on the full data (train + validation) for that number of epochs and apply it on test set. To measure feature importance, we use Gini Importance/Mean Decrease in Impurity in Random Forest and split-based feature importance measure for XGBoost and LightGBM. In all experiments, we set feature importance ratio $\alpha = 1\%$ by default. The performance evaluators we use are those common for the algorithms we chose: AUC, MAE or RMSE. All experiments are run on a Dell R920 in-memory server with 96 cores, 2 TB RAM, in Python[6].

5.3　Results

In this section, we present the results obtained by our *AFEM* approach (Table 1) and compare them to those obtained on competition sites or to other *FE* techniques in the literature [2, 6] and to the global top-down approach. The *top-down* approach is very time-consuming and our *AFEM* approach, could be faster, but hopefully not degrade performances. Features generated are shown in Table 3.

KDD Cup 2014. We have used our *AFEM* approach, generating features in families Ψ_{stat} and Ψ_{time}. In Table 4, we compare the performances of *AFEM* (LightGBM model) and the *top-down* approach. As shown, performances regularly increase with the addition of new derived features and *AFEM* outperforms the *top-down* approach.

---

[6] https://github.com/TjuJianyu/AFEM

Table 3. Number of features generated by *AFEM* for each task

Feature family	KDD 2014	WSDM 2018	Last.fm
Original	33	5	2
Statistic Ψ_{stat}	14	19	12
Time-based Ψ_{time}	384	28	–
Graph-based Ψ_{SN}	–	–	7
Representation-based Ψ_{Rep}	–	60	–
Total	431	112	21

We also compare *AFEM* to *DSM* and *OneBM* [2, 6] in Table 5 (results from *DSM* and *OneBM* are from the original papers, number of features are not available). *AFEM* outperforms the other two automatic *FE* techniques: on *DSM* our *AFEM* + LighGBM improves AUC performance by 5,41% and 130 ranks; for *OneBM* + XGBoost, resp. Random Forest our *AFEM* + LighGBM, resp. XGBoost, Random Forest improves resp. AUC by 4.35% and 66 ranks, 4,32%, 66 ranks and 0.07%, 3 ranks. In web-based applications, computing time, at the time when we apply a model, needs to be very small ensuring short latency to avoid losing customers, with a compromise between improved performance and shorter computation time. Figure 2 shows the AUC performance and time cost of feature generation and training the model with different feature importance ratios α, as well as model apply time. As we increase α, we select less and less features for training, and thus less features when we apply it. In the situation where we choose $\alpha = 0.17$ instead of 0.01 (the default used previously), we save 2/3 computation time of feature generation and model training with only 0.2% AUC reduction; with the number of features reduced from 431 to 125, the time for applying the model is reduced from 1,163 s to 264 s.

WSDM Cup 2018. In this task, there is no explicit timestamp in the dataset, but data are ordered chronologically. We use the index of examples as timestamp in *AFEM*. In Table 6, we show the performance of *AFEM* with LightGBM classifier. In our iterative approach, we can achieve 12th (top 1.1%) out of 1081 human teams. We can even get top 13.1% using statistic and time-based families only. However, in this case, *AFEM* is very slightly worse than the *top-down* method. We can also use *AFEM* with a *human-in-the-loop* approach. To illustrate this, we choose a model from an ensemble used in [10], which ranked 6th in WSDM'2018. In this single model, the authors use Ψ_{stat} features, Ψ_{Rep} features, similarity features and high-order features (a total of 186 features). We use all these features, then apply our *AFEM* approach on this initial feature set, except some duplicated features. We choose the same classification model as in [10], a LightGBM model with the same hyper-parameters. In Table 7, we show the improvement of *AFEM* for this Top model [10] in WSDM'18. *AFEM* can improve AUC by 0.43% for the top single model, which is very hard at this rank.

Last.fm. In this rating prediction task, we compare RMSE and MAE performance of our *AFEM* with BPMFSRIC [7] and other two approaches BPMF and Hao Ma`s method from BPMFSRC work. BPMF uses Bayesian probabilistic matrix factorization to predict the rating; BPMFSRIC adds social relations and item contents on top of

Table 4. Performances on private leaderboard of *AFEM* iterative and *top-down* approaches on KDD Cup 2014. FS(x) indicates features selected from feature set x; $FT_{previous}$ features from previous step (previous row). Best results are shown **in bold**.

Approach	Feature used	#features	AUC	Rank	Top%
Iterative approach	Original features	33	0.58210	169	36%
	FS$(FT_{previous})$	28	0.58271	164	35%
	$FT_{previous} + \Psi_{stat}$	70 (+42)	0.58659	142	30%
	FS$(FT_{previous})$	55	0.59463	90	19%
	$FT_{previous} + \Psi_{time}$	**411 (+356)**	**0.64290**	**15**	**3%**
	FS$(FT_{previous})$	286	0.64045	15	3%
Top-down approach	Original features + $\Psi_{stat} + \Psi_{time}$	431	0.63919	17	4%
	FS$(FT_{previous})$	296	0.63845	17	4%

Table 5. Performances on private leaderboard of *AFEM*, *DSM* and *OneBM* on KDD Cup 2014

Methods	#features	AUC	Rank	Top %
DSM without tuning	N/A	0.55481	314	66.5%
DSM with tuning	N/A	0.58630	145	30.7%
OneBM + Random Forest	90	0.58983	118	25.0%
OneBM + XGBoost	90	0.59696	81	17.2%
AFEM + Random Forest	286	0.59052	115	24.4%
AFEM + XGBoost	286	0.64014	15	3.2%
AFEM + LightGBM	**286**	**0.64045**	**15**	3.2%

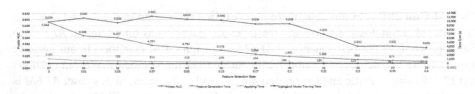

Fig. 2. AUC performance and time cost of feature generation, training model and total cost with different feature importance ratios α.

BPMF; Hao Ma's method uses social networks to regularize the process of matrix factorization. To allow for comparisons, we map listening counts into ratings of 1 to 5 in the same way as BPMFSR. We use 80% ratings data to train and validate the model and apply the model on the remaining 20% data. The whole process is repeated 10 times to calculate confidence intervals. All these settings are the same as in the literature. Furthermore, the MAE and RMSE performance of BPMFSRIC, BPMF and Hao Ma's method on Last.fm dataset are collected from BPMFSRIC [7] directly. Our *AFEM* method here derives 7 graph-based features. Table 8 shows the MAE and RMSE performance of *AFEM* and other three methods on Last.fm dataset. *AFEM* +

Table 6. Performances on private leaderboard of *AFEM* iterative approach and *top-down* approach on WSDM'18. Same notations as Table 5.

Approach	Feature used	#features	AUC	Rank	Top%
Iteratively approach	Original features	5	0.62233	882	81.6%
	$FT_{previous} + \Psi_{stat}$	24	0.67010	561	51.9%
	$FS(FT_{previous})$	22	0.67342	408	37.7%
	$FT_{previous} + \Psi_{time}$	50	0.68897	142	13.1%
	$FS(FT_{previous})$	40	0.68857	175	16.2%
	$FT_{previous} + \Psi_{Rep}$	100	0.71989	12	1.1%
	$FS(FT_{previous})$	93	0.71978	12	1.1%
Top-down approach	Original features + $\Psi_{stat} + \Psi_{time} + \Psi_{Rep}$	112	0.71917	12	1.1%
	$FS(FT_{previous})$	**94**	**0.71992**	**12**	1.1%

Table 7. Performances on private leaderboard of *AFEM* for top model on WSDM'18.

	#features	AUC	Rank
Top model	186	0.72928	7
Top model + *AFEM*	**186 + 16**	**0.73359**	**6**

Table 8. MAE and RMSE performance of Hao Ma's method, BPMF, BPMFSRIC and *AFEM* + LightGBM on Last.fm dataset.

Methods	Hao Ma's	BPMF	BPMFSRIC	AFEM + LightGBM
MAE	0.4492 ± 0.0013	0.3241 ± 0.0012	0.3244 ± 0.0014	$\mathbf{0.3131 \pm 0.0011}$
RMSE	0.6418 ± 0.0026	0.4467 ± 0.0024	0.4451 ± 0.0026	$\mathbf{0.4305 \pm 0.0023}$

LightGBM significantly outperforms all the other methods, on both performance indicators reducing regression errors RMSE by 1.5% and MAE by 1.1%.

6   Conclusion and Future Work

We propose a general feature engineering framework *AFEM* and define different complex feature families for performing automatic feature engineering on relational data. We present a block bottom-up evaluation and feature selection heuristic for selecting the most significant features. We show how to monitor the generation/ selection process to limit the number of derived features and allow computation time compatible with Web applications requirements. We evaluate the performance of *AFEM* on 3 datasets from different domains. *AFEM* outperforms most of the human solutions and literature. It gets from rank 882 with raw data to rank 12 with automatically generated features in one competition; it largely outperforms state-of-the-art

DSM and *OneBM* in another competition; it outperforms state-of-the-art on a classical dataset using social network features. But still many questions remain. In our future work, we intend to add more features and more complex feature families. The order of generating and evaluating different families is still a problem: using some algorithms such as reinforcement learning to learn the whole process is part of our future work.

References

1. Domingos, P.M.: A few useful things to know about machine learning. Commun. ACM **55** (10), 78–87 (2012)
2. Kanter, J.M., Veeramachaneni, K.: Deep feature synthesis: towards automating data science endeavors. In: 2nd International Conference on Data Science and Advanced Analytics DSAA, pp. 1–10. IEEE, Paris (2015)
3. Katz, G., Shin, E.C.R., Song, D.: ExploreKit: automatic feature generation and selection. In: 16th International Conference on Data Mining ICDM 2016, pp. 979–984, IEEE. Barcelona (2016)
4. Khurana, U., Turaga, D., Samulowitz, H., Parthasrathy, S.: Cognito: automated feature engineering for supervised learning. In: 16th International Conference on Data Mining Workshops ICDMW, pp. 1304–1307, IEEE. Barcelona (2016)
5. Khurana, U., Samulowitz, H., Turaga, D.: Feature engineering for predictive modeling using reinforcement learning. In: 32nd AAAI Conference on Artificial Intelligence AAAI-18, New Orleans, USA (2018)
6. Lam, H.T., Thiebaut, J.-M., Sinn, M., Chen, B., Mai, T., Alkan, O.: One button machine for automating feature engineering in relational databases. arXiv preprint arXiv:1706.00327 (2017)
7. Liu, J., Wu, C., Liu, W.: Bayesian probabilistic matrix factorization with social relations and item contents for recommendation. Decis. Support Syst. **55**(3), 838–850 (2013)
8. Nargesian, F., Samulowitz, H., Khurana, U., Khalil, E.B., Turaga, D.S.: Learning feature engineering for classification. In: 26th International Joint Conference on Artificial Intelligence IJCAI 2017, Melbourne, Australia, pp. 2529–2535 (2017)
9. Töscher, A., Jahrer, M.: The big chaos solution to the netflix grand prize. AT&T Labs - Research Tech report, 5 September 2009
10. Zhang, J., Fogelman-Soulié, F.: KKbox's music recommendation challenge solution with feature engineering. In: 11th ACM International Conference on Web Search and Data Mining WSDM 2018, Cup Workshop, Los Angeles, California, USA (2018)

Entity Linkage and Semantics

Entity Linking Facing Incomplete Knowledge Base

Shaohua Zhang[1,2], Jiong Lou[1,2], Xiaojie Zhou[1], and Weijia Jia[2,1(✉)]

[1] Department of Computer Science, Shanghai Jiao Tong University, Shanghai, China
{zhangsh950618,lj1994,szxjzhou}@sjtu.edu.cn
[2] Faculty of Science and Technology, University of Macau, Macau, China
jiawj@umac.mo

Abstract. Entity linking, bridging text and knowledge base, is a fundamental task in the field of information extraction. Most existing approaches highly depend on the structural features and statistics in the target knowledge base. Compared with raw text, they provide more discriminative information and make the task easier. However, in many closed domains, structural features and statistics are rarely available and the target knowledge base may be as simple and sparse as a series of separate entity records only with description. Therefore, few algorithms could work well on the incomplete knowledge base. In this paper, we propose a novel neural approach which only requires minimal text information from the knowledge base. To extract features from text effectively, we employ the co-attention mechanism to emphasize discriminative words and weaken noise. Compared with existing "black box" neural approaches, co-attention mechanism also brings better interpretability to our model. We conduct experiments on the AIDA-CoNLL benchmark and evaluate the performance with accuracy. Results show that our model achieves 82.3% in accuracy and outperforms the baseline by 1.1%.

Keywords: Entity linking · Co-attention mechanism
Neural network

1 Introduction

Given the named entity mentions in the documents, entity linking focuses on mapping each mention to the corresponding entity in the target knowledge base like Wikipedia[1], Freebase [2], and YAGO [4]. Entity linking is a fundamental task in the field of information extraction, which benefits many upstream tasks such as coreference resolution [5] and distantly-supervised relation classification [17]. Besides, it is also a significant step in knowledge base population [2,4], information retrieval [3], and question-answering [8].

The critical challenge of entity linking comes from the ambiguity of natural language, which means a named entity may have multiple aliases and different named entities could share the same name. Previous studies typically

---

[1] https://www.wikipedia.org/.

© Springer Nature Switzerland AG 2018
H. Hacid et al. (Eds.): WISE 2018, LNCS 11234, pp. 325–334, 2018.
https://doi.org/10.1007/978-3-030-02925-8_23

regard entity linking as a ranking problem, which is based on measuring the similarity between the mention and the entity. For the entity representation, a variety of structural features are explored in previous work, including entity co-occurrence [24], Wikipedia infobox, entity type [9] and Wikipedia category [22]. Besides, statistics are introduced by most of the entity linking models. Compared with raw text, such as entity description, these structural features and statistics provide more information and makes the model easier to discriminate between ambiguous entities. Unfortunately, in many closed domains, structural features and statistics are rarely available. The target knowledge base may be as simple and sparse as a series of separate entity records only with description [13], we call these incomplete knowledge bases. Therefore, these methods may not perform well when facing incomplete knowledge base. Because of the huge number of entities in the knowledge base, adding structural features and statistics for each entity manually is labor sensitive and machine learning complementing methods would inevitably introduce many errors. So we only use the entity description, which is the most common information in the knowledge base.

Therefore, we propose a more general entity linking approach for the incomplete knowledge base. Compared with previous work, we only use the context around the mention and entity description. We build the model primarily based on two assumptions: (1) most of the words in the mention context and entity description are weakly related to its disambiguation [12], and (2) dependencies between the mention context and entity description can help to extract features from text more effectively. Therefore, we employ the co-attention mechanism in our model to emphasize the discriminative words. Besides, compared with "black box" neural approaches, co-attention mechanism brings better interpretability.

Our contributions can be listed as follows:

- We propose a novel neural model for entity linking, which only requires mention context and entity description. Especially, the model can be applied to the incomplete knowledge base, which is simple and sparse.
- We employ the co-attention mechanism for efficient feature extraction from the noisy text. Also, our model has better interpretability than existing neural approaches, which benefits from the co-attention mechanism.
- We validate the model on the AIDA-CoNLL benchmark [10] and evaluate the performance with accuracy. Experiments show that our model achieves 82.3% in accuracy and outperforms the baseline by 1.1%.

The rest of the paper is organized as follows. Detailed model is presented in Sect. 2 and experiments are shown in Sect. 3. The conclusion is given in Sect. 4.

2   Methodology

2.1   Overview

As shown in Fig. 1, the overall architecture of the model contains three major components: encoder, co-attention, and decoder. Our model only takes mention

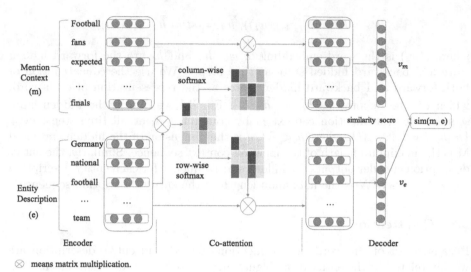

Fig. 1. Overview of our model.

context and entity description as the input, both of which are viewed as word sequences, and outputs the similarity score between each mention and candidate entity. Firstly, we put the mention context and entity description into the encoder and generate the hidden state of each word. Secondly, we generate the new hidden state sequences both for the mention context and the entity description based on the guide of the co-attention mechanism. Finally, the decoder takes the new hidden state sequences as input and generates the representations for the mention and the entity. And we use fully connected neural networks to estimate similarity score between the mention and the entity.

2.2 Encoder

In this section, we describe the encoder, which works for generating the hidden state of each word in the sequences (mention context and entity description). The encoders for the mention context and entity description share the same network structure, so we take mention context encoder as an example, which can be applied to the entity description encoder equally.

The encoder can be divided into two layers: embedding layer and Bidirectional Long Short-Term Memory (Bi-LSTM) layer. Firstly, we use the embedding layer transforms the distinct words in the mention context to the dense and low-dimensional word vectors, which are distributed in a continuous semantic space. Then we use Bi-LSTM to encode the mention context and generate the hidden states since Bi-LSTM has shown great performance in text understanding. Bi-LSTM extends the standard LSTM with an extra direction [18]. With regard to a specific time step t, Bi-LSTM considers the past states and the future states simultaneously. Given mention context sequence $w = [w_1, w_2, ...w_M]$, the process can be formulated as:

$$\overrightarrow{h}_t = lstm(\overrightarrow{h}_{t-1}, e(w_t)), \overleftarrow{h}_t = lstm(\overleftarrow{h}_{t-1}, e(w_t))$$

where $e(w_t)$ is the word embedding of w_t, $\overrightarrow{h}_t$ and $\overleftarrow{h}_t$ are the forward hidden state and backward hidden state at time step t. We use the concatenation of both forward and backward hidden states as the representation of each word, which can be denoted as $h_t = [\overrightarrow{h}_t; \overleftarrow{h}_t]$. Finally, we obtain the hidden state sequence for the mention context m by combining h_i at all time steps: $m = \{h_i | h_i, i = 1, 2, ...M\}$, where $h_i \in \mathbb{R}^d$, d is the dimension of the hidden state and M is the maximum length of the mention context sequence. Similarly, the entity description encoder outputs the hidden state sequence for each entity description $e \in \mathbb{R}^{d*N}$, where N is the maximum length of the entity description sequence.

2.3   Co-attention

Because most of the words in the mention context and entity description are weakly related to the entity disambiguation, we use the co-attention mechanism to emphasize the discriminative words and weaken the noise. In this section, we first give a brief introduction to the attention mechanism [1] and then we propose our co-attention based approach.

Generally, attention mechanism [23] use alignment function f to calculate the alignment scores between query q and each word s_i in the source sentence $s = [s_1, s_2...s_n]$. Then the normalization function, usually $softmax$, is applied to the alignment score a and transforms them into attention weights α as:

$$\alpha = softmax(a) = \left[\frac{e^{f(q,s_1)}}{\sum_{i=1}^n e^{f(q,s_i)}}, \cdots, \frac{e^{f(q,s_n)}}{\sum_{i=1}^n e^{f(q,s_i)}} \right]$$

Attention weights indicate the probability distribution of importance about the query q. Larger $f(q, s_i)$ means s_i has a higher probability of being related to the query q. In order to capture the dependence between the mention context and entity description, we choose co-attention [15], which attends to mention and entity simultaneously.

We first introduce entity-to-mention attention, which means we would like to emphasize the discriminative words in the mention context with the guide of the entity description. Specifically, we use each word in the entity description as the query q and the mention context is regarded as the source sentence s. Based on the notation in Sect. 2.2, we have the hidden state sequences of the mention context m and entity description e. Firstly, we calculate the alignment scores between each pair of context-words and description-words. We choose the inner product function to obtain the alignment score. Then we apply $softmax$ function on the each column of the alignment matrix as:

$$A_{col} = softmax_{col}(A) = softmax_{col}(m^T e)$$

where $m \in \mathbb{R}^{d*M}$, $e \in \mathbb{R}^{d*N}$, $A \in \mathbb{R}^{M*N}$ and $softmax_{col}$ is the column-wise softmax. Each column of A_{col} indicates the attention distribution of the mention context with regard to a specific word in the entity description.

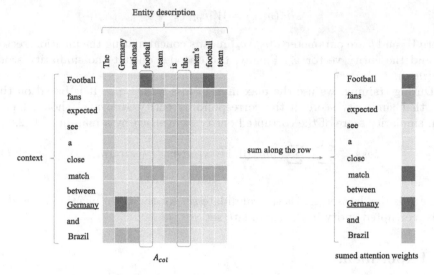

Fig. 2. The A_{col} is obtained by applying column-wise *softmax* on the alignment matrix A, and we sum along the row of A_{col} to obtain the sumed attention weight. Deeper bule in the figure means higher attention weight. (color figure online)

For example, with respect to the word "football" in the entity description, shown in the fourth column in the Fig. 2, it pays more attention to the word "Football", "fans", and "match" in the mention context. Based on the guide of the attention weights, we generate the new hidden state with weighted sum of the mention context hidden states. And the new hidden state sequence for the mention context can be obtained with $m' = mA_{col}$.

Besides, we also use the mention context to help emphasize informative words in the entity description by constructing mention-to-entity attention parallelly. Such process is similar to the entity-to-mention attention. We first obtain the alignment matrix A with the original hidden state sequences m, e. Then we employ the row-wise *softmax* on the alignment matrix. Further, we calculate the new hidden state sequence for the entity e' with the guide of the mention-to-entity attention weights.

2.4 Decoder

In this section, we give the description of the decoder in detail. We first generate the mention representation v_m and entity representation v_e. Then we measure the similarity between the v_m and v_e with $sim(m, e)$.

Based on the output from the co-attention network, we employ another Bi-LSTM to decode the new hidden state sequences for mention m' and entity e'. And we obtain the vector representation for mention v_m and entity v_e by averaging the Bi-LSTM output h_m and h_e at all time steps. Then we measure the similarity between the v_m and v_e with the fully connected layer as:

$$sim(m, e) = W[v_m; v_e] + b$$

where W and b are parameters, $[v_m; v_e]$ means concatenating the mention vector v_m and the entity vector v_e. Finally, the decoder output the similarity score $sim(m, e)$.

During training, we use the max-margin loss with Eq. 1. It is based on the idea that similarity score of the corresponding entity $sim(m, e)$ should larger than similarity score of the corrupted entity $sim(m, e')$ by a margin of 1 [22].

$$loss = \sum_{m \in T} \sum_{e' \in E_m} max(0, 1 - sim(m, e) + sim(m, e')) \tag{1}$$

where T is the dataset, E_m is the candidate set for the given mention m, and e' is the corrupted entity in the candidate set.

3   Experiment

3.1   Experiment Setting

We validate the model on the AIDA-CoNLL benchmark, which is one of the biggest manually annotated datasets [10]. AIDA-CoNLL officially provides training set, validation set and test set. Besides, all the mentions have been annotated in the documents and the corresponding entities for each mention in the Wikipedia are given in advance.

Since all the mentions have been annotated in the documents, we start from the candidate generation. For the effectiveness of the experiment, we choose the Cross-Wiki [20], which has been widely used for the candidate generation [14]. Cross-Wiki is a dictionary, which is built from the Wikipedia and a large Web corpus. Given a mention, it yields a set of candidate entities with the prior. Following the previous work [7], we add the YAGO dictionary [10], where each candidate receives a uniform prior. The number of the candidate entities may be over hundreds, so we choose the top ten entities in the candidate set according to the prior. If the corresponding entity is not in the candidate set, the model could never choose the correct entity. Therefore, we measure the performance of candidate generation with gold recall, the percentage of mentions for which the entity candidate set contains the corresponding entity. Details are give in Table 1. During the training, we skipped the samples if their corresponding entities are not in the candidate set.

As stated in Sect. 2, we only take the mention context and entity description as input. Specifically, for the mention context, we choose the windows size as 100 (50 words before mention and 50 words after the mention). For the entity description, we use the first 100 words of the corresponding Wikipedia article. If the length of the sequence is less than 100, we pad the input sequence with zero as usual practice. We remove the stop words in the sequences and pre-train the 100 dimension case-insensitive word vectors by standard tools [16] with default settings from Wikipedia articles. The word vectors are not updated during back

Table 1. Statistics of AIDA-CoNLL dataset.

Dataset	Number of mentions	Number of documents	Gold recall
AIDA-train	18448	946	96.1
AIDA-dev	4791	216	94.6
AIDA-test	4485	231	96.1

propagation. We minimize the loss function by Adam [11] with batch size of 64 and the initial learning rate is 10^{-4}. The hidden size of encoder and decoder is 100. Besides, to prevent the model from overfitting, we add the dropout [21] layer after the embedding layer with keep probability 0.75.

3.2 Performance

During testing, we choose the top-ranked entity in the candidate set as the predicted entity and evaluate the performance of all the model with micro-average accuracy. The accuracy is calculated as the number of correctly linked entity mentions divided by the total number of all entity mentions. Therefore, here $precision = recall = F_1 = accuracy$ and researchers usually use accuracy to assess the performance [19]. For fair comparison, we use the same candidate set for all the models. To demonstrate the contribution of different components, we start from the simple baseline and then add components step by step. All the models only take mention context and entity description as input and details are listed as follows:

- CNN: We use CNN to encode the context and entity description, generating the representation vectors for mention and entity. Follow the Berkely-CNN [6], we use 150 convolution kernels and set the kernel size as 5. Then we put the result through a rectified linear unit (ReLU) and combine the results with sum pooling. Finally, we measure the similarity score between the mention and entity with cosine function.
- encoder + decoder: We remove the co-attention mechanism in our model. We use the encoder to learn the hidden state sequences for mention context and entity description. Then we directly pass them to the decoder without the process of the co-attention layer. And the decoder is the same as described in Sect. 2.4.
- encoder + mention-to-entity attention + decoder: We only keep the mention-to-entity attention in our model and remove the entity-to-mention attention. We use the encoder to learn the hidden state sequences for mention context and entity description. Then we generate the new hidden state sequence for the entity based on the guide of mention-to-entity attention. And the hidden state sequence for the mention is not changed. And the decoder is the same as described in Sect. 2.4.
- encoder + entity-to-mention attention + decoder: We only keep the entity-to-mention attention in our model and remove the mention-to-entity attention.

We use the encoder to learn the hidden state sequences for mention context and entity description. Then we generate the new hidden state sequence for the mention based on the guide of entity-to-mention attention. The hidden state sequence for the entity is not changed. The decoder is the same as described in Sect. 2.4.
– our model: Our model has described in Sect. 2 in detail.

We report the accuracy in Table 2. Firstly, we compare our model with Berkely-CNN [6], which best fits our scenario. Berkely-CNN uses CNN to capture semantic correspondence between a mention and an entity. They employ CNNs at multiple granularities to exploit various kinds of topic information. Then they consider the similarity scores at different topic levels and output the final similarity score. They also combine the CNN features with sparse features, but we ignore the result for the reason that statistics are absent in the incomplete knowledge base. The accuracy of our model is higher than Berkely-CNN by 1.1%. However, Berkely-CNN uses different candidate generation methods and different input. So we conduct the experiment with CNN based on the same input and same candidate set. The result shows that the accuracy is 74.1%. We find the accuracy of simple encoder-decoder approach without co-attention is 1.9% higher than the CNN (from 74.1% to 76.0%). We speculate that LSTM has better performance of modeling the long sequence than CNN.

Table 2. Accuracy on AIDA dataset.

Model	Accuracy on test set
Berkely-CNN	81.2
CNN	74.1
encoder + decoder	76.0
encoder + mention-to-entity attention + decoder	78.1
encoder + entity-to-mention attention + decoder	79.9
encoder + co-attention + decoder (our model)	**82.3**

Further, we explore the effectiveness of the co-attention mechanism in our model. As described in Sect. 2, we employ the mention-to-entity attention and entity-to-mention attention parallelly. We remove both attentions in our model, and the accuracy dropped by 6.3% (from 82.3% to 76%). The comparison shows the great effectiveness of the co-attention. And then we try to remove the one of the attention. Compared with the no attention approach, the result shows that only mention-to-entity attention improve the accuracy by 2.1% (from 76.0% to 78.1%) and only entity-to-mention attention improve the accuracy by 3.9% (from 76.0% to 79.9%). The results show that entity-to-mention attention contributes more to the improvement of the accuracy. We believe that the mention context is noisier than the entity description. This is consistent with the fact that the topics described in the entity are more focused and the context is more dispersed.

4   Conclusion and Future Work

We have presented a new approach for entity linking. Compared with most of the existing approaches, ours requires the minimal features, and it works smoothly under the incomplete knowledge base. Our approach only takes the mention context and entity description as the input, and it introduces the co-attention mechanism to emphasize the informative words from the noisy text. We have conducted the experiments on the benchmark AIDA-CoNLL. Empirical results show that our approach outperforms the previous study and it has better interpretability. In our Future work, we will combine this approach with collective methods so that the model can consider the global topic and probably give more robust results.

Acknowledgments. This work is supported by FDCT 0007/2018/A1, DCT-MoST Joint-project No. (025/2015/AMJ) of SAR Macau; University of Macau Funds Nos: CPG2018-00032-FST & SRG2018-00111-FST; Chinese National Research Fund (NSFC) Key Project No. 61532013; National China 973 Project No. 2015CB352401; Shanghai Scientific Innovation Act of STCSM No.15JC1402400 and 985 Project of Shanghai Jiao Tong University: WF220103001.

References

1. Bahdanau, D., Cho, K., Bengio, Y.: Neural machine translation by jointly learning to align and translate. arXiv preprint arXiv:1409.0473 (2014)
2. Bollacker, K., Evans, C., Paritosh, P., Sturge, T., Taylor, J.: Freebase: a collaboratively created graph database for structuring human knowledge. In: Proceedings of the 2008 ACM SIGMOD International Conference on Management of Data, pp. 1247–1250. ACM (2008)
3. Cheng, T., Yan, X., Chang, K.C.C.: Entityrank: searching entities directly and holistically. In: Proceedings of the 33rd International Conference on Very Large Data Bases, pp. 387–398. VLDB Endowment (2007)
4. Fabian, M., Gjergji, K., Gerhard, W., et al.: YAGO: a core of semantic knowledge unifying Wordnet and Wikipedia. In: 16th International World Wide Web Conference, WWW, pp. 697–706 (2007)
5. Finin, T., Syed, Z., Mayfield, J., McNamee, P., Piatko, C.D.: Using wikitology for cross-document entity coreference resolution. In: AAAI Spring Symposium: Learning by Reading and Learning to Read, pp. 29–35 (2009)
6. Francis-Landau, M., Durrett, G., Klein, D.: Capturing semantic similarity for entity linking with convolutional neural networks. arXiv preprint arXiv:1604.00734 (2016)
7. Ganea, O.E., Hofmann, T.: Deep joint entity disambiguation with local neural attention. arXiv preprint arXiv:1704.04920 (2017)
8. Gattani, A., et al.: Entity extraction, linking, classification, and tagging for social media: a wikipedia-based approach. Proc. VLDB Endowment **6**(11), 1126–1137 (2013)
9. Gupta, N., Singh, S., Roth, D.: Entity linking via joint encoding of types, descriptions, and context. In: Proceedings of the 2017 Conference on Empirical Methods in Natural Language Processing, pp. 2681–2690 (2017)

10. Hoffart, J., et al.: Robust disambiguation of named entities in text. In: Proceedings of the Conference on Empirical Methods in Natural Language Processing, pp. 782–792. Association for Computational Linguistics (2011)
11. Kingma, D.P., Ba, J.: Adam: a method for stochastic optimization. arXiv preprint arXiv:1412.6980 (2014)
12. Lazic, N., Subramanya, A., Ringgaard, M., Pereira, F.: Plato: a selective context model for entity resolution. Trans. Assoc. Comput. Linguist. **3**, 503–515 (2015)
13. Li, Y., Tan, S., Sun, H., Han, J., Roth, D., Yan, X.: Entity disambiguation with linkless knowledge bases. In: Proceedings of the 25th International Conference on World Wide Web, pp. 1261–1270. International World Wide Web Conferences Steering Committee (2016)
14. Ling, X., Singh, S., Weld, D.S.: Design challenges for entity linking. Trans. Assoc. Comput. Linguist. **3**, 315–328 (2015)
15. Lu, J., Yang, J., Batra, D., Parikh, D.: Hierarchical question-image co-attention for visual question answering. In: Advances in Neural Information Processing Systems, pp. 289–297 (2016)
16. Mikolov, T., Sutskever, I., Chen, K., Corrado, G.S., Dean, J.: Distributed representations of words and phrases and their compositionality. In: Advances in neural information processing systems, pp. 3111–3119 (2013)
17. Mintz, M., Bills, S., Snow, R., Jurafsky, D.: Distant supervision for relation extraction without labeled data. In: Proceedings of the Joint Conference of the 47th Annual Meeting of the ACL and the 4th International Joint Conference on Natural Language Processing of the AFNLP: Volume 2-Volume 2, pp. 1003–1011. Association for Computational Linguistics (2009)
18. Schuster, M., Paliwal, K.K.: Bidirectional recurrent neural networks. IEEE Trans. Signal Proces. **45**(11), 2673–2681 (1997)
19. Shen, W., Wang, J., Han, J.: Entity linking with a knowledge base: issues, techniques, and solutions. IEEE Trans. Knowl. Data Eng. **27**(2), 443–460 (2015)
20. Spitkovsky, V.I., Chang, A.X.: A cross-lingual dictionary for English Wikipedia concepts. In: LREC, pp. 3168–3175 (2012)
21. Srivastava, N., Hinton, G., Krizhevsky, A., Sutskever, I., Salakhutdinov, R.: Dropout: a simple way to prevent neural networks from overfitting. J. Mach. Learn. Res. **15**(1), 1929–1958 (2014)
22. Sun, Y., Lin, L., Tang, D., Yang, N., Ji, Z., Wang, X.: Modeling mention, context and entity with neural networks for entity disambiguation. In: IJCAI, pp. 1333–1339 (2015)
23. Vaswani, A., et al.: Attention is all you need. In: Advances in Neural Information Processing Systems, pp. 6000–6010 (2017)
24. Yamada, I., Shindo, H., Takeda, H., Takefuji, Y.: Joint learning of the embedding of words and entities for named entity disambiguation. arXiv preprint arXiv:1601.01343 (2016)

User Identity Linkage with Accumulated Information from Neighbouring Anchor Links

Xiang Li[1,2,3(✉)], Yijun Su[1,2,3], Wei Tang[1,2,3], Neng Gao[2,3], and Ji Xiang[2,3]

[1] School of Cyber Security, University of Chinese Academy of Sciences,
Beijing, China
[2] State Key Laboratory of Information Security, Chinese Academy of Sciences,
Beijing, China
[3] Institute of Information Engineering, Chinese Academy of Sciences, Beijing, China
{lixiang9015,suyijun,tangwei,gaoneng,xiangji}@iie.ac.cn

Abstract. User identity linkage is to identify all the users belonging to the same individual in different networks and has been widely studied along with the increasing popularity of diverse social media sites. Generally, a pair of probable corresponding users on different networks may form a true "Anchor Link". Most existing methods identify a user based on unique features (username, interests, friends, etc.) and neglect the importance of users local network structure. Therefore, one challenging problem is how to address the user identity linkage problem if only structural information is available. In this paper, we explore techniques for dealing with the fundamental and accumulated information from neighbouring anchor links. Furthermore, we design a Trustworthy Predicting Approach (TPA) for computing the *authority* of an anchor link, inferring the *trustworthiness* of a candidate anchor link being true and predicting whether an anchor link is able to be veritably formed. Experiments illustrate the effectiveness of our proposed algorithm.

1 Introduction

With the vigorous development of Internet in the world, online social networks have revolutionized our daily life and brought us in a "second life". Social networks, such as Facebook, Twitter, Flickr and LinkedIn, make people easy to share their information with other familiar or unfamiliar people. According to the statistical data about Facebook, Twitter and Youtube from 2017 Pew Research Center report[1], more than half of the users tend to acquire information from multiple social media sites as shown in Table 1.

The problem of identifying users across online social networks (also known as *User Identity Linkage*) is valuable and particularly challenging. Mapping users from diverse social platforms can bring many benefits. Discrepant information

---

[1] http://www.pewresearch.org/fact-tank/2017/11/02/more-americans-are-turning-to-multiple-social-media-sites-for-news/.

© Springer Nature Switzerland AG 2018
H. Hacid et al. (Eds.): WISE 2018, LNCS 11234, pp. 335–344, 2018.
https://doi.org/10.1007/978-3-030-02925-8_24

Table 1. % of each site's news users who get news from...

	Facebook	Twitter	Youtube
Only that site	50%	18%	22%
2 sites	30%	37%	39%
3 or more	20%	45%	39%

of the same user on different social platforms helps to construct a better portrait for corresponding natural person and provide precise and personalized recommendations or advertisements [1,5].

Recent work in user identity linkage often leverages user profiles and user-generated content. However, there are several difficulties for subsequent exploitation. For example, due to the personal preferences and privacy demand, researchers have to face the dilemma that user profiles and user-generated content often behave truthless, incomplete and inconsistent. Therefore, a more challenging and interesting scenario emerges when only social circle is available.

A basic intuition to use social cycle is that when most of a person's friend say "account u on one social platform and account u' on another social platform belong to that person", it seems believable that these two accounts indeed belong to that person. Based on above intuition, a concept "shared identified friends" has been presented and widely used. In this paper, a detailed analysis on the basic concept Shared Identified Friends has been conducted. The main idea behind our method is to differentiate each identified friend according to his *authority*. As a result, we propose a key component called Authority-Trustworthiness Analysis Model to iteratively compute the *trustworthiness* and *authority* of each probable anchor link. By combining Authority-Trustworthiness Analysis Model with the process of anchor link inference, a Trustworthy Predicting Approach (TPA) is presented for solving the problem of user identity linkage purely based on structure information.

The remainder of this paper is organized as follows: Sect. 2 reviews some existing work on user identity linkage. Section 3 describes our analysis model and approach in detail. Experimental evaluation and comparison to other methods are shown in Sect. 4. Finally, Sect. 5 concludes the paper with a brief discussion.

2   Related Work

Existing approaches mainly use user's unique attributes (e.g., name, age, hometown, interests) and content (e.g., post, comment). Great efforts have been made on feature engineering [7,8,11,13,14]. For example, [11] considers distance-based profile features and neighborhood-based network features and iteratively identify unknown user identity pairs. More comprehensively, [8] models heterogeneous behaviors including distance-based profile features, style-based content features, trajectory-based content features and neighborhood-based network features in a semi-supervised manner. Other techniques such as embedding [7,13] also have been utilized to learn better features. Besides feature extraction, recent works make progress in designing better models like Energy-based model [16] and Latent User Space model [9].

Nevertheless, due to the aforementioned drawbacks (e.g., truthless, incomplete and inconsistent), a fundamental problem is how to solve user identity linkage problem by making full use of structure information. [10] proposes a propagation algorithm to find new links by computing the match score of all probable links based on degree and feedback from previously constructed anchor links. [12] firstly uses the distance vector to initial seed anchors. Then, the authors compare the local network structure by randomized spanning trees and recursive sub-graph matching. Similarly, [2] designs an Unified Similariy (US) measurement by combining the degree centrality, closeness centrality, betweenness centrality of nodes and the relative distance to the initial seed anchors. Besides, [15] presents a local degree-based method and a global embedding-based method for identifying users by only utilizing structure information. [4] gives a better search strategy for generating candidate links to be computed. In the ith phase of the alogrithm, it only allows nodes of degree roughly $D/2^i$ and above to be matched, where D is a parameter related to the largest node degree.

3 Proposed Method

A social platform can be viewed as an undirected network and each node in the network can respresent an account on the social platform. Let $\mathbf{S} = \{S_1, S_2, \ldots, S_{n_s}\}$ denotes a set of different networks and for each $S_i \in \mathbf{S}$, $S_i = (V_i, E_i)$ where $V_i = \{v_1^i, v_2^i, \ldots, v_{n_i}^i\}$ denotes the set of nodes on S_i and $E_i \subseteq (V_i \times V_i)$ denotes undirect links on S_i. Without loss of generality, we focus on two networks S_1, S_2 in this study. This is reasonable because solving the problem of two sites can be easily generalized to the problem of n_s networks in a pairwise manner.

All true anchor links between S_1 and S_2 is denoted as $T = \{(v_i^1, v_j^2)|v_i^1 \in S_1, v_j^2 \in S_2\}$ and $T^p \subseteq T$ represents prior true anchor links known in advance. For each node v_i^m, $N(v_i^m)$ represents the set of nodes linked to node v_i^m on network S_m and $F(v_i^m)$ denotes the set of matched friends among its neighbouring nodes.

3.1 Authority-Trustworthiness Analysis Model

Before presenting the ananlysis model, it is necessary to introduce the basic concept "Shared Identified Friend (SIF)". As shown in Fig. 1, we already know account pair (v_1^1, v_1^2) belong to user A, (v_2^1, v_2^2) belong to user B and so on. In this case, (v_1^1, v_1^2) (or (v_2^1, v_2^2)) is called a shared identified friend for (v_3^1, v_3^2). The set of shared identified friends for (v_3^1, v_3^2) can be represented as $SIF(v_3^1, v_3^2) = F(v_3^1) \cap F(v_3^2)$.

In this paper, only structure information can be taken into consideration. To better solve the problem, we differentiate the function of different shared identified friend. This kind of difference originates from the *authority* of distrinct

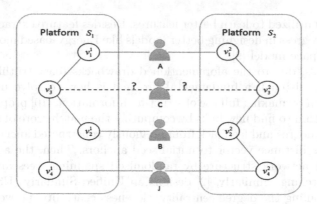

Fig. 1. An example for some notations. A solid line with an arrow on both sides denotes an true anchor link and a dash line is a probable anchor link.

people. By common sense, we have summarized two basic intuitions: (1) The authority of a person can be evaluted by the trustworthiness of judgements he has made. (2) A person that provides mostly true judgements for many objects will likely provide true judgements for another objects. Furthermore, two conclusions, which are interactively promoted, have been drawn from these two intuitions: (1) Conclusion 1: A judgement is more trustworthy if people returning this judegment are more authoritative; (2) Conclusion 2: A person is more authoritative if the judgements returned by this person are more probable to be correct. Based on above intuitions and conclusions, the definitons of two new concepts *Trustworthiness* and *Authority* naturally arises as follows:

Definition 1 *(Trustworthiness of a judgement): The trustworthiness of a judgement is the probability of this judgement being correct, according to the best of our knowledge.*

Definition 2 *(Authority of an anchor link): The authority of an anchor link (v_i^1, v_j^2) is the expected trustworthiness of the judgements provided by (v_i^1, v_j^2).*

To formulate the Authority-Trustworthiness model expediently, some notations and notions should be provided in advance. We denote $trust(v_i^1, v_j^2) \in [0, 1]$ as the trustworthiness score of a judgement for probable anchor link (v_i^1, v_j^2) and $auth(v_i^1, v_j^2) \in [0, 1]$ as the authority score of an anchor link (v_i^1, v_j^2). For a fixed order of all probable account pairs between two networks, $trust \in R^{n_1 n_2 \times 1}$ and $auth \in R^{n_1 n_2 \times 1}$ separately represent the trustworthiness and authority of all probable account pairs. In addition, a "transition" matrix $M \in R^{n_1 n_2 \times n_1 n_2}$ is defined with the same order of account pairs in $trust$ and $auth$. When $(v_p^1, v_q^2) \in SIF(v_i^1, v_j^2)$, the value of $M[v_i^1, v_j^2][v_p^1, v_q^2]$ is set to 1. Otherwise, the value is equal to 0. Naturally, a diagonal matrix D can be defined. Each of element in D is the sum of corresponding row in M.

According to above two conclusions and definitions, our iterative computation model for authority and trustworthiness can formulated as $trust =$

$D^{-1}M \cdot auth$ and $auth = (D^{-1}M)^T \cdot trust$. By viewing this iterative procedure as a HITS algorithm [3], the authroity-trustworthiness analysis model necessarily converges as the number of iteration increases arbitrarily due to the convergence proof of HITS. For each pair (v_i^1, v_j^2), its authority and trustworthiness can be computed as:

$$trust(v_i^1, v_j^2) = \frac{\sum_{(v_p^1, v_q^2) \in SIF(v_i^1, v_j^2)} auth(v_p^1, v_q^2)}{|SIF(v_i^1, v_j^2)|}$$

$$auth(v_i^1, v_j^2) = \frac{\sum_{(v_p^1, v_q^2) \in SIF(v_i^1, v_j^2)} trust(v_p^1, v_q^2)}{|SIF(v_i^1, v_j^2)|} \qquad (1)$$

From the view of weighted majority voting, above model only allows shared identified friends to vote. However, everyone identified has the right to vote. For example, for a candidate pair (v_3^1, v_3^2), we know $SIF(v_3^1, v_3^2) = \{A, B\}$ in Fig. 1. If we compute the trustworthiness of this pair as above, we ignore the matched person J, which is unreasonable. Noting that if a matched person such as J is not a shared identified friend of a certain candidate pair, it means this person disagree the corresponding account pair belongs to a real person. In this paper, we think this kind of matched person has no contribution to the trustworthiness and authority of corresponding account pairs, which means the value is zero. Therefore, our analysis model can be modified as:

$$trust(v_i^1, v_j^2) = \frac{\sum_{(v_p^1, v_q^2) \in SIF(v_i^1, v_j^2)} auth(v_p^1, v_q^2)}{|F(v_i^1) \cup F(v_j^2)|} \qquad (2)$$

$$auth(v_i^1, v_j^2) = \frac{\sum_{(v_p^1, v_q^2) \in SIF(v_i^1, v_j^2)} trust(v_p^1, v_q^2)}{|F(v_i^1) \cup F(v_j^2)|} \qquad (3)$$

3.2 Trustworthy Predicting Approach

In this paper, the analysis model described above is not able to predict. Therefore, a semi-supervised algorithm called Trustworthy Predicting Approach is designed for integrating the authority-trustworthiness anaylsis model and anchor link inference process. As shown in Algorithm 1, in each iteration, we generate the candidate set for each people in identified set. For example, assuming (v_i^1, v_j^2) has been identified, the candidate set of this person is $C(v_i^1, v_j^2) = \{(v_l^1, v_h^2)|(v_l^1, v_h^2) \in (N(v_i^1) \times N(v_j^2) - SIF(v_i^1, v_j^2)), |N(v_i^1) - N(v_j^2)| < window\})$. Then, the trustworthiness of probable anchor links in the candidate set can be computed. After computing the trustworthiness of candidate set, the authority-trustworthiness analysis model is applied to acquire the stable authority and trustworthiness for each link. During the process of analysis process, we repeat the iteration by only considering anchor links whose trustworthiness score is above a low bound.

4 Experiment Study

In this section, we compare the proposed approach with existing baseline methods. The main comparsion methods used in experiments include:

Algorithm 1. Trustworthy Predicting Approach

Input: $S_1 = (V_1, E_1), S_2 = (V_2, E_2)$, prior anchor links $T^p \neq \emptyset$, the threshold *right_low_bound*, degree window size *window*, number of iterations *iter_num*
Output: all identified anchor links T^*

1: $T^* = T^p$
2: **for** k=1,2,...,iter_num **do**
3:     $temporal\_cand = \emptyset$
4:     **for** $(v_i^1, v_j^2) \in T^*$ **do**
5:         $temporal\_cand = temporal\_cand \cup C(v_i^1, v_j^2)$
6:     $T^* = T^* \cup temporal\_cand$
7:     **while** not reach stable state **do**
8:         **for** $(v_i^1, v_j^2) \in (T^* - T^p)$ **do**
9:             compute trustworthiness score by (2)
10:        **if** $trust(v_i^1, v_j^2) < right\_low\_bound$ **then**
11:           remove (v_i^1, v_j^2) from T^*
12:        **for** $(v_i^1, v_j^2) \in T^*$ **do**
13:           compute authority score by (3)

- Local Method (LM) [15]: Nodes who has maximum number of identified friends in its network in each iteration are considered as an anchor.
- Global Method (GM) [15]: By constructing the normalized laplacian matrix for each network, the algorithm gives spectral embedding for each node and learns a linear transformation between initial seed nodes in two networks.
- UserMatch [4]: The algorithm uses the concept shared identified friends to calculate the similarity of a candidate node pair and designs a better propagation strategy for reducing the size of candidate set.
- Trustworthy Predicting Approach (TPA): Our proposed method utilize the structure information based on the Authority-Trustworthiness Analysis Model.

Experiment Setups. To evaluate the performance of comparison methods, *Recall* and *F-measure*($F1 = \frac{2*Precision*Recall}{Recall+Precision}$) are considered in this paper. Considering the labeled data required by semi-supervised model, the ratio of prior anchor links shown as (4) needs to be investigated during experiments.

$$prior\_ratio = \frac{|T^p|}{|T^a|} \tag{4}$$

4.1 Experiments on Social Networks

Datasets. Facebook is one of most popular social platforms currently. We use facebook data in Standford Large Network Dataset Collection [6] to assess the performance of different measures. Specific information about facebook dataset is shown in Table 2. Then, we generate two networks from facebook dataset by edge sampling. For each edge, a random value p with the uniform distribution

Table 2. Information about networks

Name	Nodes	Edges	Average degree
Facebook	4029	88234	43.691
Data Mining	20680	71130	6.879
Artificial Intellegence	25674	76141	5.931

Table 3. Experimental results under different prior ratio, $\alpha_o = 0.5, \alpha_s = 0.5$

Metric	Method	0.05	0.1	0.2	0.3	0.4	0.5
Recall	LM	0.0	0.00087	0.00327	0.00225	0.00351	0.00314
	GM	0.00275	0.11283	0.15389	0.17848	0.19315	0.20923
	UserMatch	0.03714	0.07114	0.13425	0.21785	0.27963	0.34399
	TPA	**0.10839**	**0.18047**	**0.28978**	**0.36145**	**0.46752**	**0.53225**
F1	LM	0.0	0.00087	0.00328	0.00225	0.00351	0.00315
	GM	0.00275	0.11283	0.15389	0.17848	0.19315	0.20923
	UserMatch	0.04963	0.08961	0.16273	0.25868	0.32558	0.40110
	TPA	**0.10928**	**0.18219**	**0.29145**	**0.36316**	**0.46875**	**0.53435**

in $[0, 1]$ is generated. Then, if $p \leq 1 - 2\alpha_s + \alpha_o\alpha_s$, the edge will be removed; if $1 - 2\alpha_s + \alpha_o\alpha_s < p \leq 1 - \alpha_s$, the edge will only be kept in first sub-network; if $1 - \alpha_s < p \leq 1 - \alpha_s\alpha_o$, the edge will only be kept in second sub-network; if $p > 1 - \alpha_s\alpha_o$, this edge will be kept in both two sub-networks. Based on above strategy, we know the overlap level is $overlap\_level = \frac{2*(1-(1-\alpha_s\alpha_o))}{1-(1-2\alpha_s+\alpha_o\alpha_s)+1-(1-\alpha_o\alpha_s)} = \alpha_o$. As a result, we call α_o "edge overlap level" and α_s "sparsity level".

Results and Comparison. To evaluate the performace of our TPA method in its entirely, we firstly compare it with baseline methods in different settings. Specifically, different prior ratios in $[0.05, 0.1, 0.2, 0.3, 0.4, 0.5]$ with same overlap level $\alpha_o = 0.5$ and same sparsity level $\alpha_s = 0.5$ are tested. From Table 3, when prior ratio is small enough such as 0.05, TPA behaves better than other methods when prior ratio increases from 0.05 to 0.5. Moreover, LM is always the poorest method under all prior ratios and UserMatch is always the best method between LM, GM and UserMatch. The deviation of Recall and F1 between TPA and UserMatch ranges from 6% to 20%.

In addition, different overlap levels $\alpha_o = [0.4, 0.5, 0.6, 0.7, 0.8, 0.9]$ with same prior ratio $prior\_ratio = 0.3$ and same sparsity level $\alpha_s = 0.5$ are tested. From Table 4, similar to Table 3, LM is always the poorest method under all overlap levels and UserMatch is always the best method between LM, GM and User-Match. However, our TPA approach still behaves better and exceeds about 10% in average than other methods.

Similarly, different sparsity levels $\alpha_s = [0.3, 0.4, 0.5, 0.6, 0.7, 0.8]$ with same prior ratio $prior\_ratio = 0.3$ and same overlap level $\alpha_o = 0.5$ are also tested.

Table 4. Experimental results under different overlap level, $\alpha_s = 0.5, prior\_ratio = 0.3$

Metric	Method	0.4	0.5	0.6	0.7	0.8	0.9
Recall	LM	0.00188	0.00224	0.00186	0.00297	0.00371	0.00258
	GM	0.1509	0.17848	0.23312	0.31501	0.43091	0.62182
	UserMatch	0.09572	0.21785	0.39910	0.55423	0.65565	0.75636
	TPA	**0.19707**	**0.36145**	**0.58859**	**0.76486**	**0.85587**	**0.93107**
F1	LM	0.00188	0.00225	0.00187	0.00297	0.00372	0.00258
	GM	0.15090	0.17848	0.23312	0.31501	0.43091	0.62182
	UserMatch	0.11371	0.25868	0.46210	0.63747	0.73927	0.83111
	TPA	**0.19890**	**0.36316**	**0.59068**	**0.77001**	**0.85874**	**0.93417**

Table 5. Experimental results under different α_s, $\alpha_o = 0.5, prior\_ratio = 0.3$

Metric	Method	0.3	0.4	0.5	0.6	0.7	0.8
Recall	LM	0.00160	0.00307	0.00224	0.00222	0.00328	0.00292
	GM	0.11658	0.14731	0.17847	0.21686	0.22608	0.26931
	UserMatch	0.25801	0.25615	0.21785	0.20015	0.18955	0.17966
	TPA	**0.45592**	**0.41500**	**0.36145**	**0.33791**	**0.32359**	**0.29155**
F1	LM	0.00161	0.00309	0.00225	0.00224	0.00329	0.00292
	GM	0.11658	0.14731	0.17847	0.21686	0.22608	0.26931
	UserMatch	0.32975	0.31246	0.25868	0.23019	0.21620	0.21467
	TPA	**0.45850**	**0.41862**	**0.36316**	**0.34185**	**0.32658**	**0.29406**

From Table 5, the performance of all methods except GM decreases when the sparsity level increases. TPA always behaves best than other methods in Table 5. In fact, when the sparsity level is greater than 0.8, GM can achieve nearly the same performance as our proposed TPA. After observing this phenomenon, we find that this phenomenon often emerges when α_o or $prior\_ratio$ is small enough and sparsity level is large enough by conducting extensive experiments.

4.2 Experiments on Co-author Networks

Datasets. Co-author networks have been widely adopted in user identity linkage problem. Firstly, 10 representative conferences on Data Mining (DM)[2] and 9 representative conferences on Artificial Intellegence (AI)[3]. Then, we crawl data from DBLP and build a co-author network by the authors of papers from January

---

[2] The conferences selected from the DM field are KDD, SIGMOD, SIGIR, ICDM, ICDE, VLDB, WWW, SDM, CIKM, and WSDM.

[3] The conferences selected from the AI field are AAAI, IJCAI, CVPR, ICML, NIPS, UAI, ACL, EMNLP and ECAI.

Table 6. Experimental results under different prior ratio

Metric	Method	0.05	0.1	0.2	0.3	0.4	0.5
Recall	LM	0.00027	0.00031	0.00056	0.00062	0.00084	0.00091
	GM	0.00027	0.00031	0.00059	0.00036	0.00042	0.00063
	UserMatch	0.00424	0.00814	0.02224	0.03321	0.04017	0.04989
	TPA	**0.02121**	**0.03314**	**0.06013**	**0.08036**	**0.11247**	**0.12627**
F1	LM	0.00028	0.00033	0.00059	0.00069	0.00088	0.00094
	GM	0.00027	0.00031	0.00059	0.00036	0.00042	0.00063
	UserMatch	0.00833	0.01553	0.04008	0.05646	0.06632	0.08029
	TPA	**0.02194**	**0.03415**	**0.06124**	**0.08221**	**0.11470**	**0.12873**

2010 to September 2017 shown as Table 2. Finally, the shared number of same users $|T^a|$ between DM and AI dataset is 4941.

Results and Comparsion. Because the overlap level of DM-AI dataset is fixed, only prior ratio needs to be considered in experiments. As shown in Table 6, TPA exhibits the best performance on predicting anchor links between the two co-author networks on AI and DM. By experiments on social networks, we know the deviation between TPA and other methods is not largely when the overlap level of datasets or the prior ratio is too small. It is shown that TPA exhibits the best performance on predicting anchor links between co-author networks between DM and AI. The deviation between TPA and other methods ranges from 1% to 4.8%. The recall and F1 of TPA raises rapidly with varying prior ratio.

5   Conclusions

In this paper, we addressed the problem of user identity linkage. Unlike user unique attributes, we explore the power of user's social circle. The heart of our idea is that if most your best friends judge the different accounts on different networks is yours, these accounts are believed to belong to you. To acquire the authority of each friend and the trustworthiness of each final judgement, an Authority-Trustworthiness Analysis Model has been presented. Finally, we design a Trustworthy Predicting Approach to resolve the problem of user identity linkage.

Acknowledgments. This work was partially supported by National Natural Science Foundation of China No. U163620068 and Strategy Cooperation Project AQ-1703 and AQ-17014.

References

1. Carmagnola, F., Cena, F.: User identification for cross-system personalisation. Inf. Sci. **179**(12), 16–32 (2009)
2. Ji, S., Li, W., Srivatsa, M., He, J.S., Beyah, R.: Structure based data de-anonymization of social networks and mobility traces. In: Chow, S.S.M., Camenisch, J., Hui, L.C.K., Yiu, S.M. (eds.) ISC 2014. LNCS, vol. 8783, pp. 237–254. Springer, Cham (2014). https://doi.org/10.1007/978-3-319-13257-0_14
3. Kleinberg, J.M.: Authoritative sources in a hyperlinked environment. In: ACM-SIAM Symposium on Discrete Algorithms, pp. 668–677 (1998)
4. Korula, N., Lattanzi, S.: An efficient reconciliation algorithm for social networks. Proc. VLDB Endow. **7**(5), 377–388 (2014)
5. Kumar, S., Zafarani, R., Liu, H.: Understanding user migration patterns in social media. In: AAAI Conference on Artificial Intelligence, pp. 1204–1209 (2011)
6. Leskovec, J., Krevl, A.: SNAP datasets: stanford large network dataset collection, June 2014. http://snap.stanford.edu/data
7. Liu, L., Cheung, W.K., Li, X., Liao, L.: Aligning users across social networks using network embedding. In: Proceedings of the Twenty-Fifth International Joint Conference on Artificial Intelligence, pp. 1774–1780 (2016)
8. Liu, S., Wang, S., Zhu, F., Zhang, J., Krishnan, R.: Hydra: large-scale social identity linkage via heterogeneous behavior modeling. In: Proceedings of the 2014 ACM SIGMOD International Conference on Management of Data, SIGMOD 2014, pp. 51–62. ACM, New York (2014)
9. Mu, X., Zhu, F., Wang, J., Wang, J., Wang, J., Zhou, Z.H.: User identity linkage by latent user space modelling. In: ACM SIGKDD International Conference on Knowledge Discovery and Data Mining, pp. 1775–1784 (2016)
10. Narayanan, A., Shmatikov, V.: De-anonymizing social networks. In: 2009 30th IEEE Symposium on Security and Privacy, pp. 173–187, May 2009
11. Shen, Y., Jin, H.: Controllable information sharing for user accounts linkage across multiple online social networks. In: Proceedings of the 23rd ACM International Conference on Conference on Information and Knowledge Management, CIKM 2014, pp. 381–390. ACM, New York (2014)
12. Srivatsa, M., Hicks, M.: Deanonymizing mobility traces: using social network as a side-channel. In: Proceedings of the 2012 ACM Conference on Computer and Communications Security, CCS 2012, pp. 628–637. ACM, New York (2012)
13. Tan, S., Guan, Z., Cai, D., Qin, X., Bu, J., Chen, C.: Mapping users across networks by manifold alignment on hypergraph. In: AAAI Conference on Artificial Intelligence (2014)
14. Zafarani, R., Liu, H.: Connecting users across social media sites: a behavioral-modeling approach. In: 19th ACM SIGKDD International Conference on Knowledge Discovery and Data Mining, KDD 2013, vol. Part F128815, pp. 41–49. Association for Computing Machinery, August 2013
15. Zafarani, R., Tang, L., Liu, H.: User identification across social media. ACM Trans. Knowl. Discov. Data **10**(2), 16:1–16:30 (2015)
16. Zhang, Y., Tang, J., Yang, Z., Pei, J., Yu, P.S.: COSNET: connecting heterogeneous social networks with local and global consistency. In: ACM SIGKDD International Conference on Knowledge Discovery and Data Mining, pp. 1485–1494 (2015)

Mining High-Quality Fine-Grained Type Information from Chinese Online Encyclopedias

Maoxiang Hao[1], Zhixu Li[1(✉)], Yan Zhao[1], and Kai Zheng[2]

[1] School of Computer Science and Technology, Soochow University, Suzhou, China
mxhao@stu.suda.edu.cn, {zhixuli,zhaoyan}@suda.edu.cn
[2] University of Electronic Science and Technology of China, Chengdu, China
zhengkai@uestc.edu.cn

Abstract. Entity typing is a necessary step in building knowledge graphs. So far, plenty of efforts have been made in mining type information for entities from online encyclopedias, but usually only coarse-grained type information could be obtained for entities, which are not fine enough for the purpose of knowledge graphs construction or query answering. The situation becomes even worse for mining type information for entities in Chinese. In this paper, we work on mining high-quality fine-grained type information for entities from not only the title-labels and info-boxes in the entity's encyclopedias page, but also the abstracts and crowd-labels in the page, which could provide a lot more candidate fine-grained type information (with noises). To maintain the high quality of the mined type information, initially we only get reliable type information from the title-labels and info-boxes. Then by putting entities, attributes, values and types into one graph, some path information can be obtained between each candidate entity-type pair, then we rely on a proposed Path-CNN binary classification model to identify more correct entity-type pairs from the graph. Compared with the previous approach and DBpedia, our work could mine a lot more high-quality fine-grained type information for entities from the online encyclopedia. By performing our approach on the largest Chinese online encyclopedia, Baidu Baike, we have generated 25,651,022 type information (with more than 80% accuracy) for the entities involved in this encyclopedia.

Keywords: Entity typing · Entity classification · Knowledge graph

1 Introduction

Nowadays, knowledge graphs such as Dbpedia [1], Freebase [2], Yago [15] and Probase [21] are widely applied in many real applications, such as knowledge reasoning [3,10,19], entity linking [14] and question answering [5], etc. Due to the large requirement to knowledge graphs of various domains and languages, many ongoing efforts are still dedicated to constructing a variety of knowledge graphs from either the world wide web or domain corpuses.

© Springer Nature Switzerland AG 2018
H. Hacid et al. (Eds.): WISE 2018, LNCS 11234, pp. 345–360, 2018.
https://doi.org/10.1007/978-3-030-02925-8_25

The task of entity typing [18] plays an import role in building high-quality knowledge graphs, which aims at deciding a certain type for given entity. For example, finding out that "New York" (entity) is an instance of "City" (type). Recently, there is an increasing interests in mining type information from Web sources. For instance, Dakka and Cucerzan et al. [6] use supervised machine learning algorithms to classify encyclopedic articles into four types, and Suzuki et al. [16] propose a neural network-based multi-task learning method to classify encyclopedic articles into 200 types. However, these traditional works could only do coarse-grained entity typing, and can not work well on fine-grained entity typing tasks as the scale of the entity types becomes relatively large (say thousands of entity types).

Some recent work focus on learning types of fine granularity for entities from the world wide web or online encyclopedias [20]. Indeed, a large number of fine-grained types could extensively enrich and energize the constructed knowledge graph, but on the other hand, they bring great challenges in filtering out the noises from the correct ones. A recent work [20] has discussed on a three-step approach to extract type information from online encyclopedias. This approach first identifies explicit InstanceOf relations (i.e. type information) and SubclassOf relations with several heuristics, and then applies an attribute propagation algorithm leveraging existing category attributes, instance attributes, identified InstanceOf and SubclassOf relations to generate new category attributes. Finally, it constructs a weighted directed graph for each instance which has been enriched with attributes and categories, and then applies a graph-based random walk method to discover more type information. Although the experimental results conducted on several large Chinese encyclopedias shows that this approach can generate large-scale type information with types of appropriate granularity, the quality of the type information is still relatively low. According to our experiments, only 70% of the type information are correct. Besides, the approach may still miss a lot of useful fine-grained types for entities.

In this paper, we propose a novel approach to mine high-quality fine-grained type information for entities from online encyclopedias. Initially, we get entity-type pairs of instanceOf relation from the title-labels and info-boxes of online encyclopedias. Then for each entity type, we extract its attributes and then build a large-scale heterogeneous graph with the existing entities, entity attributes, attribute values, types and type attributes. Based on this graph, we then find out whether an entity is a correct instance of an entity type with a proposed Path-CNN model. By using CNN as the core, the Path-CNN model could fully leverage the path information about the entity and the entity type in the constructed graph. Particularly, we use metapath2vec [7] to translate each node in the heterogeneous graph into a low-dimensional vector. Because the attribute values contain richer information, we add the attribute values when building the heterogeneous graph, so that the low-dimensional vector for each node contains more information, which can greatly improve the prediction accuracy of our model. To generate the training data for the Path-CNN model, the entity-type pairs that we collect from the structured data of encyclopedias could be used as positive samples. We also define some rules to generate some incorrect entity type pairs (i.e. without instanceOf relation) as the negative samples.

The contributions in this paper are summarized as follows:

(1) We propose a novel approach to mine high-quality fine-grained type information for entities from online encyclopedias. By putting entities, attributes, values and types into one graph, some path information can be obtained between each candidate entity-type pair. Then we rely on a proposed Path-CNN binary classification model to identify more entity-type pair from the graph.

(2) By conducting our approach on the largest Chinese online encyclopedia, Baidu Baike, we have generated 25,651,022 high-quality (more than 80% accuracy) fined-grain type information for all the entities involved in this encyclopedia.

The remainder of the paper is organized as follows. After reviewing the existing literature on entity typing in Sect. 2, we give the overview of our approach in Sect. 3. We present the Path-CNN model in Sect. 4, and then report our experiments in Sect. 5. We conclude the paper in Sect. 6.

2   Related Work

The task of entity typing [18] plays an import role in building high-quality knowledge graphs, which aims at deciding a certain type for given entity. Initially, a large number of studies focus on classifying encyclopedic articles. Toral and Mu et al. [18] propose a method of classifying encyclopedic articles into three types (Location, Organization, Person), using words included in the body of the article as well as using hypernym information of words in WordNet [8] as an external knowledge base. They finally apply a weighted voting algorithm heuristic to determine the type of each article. Dakka and Cucerzan et al. [6] use supervised machine learning algorithms, SVMs and naive Bayes, to classify encyclopedic articles into four types (PER, ORG, LOC, MISC) defined by ACE. They employ several distinct features for each article, i.e., bag-of-words, article structure, abstract, titles, and entity mentions. In a recent study, Suzuki et al. [16] construct a neural network-based multi-task learning method, which classifies encyclopedic articles into 200 types proposed by Sekine et al. [13]. They use two sets of features for building the models, one is a baseline feature set proposed by Higashinaka et al. [9], and the other is an article vector (i.e., entity embeddings), which is learned from encyclopedic hypertext structure using a Skip-gram model. In addition to these studies, there are some other studies with respect to automatic categorization [4,17]. But most of these studies can just deal with a relatively small set of coarse-grained types, e.g., at most 200 types.

Recently, there are many taxonomic hierarchies of well-known knowledge graph based on encyclopedia, which are not flawless. The entity's type in DBpedia [1] is obtained by an info-box-based method, in which the entities are classified in strict accordance with instanceOf relations. However, since the construction process is semi-manual, the coverage and granularity of types are limited with only 170 info-box-based types being used for typing millions of entities. With integrating three sources of data from Wikipedia, WordNet, and GeoNames, YAGO [15] merges the word definitions in WordNet and the Wikipedia's taxonomy tree to extract the types by a rule-based method, showing a richer entity taxonomic hierarchies. However, because of its language-dependent nature, this method can't be directly applied to our tasks. Babel-Net [11] obtains entity's types by the means of mapping Wikipedia entities and concepts to WordNet.

In addition to the above knowledge graphs, there are some Chinese knowledge graphs. Zhishi.me [12], the first Chinese knowledge graph, combines data from Baidu Baike, Hudong Baike and Chinese Wikipedia. It only uses the SKOS vocabulary to represent the category system and did not strictly define the instanceOf relation between entities and types. In order to enrich entities with types, CN-DBpedia [22], one of the most well-known Chinese knowledge graph, uses reuses the taxonomy of types in DBpedia and classify Chinese entities with DBpedia types. Among the datasets in current knowledge graph, the number of Chinese type information is limited. Very little or no effort has been devoted to obtaining type information in Chinese knowledge graph.

A very close work to our study is done by Wu et al. [20], which mines type information from abstracts, info-box and categories of article pages in Chinese encyclopedia Web sites. They present an attribute propagation algorithm to generate attributes for categories and a graph-based random walk method to infer instance types from categories of entities. Although this approach can generate the appropriate granularity of type information, the quality of the type information is still relatively low, and the approach may still drop a lot of useful fine-grained types for entities, since it only uses the attribute information in info-box and discards attribute values containing richer information. In contrast, our approach uses all the information in info-box, and constructs a heterogeneous graph including four nodes: type, attribute, attribute value and entity, which can obtain large-scale, high-quality and fine-grained types for entities.

3   Overview of the Approach

In online encyclopedias, there are basically four places that might provide us with the type information for each entity.

- In order to distinguish different entities with the same name in encyclopedia pages, there are labels for each of these entities, called **title-label**. For example, the "Harry Porter" page has the following several title-labels which are in brackets: Harry Porter (J.K. Rowling's magical serial novels), Harry Porter (a series of films made by Warner Brothers) and Harry Porter (the protagonist of the novel "Harry Porter").

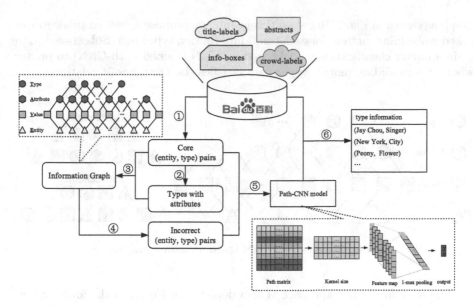

Fig. 1. The workflow of our approach.

- The **info-box** contains some structured information about the entity of this page. The format of the contents in the info-box is attribute and attribute value. For example, the info-box of "Jay Chou" contains information like (Nationality, China), (Occupation, Singer), (Representative work, Nunchakus) etc.
- The **abstract** is a brief description about the entity of the page in natural language. Usually, the first sentence in the abstract contains the type information of the entity. For example, the first sentence in Jay Chou's abstract is "Jay Chou (born 18 January 1979) is a Taiwanese musician, singer, actor, and director". We may get some fine-grained type information about "Jay Chou" from the abstract, by taking the risk that we may involve some noises which are caused by erroneous extraction results.
- There are also a large number of labels given by volunteers to the entity of the page. For instance, the page "Jay Chou" has labels given by volunteers, such as "Singer", "Musician", "Music", etc. We call these manually-given labels as **crowd-labels**. Obviously, crowd-labels provide a large number of candidate type information for entities, but many of these labels are not type information which should be removed, such as the label "Music" for "Jay Chou".

Given all the resources above, we work on getting the type information for entities from the online encyclopedias step by step. Initially, we can easily get a number of (entity, type) pairs satisfying the `instanceOf` relation from the info-boxes and title-labels. Then, we organize each entity with its attributes values and all the candidate types from either its crowd-labels or its abstract into a graph as shown in Fig. 2. Based on this graph, we propose a custom-made graph-based embedding method for each candidate (entity, type) pair. Subsequently, we train a binary classification model based on CNN (named Path-CNN) to predict whether a candidate (entity, type) pair satisfies the `instanceOf` relation.

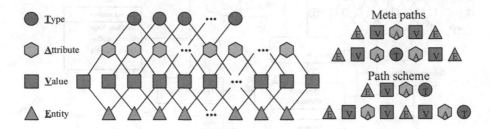

Fig. 2. The structure of the information graph.

The workflow of our approach is also depicted in Fig. 1. In the following, we discuss on the details of each step.

STEP 1: Getting Core Entity-Type Pairs. We first extract the (entity, type) pairs that satisfy the `instanceOf` relation separately from info-boxes and title-labels in encyclopedia. Here we first collect all crowd-labels with a relatively high frequency (say > 3) into a candidate type label set. Then for those (attribute, value) pairs in info-boxes, if the "attribute" here does exist in the candidate type label set and the entity set contains the "value", then this (attribute, value) pair will be added to the core (entity, type) pairs set. Similarly, if an entity has title-label and the lexical head extracted from it is within the candidate type label set, then we also add this (title, lexical head) to the core (entity, type) pairs set.

STEP 2: Type Attributes Inference. The attributes in the article's info-boxes are called entity attributes. In the core (entity, type) pairs set we extracted in previous step, a type has some entities with attributes, when more than one-third of the entities share some attributes, these attributes should belong to this type, which we call type attributes.

STEP 3: Building Information Graph. After generating attributes for types, we organize each entity, its attributes and values, each type and its attributes into a so-called *Information Graph*, such as the one shown in Fig. 2.

Definition 1. *An **Information Graph** is a heterogeneous graph $G = (N, E)$, where*

- $N = \{N_e \cup N_v \cup N_a \cup N_t\}$ *is the set of nodes, consisting of N_e, the set of all entities, N_v, the nodes representing values of entities, N_a, the nodes representing attributes of entities and types, and N_t, the type nodes.*

– $E = \{E_{ev} \cup E_{va} \cup E_{at}\}$ is the set of edges, where E_{ev} represents edges between the entity and its values, E_{va} indicates that there are multiple values corresponding to the same attribute name, E_{at} stands for edges between attributes and types.

STEP 4: Getting Incorrect Entity-Type Pairs. We defined some rules to generate the incorrect entity type pairs(i.e. the pairs without instanceOf relation) from Information Graph. The attributes of an entity and a type may overlap with each other. If there are more overlapping parts, the relationship between them should also be closer. Therefore, once the number of overlapping attributions between an entity and a type is 30%-50% of the number of the type attributes, we take this (entity, type) pair as a incorrect entity type pair.

STEP 5: Training Path-CNN Model. We use the core entity type pairs that are extracted from the title-labels, info-boxes and the incorrect entity type pairs that are generated by rules as the training data to train a so-called Path-CNN model. More details of the model will be given in Sect. 4.

STEP 6: Identifying More Type Information. Finally, we use the Path-CNN model to identify more types for each entity from the Information Graph. For example, given an entity "Jay Chou" with a number of candidate types such as "Person", "Musician", "Music" "Singer" and so on, we may identify that (Jay Chou, Person), (Jay Chou, Musician) and (Jay Chou, Singer) are the correct type information with the Path-CNN model.

4   Path-CNN Model

In this section, we propose the binary classification CNN model, namely Path-CNN, for mining entity types. We first introduce how to use the metapath2vec [7] model to transform all the nodes in the Information Graph into low-dimensional vectors and then we introduce the input data of the model. Finally, we give the details of the Path-CNN model.

4.1   Node Embedding Based on Meta-Path

The metapath2vec model [7] proposed by Dong et al. formalizes meta-path-based random walks to construct the heterogeneous neighborhood of a node and then leverages a heterogeneous skip-gram model to perform node embeddings. A meta-path is a path that connects multiple node types through a set of relations and can be used to describe the different semantic relationships of various connections between different types of nodes in graph.

Here we adopt two kinds of meta-path (i.e. "EVAVE" and "EVATAVE") in Information Graph as given in Fig. 2. For example, the meta-path "EVAVE" represents that two entity (E) have attribute values (V) and the corresponding attributes (A) are the same.

Algorithm 1: Getting path set $\mathcal{P}$

    Input: entity type pair (e, t) and Information Graph G
    Output: path set $\mathcal{P}$

1 **begin**
2     $set_{a_e} \leftarrow \text{getEntityAttributes}(e, G)$;
3     $set_{a_t} \leftarrow \text{getTypeAttributes}(t, G)$;
4     $set_c \leftarrow set_{a_e} \cap set_{a_t}$;
5   **for** $a \in set_c$ **do**
6         $path \leftarrow (e, v, a, t)$;
7   put $path$ in $\mathcal{P}$;
8   **end**
9   **for** $a_t \in set_{a_t}$ **do**
10   **for** $a_e \in set_{a_e}$ **do**
11   **for** $e_c \in (FindEntitiesByAttribute(a_t, G) \cap$
                $FindEntitiesByAttribute(a_e, G))$ **do**
12                 $path \leftarrow (e, v_1, a_e, v_2, e_c, v_3, a_t, t)$;
13   put $path$ in $\mathcal{P}$;
14   **end**
15   **end**
16   **end**
17   **return** $\mathcal{P}$
18 **end**

4.2 Input of the Model

We also adopt two kinds of path schemes in Information Graph as given in Fig. 2, where $path_1$ scheme, denoted in the form of $Entity(e) \rightarrow Value(v) \rightarrow Attribute(a) \rightarrow Type(t)$, means that e and t have the same a, and $path_2$ scheme, denoted in the form of $Entity(e_1) \rightarrow Value(v_1) \rightarrow Attribute(a_1) \rightarrow Value(v_2) \rightarrow Entity(e_2) \rightarrow Value(v_3) \rightarrow Attribute(a_2) \rightarrow Type(t)$, represents that e_1 and e_2 have the same attribute a_1, as well as e_2 and t share the same attribute a_2.

Given an (entity, type) pairs (e, t) where e is an entity and t is a type. According to $path_1$ and $path_2$ schemes, it's possible to obtain a set of paths $\mathcal{P} = \{p_1, p_2, \ldots, p_n\}$ from e to t. Algorithm 1 shows how to get the path set $\mathcal{P}$ in Information Graph G. We get the paths from e to t based on $path_1$ scheme, and add paths to $\mathcal{P}$ (lines 5–8), v is the attribute value corresponding to a in e. The paths from e to t are obtained according to $path_2$ scheme, and are put into $\mathcal{P}$ (lines 9–16), v_1 is the attribute value of e about a_e, v_2 is the attribute value of e_c about a_e, and v_3 is the attribute value of e_c about a_t.

The input data of our model is the path set $\mathcal{P}$ corresponding to each (entity, type) pair. After transforming all the nodes in Information Graph into low-dimensional vectors (see Sect. 4.1), we convert each node in $\mathcal{P}$ to a vector via the node embedding matrix so that all the paths in $\mathcal{P}$ can be transformed into a two-dimensional matrix. As shown in Fig. 3, let $\mathbf{x}_i \in \mathbb{R}^d$ be the d-dimensional node vector corresponding to the i-th node in the path. A path with n nodes could be represented as

$$\mathbf{x}_{1:n} = \mathbf{x}_1 \oplus \mathbf{x}_2 \oplus \ldots \oplus \mathbf{x}_n, \tag{1}$$

where $\oplus$ is the concatenation operator. In general, let $\mathbf{x}_{i:i+j}$ refer to the concatenation of nodes $\mathbf{x}_i, \mathbf{x}_{i+1}, \ldots, \mathbf{x}_{i+j}$. However, there are two kinds of path with different lengths in $\mathcal{P}$. For the convenience of input, we will complement zero in the paths of shorter length and convert the two paths to the same length. For example, the path $E \to A \to V \to T$ will be processed into the $E \to A \to V \to 0 \to 0 \to 0 \to 0 \to T$. Therefore, the input to this model becomes a 3-dimensional matrix.

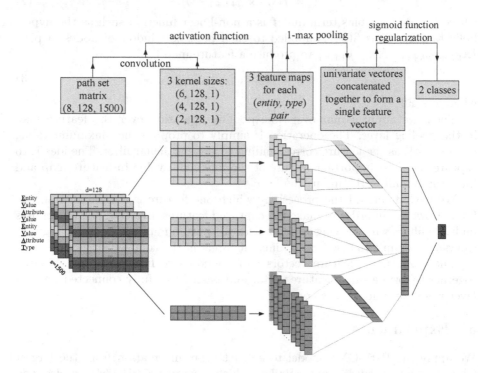

Fig. 3. The architecture of the path-CNN model.

4.3   Details of the Path-CNN Model

The architecture of the Path-CNN model is given in Fig. 3. In general, the Path-CNN model uses three convolution kernels of different size to extract features from $\mathcal{P}$, merges all the features into one vector and finally outputs the result. We will present the details of this model in the following of this subsection.

Basically, the information that could be leveraged for predicting whether an (entity, type) pair satisfies the `instanceOf` relation is included in $\mathcal{P}$. Thus, we should utilize all local features to perform prediction globally, which is the task of the convolutional layer. First, the convolutional layer extracts local features with a sliding window of length h over the path in $\mathcal{P}$, and then merges all local features into a vector of the path by max-pooling operation.

A number of feature maps are obtained by doing convolution operation on input layer. A convolution operation involves a *filter* $\mathbf{w} \in \mathbb{R}^{hd}$, where h represents the length of the sliding window, and d represents the dimension of the node vector. Through such a large convolution window, a column of feature maps will be obtained. For example, a feature c_i is generated from a window of nodes $\mathbf{x}_{i:i+h-1}$ by

$$c_i = f(\mathbf{w} \cdot \mathbf{x}_{i:i+h-1} + b), \tag{2}$$

where $b \in \mathbb{R}$ is a a bias term and f is a non-linear function such as the hyperbolic tangent. This filter is applied to each possible window of nodes in path $\{\mathbf{x}_{1:h}, \mathbf{x}_{2:h+1}, \ldots, \mathbf{x}_{n-h+1:n}\}$ to produce a feature map

$$\mathbf{c} = [c_1, c_2, \ldots, c_{n-h+1}], \tag{3}$$

where $\mathbf{c} \in \mathbb{R}^{n-h+1}$.

Then, we apply a max-over-time pooling operation over the feature map in the pooling layer. The operation is simply to propose the maximum value, $\hat{c} = max\{\mathbf{c}\}$, as the feature corresponding to this particular filter. The idea is to capture the most important feature of the maximum value in feature map and filter the zerofill in the input.

We have described the process by which one feature is extracted from one filter. However, in order to make the captured features diversity, our model uses multiple filters with varying window sizes to capture multiple features. Figure 3 shows an example that we use 3 different filters in the convolution layer.

These univariate feature vectors corresponding to filters are concatenated together to form a single feature vector and passed to a fully connected sigmoid layer whose output is *"Yes"* or *"No"*.

5   Experiments

We apply the Path-CNN model to extract type information from the largest Chinese encyclopedia, Baidu Baike, which contains 7,976,064 articles and 120,540,204 info-boxes. In this section, we first determine the parameters used in metapath2vec and our model. Then, we evaluate the accuracy of the generated type information. Finally, we compare the type information obtained by our approach with those generated by baseline and an existing knowledge graph.

5.1   Experimental Setting

Metapath2vec Model Parameters. In Sect. 4.1, we apply the metapath2vec model to perform nodes embedding in Information Graph, which can guide random walk through the meta-path to construct the node's neighbor node set. For node embeddings, we specify two meta-path schemes: "EVATAVE" and "EVAVE" (as shown in Fig. 2) to guide random walks. The values of these parameters are listed below.

(1) The number of walks per node w: 100;
(2) The walk length l: 500;
(3) The vector dimension d: 128;
(4) Other parameters by default.

Path-CNN Model Parameters. We use the core (entity, type) pairs extracted from Sect. 3 STEP 1 as the positive samples and take the incorrect (entity, type) pairs generated from Sect. 3 STEP 4 as the negative samples. From the dataset, we randomly select 80% instances as training samples and the rest as validation samples. We use empirical values to determine the optimal parameters for each data set, including path embeddings size s, filter windows size h, number of filters n, optimizer f, learning rate λ and mini-batch size b. The values of these parameters are listed below.

(1) The path embeddings size s: 1500;
(2) The filter windows size h: 6, 4, 2;
(3) The number of filters n: 32;
(4) The optimizer f: Adam;
(5) The learning rate λ: 1e$-$5;
(6) The mini-batch size b: 64.

Table 1. Comparison between the existing methods and our approach.

Model	Type number	Source	Type information number	Precision
Instance Type Ranker	1078	info-boxes	191,770	90.51%
		abstracts	732,352	73.39%
		crowd-labels	3,159,482	68.95%
Path-CNN model	4,518	title-labels	300,315	92.43%
		info-boxes	191,770	90.51%
		abstracts	9,008,445	84.79%
		crowd-labels	18,990,844	80.46%

5.2 Comparing with Existing Approach

We obtain 25,651,022 distinct type information, which are derived using heuristic rules from title-labels and info-boxes, and inferred using the Path-CNN model from abstracts and crowd-labels. According to the statistics, 7,976,064 distinct entities have been typed with 4,518 different types, and each entity has 3 type information on average.

Here we adopt the same labeling process used in Yago to build the ground-truth for our data set. More specifically, we invite 3 volunteers to partici-pant in the labeling process, and each volunteer selects "Correct", "Wrong" or

"Unknown" to label each (entity, type) pair. For the existing method: Instance Type Ranker, we use the same labeling process to evaluate the accuracy.

We mainly compare our approach with a state-of-the-art existing approach, **Instance Type Ranker** [20], which also mines type information from abstracts, info-box and categories of article pages in Chinese encyclopedia Web sites. The details of this approach is already presented in the Related work section.

As shown in Table 1, our model has learned 4518 different types for all the entities, which is much more than the 1078 different types learned by the Instance Type Ranker approach. From each particular resource, we also learns a lot more type information than the Instance Type Ranker approach. More importantly, the precision of our type information is also higher than the Instance Type Ranker approach. The experimental results prove that our approach could mine higher quality and finer grained type information for entities than the previous approach.

5.3   Comparison with DBpedia

We also compare the type information generated by our approach with an existing well-known knowledge graph, that is, the Chinese version of DBpedia. Table 2 not only give the number of types, entities and type information in our Data and DBpedia, but also show the overlap of the type information between our data and DBpedia. As can be observed, the number of the type information obtained by our model is significantly larger than DBpedia, and the overlapped part are very few.

Table 2. Number comparison and the overlap between DBpedia and our data.

(a) Number of the Data			(b) Overlap of the Data	
	Our data	DBpedia		Overlap
Type number	4,518	170	Type overlap	82
Entity number	7,976,064	876,725	Entity overlap	425,422
Type information number	25,651,022	1,534,268	Type information overlap	156,427

We finally compare the granularity of the type information between our data and DBpedia. here we randomly select 10,000 entities from overlapping entities, and take these entities and their corresponding types as samples to ask three volunteers to do manual comparison between our data and the DBpedia data. For each sample, volunteers label the comparison result with "Better", "Poorer" and "Similar". For example, for the entity "Jay Chou", the types we obtained are "Person", "Actor", "Film Actor", "Singer", "Director" and "Musician", while in DBpedia there are the following types: "Person" and "Actor" (volunteers chose the "Similar" label), "Artist" (volunteers chose the "Poorer" label) and all volunteers chose the "Better" label in the remain cases. Figure 4 shows the comparison results and some examples of the comparison results are also listed

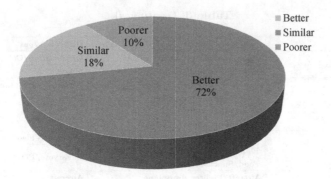

Fig. 4. Compared the granularity with DBpedia.

in Table 3. Apparently, most of our entities have more fine-grained types than those in DBpedia.

Table 3. Example type information comparison between our results and DBpedia.

Entities	Path-CNN Types	DBpedia types
Redis	Software/Database	Software
Linux	Software/Operating System	Software
Peony	Plant/Flower	Eukaryote/Plant
Pitaya	Plant/Fruit Food/Fruit	Eukaryote/Plant
Beijing	Place/City/Capital	Place/Settlement
Mencius	Person/Ideologist Person/educationist	Person/Philosopher
YouTube	Website/Video Website	Organisation/Company
Jay Chou	Person/Actor/Film Actor Person/Musician/Singer Person/Director	Person/Actor
Microsoft	Company/Software Company Company/Technology Company	Organisation/Company
Christmas	Festival/Traditional Festival	Holiday
Coca-Cola	Beverage	Food/Beverage
Bill Gates	Person/Enterpriser/Chairman Person/Philanthropist	Person
WhatsApp	Software/Application	Software
Hello Venus	Group/Girl Group Band	Organisation/Music Group/Band
Caspian Sea	Place/Lake/Saline Lake	Place/Lake
Giant Panda	Animal/Mammal Animal/Endangered Animal	Eukaryote/Animal/Mammal
MacBook Pro	Computer/Laptop	Device
Spirited Away	Film/Animated Film Film/Fantasy Film	Film
Glorious Mind	Music/Opening Theme Song	Single
Egg Srop Soup	Food/Soup	Food

(continued)

Table 3. (*continued*)

Entities	Path-CNN Types	DBpedia types
Random House	Company/Publishing Company	Organisation/Company
Moscow Nights	Music/Song/Single	Single
Arsene Wenger	Person/Sport Person/Soccer Mannager	Person/Soccer Manager
Spratly Islands	Place/Island/Archipelago	Place
The Lazy Song	Music/Song/Single Music/Pop Music	Single
Syobon Action	Game/Singe-Player Game	Software/Video Game
Sunflower Seed	Plant/Seed Food/Snacks	Eukaryote/Plant
Nakajima Ki-27	Aircraft/Fighter Aeroplane	Aircraft
Monocled Cobra	Animal/Snake	Eukaryote/Animal
Earth Simulator	Computer/Super Computer	Software
Michael Jackson	Person/Musician/Singer Person/Dancer Person/Actor	Person/Artist Music Group
Time (magazine)	Magazine/News Magazine	Magazine
The Shawshank Redemption	Film/Inspirational Film Film/Drama Film Film/Prison Film	Film
Stephen Hawking	Person/Scientist/Physicist	Person/Scientist
IMI Desert Eagle	Weapon/Firearms/Pistol	Weapon
Magnesium Oxide	Chemical Compound/Oxide Solid	Chemical Substance Chemical Compound
Beep Media Player	Software/Player/Music Player Software/Freeware	Software
League of Legends	Game/Video Game/MOBA Game	Software/Video Game
Sputnik Sweetheart	Book/Novel	Book
Grand Theft Auto IV	Game/Video Game/Action Game	Software/Video Game
Brandenburg Concertos	Music/Orchestral Music	Music
Mikoyan-Gurevich MiG-9	Aircraft/Fighter Aeroplane	Aircraft
Warcraft III: Frozen Throne	Game/Video Game/RTS Game	Software/Video Game
The Adventures of Little Carp	Television Show/Cartoon	Television Show
Humble Administrator Garden	Place/Tourist Attraction/Garden	Place/World Heritage Site

6   Conclusions

In this paper, we propose a novel approach to mine high-quality fine-grained type information for entities from online encyclopedias. By putting entities, attributes, values and types into one graph, some meta-path information can be obtained between each candidate entity-type pair. Then we rely on a proposed Path-CNN binary classification model to identify more entity-type pairs from the graph. Compared with the previous approach and DBpedia, the approach could mine a lot more fine-grained type information for entities from the online encyclopedia. Besides, the quality of our type information is also better than that obtained by previous approach. As for future work, we would like to mine more type information for entities from the article text of the online encyclopedias.

Acknowledgments. This research is partially supported by National Natural Science Foundation of China (Grant No. 61632016, 61402313, 61472263), and the Natural Science Research Project of Jiangsu Higher Education Institution (No. 17KJA520003).

References

1. Auer, S., Bizer, C., Kobilarov, G., Lehmann, J., Cyganiak, R., Ives, Z.: DBpedia: a nucleus for a web of open data. In: Aberer, K., et al. (eds.) ASWC/ISWC -2007. LNCS, vol. 4825, pp. 722–735. Springer, Heidelberg (2007). https://doi.org/10.1007/978-3-540-76298-0_52
2. Bollacker, K., Evans, C., Paritosh, P., Sturge, T., Taylor, J.: Freebase: a collaboratively created graph database for structuring human knowledge. In: SIGMOD 2008, pp. 1247–1250 (2008)
3. Bordes, A., Usunier, N., Garcia-Duran, A., Weston, J., Yakhnenko, O.: Translating embeddings for modeling multi-relational data. In: ICONIP, pp. 2787–2795 (2013)
4. Chang, J.Z., Tsai, R.T., Chang, J.S.: Wikisense: supersense tagging of Wikipedia named entities based wordnet. In: PACLIC 23, pp. 72–81 (2009)
5. Cui, W., Wang, H., Wang, H., Song, Y., Hwang, S.W., Wang, W.: KBQA: learning question answering over QA corpora and knowledge bases. PVLDB **10**(5), 565–576 (2017)
6. Dakka, W., Cucerzan, S.: Augmenting Wikipedia with named entity tags. In: IJCNLP, pp. 545–552 (2008)
7. Dong, Y., Chawla, N.V., Swami, A.: metapath2vec: scalable representation learning for heterogeneous networks. In: KDD, pp. 135–144 (2017)
8. Fellbaum, C., Miller, G.: WordNet: An Electronic Lexical Database. MIT Press, Cambridge (1998)
9. Higashinaka, R., Sadamitsu, K., Saito, K., Makino, T., Matsuo, Y.: Creating an extended named entity dictionary from Wikipedia. In: COLING, pp. 1163–1178 (2012)
10. Lin, Y., Liu, Z., Zhu, X., Zhu, X., Zhu, X.: Learning entity and relation embeddings for knowledge graph completion. In: AAAI, pp. 2181–2187 (2015)
11. Navigli, R., Ponzetto, S.P.: BabelNet: the automatic construction, evaluation and application of a wide-coverage multilingual semantic network. Artif. Intell. **193**(6), 217–250 (2012)
12. Niu, X., Sun, X., Wang, H., Rong, S., Qi, G., Yu, Y.: Zhishi.me - weaving Chinese linking open data. In: Aroyo, L., et al. (eds.) ISWC 2011. LNCS, vol. 7032, pp. 205–220. Springer, Heidelberg (2011). https://doi.org/10.1007/978-3-642-25093-4_14
13. Sekine, S., Sudo, K., Nobata, C.: Extended named entity hierarchy. In: LREC (2002)
14. Shen, W., Han, J., Wang, J., Yuan, X., Yang, Z.: Shine+: a general framework for domain-specific entity linking with heterogeneous information networks. IEEE Trans. Knowl. Data Eng. **30**(2), 353–366 (2018)
15. Suchanek, F.M., Kasneci, G., Weikum, G.: Yago: a large ontology from Wikipedia and wordnet. Web Semant.: Sci. Serv. Agents World Wide Web **6**(3), 203–217 (2008)
16. Suzuki, M., Matsuda, K., Sekine, S., Okazaki, N., Inui, K.: Neural joint learning for classifying Wikipedia articles into fine-grained named entity types. In: PACLIC 30 (2016)

17. Tardif, S., Curran, J.R., Murphy, T.: Improved text categorisation for Wikipedia named entities. In: ALTA, pp. 104–108 (2009)
18. Toral, A., Mu, R.: A proposal to automatically build and maintain gazetteers for named entity recognition by using Wikipedia. In: EACL, pp. 56–61 (2006)
19. Wang, Q., Liu, J., Luo, Y., Wang, B., Lin, C.Y.: Knowledge base completion via coupled path ranking. In: ACL, pp. 1308–1318 (2016)
20. Wu, T., Ling, S., Qi, G., Wang, H.: Mining type information from Chinese online encyclopedias. In: Supnithi, T., Yamaguchi, T., Pan, J.Z., Wuwongse, V., Buranarach, M. (eds.) JIST 2014. LNCS, vol. 8943, pp. 213–229. Springer, Cham (2015). https://doi.org/10.1007/978-3-319-15615-6_16
21. Wu, W., Li, H., Wang, H., Zhu, K.Q.: Probase: a probabilistic taxonomy for text understanding. In: SIGMOD, pp. 481–492 (2012)
22. Xu, B., et al.: CN-DBpedia: a never-ending Chinese knowledge extraction system. In: Benferhat, S., Tabia, K., Ali, M. (eds.) IEA/AIE 2017. LNCS (LNAI), vol. 10351, pp. 428–438. Springer, Cham (2017). https://doi.org/10.1007/978-3-319-60045-1_44

Semantics-Enabled Personalised Urban Data Exploration

Devis Bianchini[✉], Valeria De Antonellis, Massimiliano Garda,
and Michele Melchiori

Department of Information Engineering, University of Brescia,
Via Branze 38, 25123 Brescia, Italy
{devis.bianchini,valeria.deantonellis,m.garda001,
michele.melchiori}@unibs.it

Abstract. Research challenges for Smart Cities concern the study of
methods and techniques to help citizens and administrators to effectively
obtain information of interest from the large amounts of data in multiple,
heterogeneous sources. Accessing heterogeneous data sources and aggregating urban data according to several perspectives can be achieved by
defining proper indicators. In addition, semantic web technologies may
be used to enable interoperability and improve data access. In this paper,
we propose an ontology-based framework to support personalised urban
data exploration. The framework is composed of: (i) the so-called Smart
Living Ontology, providing a semantic representation of city indicators;
(ii) a Semantic Layer, exploiting the ontology and user characterisation
to enable personalised access to urban data.

Keywords: Urban data exploration · Semantic web
Web applications for smart cities

1 Introduction

Exploring large collections of Smart City data from multiple and heterogeneous
sources is a challenging issue, that can be achieved by defining proper indicators about energy consumption, garbage collection, level of pollution, citizens'
safety and security [13]. Indicators are able to aggregate urban data according to
several perspectives and to provide a comprehensive view over underlying data
without being overwhelmed by the data volume [9]. Citizens receive information about their city. Public Administration (PA), utility and energy providers
have at their disposal new tools to take actions that might improve citizens'
daily life. Building managers may use urban data to take decisions about their
administered buildings.

This scenario is characterised by an increasing data variety and heterogeneity
of data sources and involved actors. Indicators designed on top of these sources,
computed at building, district or city level, aggregate data that can have different relevance with respect to the target users and the conditions in which

© Springer Nature Switzerland AG 2018
H. Hacid et al. (Eds.): WISE 2018, LNCS 11234, pp. 361–376, 2018.
https://doi.org/10.1007/978-3-030-02925-8_26

users act (e.g., activities that users are performing). Semantic web technologies may be used to enable interoperability and improve data access also in the Smart City context. In recent research efforts, semantic web technologies have been proposed to develop semantic-driven applications [12], such as Smart Urban Cockpits and dashboards [19], and tools to support smart city ontology building and maintenance [15]. In this paper, we propose the use of semantic web technologies from a data exploration viewpoint. Ontologies are used to properly represent knowledge structure in terms of concepts, hierarchies and semantic relationships, thus they can be used to facilitate exploration by exploiting the knowledge structure. In particular, the inheritance nature of ontology hierarchies allows exploration at different granularity levels with one single request. Based on these considerations, we propose a novel semantics-enabled framework for personalised urban data exploration, apt to support users in finding indicators of interest, exploiting the ontology in which they are formally represented and taking into account the defined users' profiles. The proposed data exploration approach is based on knowledge both on the Smart City relevant indicators and their context, modelled as an ontology, and associated with user's profile, which permit personalization of the exploration experience for the Smart City users. The framework is particularly useful for decision makers, who need to know the different performances of monitored systems in the Smart City and have a view on urban data at different aggregation levels. For example, the framework may allow building managers to monitor electrical consumption of administered buildings, by exploiting the indicators hierarchy in the ontology to distinguish electrical consumption according to different perspectives (e.g., consumption in common spaces, consumption of elevators), and to compare average values of consumption with other buildings at district or city level. Furthermore, the framework may enable citizens to make decisions about their activities by observing specific indicators (e.g., to avoid sport activities when pollution levels overtake tolerance thresholds). The framework is composed of: (a) the so-called Smart Living Ontology[1], to provide a formal representation of Smart City indicators, with reference to the kinds of activities and users' categories for which indicators can provide relevant information; (b) a Semantic Layer, developed on top of the ontology, to enable personalised exploration of urban data.

This work was performed in the context of the Brescia Smart Living (BSL) Italian project[2], that promotes a holistic view of the city, where different types of data have to be collected and properly explored to provide new services to both citizens and PA.

This paper is organized as follows: in Sect. 2, motivations and web-based architecture are presented; Sects. 3 and 4 present the Smart Living Ontology and the users' profiling, respectively; Sect. 5 describes Semantic Layer; in Sect. 6 we discuss preliminary experiments on ontology-based indicators selection; in

---

[1] The TBox of the ontology can be found at https://tinyurl.com/onto-schema (a free Web Protégé account is required).

[2] http://www.bresciasmartliving.eu.

Sect. 7 we highlight the cutting-edge features of our approach compared to the literature; finally, Sect. 8 closes the paper.

2 Motivating Scenario

Several categories of users may have access to urban data: citizens, mobility managers, energy managers, utility and energy providers, building managers. Different categories may be interested to different kinds of data to serve specific purposes. As a motivating example, let's consider John, the manager of several buildings located in different districts of the Smart City. John monitors the electrical consumptions of the buildings, in order to implement energy saving policies (e.g., introduction of LED lamps in common spaces of the buildings or planning renovation work to increase the energy efficiency class). Challenging issues are related to the capability of enabling John to fruitfully exploit available information. In our framework, we considered the following issues.

Semantic specification of city indicators. In the Smart City context, indicators are used as measurement tools that, aggregating urban data according to several dimensions, can provide useful information to different categories of users (e.g., consumption-based indicator of electrical energy use). In our framework, indicators are defined over the BSL platform, containing data coming from heterogeneous sources. The Smart Living Ontology is defined to provide semantic specification of the indicators. An ontology-based framework is proposed to enable users, with no previous knowledge about the data, to leverage ontology concepts and semantic relationships to find the information of interest.

Personalised data exploration. Given the wide variety of urban data that can be explored in the BSL platform, we propose that the selection of proper indicators can be personalised taking into account users' profiles, composed of user's category, activities and preferential indicators. The profile is initialised during the registration to the BSL platform and can be updated by the user. Preferential indicators are stored during user's interactions with the platform.

Indicators recommendation to support decision making. Indicators recommendation is provided in order to help users to take decisions in their daily life, by visualising needed aggregated data. Indicators are suggested during data exploration to the user, considering also his/her profile. For example, John is provided with suggestions about indicators on electrical consumption of the administered buildings. Comparison against values of consumptions according to different perspectives (e.g., consumption in common spaces or elevators) and of similar buildings in the same district may validate John's decision to start building energy saving activities. Therefore, indicators selection may be followed by an exploration of urban data starting from selected indicators, to enable users to find desired information.

2.1   Web-based Architecture for the Personalised Approach

Figure 1 shows an architecture overview of the semantics-enabled data exploration framework proposed in this paper. The framework is developed with web-based technologies and is organised over multiple layers. Data on field, collected from domain-specific platforms through IoT technologies, as well as data from sources external to the BSL project (weather data, pollution data, etc.) are loaded into the BSL platform (*BSL platform Layer*). Data is transferred on the BSL platform using RESTful services, SOAP-based services and MQTT Agents. JSON has been adopted as data format. Data is aggregated into smart city indicators, that are semantically specified in the *Semantic Layer* to enable personalised urban data exploration (as described in Sects. 3 and 5). The *User Access Layer* includes a web-based Smart City Dashboard to be used by citizens, PA and other users to explore data and take decisions.

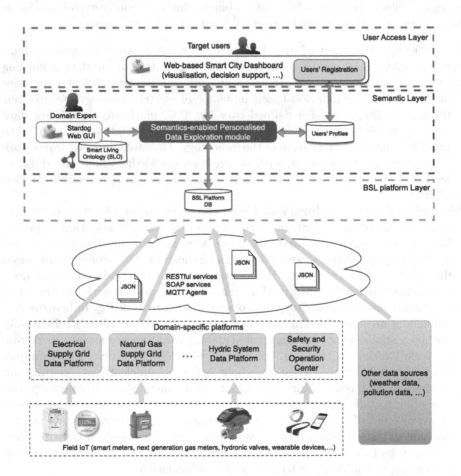

Fig. 1. Web-based architecture of the semantics-enabled data exploration framework

Using the web browser, users can register themselves and can update their profile. Profiles are used by the Semantics-enabled Personalised Data Exploration module as described in Sect. 5. This module is implemented in Java and deployed under the Apache TomEE[3] application server. The Smart Living Ontology is deployed in OWL using the Stardog[4] Triplestore. The Stardog module supports domain experts to maintain the ontology (concepts, relationships and individuals), interacting with the web-based administration console provided by the module.

3 The Smart Living Ontology

Figure 2 reports the main concepts and relationships of the Smart Living Ontology. We will describe these concepts in the following. In order to face the heterogeneity and complexity of the Smart City domain, we rely on some well-known general purpose ontologies apt to cover a set of required pivotal concepts as specified in the following.

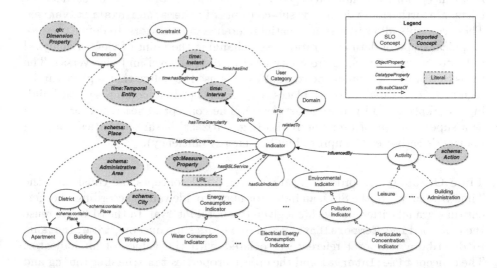

Fig. 2. A portion of the Smart Living Ontology, containing main concepts, object properties, datatype properties and `subClassOf` relationships

3.1 Reference Ontologies

Reference ontologies are exploited to define: (i) concepts that enable a geospatial mapping of the main structures of the city (e.g., buildings, streets, areas) as

[3] http://tomee.apache.org/.
[4] https://www.stardog.com/.

well as the possibility of expressing inclusion relationships (e.g., to state that a district is physically located within a specific geographic area); (ii) temporal entities, that we used as analysis dimensions; (iii) other high level concepts, that have been specialised to define the hierarchy of indicators and activities. A thorough documentation of the concepts imported from foundation ontologies can be found following the link associated with each ontology in the following descriptions.

Data Cube. The Data Cube vocabulary[5] contains the main definitions for multi-dimensional data analysis, such as dimensions and measures. In the SLO, indicators and analysis dimensions are considered as sub-concepts of measure (`qb:MeasureProperty`) and dimension (`qb:DimensionProperty`) concepts in the Data Cube vocabulary, respectively.

Schema.org. Schema.org[6] is a general-purpose ontology, that collects several concepts belonging to different domains. The concept `schema:Place` is used as a root concept for the definition of geographical entities. Some of them, such as districts, are furthermore defined as sub-concepts of `schema:AdministrativeArea`. These concepts allow to assign a spatial reference to indicators, to define the geographical scope of data exploration (e.g., changing the point of view of analysed indicator values, switching from city level to district level and vice versa). The transitive property `schema:containsPlace` is used to define that a place can be seen as a "container" of other places (e.g., a district may contain several buildings, workplaces and private apartments). The concept `schema:Action` is used as a super-concept of user's activities (e.g., citizens' leisure activities, building manager's activities to improve building energy efficiency).

Time Ontology. In order to deal with temporal concepts, we considered the Time Ontology[7]. In the SLO, an indicator may provide values according to different time granularities and within a given temporal interval. In the SLO, we reuse the concept `time:TemporalEntity` and one of its sub-concepts, `time:Instant`, to describe a temporal reference associated with the values of an indicator. The concept `time:Interval` and the object properties `time:hasBeginning` and `time:hasEnd` are used to define time intervals on which indicators values are available.

3.2   Semantic Description of Smart City Indicators

The Smart Living Ontology contains the formal definitions of indicators, that aggregates and summarize urban data of interest for the roles operating in the Smart City actors (e.g., citizens, building managers, PA).

---

[5] http://purl.org/linked-data/cube ("qb:" is the ontology prefix).
[6] http://schema.org/ ("schema:" is the ontology prefix).
[7] http://www.w3.org/2006/time ("time:" is the ontology prefix).

Indicators are specified as individuals of the `Indicator` concept or one of its sub-concepts in the indicators hierarchy (e.g., `ElectricalEnergyConsumption-Indicator`, that is a sub-concept of `EnergyConsumptionIndicator`). Moreover, an indicator i is further specified by a set of domain individuals D_i (e.g., environment, safety, energy, mobility), through the `relatedTo` relationship, and a set of constraints C_i. As shown in Fig. 2, in the SLO a constraint $c_i \in C_i$ can be either a dimension (time and space) or a user's category (e.g., citizen, building manager). An indicator can be `boundTo` a time interval (e.g., values of electrical consumption available for the year 2017), may have a time granularity (`hasTimeGranularity` relationship), may be defined at city, street, district or more specific levels, such as buildings, workplaces and private apartments (`hasSpatialCoverage` relationship), is designed for specific users' categories (`isFor` relationship). An indicator i is linked to a web-based service of the BSL Platform (`hasBSLService` property in Fig. 2), whose execution allows to display the indicator values on the Smart City Dashboard. Finally, indicators are related to activity types (e.g., leisure, building administration), in order to state that an activity can be influenced by the knowledge provided by the indicator (`influencedBy` property).

We can summarize an indicator i by means of a tuple: $\langle ID_i, T_i, D_i, C_i, r_i \rangle$ ($\forall i = 1, \ldots, N_\mathcal{I}$), where ID_i is a unique identifier (i.e., an URI), T_i is the indicator type (e.g., `ElectricalEnergyConsumptionIndicator`), D_i is the set of domains individuals, C_i is the set of constraints, r_i is the link to the web-based service of the BSL Platform whose execution provides indicator values.

4   Users' Profiles

BSL users are profiled (described) according to their category (e.g., citizen, building manager), their activities, defined as individuals of the concept `Activity` or its sub-concepts, the types of indicators explored by the user through the interactions with the framework, referring to the concept `Indicator` or its sub-concepts. Moreover, different users can have access to different indicators up to the finest level of aggregation that is allowed depending on the user's category: a citizen can select indicators concerning his/her apartment(s) only, building managers can select indicators on their administered buildings only, energy managers can select indicators that only concern the workplaces they are responsible for, etc.

When the user moves from apartment, building or workplace levels up to district or city levels, only aggregated data is displayed. Therefore, during registration, citizens, building managers and other categories of users also specify the place (e.g., apartment or building) they act in. This has a two-fold advantage: (i) it enables data privacy preservation, for instance preventing building managers to visualize data on buildings they do not administer; (ii) it will be used to personalise indicators selection and data exploration, as explained in the next sections.

Based on these considerations, the profile $p(u)$ of a user $u \in \mathcal{U}$ stored within the BSL Platform is defined as $p(u) = \langle ID_u, cat_u, \mathcal{I}_u, A_u, P_u \rangle$, where ID_u is the identifier associated to user's account, cat_u is the user's category, $\mathcal{I}_u$ is the set of individuals representing indicators that have been selected by u in previous interactions with the framework, A_u is the set of activities individuals as defined above, P_u is the list of individuals of concepts representing places where the user acts, namely `Building` for building managers, `Apartment` for citizens, `Workplace` for energy managers, etc. The registration wizard, starting from the user's category, prompts to the user proper masks to insert P_u instances and connect them to districts. Such instances will be also inserted in the SLO to enable semantics-enabled urban data exploration, while only districts and upper level places are inserted in the ontology by the domain experts. This reduces the complexity of ontology population and maintenance for domain experts.

5 The Ontology-based Semantic Layer

5.1 Formulating a Request for Personalised Data Exploration

Personalised data exploration for each user can be achieved by effectively exploiting city indicators, properly selected according to the indicators domains, constraints and profiles of users who are looking for indicators values, thus performing personalised data exploration. The user starts data exploration by specifying domains (e.g., pollution, energy) and/or indicators of interest, that are searched among the individuals of `Domain` and `Indicator` concepts or sub-concepts, respectively. Nevertheless, to support the user in the request formulation without requiring a detailed knowledge of ontology concepts and individuals, the framework enables the user to specify a set of keywords $K_r = \{k_{r1}, k_{r2}, \ldots, k_{rn}\}$. The set K_r is processed according to techniques aimed to match the keywords with ontology terms [18]. The adopted procedure relies on WordNet[8], to retrieve synonyms, hypernyms and hyponyms of the keywords, and identifies a mapping between the input list of keywords and ontology individuals using probabilistic techniques. Following this approach, user's requests are processed in a more flexible way, to deal with the different levels of expertise (i.e., knowledge of the terminology and lexicon) users have. WordNet adoption increases the recall and precision of indicators selection algorithm, at the cost of a slight increment in terms of response time (see experiments in Sect. 6), that is still acceptable for the considered exploration scenarios.

Beyond the desired domains and/or indicators expressed in K_r, the user's profile is exploited to take into account other elements for indicators selection, namely the user's category cat_u, the activities $\mathcal{A}_u$ and the set $\mathcal{I}_u$ of indicators explored in the past by the user. Formally, we represent the request submitted to the framework as follows: $r(u) = \langle D_r, \mathcal{I}_r, p(u) \rangle$, where D_r is the set of desired domains, $\mathcal{I}_r$ is the set of indicators of interest, $p(u)$ is the user's profile. D_r and $\mathcal{I}_r$ are the output of the WordNet-based disambiguation procedure.

---

[8] https://wordnet.princeton.edu/.

5.2   Semantics-Enabled Data Exploration

The semantics-enabled approach proposed here for urban data exploration is articulated over a set of steps, that are summarized in Fig. 3:

- *candidate indicators selection* - the overall set of available indicators $\mathcal{I}$ is properly pruned taking into account the desired domains and indicators specified in the request $r(u)$ and the user's profile; let's denote with $\mathcal{I}_{cand}$ the set of generated candidate indicators;
- *semantics-enabled personalised data exploration* - candidate indicators are used as starting point to enable personalised exploration over the hierarchy of smart city indicators and over the semantic relationships between analysis dimensions;
- *data visualisation on the web-based dashboard* - selected indicators are used to visualise urban data on the web-based Smart City Dashboard, through the invocation of BSL platform Web services; this step may motivate further refinement of candidate indicators, thus implementing an iterative and personalised exploration of urban data.

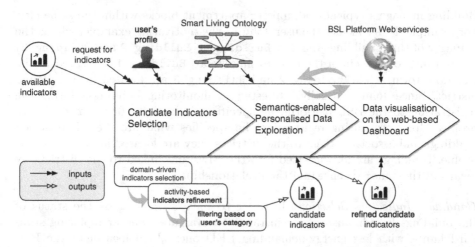

Fig. 3. The steps of semantics-enabled data exploration

5.3   Candidate Indicators Selection

Indicators selection takes as input the set $\mathcal{I}$ of all available indicators, the request $r(u)$, that includes the user's profile, and the SLO. The outcome of this step is a set of candidate indicators, namely $\mathcal{I}_{cand}$, containing indicators that are compliant with the request. The user may start from indicators in $\mathcal{I}_{cand}$ to further perform data exploration as described in the next section. The selection step is further organised in three sub-steps, corresponding to candidate indicators selection based on domains, indicators enrichment based on activities extracted from user's profile, indicators filtering according to user's category.

In candidates selection based on domains specified in D_r, the individuals of Domain concepts linked to each indicator $i \in \mathcal{I}$ (if any), denoted with d_i, are retrieved for each i (by considering the relatedTo relationship in the SLO). If there is an overlapping between d_i and D_r, then i is added to $\mathcal{I}_{cand}$.

In activity-based candidate indicators refinement, for each activity extracted from the user's profile, the influencedBy relationship in the SLO is used to retrieve additional candidate indicators. At this point of the selection step, the set $\mathcal{I}_u$ from the user's profile $p(u)$ is considered adding the elements of the $\mathcal{I}_u$ set to the indicators in $\mathcal{I}_{cand}$. In fact, the set $\mathcal{I}_u$ traces past exploration history of the user providing worthy, albeit not novel, candidate indicators.

Finally, in the indicators filtering based on user's category, each indicator $i \in \mathcal{I}_{cand}$ is analysed to filter out indicators that are not compliant with the user's category. The isFor relationship in the SLO is considered to get the individuals of concept UserCategory, that are semantically related to i.

5.4   Personalised Data Exploration Scenarios

Building managers typically administer apartment blocks within the same city. For example, let's consider the user John in the motivating example, who is the manager of three buildings (namely Building 1, Building 2 and Building 3) located in two districts of the city. In particular, Building 1 is located in the city downtown, while Building 2 and Building 3 are located in San Polino district. Since John is usually interested in monitoring buildings, during the registration to the BSL platform, he specifies the activity Monitoring in his profile. Moreover, during registration, he specifies what are the administered buildings and associates them to the districts they are located in, as part of his profile. Buildings are also inserted into the SLO and linked to the districts by means of the schema:containsPlace relationship.

Candidate Indicators Selection. In order to have an insight on the status of the buildings he administers, for instance to evaluate whether replacing standard lamps with less energy-demanding LED ones, John logs in to the BSL platform and asks for consumption indicators, specifying the keywords $K_r =$ {energy, consumption}. The platform processes the request as explained in Sect. 5.3 and returns, among the others, the indicator NormalizedElectrical-EnergyConsumption (NEEC), which reports electrical consumption normalised with the number of apartments in the building. The indicator is selected because it is both compatible with the keywords given in the request and it is associated with the activity Monitoring in the ontology.

Exploration of Indicators. Semantic description of the indicator NEEC is shown in Fig. 4. Starting from NEEC indicator, John can further explore other indicators being guided by the semantic relationships in the SLO. Exploration can be performed according to different perspectives, given the knowledge structure

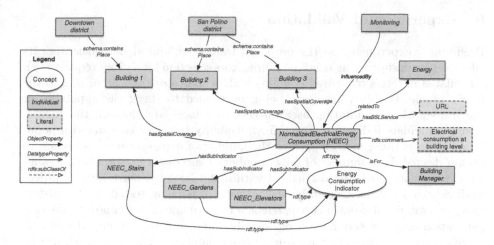

Fig. 4. Example of candidate indicator and related properties

in the ontology: (a) exploration over the indicators hierarchy; (b) personalised exploration over the indicators dimensions.

In the first exploration scenario, John selects the NEEC indicator and the framework suggests him more specific indicators NEEC_Stairs, NEEC_Elevators and NEEC_Gardens, which are related to the NEEC indicator through the hasSub-Indicator relationship in the ontology. Since John's focus is on evaluating the electrical consumption of the lighting plants of stairs, he selects NEEC_Stairs.

Personalised exploration over the indicators dimensions exploits both the semantic relationships that relate individuals of schema:Place concept or its sub-concepts and the information stored in the user's profile. In particular, knowledge on the spatial coverage of indicators is obtained through the hasSpatialCoverage relationship. Starting from indicators previously selected for the John's building, either NEEC or one of its sub-indicators, the containment relationship that relates John's buildings with districts is exploited. Therefore, John could choose to visualize the average consumption provided by the indicators for the buildings of the districts, in order to compare his buildings against others having similar characteristics or using different lighting solutions. Similarly, indicators for John's buildings could also be suggested over several years (boundTo relationship) or over different time granularities (e.g., years, months, days), according to the hasTimeGranularity relationship. Comparison between indicators may stimulate John to consider the replacement of energy consuming light bulbs with modern LED lamps in shared spaces, after analysing the affordability of the expenditures, with respect to the ones sustained by other similar buildings.

6   Experimental Validation

Preliminary experiments on the proposed framework aim at demonstrating its effectiveness in supporting candidate indicators selection for a given request $r(u)$, formulated in terms of desired domains and/or indicators and according to the user's profile. Usability tests are being performed to check the capability of the framework in facilitating user's access to urban data through the suggestion of candidate indicators. To perform usability tests, we considered a population of users using metrics such as the number of exploration steps needed to obtain desired data, number of fails, number of successful explorations. Usability experiments are being carried on within the Brescia Smart Living project until September 2018. Currently, the framework is being tested, with satisfaction, by a sample of users in two districts, a modern one (San Polino), where new generation smart meters have been installed, and a district in city downtown, more densely populated and presenting older buildings. The framework will be also used by other partners involved in the project as representatives of PA (in particular, the Municipality of Brescia, Italy), utility and energy providers.

For a more technical evaluation of our approach effectiveness, the recall and precision metrics are considered. In particular, given a request $r(u)$, precision $P(r(u))$ (i.e., the number of relevant indicators returned among the search results, compared to the number of returned indicators) and recall $R(r(u))$ (i.e., the number of relevant indicators returned among the search results, compared to the total number of relevant indicators) of the candidate indicators selection step must be as closer to 1.0 as possible. To quantify precision and recall, we used a SLO composed of 57 concepts (and, among them, 30 indicators), 104 individuals, 207 object and datatype properties. Table 1 reports precision and recall values of our approach compared to a keyword-based search, where keywords have been properly expanded with synonyms using WordNet lexical system. Precision ($\overline{P}$) and recall ($\overline{R}$) values have been computed on two kinds of requests: (A) requests where the user specified a set of keywords to identify desired domains and indicators, and the user's profile does not contain any activity or preferential indicator; (B) requests where the user presents a richer profile (containing category, activities and preferential indicators), but specifies a few keywords in the keyword set K_r, that only correspond to individuals of the Domain concept. Five requests for each type have been issued and average values have been computed.

Table 1. Average precision and recall values obtained for the preliminary evaluation

	Keyword-based search	Ontology-based search
$\overline{P}_A$	0.49	0.99
$\overline{R}_A$	0.97	0.98
$\overline{P}_B$	0.33	0.94
$\overline{R}_B$	0.27	0.93

The second type of request is used to demonstrate how relationships within the SLO and the knowledge structure in the ontology are effective in improving precision and recall of indicators selection. In fact, with respect to the keyword-based approach, our framework enables a better precision by refining the set of candidate indicators based on the user's category, the specified domain(s) and other ontological relationships. On the other hand, recall is increased by exploiting the relationships between other elements of the user's profile (i.e., activities and preferential indicators) and the available indicators in the ontology, thus including among search results candidate indicators that are not described with the keywords specified in the user's request or with keyword synonyms as extracted from WordNet. Since both compared approaches use WordNet to perform keywords disambiguation and the same keywords across the approaches have been used during tests, difference in average precision and recall is not due to keyword homonymy or polysemy, but it is influenced by the knowledge structure in the ontology.

The formulation of the request as a set of keywords, instead of asking the user to specify required properties and constraints, enables more flexibility, since it does not require a detailed knowledge of the ontology, its concepts and relationships. Furthemore, processing time required to expand keywords sets with the use of WordNet is affordable and acceptable for the considered exploration scenarios. Response times required by such a procedure has been already tested in other approaches in different contexts, such as [18]. Table 2 contains average execution time (in msec) of the candidate indicators selection algorithm (with and without the time spent for the WordNet-based disambiguation of keywords) for the two kinds of requests we considered in the preliminary evaluation. Tests have been performed on a Windows-based machine equipped with an Intel i7 2.00GHz CPU, 8GB RAM, SSD storage.

Table 2. Average execution time (in msec) for the candidate indicators selection

Request type	Indicators selection	Keyword disambiguation	Total
A	1454	1105	2559
B	269	1056	1325

7 Related Work

In literature, the adoption of ontologies in Smart City projects targets *energy management*, where diagnostic models are built to discover energetic losses [20] or to perform optimisation for cost saving [2,5,8,17]; *facility discovery*, to search for city facilities and services [4,8]; *events monitoring and management* [3,16]; *Ontology-Based Data Access* (OBDA), to cope with heterogeneous data sources inside the Smart City [1,6,7,14]. Here, we focus on a different perspective, by proposing an ontology-based data exploration approach on top of the Smart Living Ontology. Compared to recent efforts, the Smart Living Ontology enables a personalised exploration of data, where semantic relationships and knowledge structure in the ontology are properly exploited. Urban data is aggregated through smart city indicators, that are semantically described to enable data exploration according to users' category (e.g., citizens, building managers, PA) and activities. With respect to an OBDA perspective, our ontology-based Semantic Layer makes possible the exploration of urban data through indicators, also considering the influence that they might have on users' activities. The semantic modeling we pursue also reinforces the characteristics of the BSL project, that if compared to other Smart City projects [2–5] provides a wider spectrum of urban data, considering data on energy consumption, environmental conditions, safety and security. Compared to approaches focused on Ontology-Based Data Warehouses (OBDW), that store analytical data, indicators, requirements and their semantics [10,11,19], our data exploration framework exploits indicators hierarchy and considers users' profiles to enrich exploration over dimensions hierarchies.

The approach presented in [12] focuses on the ontology development phase. It defines a set of high level concepts, mapped to the ones of ontologies underneath twenty Smart City applications. In [15] an ontology that models a Smart City as a composition of information objects, agents and measures is proposed. In [19] a semantic characterisation of Smart City indicators is provided. Differently from the aforementioned solutions, our approach is designed for multiple categories of users, introducing a semantic relationship between users' profiles and urban data collected from the Smart City to foster personalised urban data exploration.

8   Conclusions

In this paper, we described a semantics-enabled framework composed of: (i) a so-called Smart Living Ontology, apt to provide a flexible representation of Smart City indicators; (ii) a Semantic Layer, to enable personalised exploration of urban data for different categories of users. Ontologies represent knowledge structure in terms of concepts, hierarchies and semantic relationships and can be used to facilitate exploration by exploiting the knowledge structure. In particular, the inheritance nature of ontology hierarchies allows exploration at different granularity levels and according to different exploration scenarios. Future efforts will be devoted to extend the set of semantic relationships in the SLO as follows: (a) further relationships between indicators will be identified (e.g., to assert that two or more environmental indicators must be jointly monitored due to their harmful impact on the ecosystem); (b) strategies to promote the users' virtuous behaviours will be studied and implemented on top of the relationships, providing advices for healthy activities that should be practised by users. This will be accomplished by collecting and formalizing additional knowledge about users' lifestyle, and then enriching the SLO with specific background semantics.

Acknowledgments. The BSL consortium is leaded by A2A and includes as partners: Beretta Group, Cauto, Cavagna Group, the Municipality of Brescia, University of Brescia, Enea, STMicroelectronics and an association of private companies (for more information, https://www.bresciasmartliving.eu/).

References

1. The GrowSmarter project. http://www.grow-smarter.eu/home/
2. OPTIMising the energy USe in cities with smart decision support systems. http://optimus-smartcity.eu/
3. The Res Novae project. http://resnovae-unical.eu
4. The San Francisco Park project. http://sfpark.org
5. The BESOS project. Building Energy decision Support systems for Smart cities. http://besos-project.eu
6. The ROMA project. Resilience enhancement of a Metropolitan Area. http://www.progetto-roma.org
7. Bellini, P., Benigni, M., Billero, R., Nesi, P., Rauch, N.: Km4city ontology building vs data harvesting and cleaning for smart-city services. J. Vis. Lang. Comput. **25**(6), 827–839 (2014)
8. Brizzi, P., Bonino, D., Musetti, A., Krylovskiy, A., Patti, E., Axling, M.: Towards an ontology driven approach for systems interoperability and energy management in the smart city. In: International Conference on Computer and Energy Science (SpliTech), pp. 1–7 (2016)
9. Chauhan, S., Agarwal, N., Kar, A.: Addressing big data challenges in smart cities: a systematic literature review. Info **18**(4), 73–90 (2016)
10. Fox, M.S.: PolisGnosis project: representing and analysing city indicators. In: Enterprise Integration Laboratory, University of Toronto Working paper (2015)
11. ISO: Sustainable development of communities - Indicators for city services and quality of life. Standard, International Organization for Standardization (2014)
12. Komninos, N., Bratsas, C., Kakderi, C., Tsarchopoulos, P.: Smart city ontologies: improving the effectiveness of smart city applications. J. Smart Cities **1**(1), 1–16 (2015)
13. Lanza, J., et al.: Managing large amount of data generated by a smart city internet of things deployment. Int. J. Semant. Web Inf. Syst. **12**(4), 22–42 (2016)
14. Lopez, V., Stephenson, M., Kotoulas, S., Tommasi, P.: Data access linking and integration with DALI: building a safety net for an ocean of city data. In: Arenas, M., et al. (eds.) ISWC 2015. LNCS, vol. 9367, pp. 186–202. Springer, Cham (2015). https://doi.org/10.1007/978-3-319-25010-6_11
15. Psyllidis, A.: Ontology-based data integration from heterogeneous urban systems: a knowledge representation framework for smart cities. In: Proceedings of the 14th International Conference on Computers in Urban Planning and Urban Management (2015)
16. Rani, M., Alekh, S., Bhardwaj, A., Gupta, A., Vyas, O.P.: Ontology-based classification and analysis of non-emergency smart-city events. In: 2016 International Conference on Computational Techniques in Information and Communication Technologies (ICCTICT), pp. 509–514 (2016)
17. Rossello-Busquet, A., Brewka, L.J., Soler, J., Dittmann, L.: Owl ontologies and SWRL rules applied to energy management. In: 2011 UkSim 13th International Conference on Computer Modelling and Simulation, pp. 446–450 (2011)

18. Royo, J.A., Mena, E., Bernad, J., Illarramendi, A.: Searching the web: from keywords to semantic queries. In: Third International Conference on Information Technology and Applications (ICITA 2005), pp. 244–249 (2005)

19. Santos, H., Dantas, V., Furtado, V., Pinheiro, P., McGuinness, D.L.: From data to city indicators: a knowledge graph for supporting automatic generation of dashboards. In: Blomqvist, E., Maynard, D., Gangemi, A., Hoekstra, R., Hitzler, P., Hartig, O. (eds.) ESWC 2017. LNCS, vol. 10250, pp. 94–108. Springer, Cham (2017). https://doi.org/10.1007/978-3-319-58451-5_7

20. Tomašević, N.M., Batić, M.Č., Blanes, L.M., Keane, M.M., Vraneš, S.: Ontology-based facility data model for energy management. Adv. Eng. Inform. 29(4), 971–984 (2015)

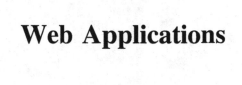

Web Applications

Recommendation for MOOC with Learner Neighbors and Learning Series

Yanxia Pang[1,2(✉)], Chang Liao[3], Wenan Tan[2], Yueping Wu[2], and Chunyi Zhou[1]

[1] East China Normal University, Shanghai, China
yxpang@sspu.edu.cn
[2] Shanghai Polytechnic University, Shanghai, China
[3] Fudan University, Shanghai, China

Abstract. MOOCs (Massive Open Online Courses) have become increasingly popular in recent years. Learning item recommendation in MOOCs is of great significance, which can help learners select the best contents from the huge overloaded information. However, the recommendation is challenging, since there's a high percentage of drop-out due to low satisfaction. Not like traditional recommendation task, learner satisfaction plays an important role in course engagement. The lower the satisfaction is, the higher possibility the learner would drop out the course. Aiming at this, we propose a new recommendation model-Recommendation with learner neighbors and learning series, called RLNLS. It takes achievement motivation on satisfaction into account by exploiting and predicting learning features. A new feature model aiming at satisfaction is proposed according to Expectancy-value Theory. More specifically, knowledge distance is presented to prediction of learning features with learner neighbors and learning series. Hawkes process is modified and utilized for learning intensity prediction. The experimental results on real-world data show the effectiveness of the proposed model in recommending courses and reducing drop-out rate by a large margin.

Keywords: Recommendation · MOOC · Learner neighbors · Learning series Drop-out

1 Introduction

With the explosive use of Internet connection, more learners prefer to learn MOOCs (Massive Open Online Courses). 2012 is known as "the year of MOOC" [1]. Coursera (http://www.coursera.org) is one of the most popular Massive Open Online Course (MOOC) platforms. By January 2018, it has got more than 2000 courses and 25 million learners. Massive educational data is generated for the fast development of MOOCs [2]. MOOC platforms collect data of learning process including learning duration, test scores, remarks, etc. That facilitates further research on learning process.

Meanwhile, it is hard for learners to find suitable courses among massive MOOCs. MOOCs have a high drop-out rate about 90% [3]. Figure 1 is an example of a course on edX (one of the most popular MOOC platform). The vertical ordinate is about the

© Springer Nature Switzerland AG 2018
H. Hacid et al. (Eds.): WISE 2018, LNCS 11234, pp. 379–394, 2018.
https://doi.org/10.1007/978-3-030-02925-8_27

ratio of active learners compared with the number of all learners enrolled in the course. The horizontal ordinate is the week number in the learning process of the course. With the increasing of week number, the ratio value goes down rapidly. At the end of the courses, there are only 10% active learners (http://hdl.handle.net/1842/6683).

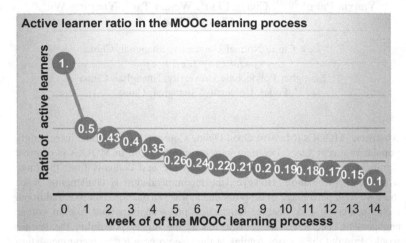

Fig. 1. Ratio of active learners

Therefore, course recommendation is of great significance. The recommendation provides suitable learning items further navigation with MOOCs [4]. MOOC recommendation improves learning insistence. It recommends learning items like courses or videos. Suitable learning items help for better learning performance and learning experience of satisfaction. That encourages learners to keep active and insist on further learning. It makes learning more adaptive and helpful for better learning experience [5].

MOOC platforms usually recommend according to fields of engaged learning objects. However, it is coarsely granular. Meanwhile, Collaborative filtering (CF) is widely used for recommendation. Interest of learner neighbors is usually the main factor in CF [6, 7]. *Learner neighbors* mean learners who are similar to the target learner. CF recommends learning items that learner neighbors engaged in. It supports collaborative learning [8]. Collaborative learning is a learning mode for learners to learn together for better performance.

However, interest is not the only factor influencing learners' insistence on MOOC learning. The high drop-out rate in MOOCs owes a lot to satisfaction lack [9]. Jiang thinks satisfaction is important in recommendation [10]. The achievement motivation is frustrated by low satisfaction.

When a learner starts a MOOC, he usually works well at the beginning. The learner does a good job on video watching, homework and test. The expectation for course completion is high. But with the learning going on, the difficulty value increases and learners tend to feel more frustrated [11]. Frustration leads to lower scores, while low scores lead to less satisfaction and more frustration. That will decrease the learner' insistence on learning. Therefore, interest is not enough for learners to complete the whole course. So, recommendation should take achievement motivation into account for satisfaction.

To fill this gap, this paper builds a new feature model for satisfaction. It is used for recommendation for MOOCs with more features with consideration of satisfaction. Feature prediction is conducted in 2 views for better precision. One is by the view of learning series of the target learner. The other is by the view of learner neighbor. Learning behavior is the learning activity by a learner on the learning item with properties including learning score (learning score is the grade got by the learner in the test of learning item), learning time, etc. Existing prediction methods on learning score are usually content-based. They cannot reflect the evolution of features on learning series. With learning series and learning neighbors, different models are improved according to the features. The self-motivate or self-demotivate effect between learning behaviors are creatively modeled. The detailed technique contributions are as follows:

1. Learning series is formally defined as a series of learning behaviors in sequence of time. Learning series enriches the information for MOOC recommendation. More specifically, learning process is different from commercial item consuming. Past learning behaviors provide prerequisite, interest and performance expectation for future learning. Moreover, improved Hawkes point process is creatively applied to model the learning intensity (*learning intensity* reflects the intensity of learning behaviors).
2. Learner neighbors correspond to the neighbors in collaborative recommendation. Combination of feature prediction with learner neighbors and learning series can improve the accuracy of recommendation result. *Knowledge distance* and *time distance* are combined for feature prediction. To the best of our knowledge, knowledge distance was not considered for MOOC feature prediction before.

The rest of the paper is organized as following: Sect. 2 describes related work on recommendation and score prediction. Section 3 is about preliminary including symbols description and distance definition. Section 4 introduces RLNLS in detail. It is introduced by prediction and combination of predicted features. It includes feature modeling, feature prediction and course recommendation. In Sect. 5, experiment data and result analysis are discussed. Section 6 is conclusion and future work.

2 Related Work

Many real-world MOOC recommender systems are based on the feature of interest [12, 13]. They are usually knowledge-based or machine learning-based. Knowledge-based recommendation makes use of meta information, such as age or gender, to design rules or patterns for recommendations [14]. Machine learning-based recommendation builds models from historical behaviors [15].

Collaborative Filtering is widely used in MOOC recommendation on learner neighbors' interest [8]. It is executed by KNN, SVD, factor model and so on. Interest feature is usually valued with preference 0/1 or favorite stars indicating the interest [13].

Tang recommends on learning objects with an evolving system [16]. Lu analyses the evolution of user profiles for recommendation, because he thinks the interest of users changes with time [17]. But under the background of MOOCs, once a learner starts a course, he will go on with the course until dropping out or completing. The interest of one course doesn't change so fast. For MOOC related recommendation interest does not play such a big role as it does in other fields.

Interest is not enough to profile the learner. Side information or context enriches the data for recommendation. Sun applies association rule mining to recommend courses. Other learners' learning paths are referred for recommendation [18]. Bendakir adds side information like reviews to improve CF recommendation [19]. Mi recommends with Context Trees to transmit the solution into a sequential decision problem [20]. Zhao recommends for MOOC on sequential inter-topic relationships [21]. Chen specifies local similarity between learners to improve collaborative filtering recommendation [22]. This paper combines learner neighbors with learning series to improve collaborative filtering recommendation.

Elbadrawy predicts learning scores by learner grouping [23]. The method in [24] estimates course recommendation values by accumulating weights of subject importance within the study field. Wu predicts subjective problem scores with a fuzzy framework to match learning skills requirements of the problem and skills of the learner [25]. Existing score prediction research does not consider on the knowledge accumulation of past learning behaviors.

By contrast, this paper combines features of learning scores, learning duration and learning intensity for a new feature model for recommendation. They represent more about satisfaction. Features are predicted in views of learner neighbors and learning series. Knowledge distance and time distance are combined into the prediction for better result.

3   Preliminary

In this section, we will describe symbols related and heuristic definition of different types of distance. Related symbols are listed in Table 1.

3.1   Distance Definition

In RLNLS, recommendation value is generated on features. The feature values refer to those of learner neighbors and learning series. Knowledge distance and time distance cause decay of the referred feature values. Related distances include knowledge distance between learning items, correlation distance between a learner and a learning item, and time distance between learning behaviors. According to the experience, they are defined as follows.

Knowledge distance between learning items is defined in hierarchy on related properties. On MOOC platforms, the leaf item of knowledge is usually in the form of video. Each video has properties including knowledge point, subject and stage.

Video is usually the minimum learning objects. It usually focuses on a specific topic point. Knowledge point is the topic of the video. It may be "area of a circle", or

Table 1. Symbols.

Symbol	Description
l	Learner l
i	Learning item i
r	Recommendation value
$time_{l_1 i_1}$	Time of the learning behavior by learner l_1 on item i_1
$rtime$	Recommendation time
w	Weight parameter
kd_{vi}	Knowledge distance between item v and item i
cd_{li}	Correlation distance between learner l and item i
$td_{lv,li}$	Time distance between learning behaviors of learner l on item v and item i
sim_{lu}	Similarity between learner l and u
s_{li}	Learning score of learner l on item i
dv_{li}	Learning devotion of learner l on item i
t_{li}	Learning duration of learner l on item i
$\hat{r}_{li}$	Recommendation value for learner l on item i
$\widehat{cs}_{li}$	Predicted score of learner l on item i with learner neighbors
$\widehat{qs}_{li}$	Predicted score of learner l on item i with learning series
$\widehat{ct}_{li}$	Predicted duration of learner l on item i with learner neighbors
$\widehat{qt}_{li}$	Predicted duration of learner l on item i with learning series
$\hat{h}_l$	Predicted learning intensity of learner l

"application fractions". Subject is the field that the video belongs to like "math". Stage is the learning period of the target learner, like "Grade 4".

For a learner, learning features like learning score and learning devotion between items tend to be similar. Knowledge distance between items influences their similarity. It is defined in different levels with properties considering the influence on learning features. Properties have different influence on similarity. It does not obey the traditional distances. According to experience, knowledge distance is defined in levels according to the influence of properties. In Fig. 2 each square indicates a property. Same square means same related property value. Area outside the square means different property values. Distance between same videos is defined as 0. If two different videos i and i_1 are in the same knowledge point, the knowledge distance between them kd_{i,i_1} is defined as 1. If two videos from different knowledge points are in the same subject and the same stage, their distance is defined as 2. Or else, if they are in the same subject but different stages, the distance between them is defined as 3. Otherwise, two videos belong to different knowledge points. Then if they are in the same stage but different subjects, their distance is defined as 4. If they are neither in the same subject nor the same stage, the distance is defined as 5. Stage influences learning features including learning devotion and learning performance. Lower stage learners tend to get high scores. Higher stage learners tend to get lower scores.

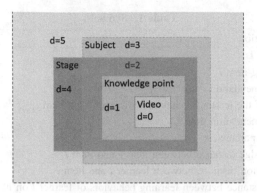

Fig. 2. Knowledge distance

Correlation distance between a learner and a learning item shows how near is the learner's learning pace to the learning item. Correlation distance between a learner l and an item i is defined as the minimal knowledge distance between item i and all items enrolled in by learner l as Eq. (1). $kd_{i,i1}$ is the knowledge distance between learning items i and i_1. $\{i_1, i_2, \ldots\}$ are learning items engaged in by learner l.

$$cd_{li} = \min\left(kd_{i,i_1}, kd_{i,i_2}, \ldots\right) \tag{1}$$

Time distance between learning behaviors is the time span between them, as Eq. (2) shows. $td_{l_1 i_1, l_2 i_2}$ is the time distance between the learning behavior of learner l_1 on item i_1 and the learning behavior of learner l_2 on item i_2. $time_{l_1 i_1}$ is the time of learning behavior by learner l_1 on item i_1.

$$td_{l_1 i_1, l_2 i_2} = |time_{l_1 i_1} - time_{l_2 i_2}| \tag{2}$$

4   Recommendation on Improved Features Predicted with Learner Neighbors and Learning Series

To recommend for better satisfaction, features are modeled in Sect. 4.1. Before recommendation, the features are predicted in Sect. 4.2. On the prediction results of Sects. 4.2, 4.3 combines different features for recommendation.

4.1   Feature Modeling

Recommendation for MOOCs usually adopts the feature of interest [26]. However, that is not enough. Anggraini thinks satisfaction is important for MOOC selection [27]. For better satisfaction, the achievement motivation should be considered in recommendation. Jacquelynne Eccles suggests learners' achievement motivation is most proximally determined by two factors: expectancies for success and subjective task value of the learning item. The achievement motivation can be calculated as (3) shows [28].

$$Motivation = Expectancy * Value \tag{3}$$

Expectancy refers to how confident an individual is in his or her ability to succeed in a task whereas task values refer to how important, useful, or enjoyable the individual perceives the task. That is usually influenced by the learning scores of past experiences. When a learner thinks a task is useful, or enjoyable, he will devote more on the task. The devotion indicates task value cherished by the learner. Applying the selected features to (3), we get Eq. (4). The recommendation value $\hat{r}_{li}$ is calculated with learning score s_{li} and learning devotion dv_{li}.

$$\hat{r}_{li} = s_{li} * dv_{li} \tag{4}$$

Devotion means the cost one pays on the task. Learning duration t_{li} and learning intensity h_{li} can be multiplied to indicate the devotion as (5). dv_{li} in (4) is replaced with (5). The recommendation value can be calculated as (6)

$$\widehat{dv_{li}} = t_{li} * h_{li} \tag{5}$$

$$\hat{r}_{li} = s_{li} * t_{li} * h_{li} \tag{6}$$

As Fig. 3 shows, learning score and learning devotion are adopted as features for recommendation on satisfaction. The devotion can be calculated by learning intensity and learning duration on time. Features of *learning score, learning duration and learning intensity* reflect more on achievement motivation than interest only. RLNLS (Recommendation with Learner Neighbors and Learning Series) predicts achievement motivation of learners' next behavior on the feature model. It reflects both learners' interest and capacity. To the best of our knowledge, current work on recommendation has not considered the capacity of learners. The achievement motivation has not been discussed for better learning experience in recommendation.

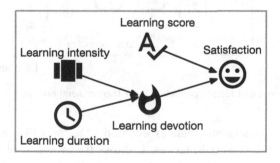

Fig. 3. The feature model of RLNLS

4.2   Feature Prediction

Before recommending on the learning features, RLNLS needs predict the features first. According to feature model of Sect. 4.1, learning score, learning duration and learning intensity are predicted. Section 4.2.1 introduces the framework of the 2 views of prediction of learner neighbors and with learning series.

4.2.1   Framework of Prediction with Learner Neighbors and Learning Series

RLNLS predicts in 2 views. One is the view of time. It is by the target learner's learning series. With learning series, the evolution of learning features on time can be modeled. The interaction between learning behaviors is combined. The other is the view of neighbors. It is performed with CF on similar learners. Similar neighbors' preference reflects more on the difference between learning objects.

Figure 4 demonstrates the idea of 2 views. Circles indicate learning behaviors. Features of the target learning behavior on time t of learner l_2 is to be predicted. Solid line and dashed line arrows point at the target behavior. The solid line comes from previous learning behavior of the same learner. The dashed lines come from learning behaviors of learner neighbors.

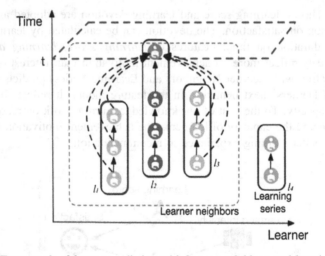

Fig. 4. Framework of feature prediction with learner neighbors and learning series

Past learning behaviors contribute expectation of features like learning score and learning devotion to following behaviors. It shows the evolution of learning features with time. Because learning series focuses more on the target learner, the predicted values by learning series do not differentiate well between learning items. Meanwhile, prediction on learner neighbors fixes this problem. Learner neighbors help differentiate feature values between learning items.

For all the predictions, the contribution from learner neighbors or learning series may decay with the knowledge distance or time distance. The influence of past learning behaviors varies with distance between them [29]. Especially contribution of past learning behaviors is time-dependent. Some features' weight of contribution varies with the knowledge distance between learning items.

In the following, Sect. 4.2.2 is about prediction for learning score and learning duration on learning series. Section 4.2.3 is about prediction on learner neighbors for features including learning score and learning duration. Section 4.2.4 is about calculation for learning intensity.

4.2.2 Prediction for Learning Score and Learning Duration with Learning Series

Past learning behaviors of the learning series contribute expectation of features to the following learning behavior. To predict features of learning score and learning duration with learning series, RLNLS improves different models of time series for the features.

Prediction on learning score with Inverse Distance Weighting (IDW) is conducted as follows. Past learning scores indicate the ability of learning for the target learner. They can be linearly combined to predict the target behavior's score. Past learning scores' contribution decays with the knowledge distance. If the knowledge distance between the related item and the target item is far, the score will have little weight for contribution. So, knowledge distance is adopted as the inverse weight of score.

Inverse Distance Weighting (IDW) [30] is usually employed to model the inverse correlation between distance and weight. Therefore, RLNLS adopts IDW for learning score prediction with learning series. IDW assigns weights to learning behaviors of the learning series. Learning scores are aggregated with IDW as Eq. (7) shows. kd_{vi} is the knowledge distance between item v and item i, and α is a positive power parameter that controls the decay rate of weight with kd_{vi}. Bigger α denotes a faster decay of weight by knowledge distance. $kd_{vi}^{-\alpha}$ assigns a bigger weight to closer item's score.

$$\widehat{s_{li}} = \frac{\sum_{v=0}^{V} s_{lv} * kd_{vi}^{-\alpha}}{\sum_{v=0}^{V} kd_{vi}^{-\alpha}} \tag{7}$$

Prediction on learning duration with Simple Exponential Smoothing (SES) is applied with time distance consideration. SES is frequently used for time series as an exponential moving average model [31]. Contribution of past learning durations decays with time. So, SES is adopted to model the contribution of past decayed learning durations. Equation (8) shows learning duration calculation with SES. td_{lv} is the time distance between the recommendation time and the time of past behavior of learner l on item v. β is a smoothing parameter between (0, 1). The bigger β is, the less weight the farther behavior will have. Usually in SES the time distance is set by difference between indexes of the behaviors on the series. Here Time distance is modified as real day difference for better representation of the time distance. Finally, normalization is performed by dividing the weight sum as Eq. (8) shows.

$$\widehat{qt}_{li} = \frac{\sum_{v=0}^{vs} t_{lv} * \beta * (1 - \beta)^{-td_{lv}}}{\sum_{v=0}^{vs} \beta * (1 - \beta)^{-td_{lv}}} \tag{8}$$

The algorithm is illustrated as follows. For each behavior of the target learner, the related time duration is aggregated weightly.

Algorithm 1. SES for learning feature prediction with learning series

---

Input: learning durations t, time points $time$, learner l, item v;
Output: predicted learning duration qt;

1 $qt = 0; den = 0$;
2 For each v among learning items of the target learner l
3   Get t_{lv} of the target learner l on learning item v;
4   Get $time_{lv}$ of learner l on item v:;
5   Get the recommendation time $rtime$;
6    $td_{lv}=rtime - time_{lv}$; //time distance
7    $qt += t_{lv} * \beta * (1 - \beta)\wedge(-td_{lv})$;
8    $den += \beta * (1 - \beta)\wedge(-td_{lv})$;
9 End for
10 $qt = qt/den$;
11 Return qt;

4.2.3 Prediction for Learning Score and Learning Duration with Learner Neighbors

Prediction with learning series shows the feature expectation evolution of the target learner. History items of the target learner are often in same knowledge section, so they have the same knowledge distance with the candidate item. Difference between candidates is not obvious. Learner neighbors' behaviors enrich additional context for that. Collaborative filtering (CF) is proved an effective prediction method on learner neighbors.

CF based learning duration prediction with learner neighbors is conducted to enrich information about different items. CF reflects different features on items according to those of learner neighbors. The learning duration of the target behavior $\widehat{ct}_{li}$ is predicted by summing up learner neighbors' learning duration. Neighbors' learning durations are weighted by similarity between the target learner and the learner neighbors.

Learning duration t_{ui} reflects the interest or the value held by learner u on item i. Learning interest decays with time on. The time distance makes the contribution of the related learning duration decay. It is inversely proportional to the weight of the contribution. The longer the time distance is, the more the contribution decays, and the less weight it should have for prediction. td_{ul} is the time distance of learning behaviors between the target learner l and the similar learner u on item i. Learning duration is predicted with CF as the following equation shows.

$$\widehat{ct_{li}} = \frac{\sum_{u=0}^{U} t_{ui} * sim_{lu}/td_{ul}}{\sum_{u=0}^{U} sim_{lu}/td_{ul}} \tag{9}$$

CF based learning score prediction with learner neighbors is conducted with knowledge distance consideration. Similar capacity learners tend to have similar learning scores $\widehat{cs_{li}}$. CF is adopted to predict the score by aggregating learner neighbors' scores with similarity as the weight.

Correlation distance cd_{ui} between the learner neighbor u and the candidate item i is adopted as weight punishment. It is inversely proportional to the contribution of learning scores of learner neighbors. Correlation distance is combined for score predication as following equation shows.

$$\widehat{cs_{li}} = \frac{\sum_{u=0}^{U} s_{ui} * sim_{lu}/cd_{ui}}{\sum_{u=0}^{U} sim_{lu}/cd_{ui}} \tag{10}$$

4.2.4 Hawkes Process for Learning Intensity Calculation

Learning intensity reflects the number of learning behaviors in a unit time. Hawkes process is used for discrete time series [32]. It has the characteristic of self-motivate. Event points motivate more following event. The motivate effect decays with time.

Learning score also has the effect of motive or demotivate on following learning. Better performance leads to more learning behaviors. Learning intensity will be motivated. On the contrary, worse performance will decrease the learning intensity. Learning score has the effect of self-demotivate on learning series. The influence of learning score usually goes down with time on. So, learning process obeys the rule of Hawkes process. Hawkes process is creatively improved to model the motivate and demotivate effect of learning score.

Equation (11) improves Hawkes process for learning intensity calculation. s_{lv} is the score of learner l on item v. If it is more than 50 points, the learner may feel encouraged to try more learning behaviors. On the contrary, it is demotivating for further learning behaviors. td_{lv} is the time distance between the past learning behavior about item v and the recommendation. Exponential function is adopted as the kernel function to model the decay with time. Initial intensity is set zero without loss of generality.

$$\widehat{h_l} = \sum_{v=0}^{V} (s_{lv} - 50) * e^{-|td_{lv}|} \tag{11}$$

The algorithm is illustrated as following. For each learning behavior of the target learner, the time distance and learning score are used for intensity calculation.

Algorithm 2. Improved Hawkes Point Process for learning
intensity

Input: learner score *s*, time point *time*, learner *l*, item *v*;
Output: learning intensity h_l;

1 $h_l = 0$;
2 For each *v* in learning items of the target learner *l*
3   Get s_{lv} of the target learner *l* on item *v*;
4   Get $time_{lv}$ of the target learner *l* on item *v*;
5   Get the recommendation time *rtime*;
6   td_{lv}=rtime-$time_{lv}$;//time distance
7   h_l+= (s_{lv}-50) *exp (-td_{lv});
8 End for
9 Return h_l;

4.3   Recommendation on Predicted Features in Different Views

For learning score s_{li} and learning duration t_{li}, they are both predicted in the 2 views. The predicted values of the 2 views are regressed linearly as Eqs. (12) and (13). The weight parameters are deduced by minimizing the ordinary least square (OLS) [33] between the prediction and the ground truth. For learning score prediction, the optimization function is listed in Eq. (14). The values of w_1 and w_2 are achieved by minimizing $J(w)$. So are values of w_3 and w_4.

$$\widehat{s_{li}} = w_1 * \widehat{cs_{li}} + w_2 * \widehat{qs_{li}} \tag{12}$$

$$\widehat{t_{li}} = w_3 * \widehat{ct_{li}} + w_4 * \widehat{qt_{li}} \tag{13}$$

$$J(w) = \sum_{all(l,i)} \left(\widehat{s_{li}} - s_{li} \right)^2 \tag{14}$$

On the feature model, RLNLS calculates achievement motivation for recommendation as Eq. (6) shows. With Eqs. (12) and (13), Eq. (6) is transmitted to Eq. (15).

$$\widehat{r_{li}} = s_{li} * t_{li} * h_{li} = \left(w_1 * \widehat{cs_{li}} + w_2 * \widehat{qs_{li}} \right) * \left(w_3 * \widehat{ct_{li}} + w_4 * \widehat{qt_{li}} \right) * \widehat{h_l} \tag{15}$$

5   Experiment

5.1   Dataset

RLNLS is tested on learning data from the Mic-video Platform of ECNU (http://jclass. pte.sh.cn). For the consideration of privacy, most MOOC platforms do not provide their data for the public. Mic-video Platform records learning data in detail. It provides information necessary for the experiment. The experiment is finally on 7163 learning behavior records, 686 learners and 136 video items. The dataset is split into train set and test set by time sequence to model the real situation. The earlier part is adopted as train set. The recommendation time adopts a little later than last recorded behavior in the train set.

Figure 5 shows the distribution of learning behaviors on learners as horizontal axis and videos as vertical axis. Points indicate the learning behaviors. In the left part of Fig. 5, different colors of the left part are the clustering results. In the right part different colors indicate actual subjects.

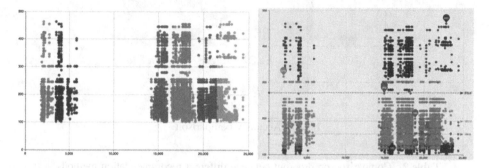

Fig. 5. Distribution of learning behaviors

5.2 Comparison on Accuracy of Recommendation

Figure 6 is about recommendation accuracy comparison between different methods. It is measured by Precision, Recall and F1-score. IDWS has good performance on accuracy than SEST. IDWS is recommendation with IDW method on learning score. SEST is recommendation with SES on learning duration. CF Recommendation on learning score CFS shows better performance than that on learning duration CFT. HawInt is the recommendation on learning intensity. RLNLS outperforms other methods in Precision, Recall and F1-score.

5.3 Comparison on Improvement of Drop-Out

Table 2 measures the drop-out rate after recommendation. It is hard to make sure that a learner has dropped out, because the learner may be back long time later. We adopt timeDrop, rateDrop, timeSpan and rateSpan as measurements of drop-out. timeDrop and rateDrop are about time length after the last recorded learning behavior of the target learner. timeDrop is about the time length. And timeRate is the rate of improvement compared with average timeDrop of all learners. timeSpan and rateSpan are about the time span between the last learning behavior before recommendation and the first behavior after recommendation. rateSpan is the rate of improvement compared with average timeSpan of all learners. RCFS shows obvious improvement on drop-out.

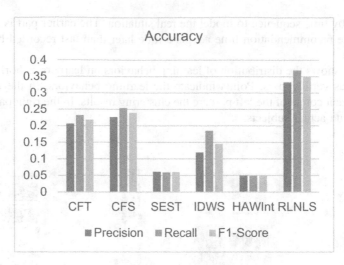

Fig. 6. Accuracy comparison

Table 2. Comparison on drop-out between different recommendation methods.

Recommendation	CFS	CFT	RLNLS
timeDrop	1368.13	1470.5	1309
rateDrop	0.2607	0.2053	0.2410
timeSpan	113.65	105.64	40
rateSpan	0.0607	0.1269	0.6694

6 Conclusion

In summary, this paper proposes a recommendation solution for MOOCs called RLNLS. It aims to decrease the drop-out rate with recommendation. Compared with traditional MOOC recommendation based on interest of learners, RLNLS builds a feature model combining satisfaction. According to Expectancy-value theory, interest is not enough for satisfaction motivation. Learning score, learning duration and learning intensity are adopted as features for recommendation. They reflect more on achievement satisfaction than interest. It helps for less frustration in case of drop-out.

From technique perspective, features are predicted in views of learner neighbors and learning series. Learner neighbors' features reflect more on difference between learning items. Learning series' behaviors reflect the evolution of learning features about the target learner. The combination of the 2 views' prediction helps for better result of recommendation. Traditional MOOC recommendations are usually content-based or neighbor-based. Few of them consider the evolution of features on learning series. Time series models are improved for different feature prediction. Especially Hawkes point process is improved to model the self-motivate and self-demotivate effect for future learning behaviors. It is a creative work for educational recommendation.

Knowledge distance and time distance are creatively combined for feature weight decay. That increases the accuracy of recommendation.

From empirical study on real dataset, it verifies the effectiveness of RLNLS. Accuracy comparison on recommendation is conducted. The result shows the combination of predicted features with learner neighbors and learning series is closer to ground truth. And RLNLS has better accuracy in recommendation than other methods. Especially in drop-out rate decreasing, RLNLS makes obvious improvement.

As for future work, learning is a complicated activity. Lots of factors may influence it. Limited by the data, only some of the most related features for recommendation are selected. More features will enrich the feature model and improve the recommendation further.

References

1. Pappano, L.: The year of the MOOC-the New York Times (2012). http://www.nytimes.com/2012/11/04/education/edlife/massive-open-online-courses-are-multiplying-at-a-rapid-pace.html
2. The White House: Challenges and Opportunities with Big Data, a Community White Paper (2013)
3. Breslow, L., Pritchard, D.E., DeBoer, J., et al.: Studying learning in the worldwide classroom: research into edX's first MOOC. Res. Pract. Assess. **8**, 13–25 (2013)
4. Didou, Y., Khaldi, M.: Using recommendation systems in MOOC. Enhancing Knowl. Discov. Innov. Digit. Era 176 (2018)
5. Onah, D.F.O., Sinclair, J.: Massive open online courses: an adaptive learning framework. In: INTED2015 Proceedings, pp. 1258–1266 (2015)
6. Konstan, J.A., Miller, B.N., Maltz, D., et al.: GroupLens: applying collaborative filtering to usenet news. Commun. ACM **40**(3), 77–87 (1997)
7. Goldberg, D., Nichols, D., Oki, B.M., et al.: Using collaborative filtering to weave an information tapestry. Commun. ACM **35**(12), 61–70 (1992)
8. Toscher, A., Jahrer, M.: Collaborative filtering applied to educational data mining, KDD cup (2010)
9. Halawa, S., Greene, D., Mitchell, J.: Dropout prediction in MOOCs using learner activity features. Exp. Best Pract. MOOCs 7 (2014)
10. Jiang, Y., Shang, J., Liu, Y.: Maximizing customer satisfaction through an online recommendation system: a novel associative classification model. Decis. Support Syst. **48**(3), 470–479 (2010)
11. Kizilcec, R.F., Piech, C., Schneider, E.: Deconstructing disengagement: analyzing learner subpopulations in massive open online courses. In: Third International Conference on Learning Analytics and Knowledge, LAK 2013 Leuven, Belgium (2013)
12. Ray, S., Sharma, A.: A collaborative filtering based approach for recommending elective courses. CoRR, abs/1309.6908 (2013)
13. Denley, T.: Course recommendation system and method. US Patent Application 13/441,063 (2013)
14. Trewin, S.: Knowledge-based recommender systems. Encycl. Libr. Inf. Sci. **69**(Suppl. 32), 180 (2000)

15. Adomavicius, G., Tuzhilin, A.: Toward the next generation of recommender systems: a survey of the state-of-the-art and possible extensions. IEEE Trans. Knowl. Data Eng. **17**(6), 734–749 (2005)
16. Tang, T.Y., McCalla, G.: Smart recommendation for an evolving e-learning system: architecture and experiment. Int. J. Elearning **4**(1), 105 (2005)
17. Lu, Z., Pan, S.J., Li, Y., et al.: Collaborative evolution for user profiling in recommender systems. In: IJCAI pp. 3804–3810 (2016)
18. Sun, Y., Han, J., Yan, X., et al.: PathSim: meta path-based top-k similarity search in heterogeneous information networks. Proc. VLDB Endow. **4**(11), 992–1003 (2011)
19. Bendakir, N., Aïmeur, E.: Using association rules for course recommendation. In: Proceedings of the AAAI Workshop on Educational Data Mining, vol. 3 (2006)
20. Mi, F., Faltings, B.: Adaptive sequential recommendation using context trees. In: IJCAI, pp. 4018–4019 (2016)
21. Zhao, J., Bhatt, C., Cooper, M., et al.: Flexible learning with semantic visual exploration and sequence-based recommendation of MOOC videos. In: Proceedings of the 2018 CHI Conference on Human Factors in Computing Systems, p. 329. ACM (2018)
22. Chen, Y., Zhao, X., Gan, J., Ren, J., Hu, Y.: Content-based top-N recommendation using heterogeneous relations. In: Cheema, M.A., Zhang, W., Chang, L. (eds.) ADC 2016. LNCS, vol. 9877, pp. 308–320. Springer, Cham (2016). https://doi.org/10.1007/978-3-319-46922-5_24
23. Elbadrawy, A., Karypis, G.: Domain-aware grade prediction and top-N course recommendation. In: Proceedings of the 10th ACM Conference on Recommender Systems, pp. 183–190. ACM (2016)
24. Polyzou, A., Karypis, G.: Grade prediction with course and student specific models. In: Bailey, J., Khan, L., Washio, T., Dobbie, G., Huang, J.Z., Wang, R. (eds.) PAKDD 2016. LNCS (LNAI), vol. 9651, pp. 89–101. Springer, Cham (2016). https://doi.org/10.1007/978-3-319-31753-3_8
25. Wu, R., Liu, Q., Liu, Y., et al.: Cognitive modelling for predicting examinee performance. In: IJCAI, pp. 1017–1024 (2015)
26. Lee, Y., Cho, J.: An intelligent course recommendation system. Smart CR **1**(1), 69–84 (2011)
27. Anggraini, A., Tanuwijaya, C.N., Oktavia, T., et al.: Analyzing MOOC features for enhancing students learning satisfaction. J. Telecommun. Electron. Comput. Eng. (JTEC) **10**(1–4), 67–71 (2018)
28. Eccles, J.: Expectancies, values, and academic behaviors. In: Spence, J.T. (ed.) Achievement and Achievement Motives: Psychological and Sociological Approaches, pp. 75–146. W. H. Freeman, San Francisco (1983)
29. Husman, J., Lens, W.: The role of the future in student motivation. Educ. Psychol. **34**(2), 113–125 (1999)
30. Lu, G.Y., Wong, D.W.: An adaptive inverse-distance weighting spatial interpolation technique. Comput. Geosci. **34**(9), 1044–1055 (2008)
31. Gardner, E.S.: Exponential smoothing: the state of the art - Part II. Int. J. Forecast. **22**(4), 637–666 (2006)
32. Hawkes, A.G.: Spectra of some self-exciting and mutually exciting point processes. Biometrika **58**(1), 83–90 (1971)
33. Li, D., et al.: ECharts: a declarative framework for rapid construction of web-based visualization. Vis. Inform. (2018)

In-depth Exploration of Engagement Patterns in MOOCs

Lei Shi[1(✉)] and Alexandra I. Cristea[2]

[1] University of Liverpool, Liverpool, UK
lei.shi@liverpool.ac.uk
[2] Durham University, Durham, UK
alexandra.i.cristea@durham.ac.uk

Abstract. With the advent of 'big data', various new methods have been proposed, to explore data in several domains. In the domain of learning (and e-learning, in particular), the outcomes lag somewhat behind. This is not unexpected, as e-learning has the additional dimensions of learning and engagement, as well as other psychological aspects, to name but a few, beyond 'simple' data crunching. This means that the goals of data exploration for e-learning are somewhat different to the goals for practically all other domains: finding out what students do is not enough, it is the means to the end of supporting student learning and increasing their engagement. This paper focuses specifically on student engagement, a crucial issue especially for MOOCs, by studying in much greater detail than previous work, the *engagement of students based on clustering students according to three fundamental* (and, arguably, comprehensive) *dimensions: learning, social and assessment*. The study's value lies also in the fact that it is among the few studies using *real-world longitudinal data* (6 runs of a course, over 3 years) from a *large number of students*.

Keywords: E-learning · Learning analysis · Behavioral analysis
Clustering analysis · K-means · MOOCs

1 Introduction

"Educators, theorists, and policymakers alike tout engagement as a key to addressing educational problems, such as low achievement and escalating dropout rates" [16]. However, it is a known fact that, in many online environments, and especially in MOOCs, due to "the absence of teacher supervision and opportunities to provide direct feedback, students may lack opportunities to control and interact with a learning environment" [12]. Thus, engagement of the students is a vital target for analysis as well as enhancement, especially for MOOCs. This paper proposes an *in-depth method for exploration of engagement patterns in MOOCs*, based on clustering of students according to *three fundamental* (and, arguably, comprehensive) *dimensions: learning, social and assessment*. This is, to our knowledge, the first study proposing such an in-depth engagement exploration based on clustering (here, the widely popular k-means clustering method is used). Additionally, the study analyses *real-world data* from a *longitudinal data collection* of 6 runs of a course, between 2015–2017, with a *large*

© Springer Nature Switzerland AG 2018
H. Hacid et al. (Eds.): WISE 2018, LNCS 11234, pp. 395–409, 2018.
https://doi.org/10.1007/978-3-030-02925-8_28

number of students (48,698), on a MOOC platform that has seen less exploration, albeit rooted in pedagogical principles, unlike some of its competitors, namely, *FutureLearn* (www.futurelearn.com).

The remainder of the paper contains first related work, after which Sect. 3 presents our methodological approach; next, Sect. 4 reports the results of the in-depth analysis. We briefly discuss findings and conclude in Sect. 5.

2   Related Work

Student engagement has many definitions, depending on the perspective it is employed from. Similarly, there is no clear collective understanding on the methods of monitoring and measuring student engagement [7, 16]. An interesting way of defining student engagement is that based on the Flow theory [4]. Recent highly-cited work under this umbrella [13] analyses student engagement defined in terms of concentration, interest and enjoyment. The work however focusses on self-reporting from a relatively small number of US high-school students, unlike our study. Recent research on engagement in MOOCs [6] defines engagement in terms of length of time of video-watching, as well as existence of problem solving attempts, parameters which are related to our current study.

A special issue on the subject discusses also the variety in grain-size of the measurement tools [2, 3, 15], starting from microlevel (e.g., individual's engagement in the moment, task or learning activity) versus macrolevel (as in groups of learners), with measurements for the former such as brain imaging, eye tracking, etc., and for the latter, discourse analysis, observations, ratings, etc. [16] They also note that, whilst they categorize engagement as behavioral, cognitive, emotional and agentic, the motivational and self-regulated constructs run through each of these dimensions. Behavioral engagement, targeted also by the current study, has been shown in the past to be related to achievement in learning [11, 14]. [16] cautions however that higher-order processing (such as exams or strategic thinking tasks) might not be well caught by behavioral engagement. However, we argue here that MOOCs don't provide normally exams per se, and that the type of tests at the end of a MOOC often emulate the level of the tasks given during the MOOC (including the inclusion or exclusion of higher-order processing tasks).

A relatively recent review of measurement methods for student engagement [7] concludes that, whilst most technology-mediated learning research uses self-report measures of engagement, in fact, physiological and systems data offer an alternative method to measuring engagement, and that more research is needed in this area, including determining the system data and values needed for the engagement evaluation. Our in-depth longitudinal study builds upon these recommendations and proposes novel ways of analyzing and measuring the student engagement in MOOCs.

3   Method

3.1   MOOC Settings and the Dataset

Each MOOC on FutureLearn is hierarchically structured into *weeks*, *activities* and *steps*. A *week* may contain several *activities* and an *activity* may contain several *steps*. A *step* is a basic learning unit which may be an article, a video with or without a discussion (comment) list. A *step* may also be a quiz which consists of a set of questions.

The MOOC presented in this study consists of 4 *weeks*. Each *week* contains 4 *activities* and each *activity* contains between 2 and 8 *steps*. In total, there are 18 *steps* in Week 1, 22 *steps* in Week 2, 15 *steps* in Week 3 and 19 *steps* in Week 4. Thus, in total, there are 74 *steps* in the MOOC. The last *activity* of each *week* contains a 'quiz type' of *step*. Each quiz has 5 questions, so there are 20 questions in total in the MOOC. Each *step*, except the 'quiz type' of *step*, provides a discussion board where students can submit comments, and 'like' (as in social network apps, e.g. Weibo) each other's comments. Each *step*, except the 'quiz type' of *step*, also provides a "Mark as complete" button for students to claim that they have learnt the *step*.

The MOOC ran 6 times between 2015 and 2017. The total number of students we analyzed is of 45,321, divided between the six runs as 12,628, 9,723, 7,755, 6,218, 8,432 and 3,942. However, 3,377 students unenrolled from the MOOC. Next, 20,532 did not visit any *step*, being thus passive. Therefore, after filtering these out, in total, 24,798 students are considered in this study.

The dataset collected on FutureLearn platform contains behavioral information including visiting a *step*, marking a *step* as completed, submitting a comment, 'like'ing a comment, and attempting to answer a question in a quiz. These 5 types of behavioral information are all considered in this study. Besides, each time a student attempts to answer a question, FutureLearn records if the answer is correct or incorrect. This is the additional data used in this study.

3.2   Clustering and Fundamental Dimensions

With complex dataset, we empirically explore student engagement patterns without relying on predefined classes. Clustering, an unsupervised machine learning method, can uncover new relationships in a complex dataset, and has been used to develop profiles that are grounded in student behaviors [1]. In this study, k-means [10], a well-known non-hierarchical clustering method, is used to partition students into different clusters. This is essential, as it provides insights into engagement patterns caused by the diversity of students, as well as opportunities to compare these patterns and predict behaviors. K-means requires k, the number of clusters, as a parameter, but determining this parameter is known to be a challenging issue. One way to determine an optimal k is the "elbow method", which relays on visually identifying the "elbow point" of a curve drawn on a line chart, but the problem is that this "elbow" cannot always be unambiguously identified, and sometimes there is no elbow or are several elbows [8]. In our case, indeed, we were unable to identify a conclusive k, but instead obtained several interesting clustering options. Besides, a k-means algorithm normally favors higher values of k, but the latter is not necessarily desirable, as it is very important to consider a more sensible k for the nature of the dataset. It is common to run k-means clustering a few times with 3, 4 or 5 as the k, and compare the results, to determine which is the "final optimal" k to use [1]. In this study, we start with $k = 2$, with further increments of 1.

The clustering is based on three fundamental dimensions, namely, *learning*, *social* and *assessment*. In terms of how we determine these three fundamental dimensions, firstly, the core of using FutureLearn (or any other e-learning platforms) is to learn. Thus, to explore engagement patterns, it is essential to investigate how students access learning content. On FutureLearn, basic learning units of a MOOC are *steps*. Therefore, we consider how students visit *steps* as the first dimension, and we label it as 'learning'. Secondly, FutureLearn employs a social constructivist approach inspired by Laurillard's Conversation Framework [9], which describes a general theory of effective learning through conversation (or social interaction). Therefore, we consider how students interact with each other as the second dimension, and label it as 'social'. Thirdly, FutureLearn, as an xMOOCs platform, consider both content and assessment as essential elements of the teaching and learning process [5]. Therefore, we consider how students attempt to answer questions in quizzes as the third dimension, and label it as 'assessment'. Regarding the parameters for the k-means algorithm, we use: (1) (the number of visited) *steps* to represent the first dimension – learning; (2) (the number of submitted) comments to represent the second dimension – social; and (3) (the number of) attempts (to answer questions in quizzes) to represent the third dimension – assessment.

4   Results

4.1   Clustering and Validation

Firstly, we tested $k = 2$ (two clusters) for the k-means analysis. The convergence was achieved in the 17th iteration. The final cluster centers (Fig. 1) show that on the standardized scale, all the three variables, i.e., Zscore(steps), Zscore(comments) and Zscore(attempts), of cluster I are higher than those of cluster II. This suggests that students who were allocated in cluster I may be more engaged in the learning, in terms of visiting steps, submitting comments (discussions) and attempting to answer questions in quizzes, than those who were allocated in cluster II.

A one-way ANOVA test was conducted to compare the relative weight of Zscore(steps), Zscore(comments) and Zscore(attempts) in the clustering process. The result shows very large F scores ($F_{steps} = 85,155.321$; $F_{comments} = 4,441.474$; $F_{attempts} = 90,965.767$), and very small p values ($p_{steps} < .001$, $p_{comments} < .001$, $p_{attempts} < .001$) indicating that all three variables have a statistically significant impact on determining the clustering, and that the variables of Zscore(steps) and Zscore(attempts) have stronger impact than the variable of $F_{comments}$ ($F_{steps} \approx F_{attemps} \gg F_{comments}$, $F_{steps} \approx F_{attemps} \approx 22 \times F_{comments}$).

Secondly, we tested $k = 3$ (three clusters) for the k-means analysis. The convergence was achieved in the 16th iteration. We can see from Fig. 2, the final cluster centers, that similar to the k-means analysis result using $k = 2$, the majority of students

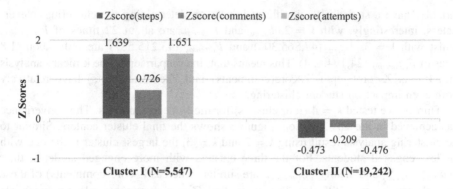

Fig. 1. Comparisons of standardized numbers of steps, comments and attempts between two clusters (k-means analysis, when k = 2)

(18,998, 69.52%) were allocated in cluster II, and they were less engaged in terms of visiting steps, submitting comments (discussions) and attempting to answer questions in quizzes, than those who were allocated in cluster I and cluster III. Interestingly, Fig. 2 also shows that despite $mean_{Zscore(steps)}$ and $mean_{Zscore(attempts)}$ of cluster I and cluster III are similar, $mean_{Zscore(comments)}$ of cluster I is much smaller than that of cluster III. This suggests that among the students who were more engaged, those who were allocated in cluster II might submit much more comments (discussions) than those who were allocated in cluster III.

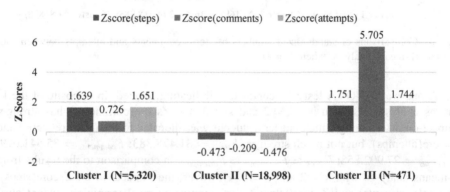

Fig. 2. Comparisons of standardized numbers of steps, comments and attempts between three clusters (k-means analysis, when k = 3)

A one-way ANOVA test was performed to compare the relative weight given to Zscore(steps), Zscore(comments) and Zscore(attempts) in order to determine which cluster a student was allocated to. We find from the ANOVA test result that, similar to k-means analysis with $k = 2$, the F scores of all three variables are very large ($F_{steps} = 47,865.306$; $F_{comments} = 24,194.624$; $F_{attempts} = 43,215.587$, and their p values are very small ($p_{steps} < .001$, $p_{comments} < .001$, $p_{attempts} < .001$) indicating all these three variables have a statistically significant impact on determining the clustering. Nevertheless, interestingly, with k = 2, F_{steps} and $F_{attemps}$ are about 22 times of $F_{comments}$; whilst with k = 3, F_{steps} (47,865.306) and $F_{attemps}$ (43,215.587) are only about 1.8 times of $F_{comments}$ (24,194.624). This means that, in comparison to the k-means analysis with k = 2, Zscore(steps), Zscore(comments) and Zscore(attempts) have relatively more even impact on student clustering.

Thirdly, we tested $k = 4$ (four clusters) for the k-means analysis. The convergence was achieved in the 18th iteration. Figure 3 shows the final cluster centers. Similar to the clustering analysis result using $k = 2$ and $k = 3$, the largest cluster is the one with the less engaged students. For the three clusters with more engaged students, their $mean_{Zscore(steps)}$ and $mean_{Zscore(attempts)}$ are similar, whereas Zscore(comments) of these three clusters are very different. This means that, Zscore(comments) plays a major role of allocating engaged student to different clusters.

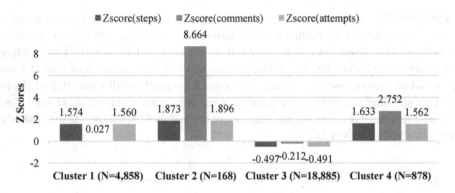

Fig. 3. Comparisons of standardized numbers of steps, comments and attempts between four clusters (k-means analysis, where k = 4)

A one-way ANOVA test was conducted, indicating, indeed, in opposite of the k-means analyses results using $k = 2$ and $k = 3$, the Zscore(comments) has stronger impact on determining which cluster a student was allocated to than Zscore(steps) and Zscore(attemps), but not much stronger ($F_{steps} = 31,478.383$; $F_{comments} = 35,541.079$; $F_{attempts} = 27,902.858$; $F_{steps} \approx F_{comments} \approx F_{attempts}$), in comparison to the results from k-means analyses with $k = 2$ and $k = 3$. However, the impacts of all Zscore(steps), Zscore(comments) and Zscore(attempts) are significant on determining student clustering ($p_{steps} < .001$, $p_{comments} < .001$, $p_{attempts} < .001$).

We also tested $k \in \{n|5 \le n \le 8\}$. However, the convergence could not be achieved within 20 iterations. Because there are only three variables, i.e. Zscore(steps), Zscore(comments) and Zscore(attempts), we determined that the largest possible number of clusters should be $k \in \{n|n \le 2^3 = 8\}$. Therefore, we decided to discard the options where k ≥ 5.

Overall, with $k \in \{n|2 \le n \le 4\}$, all k-means analysis results suggest a division between more engaged students and less engaged students; whilst with $k \in \{3, 4\}$, the k-means analysis results reveal more information about how the students were engaged in learning. Additionally, we conducted two Tukey's honestly significant difference (HSD) post hoc tests (at 95% confidence interval): the first test was to compare the differences of Zscore(steps), Zscore(comments) and Zscore(attempts) between these three clusters under the condition of $k = 3$ (Table 1 shows the result); the second test was to compare these three variables between four clusters under the condition of $k = 4$ (Table 2 shows the result). We can see from Table 1 that, when $k = 3$, all the variables, i.e., Zscore(steps), Zscore(comments) and Zscore(attempts), are significantly different ($p < .001$) between these three clusters. However, as shown in Table 2, when $k = 4$, whilst Zscore(steps) and Zscore(attempts) are significantly different ($p < .001$) between those four clusters, the Zscore(comments) does not significantly differ from cluster I and cluster IV. Therefore, we determine that only when $k = 3$, we can obtain strong and stable clusters.

Table 1. Tukey HSD test result - multiple comparisons ($k = 3$)

Dependent variable	Cluster number of case	Cluster number of Case	Mean difference	Std. error	Sig.	95% Confidence interval	
						Lower bound	Upper bound
Z score (steps)	I	II	2.094*	.007	.000	2.077	2.110
		III	−.149*	.022	.000	−0.200	−0.098
	II	I	−2.094*	.007	.000	−2.110	−2.077
		III	−2.243*	.021	.000	−2.292	−2.193
	III	I	.149*	.022	.000	0.098	0.200
		II	2.243*	.021	.000	2.193	2.292
Z score (comments)	I	II	.429*	.009	.000	0.408	0.450
		III	−5.480*	.028	.000	−5.546	−5.414
	II	I	−429*	.009	.000	−0.450	−0.408
		III	−5.909*	.027	.000	−5.973	−5.846
	III	I	5.480*	.028	.000	5.414	5.546
		II	5.909*	.027	.000	5.846	5.973
Z score (attempts)	I	II	2.070*	.007	.000	2.052	2.087
		III	−.161*	.023	.000	−.214	−.108
	II	I	−2.070*	.007	.000	−2.087	−2.052
		III	−2.231*	.022	.000	−2.283	−2.179
	III	I	.1613*	.023	.000	.108	.214
		II	2.231*	.022	.000	2.179	2.283

*. The mean difference is significant at the 0.05 level.

Table 2. Tukey HSD test result - multiple comparisons (k = 4)

Dependent variable	Cluster number of case	Cluster number of case	Mean difference	Std. error	Sig.	95% Confidence interval	
						Lower bound	Upper bound
Z score (steps)	I	II	−.299*	.036	.000	−.391	−.207
		III	2.071*	.007	.000	2.052	2.090
		IV	−.059*	.017	.002	−.102	−.016
	II	I	.299*	.036	.000	.207	.391
		III	2.370*	.035	.000	2.279	2.461
		IV	.240*	.038	.000	.141	.338
	III	I	−2.071*	.007	.000	−2.090	−2.052
		II	−2.370*	.035	.000	−2.461	−2.279
		IV	−2.130*	.016	.000	−2.171	−2.090
	IV	I	.059*	.017	.002	.016	.102
		II	−.240*	.038	.000	−.338	−.141
		III	2.130*	.016	.000	2.090	2.171
Z score (comments)	I	II	−8.636*	.034	.000	−8.724	−8.549
		III	.239*	.007	.000	.221	.257
		IV	−2.725*	.016	.000	−2.766	−2.684
	II	I	8.636*	.034	.000	8.549	8.724
		III	8.876*	.034	.000	8.789	8.962
		IV	5.911*	.037	.000	5.817	6.005
	III	I	−.239*	.007	.000	−.257	−.221
		II	−8.876*	.033	.000	−8.962	−8.789
		IV	−2.964*	.015	.000	−3.003	−2.926
	IV	I	2.725*	.016	.000	2.684	2.766
		II	−5.911*	.037	.000	−6.005	−5.817
		III	2.964*	.015	.000	2.926	3.003
Z score (attempts)	I	II	−.336*	.038	.000	−.432	−.240
		III	2.051*	.008	.000	2.032	2.071
		IV	−.002	.018	**1.000**	−.047	.043
	II	I	.336*	.038	.000	.240	.432
		III	2.387*	.037	.000	2.292	2.483
		IV	.334*	.040	.000	.231	.438
	III	I	−2.051*	.008	.000	−2.071	−2.032
		II	−2.387*	.037	.000	−2.483	−2.292
		IV	−2.053*	.017	.000	−2.095	−2.011
	IV	I	.002	.018	**1.000**	−.043	.047
		II	−.334*	.040	.000	−.438	−.231
		III	2.053*	.017	.000	2.011	2.095

*. The mean difference is significant at the 0.05 level.

4.2   Comparisons Between Clusters

Table 3 summarizes descriptive statistics of steps, comments and attempts of these three clusters. Cluster II has the most students i.e. 18,998, followed by Cluster I with 5,320 students; whilst Cluster III is the least represented, with only 471 students.

Table 3. Descriptive statistics of steps, comments and attempts of three clusters

		N	Min	Max	Median	Mean	SD
Cluster I	Steps	5,320	6	6	74	64.95	12.241
	Comments	5,320	0	0	1	3.81	5.403
	Attempts	5,320	0	0	23	22.79	7.228
Cluster II	Steps	18,998	1	1	10	12.34	11.147
	Comments	18,998	0	0	0	.75	2.023
	Attempts	18,998	0	0	0	2.21	3.618
Cluster III	Steps	471	20	20	74	68.69	11.373
	Comments	471	24	24	36	42.78	20.198
	Attempts	471	5	5	25	24.39	6.446

The 1st Dimension – Learning: Step Visits and Completion Rate

In Cluster I (5,320 students), a very large amount of the students visited a large percentage of steps. In particular, more than half, i.e., 3,079 (57.88%), students visited more than 90% of the steps; 5,278 (99.21%) students visited more than half, i.e., 50%, of the steps. Although very few, there were still some, i.e., 2 (0.04%), students visited less than 10% of the steps; and 5 (0.09%) students visited 10%~20% of the steps. Nevertheless, the smallest percentage of steps visited by the students was 8.0% (6 steps), and there were 912 (17.14%) students who visited all, i.e., 100%, of the steps. In terms of completion rate, 4,920 (92.48%) students marked more than 90% of the steps they visited as 'complete', by clicking on the button "Mark as complete". There wasn't any student who did not mark any step that they visited.

In Cluster II (18,998 students), a very large amount of the students visited only a very small percentage of steps, and the more percentage of the steps, the fewer the students. In particular, 8,032 (42.28%) students visited less than 10% of the steps; 4,231 (22.27%) students visited 10%~20% of the steps; 3,393 (17.86%) students visited 20%~30%; 1,789 (9.42%) students visited 30%~40%; and 1,087 (5.72%) students visited 40%~50%. In total, 18,532 (97.55%) students visited less than 50% of the steps. However, there were 9 (0.05%) students visited more than 90% of the steps, and 31 (0.16%) students visited 80%~90%. Nevertheless, the largest percentage of the steps visited was 94.67% (71 steps), i.e., there wasn't any student who visited all the steps. Regarding the completion rate, 5,493 (28.91%) students marked less than 10% of the steps that they visited as 'complete', and 5,702 (30.01%) students marked more than 90% of the steps that they visited as 'complete'. 11,845 (62.35%) students marked more than 50% of the steps they visited as 'complete'. Comparing to 'visit rate', the completion rate was very high. Interestingly, albeit 5,339 (28.10%) students did not mark any step that they visited as 'complete', there were still 1,541 (8.11%) students who marked all the steps that they visited as 'complete'.

In Cluster III (471 students), similar to Cluster I, a very large number of students visited a large percentage of steps. In particular, 332 (70.49%) students visited more than 90% of the steps; 53 (11.25%) students visited 80% ~ 90%; 30 (6.37%) students visited 70% ~ 80%. Interestingly, only 1 (0.21%) student visited 20% ~ 30% of the steps, and only 1 (0.21%) student visited 10% ~ 20%. Surprisingly, no student visited less than 10% of the steps. Regarding completion rate, surprisingly, 457 (97.03%) students marked more than 90% of the steps they visited as 'complete'; 295 (62.63%) students marked all of the steps that they visited as 'complete'; apart from only 1 (0.21%) student marking only 13.51% of the steps as 'complete', all the rest, 471 (99.79%) students, marked more than 70% steps as 'complete'. The completion rate in Cluster III was surprisingly very high.

Figure 4 compares the percentage of steps visited by students for these three clusters. Cluster I and Cluster III are similar – the larger the percentage of the step being visited, the more the students; whereas Cluster II shows an opposite trend. A Kruskal-Wallis H test was conducted. The result shows that there is a statistically significant difference in the number of steps being visited between these three clusters, $\chi^2(2) = 1,318,881.844$, p < .001, with a mean rank correct answers rate of 21,773.63 for Cluster I, 9,520.72 for Cluster II, and 22,397.58 for Cluster III. Further Mann-Whitney U tests show that there are statistically significant differences between Cluster I and Cluster II (Z = −111.106, U = 365,933.5,796, p < .001), between Cluster I and Cluster III (Z = −8.086, U = 978,445.5, p = < .001), and between Cluster II and Cluster III (Z = −36.97, U = 37,226, p < .001).

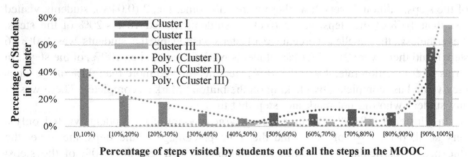

Percentage of steps visited by students out of all the steps in the MOOC

Fig. 4. Comparison of the number of steps visited by students between three clusters

Figure 5 shows comparisons of the completion rate between clusters. Again, Cluster I and Cluster III share a similar trend, whilst Cluster II is very different. A Kruskal-Wallis H test was performed to examine these differences. The result shows that there is a statistically significant difference in completion rate between these three clusters, $\chi^2(2) = 8,461.923$, p < .001, with a mean rank correct answers rate of 19,830.14 for Cluster I, 10,106.02 for Cluster II, and 20,741.28 for Cluster III. Further Mann-Whitney U tests show statistically significant differences between Cluster I and Cluster II (Z = −88.502, U = 10,796,504, p < .001), between Cluster I and Cluster III (Z = −5.635, U = 1,069,606, p < .001), and between Cluster II and Cluster III (Z = −31.447, U = 726,183, p < .001).

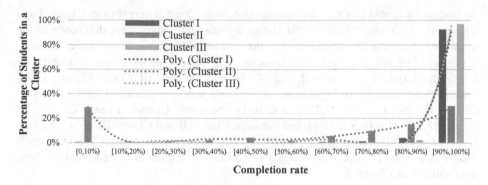

Fig. 5. Comparison of completion rate between three clusters

The 2nd Dimension – Social: Comments and 'Likes'

In Cluster I (5,320 students), 2,959 (55.62%) students submitted at least one comment; they submitted 20,254 comments in total, with an average of 6.84, standard deviation of 5.63, and median of 5. Overall (all 5,320 students together), the average number of comments was 3.81 with standard deviation of 5.40. The median number of comments was 1. There were 26 students who submitted the largest number of comments, i.e., 23. Regarding 'likes', all those 2,959 students, who submitted at least one comment, received 30,044 'likes', in total. The average number of 'likes' was 10.15 with standard deviation of 12.34. The median number of 'likes' was 5. There were 409 students who submitted at least one comments but did not receive any 'likes'. The most popular student received the largest number, 81, of 'likes'.

In Cluster II (18,998 students), only 5,007 (26.36%) students submitted at least one comment; they submitted 14,318 comments in total, with an average of 2.86, standard deviation of 3.08, and median of 2. Overall (all 18,998 students together), the average number of comments was 0.75 with standard deviation of 2.02. The median number of comments was 0. There was 1 student who submitted the largest number of comments, i.e., 26. Regarding 'likes', all those 5,007 students, who submitted at least one comment, received 14,979 'likes', in total. The average number of 'likes' was 2.99 with standard deviation of 5.43. The median number of 'likes' was 1. There were 1,773 students who submitted at least one comments but did not receive any 'likes'. The most popular student received the largest number, 86, of 'likes'.

In Cluster III (471 students), all (100%) students submitted comments: 20,151 in total, with an average of 42.78, standard deviation of 20.20, and median of 36. 24 students submitted the smallest number, 24, of comments. One student submitted the largest number, 179, of comments. Regarding 'likes', in total, they received 30,164 'likes'. The average number of 'likes' was 64.04 with standard deviation of 50.38. The median number of 'likes' was 52. Only 1 student did not receive any 'likes'. The most popular student received the largest number, 441, of 'likes'.

Figure 6 (left) compares the percentage of students submitting comments in the three clusters. Cluster I and Cluster II are similar– a very large percentage of students submitted very few comments; although a larger percentage of students did not submit any comments in Custer II. As for Cluster III, the peak appears between 20 and 30, and no student submitted less than 24 comments, which is very different from Cluster I and Cluster II. A Kruskal-Wallis H test shows a statistically significant difference in the number of comments submitted by the students between these three clusters, $\chi^2(2) = 4,168.348$, p < .001, with a mean rank correct answers rate of 15,605 for Cluster I, 11,194.54 for Cluster II, and 24,553.76 for Cluster III. Mann-Whitney U test results show statistically significant differences between Cluster I and Cluster II (Z = − 48.616, U = 322,02,216, p < .001), between Cluster I and Cluster III (Z = −37.337, U = 0, p < .001), and between Cluster II and Cluster III (Z = −46.886, U = 112, p < .001). Interestingly, the Mann-Whitney U test for Cluster I and Cluster II results is $U = 0$. Thus, all the students in Cluster III submitted more comments than any students in Cluster I.

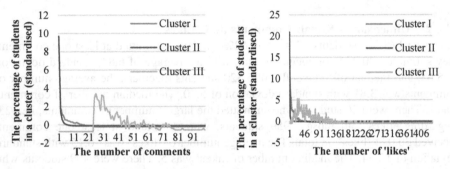

Fig. 6. Comparison of the number of comments (on the left) and 'likes' (on the right) vs the percentage of students between three clusters

Figure 6 (right) shows the comparison of the number of 'likes' received by certain percentage of students between three clusters. Similar to the comparison of the number of comments, Cluster I and Cluster II share a similar trend, but Cluster III has a very different trend: the peak of Cluster I and Cluster II appear in the very left end of the horizontal axis i.e. the majority of students in Cluster I and Cluster II received very few, if not zero, comments; whereas Cluster III has a peak at around 30 'likes'.

The 3rd Dimension – Assessment: Attempts and Correct Answers Rate
In Cluster I (5,320 students), only 51 (0.96%) students did not attempt to answer any question; the majority (5,269, 99.04%) students attempted between 2- and 88-times answering questions. Overall (all 5,320 students together), the average number of attempts was 22.79 with standard deviation of 7.23. The median number of attempts was 23. Regarding the correct answers rate, all those 5,269 students who attempted answering questions correctly answered at least one question. The average correct answers rate was 72.48% with standard deviation of 11.15%. The median correct answers rate was 71.42%, and the lowest correct answers rate was 9.52%. Gratifyingly, 43 (0.82%) students' correct answers rate was 100%.

In Cluster II (18,998 students), 13,477 (70.94%) of them did not attempt to answer any question, and the rest, 5,521 (29.06%), attempted to answer at least one question.

Overall (all 18,998 students together), the average number of attempts was 2.21 with standard deviation of 3.62, but the median number of attempts was 0. In terms of the correct answers rate, among those 5,521 (29.06%) students who attempted at least once to answer a question, only 14 (0.25%) of them did not correctly answered any question, even though they attempted between 1 and 6 (mean = 1.86, SD = 1.61) times to answer a question. Excluding those 13,477 students who did not attempted to answer any questions, the overall average correct answers rate was 67.84% with standard deviation of 14.52%; and the median correct answers rate was 62.50%. Surprisingly, there were 372 (6.74%) students whose answer was 100% correct.

In Cluster III (471 students), every student attempted at least 5 times to answer a question. The maximum number of attempts was 45, the median was 25. The average number of attempts was 24.39 with standard deviation of 6.45. With regards to correct answers rate, the lowest was 43.75%, and the median was 74.07%. Overall, the average correct answers rate was 73.67% with standard deviation of 10.06%. However, only 2 (0.42%) students' correct answers rate was 100%.

The 100% stacked column chart (Fig. 7 left) suggests that the pattern of attempting answering questions between three clusters are very different: the majority of students (13,477; 70.94%) in Cluster II did not attempted to answer any question; whilst almost all the students in Cluster I (5,269; 99.04%) and Cluster III (471; 100%) had attempted answering questions. A Kruskal-Wallis H test shows a statistically significant difference in the number of attempts, $\chi^2(2) = 15,326$, p < .001, with a mean rank attempts of 21,678.18 for Cluster I, 9,552.93 for Cluster II, and 22,176.74 for Cluster III. Further Mann-Whitney U tests show statistically significant differencesbetween Cluster I and Cluster II (Z = −120.445, U = 949,826.5, p < .001), between Cluster I and Cluster III (Z = −5.709, U = 1,054,516, p < .001), between Cluster II and Cluster III (Z = −44.784, U = 965,172.5, p < .001).

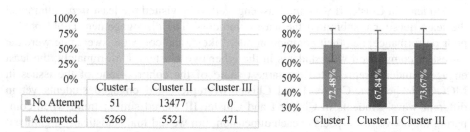

Fig. 7. Comparisons of numbers of attempts (left); correct answers rates (right)

Although the mean correct answers rates of these three clusters are similar, as shown in Fig. 7 on the right, a Kruskal-Wallis H test shows that there is a statistically significant difference in correct answers rate between these three clusters, $\chi^2(2) = 499.995$, p < .001, with a mean rank correct answers rate of 6,268.20 for Cluster I, 4,939.29 for Cluster II, and 6,610.90 for Cluster III. Further Mann-Whitney U tests show statistically significant differences between Cluster I and Cluster II

(Z = −21.323, U = 11,113,796, p < .001), between Cluster I and Cluster III (Z = −2.148, U = 1,166,974.5, p = .032 < .05), and between Cluster II and Cluster III (Z = −10.894, U = 912,535.5, p < .001).

5 Discussions and Conclusions

In this study, we first defined 3-dimensional metrics to measure engagement patterns. The *fundamental* (and, arguably, comprehensive) *dimensions include learning, social* and *assessment*. We then employed *k-means analysis* to cluster students based on the metrics. Our clustering and validation approach resulted in very strong and stable 3 clusters. We further applied statistical models to explore the differences of engagement patterns between clusters, in 3 dimensions, from 2 aspects (Table 4):

Table 4. Three dimensions and two aspects explored in the study

Dimension	Determining aspect	Performance aspect
Learning	Steps visited	Completion rate
Social	Comments submitted	'Likes' received
Assessment	Attempts to answer questions	Correct answers rate

The statistical analysis further supported that our clustering results were very strong and stable, as all three dimensions defined in the metrics (determining aspect) had statistically significant impact on determining clusters, with very large F values, at a $p < .001$ level. Moreover, all the above three performance aspects were statistically significantly different between those three clusters. This allowed exploring in depth how students were engaged in learning in the MOOC.

Students in Cluster II were the least engaged: they visited the least *steps*, attempted the least questions, submitted the least comments. The disengagement could predict poor performance, i.e., their completion rate, 'likes', and correct answers rate were the lowest in comparison with students in the other two clusters. Unfortunately, the least engaged students represented the largest share of the cohort – one of the issues in MOOCs in general. Cluster I and Cluster III represented engaged students yet in different ways. Students in Cluster I and Cluster III shared similar trend when comparing the above two aspects in each dimension, but we did find statistically significant differences at a $p < .001$ level. Yet, U values from Mann-Whitney U tests suggested that social dimension (comments) was the most differentiating aspect (only $U_{comments} = 0$, meaning all students in Cluster III submitted more comments than all students in Cluster I). Importantly, Cluster III received higher scores than Cluster I in terms of for all performance aspects. This interesting result suggests that socially engaged students would also be more engaged in the various learning activities, and their performance could be better than of those who are less socially engaged. Thus, recommender systems could support students in the social exchange, in order to enhance the learning – unlike it was considered in the past, that social exchange could only be distracting from the mainstream learning activity. This, we believe, is an important characteristic of MOOCs in particular, which may not be shared with other type of learning environments. Further work is necessary to refine the recommendations.

References

1. Antonenko, P.D., et al.: Using cluster analysis for data mining in educational technology research. Educ. Tech. Res. Dev. **60**(3), 383–398 (2012)
2. Cristea, A.I., et al.: Earliest predictor of dropout in MOOCs: a longitudinal study of FutureLearn courses. Presented at the 27th International Conference on Information Systems Development (ISD2018), Lund, Sweden, 22 August 2018
3. Cristea, A.I., et al.: How is learning fluctuating? FutureLearn MOOCs fine-grained temporal analysis and feedback to teachers and designers. Presented at the 27th International Conference on Information Systems Development (ISD2018), Lund, Sweden, 22 August 2018
4. Csiksczentmihalyi, M., et al.: Flow: the psychology of optimal experience. Aust. Occup. Ther. J. **51**(1), 3–12 (2004)
5. Ferguson, R., Clow, D.: Examining engagement: analysing learner subpopulations in massive open online courses (MOOCs). Presented (2015)
6. Guo, P.J., et al.: How video production affects student engagement: an empirical study of MOOC videos. Presented (2014)
7. Henrie, C.R., et al.: Measuring student engagement in technology-mediated learning: a review. Comput. Educ. **90**, 36–53 (2015)
8. Kodinariya, T.M., Makwana, P.R.: Review on determining number of cluster in K-means clustering. Int. J. **1**(6), 90–95 (2013)
9. Laurillard, D.: Rethinking University Teaching: A Conversational Framework for the Effective Use of Learning Technologies. London, RoutledgeFalmer (2002)
10. MacQueen, J., et al.: Some methods for classification and analysis of multivariate observations. In: Proceedings of the Fifth Berkeley Symposium on Mathematical Statistics and Probability, Oakland, CA, USA, pp. 281–297 (1967)
11. Marks, H.M.: Student engagement in instructional activity: patterns in the elementary, middle, and high school years. Am. Educ. Res. J. **37**(1), 153–184 (2000)
12. Al Mamun, M.A., et al.: Factors affecting student engagement in self-directed online learning module. In: The Australian Conference on Science and Mathematics Education (Formerly UniServe Science Conference), p. 15 (2017)
13. Shernoff, D.J., et al.: Student engagement in high school classrooms from the perspective of flow theory. In: Csikszentmihalyi, M. (ed.) Applications of Flow in Human Development and Education: The Collected Works of Mihaly Csikszentmihalyi, pp. 475–494. Springer, Dordrecht (2014). https://doi.org/10.1007/978-94-017-9094-9_24
14. Shi, L., et al.: Towards understanding learning behavior patterns in social adaptive personalized E-learning systems. In: The 19th Americas Conference on Information Systems, pp. 1–10. Association for Information Systems, Chicago (2013)
15. Shi, L., Cristea, A.I.: Demographic indicators influencing learning activities in MOOCs: learning analytics of FutureLearn courses. Presented at the 27th International Conference on Information Systems Development (ISD2018), Lund, Sweden, 22 August 2018
16. Sinatra, G.M., et al.: The challenges of defining and measuring student engagement in science. Educ. Psychol. **50**(1), 1–13 (2015)

Topic Evolution Models for Long-Running MOOCs

Arti Ramesh[1](✉) and Lise Getoor[2]

[1] SUNY Binghamton, Binghamton, USA
artir@binghamton.edu
[2] University of California, Santa Cruz, USA
getoor@ucsc.edu

Abstract. Massive open online courses (MOOCs) have emerged as a powerful platform for imparting education in the last few years. Discussion forums in online courses connect various geographically separated MOOC participants and serve as the primary means of communication between them. The text in the forums reflects many important aspects of the course such as the student population and their changing interests, parts of the course that were well received and parts needing attention, and common misconceptions faced by students. In order to improve the quality of online courses and students' interaction and learning experience, instructors need to actively monitor and discern patterns in previous iterations of the course and mold the course to suit the needs of the ever-changing student population. To enable this, in this work, we perform a systematic detailed analysis of the evolution of fine-grained topics in online course discussion forums across repeated MOOC offerings using seeded topic models and draw important insights on the nature of students, types of issues, and student satisfaction. We present topic evolution results on two successful long-running MOOCs: (i) a business course, and (ii) a computer science course. Our models uncover interesting topic trends in both courses including the decline of logistic issues in both courses as iterations unfold, decline in grading related issues when automatic grading is adopted in the business course, and prevalence of technical issues in the computer science course in comparison to the business course. Our models throw light on the different ways students interact on MOOCs and their changing needs, and are useful for instructors to understand the progression of courses and accordingly fine-tune courses to meet student expectations.

1 Introduction

Massive open online courses (MOOCs) are increasingly becoming a powerful educational platform, providing students from all over the world access to high quality education. As MOOCs continue to grow, instructors are faced with the problem of understanding the needs and expectations of the ever-changing student population, molding the course to better suit their interests, identifying

© Springer Nature Switzerland AG 2018
H. Hacid et al. (Eds.): WISE 2018, LNCS 11234, pp. 410–421, 2018.
https://doi.org/10.1007/978-3-030-02925-8_29

issues in past iterations, and addressing them in future iterations. This endeavor ensures a smoother delivery of the course and helps in fostering a superior learning experience.

With MOOCs, there is tremendous opportunity to develop models that automatically gauge feedback by interpreting textual content in the discussion forums. The text in the forums reflects many important aspects of the course such as the student population and their changing interests, parts of the course that were well received and parts needing attention, and common misconceptions faced by students. In the existing framework, online instructors often resort to manually poring through posts in the forums to determine improvements for future iterations. And, when they make improvements, they again have to rely on manual effort to determine whether the improvements have been helpful. An automatic way to quantitatively measure how the health of the course and the interests of the student population are evolving over time will be helpful for instructors to mold their course to fit student needs better and help improve their online interaction and learning experience. While most previous work in this space interpret text in the discussion forums of individual courses [4–6, 11, 13], analyzing text across repeated offerings of a course provides a panoramic view of course progression. This analysis can potentially help instructors discern topic patterns corresponding to relevant topics such as course materials and issues, and focus limited instructor resources on addressing the most prevalent and relevant problems.

In this work, we develop weakly-supervised topic evolution models to analyze posts in online course discussion forums across repeated offerings. We leverage a seeded variant of topic modeling, seeded latent dirichlet allocation (seeded LDA) [9], to induce and track evolution of specific topic clusters relevant to online courses across iterations. The large number of posts in MOOC discussion forums in each iteration of the course and privacy issues surrounding the creation and distribution of labeled data make weakly-supervised seeded topic evolution models attractive for understanding forum posts and improving future course offerings.

We present a fine-grained analysis of discussion forums in two successful and popular MOOCs that have run for several iterations. We first identify important topic themes relevant in online courses. Next, we track the rise and decline of these topics as the iterations unfold. This analysis reveals how the focus of the course, the student population, and their needs are evolving with time. Our models and analysis are useful for educators, MOOC practitioners, and instructors to understand the longitudinal evolution of online courses. Our analysis is also helpful to instructors to fine-tune their courses to meet student expectations, which subsequently helps in achieving superior interaction and learning experience, when students interact on the MOOC. To the best of our knowledge, ours is the first work that models evolution of topics across repeated offerings of online courses.

Our main contributions in this work are as follows:

– We show how to use seeded LDA to categorize discussion forum posts in online courses. We construct *four* different seeded LDA models for each of the two courses using a common set of seed words. Using our models, we track the progression of seeded topics and draw important insights on topic patterns of two successful long-running MOOCs: (i) *thirty four* iterations of a business course (BUSINESS), and (2) *fifteen* iterations of a computer science course (CS).
– We identify the three primary purposes of forums in online courses upon carefully mining posts across different courses: (i) socializing with fellow classmates (*social*), (ii) reporting issues in the course (*issue*), and (iii) discussing course related material (*technical*). We categorize the posts according to these three primary purposes of the forums. Our temporal analysis uncovers changes in topic patterns such as an increase in issue posts after the 4th iteration in the CS course, when the course splits into two parts. Our temporal analysis is helpful in understanding how big course changes such as splitting the course affects the students.
– Next, we categorize the posts referring to the three most important *course elements* in online courses: (i) lectures, (ii) quizzes, and (iii) certificate, to understand the emphasis on each of them, respectively. While lectures are the most popular course element in the BUSINESS course, quizzes surpass lectures in the CS course. We also observe that certificate receives more attention in the BUSINESS course in comparison to the CS course. This analysis provides insight on which course elements get more attention in the course and subsequently, the nature of students in both the courses, and their evolution with time.
– We then show how to use a combination of seeded topic models to perform a finer-grained analysis of issue posts and study their distribution across: (i) lecture, quiz, and certificate course elements, and (ii) fine-grained lecture and quiz sub-topics. We find that though lectures are the most dominant course element in the BUSINESS course, most issues are reported on quizzes. We also observe that while grading issues tend to occur across both the courses, submission issues predominate the forums in the CS course. Interestingly, we notice that grading issues decline in the BUSINESS course after peer grading is replaced by automatic grading, indicating a general preference for the latter. Our fine-grained analysis sheds light on issues faced by students across iterations that could negatively impact student satisfaction and enrollment in future iterations.
– In the CS course, we observe another important dimension, *technical* posts, dominating the issue posts. Technical issues in software installation and code compilation are unique to computer science courses. We use a combination of seeded LDA models to separate logistic issues from technical issues and find that logistic issues follow a similar trend to the BUSINESS course, declining with time, indicating the need to focus on technical issues in the CS course.

2   Related Work

Recently, there has been a growing interest in understanding text in online course discussion forums. Previous work in this space focus on understanding forum content in individual courses to improve the experience of MOOC participants [1,4–6,11,13]. To the best of our knowledge, ours is the first work studying the evolution of forum content in online courses over time. Topic evolution has been studied previously on scientific research papers [7,8]. Temporal variants of LDA such as Dynamic Topic Model (DTM) [3] and Topics over Time (ToT) [12] model LDA topic and word distributions over time. Both these models do not allow tracking specific topics of interest to the user and only model the evolution of topics that are more dominant in the data and tend to ignore rarer topics. Hence, topic evolution models such as DTM and ToT are not effective in this setting. Hall et al. [8] analyze the history of ideas in NLP conferences using topic modeling. They use LDA to model topic evolution across years and use hand-selected seed words to track evolution of specific topics. They also note that the temporal variants of LDA (DTM and ToT) impose constraints on the time periods, rendering them inflexible and unsuitable for modeling documents that can change dramatically from one time period to another. In online courses, each course iteration attracts an entirely new cohort of students and the content in the forums can potentially vary significantly according to their interests and backgrounds. Hence, seeding topic models with words is a simple, yet effective means to track evolution of specific topics of interest. In our work, we leverage a seeded variant of LDA, Seeded LDA [9] to track specific topics of conversation in the forums across iterations. Seeded LDA guides topic discovery to learn specific topics of interest by allowing the user to input a set of seed words that are representative of the underlying topics in the corpus. Seeded LDA uses these seed words to improve topic-word distribution by inducing topics to obtain a high probability mass for the given seed words. Similarly, it also improves the document-topic distribution in online courses by biasing documents to select topics related to the seed words. The seed set need not be exhaustive as the model gathers related words based on co-occurrence of other words with the specified seed words in the documents. We refer the reader to [9] for more details.

3   Data

We analyze data from two popular long running Coursera MOOCs: (a) a business course, and (b) a computer science course. We refer to these courses as BUSINESS course and CS course, respectively. Both courses are active courses attracting thousands of students every iteration. We analyze 34 iterations of the BUSINESS course, each iteration spanning 6 weeks. The BUSINESS course has on an average is greater than $50,000$ users and $5,000$ posts per iteration. The highest number of students registered is approximately $100,000$ students and the corresponding discussion forum has $10,000$ posts. Hence, in total, we analyze approximately $200,000$ posts across all iterations of BUSINESS course.

The CS course spans 8 weeks for the first three iterations. From the fourth iteration, the course splits into two parts spanning 6 weeks each, which we refer to as CS-1 and CS-2. In all, we analyze data from 15 course offerings, 3 iterations of the original course, and 6 iterations each of CS-1 and CS-2. The CS course has on an average is greater than $100,000$ users and $13,000$ posts per iteration. The highest number of students registered is approximately $250,000$ students and the corresponding discussion forum has $47,000$ posts. Hence, in total, we analyze approximately $110,000$ posts across all iterations of CS course.

4   Topic Discovery in Online Courses

In this section, we build the seeded topic models for discovering topics in forum posts. Due to the absence of labeled data, the seed words are hand-selected. We construct *four* seeded LDA models using four different combinations of seeded topics. Ramesh et al. [10,11] give the seed words for the different coarse and fine-grained topics. For each seeded LDA model, we include an additional k unseeded topics in the seeded LDA models to capture topics that do not fall under the seeded topic categories. After experimenting with different values of k, we choose $k = 5$. Hence, the number of topics for each model is the sum of number of seeded topics and 5 unseeded topics. We train all our seeded LDA models for 1000 iterations. We use $\alpha = 0.0001$ and $\beta = 0.0001$ to create a sparse topic distribution so that fewer topics with high values emerge.

4.1   Primary Purpose of Forums

In the first categorization, we classify the posts into three categories: (i) social, (ii) issues, and (iii) course content topics. These three topic categories reflect the three primary purposes in which the forums are utilized in online courses. Social posts capture the social aspect of forums, where students can e-socialize with fellow classmates. These posts usually fall into one of the following subcategories: (a) student introductions, and (b) formation of study groups. Issue posts are posts that intend to bring issues in the course to the attention of the instructor and fellow classmates and/or ask for their help in solving them.

4.2   Course Elements

Unlike most classroom courses, online courses attract a diverse set of students with varied interests and expectations from the course. Anderson et al. [2] classify students according to their interaction on the MOOC. Of these, three most common types of students include: (1) *viewers*: students interested in *viewing* video lectures, (2) *solvers*: students interested in *solving* assignments, and (3) students interesting in obtaining a certificate. These three types of students map to the three corresponding course elements: (i) lectures, (ii) quizzes/assignments, and (iii) certificate. In the second categorization, we identify posts corresponding to these three important course elements. Analyzing references to these elements in the forums helps us understand the different types of students in the course and which course elements to focus on for future iterations.

4.3   Fine-Grained Analysis of Issue Posts

In the third categorization, we further drill down on issue posts to understand how they are distributed across course elements. For this, we combine the two seeded LDA models to categorize issue posts further across the three course elements. We label posts for which *issue* topic has the highest value in the document-topic distribution as *issue* posts and further categorize these posts across the three course elements.

We identify fine-grained sub-topics for *lecture*, *video* and *subtitles*, and *quiz*, *submission*, *grading*, and *deadline*, and categorize the logistic issue posts into these fine-grained topics. Similarly, in the fourth categorization, we combine the seeded LDA models for *issue* with fine-grained seeded LDA models to understand how *issue* posts are distributed across fine-grained lecture and quiz topics.

5   Topic Trends in Online Courses

In this section, we present an in-depth analysis of topic evolution across iterations of the BUSINESS course and the CS course. For each course, we conduct experiments to answer the following questions:

1. How are posts distributed across topics constituting the three primary purposes of forums: (a) social, (b) issues, and (c) technical topics, and how is that evolving with time?
2. Next, we answer the question of which course elements are most popular in the course and how is the emphasis on each of them changing with time?
3. Finally, we drill down deeper on issue posts and analyze which topics constitute the focus of issue posts and how are they changing as iterations unfold?

We run seeded LDA models described in Sects. 4.1, 4.2, and 4.3 across iterations of the BUSINESS course and the CS course to answer the above three questions.

5.1   BUSINESS course

Here, we present topic evolution analysis of posts in the BUSINESS course.

Primary Purpose of Forums. In our first set of experiments, we study the evolution of social, issue, and technical topics across iterations. For each iteration of the course, we add the topic distribution values for each topic across all the posts to get the total number of posts in each topic category. We then plot the number of posts in each topic category across iterations. Figure 1(a) gives the number of *social*, *issue*, and *technical* posts across iterations. In the BUSINESS course, we observe that social posts contribute to a significant number of posts in the forum, emphasizing the importance of forums as a socializing platform. This is closely followed by technical posts. Issue posts are fewer in number in comparison to social and technical posts and decline to negligible numbers in the

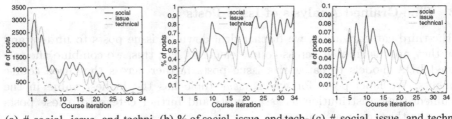

(a) # social, issue, and techni- (b) % of social, issue, and tech- (c) # social, issue, and technical posts   nical posts   cal posts per student

Fig. 1. BUSINESS course: evolution of social, issue, and technical posts across iterations

later iterations. Social and issue posts also decline over time, but always remain higher than issue posts in the later iterations.

Analyzing the percentage of social, issue, and technical posts in the total number of posts in each iteration (Fig. 1(b)), we observe that social and technical topics together constitute around 80% of posts. The percentage of social posts continuously increases, emphasizing the importance of the social aspect in learning. The percentage of technical discussions follows a steady path of evolution across iterations and they constitute a significant percentage of forum discussions even in the later iterations, which helps us to understand that there is a significant amount of interest in the technical course content. Issues contribute to less than 20% of posts in the early iterations, declining steadily, dropping to less than 10% after 30 iterations. Analyzing the number of social and issue posts per student (Fig. 1(c)), we observe that an increase across all three categories steadily in the initial iterations followed by a steady decline, indicating that fewer students tend to post in the forums as the course stabilizes.

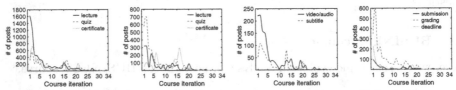

(a) # lecture, quiz, and (b) # lecture, quiz, and (c) # lecture issue posts (d) # quiz issue posts
certificate posts   certificate *issue* posts

Fig. 2. BUSINESS course: distribution of posts and *issue* posts across three course elements: lecture, quiz, and certificate, and lecture and quiz sub-topics

Course Elements. In our second set of experiments, we analyze the emphasis on different course elements across iterations. This analysis is helpful to understand what are the most sought after course elements in this course. Figure 2(a)

gives the number of posts in the three course elements across iterations. We observe that lectures emerge as the most dominant course element in BUSINESS course across all iterations, followed by quiz and certificate, in that order. We observe that certificates are a popular topic category in BUSINESS course, consistently attracting posts throughout all iterations.

Fine-Grained Analysis of Issue Posts. In our third set of experiments, we analyze how issue posts are distributed across the course elements. Figure 2(b) gives the distribution of issues across the course elements. It is interesting to note that while lectures are the most discussed course element, most issues are reported on quizzes in the initial iterations.

While we observe a consistent interest in certificates across all iterations, there are two periods which show an increased incidence of certificate issues: around 6th iteration and 16th iteration. Analyzing the certificate issue posts in these iterations, we find that there was a delay in dispatching certificates in both these periods, causing a flurry of certificate issue posts.

Next, we perform a finer-grained analysis of issue posts across fine-grained lecture and quiz topics. Figure 2(c) gives the distribution of issues across lecture sub-topics: video/audio and subtitles. We notice that video/audio issues are more prominent in the earlier iterations. Both video/audio and subtitle issues decline and contribute almost equally to lecture issues in the middle iterations before declining to negligible number of posts in the later iterations.

Figure 2(d) gives the distribution of issues across quiz sub-topics. We observe that a major proportion of quiz issues fall under grading, with submission and deadline hardly contributing to quiz-related issues. Grading issues follow a steep decline from the third iteration after peer grading was replaced with automatic grading in the course, indicating a preference among students for the latter. Often instructors make modifications to the course responding to feedback from students. Our analysis not only helps them identify the issues but also provides them with a simple and effective tool to evaluate the success of their improvements.

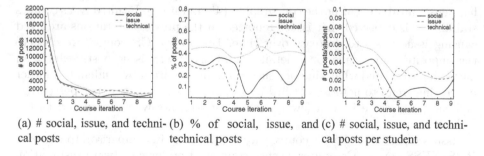

(a) # social, issue, and technical posts   (b) % of social, issue, and technical posts   (c) # social, issue, and technical posts per student

Fig. 3. CS course: evolution of social, issue, and technical posts across iterations

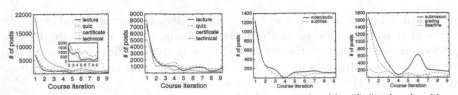

(a) # lecture, quiz, cer- (b) # lecture, quiz, cer- (c) # lecture related is- (d) # quiz related issue
tificate, and technical tificate, and technical sue posts   posts
posts   *issue* posts

Fig. 4. CS course: distribution of *issue* posts across fine-grained topics

5.2 CS Course

Here, we present topic evolution results for the CS course. In all, we analyze data
from 15 course offerings, 3 iterations of the original course and 6 iterations each
of CS-1 and CS-2, respectively. We coalesce CS-1 and CS-2 for each iteration
and treat that as a single course, giving us *nine* iterations of the CS course.

Primary Purpose of Forums. Figure 3(a) gives the number of posts in the
social, issue, and technical topics across iterations in the CS course. For the
technical topic, we add the values in the document-topic multinomial distribution given by seeded LDA across CS technical seeded topics and present their
combined evolution over time. We notice a different trend in the CS course when
compared to the BUSINESS course. While social posts dominate the forums in the
initial iterations, they slowly decline from the 4th iteration. Technical and issue
posts primarily dominate the forums from the 5th iteration, with issue posts
being more predominant when compared to technical posts.

Figure 3(b) gives the percentage of social and issue posts in the forums. We
again observe that as iterations unfold, the percentage of technical posts remains
the same, while there is a marked increase in issue posts. Percentage of issue
posts peaks in the 5th iteration and declines thereafter, but still remains higher
than social and technical posts. It is also interesting to note that this increase
happens immediately after the course splits (4th iteration). Intuitively, as courses
stabilize, it is expected that issues reported in the previous iterations are fixed
causing issue posts to decrease over time. But in the CS course, we observe
the opposite, which calls for a detailed analysis of why issue posts exhibit an
increasing trend and what kind of issues are being reported by students. Another
interesting trend to note is that in Fig. 3(c), we observe that higher percentage
of issue posts come from a fewer number of students when compared to the
technical posts.

Issues reported in the CS course vary significantly in comparison to issues in
BUSINESS course. Computer science courses often have software installation
prerequisites that could potentially trigger a large number of posts around errors
in installing/compiling software. Unlike logistic issues, these issues are inherently

different in nature and in most cases cannot be easily fixed by the instructor, especially in an online setting.

Course Elements. In our next set of experiments, we analyze the evolution of topics corresponding to the three course elements. We add the *technical* topic to this classification for readability, as we will be drilling deeper into the technical issues along with issues in course elements in Sect. 5.2.

Figure 4(a) gives the evolution of course elements as iterations unfold. Technical topic attracts the most number of posts in the CS course across all iterations. Concentrating on the zoomed in portion of Fig. 4(a), we observe that among the three course elements, quizzes emerge as the most dominating course element in the CS course. While lectures are the most sought after course element in the BUSINESS course, they rank second in popularity in the CS course, and this is followed by certificate. This analysis helps instructors prioritize and focus limited resources on course elements that students care about the most.

Fine-Grained Analysis of Issue Posts. Next, we investigate how issues are distributed across the course elements. We add the *technical* topic category to the list of course elements to model the evolution of technical issue posts. Figure 4(b) gives the evolution of issues across lecture, quiz, certificate, and technical topics. We find that technical issues dominate issue posts across all iterations, followed by quiz related issues. At iteration 5, where we observe an overall increase in issues in Fig. 3, we observe a similar spike in the quiz and technical issue posts in Fig. 4(b) as well. A plausible reason for this increase is that when the course splits into two courses in the 4th iteration, more programming assignments were added, which led to more technical and quiz issues to be reported. Lecture and certificate topics hardly contribute to the issue posts and decline to small numbers as iterations progress.

We further break down issues across lecture and quiz sub-topics in Fig. 4. Figure 4(c) gives the distribution of issue posts across lecture sub-topics: video/audio and subtitles. As we observed in Fig. 4(b), there are only a few lecture issue posts in each iteration and this reflects in the finer analysis as well. Between the lecture subtopics, video/audio is the most contributing sub-category. Performing a similar analysis on quiz sub-topics (Fig. 4(d)), we find that most of the quiz issue posts fall under the submission category, which is followed by grading, and deadlines. While grading consistently remains a contributing issue category across both the courses, we note that the structure of CS course requires submitting computer programs in an online platform inciting a significant number of issue posts in the submission category. Isolating logistic issues in the CS course and comparing this to Fig. 1(a), we find they follow a similar pattern to the BUSINESS course, declining over time. This supports our hypothesis that as courses stabilize lesser logistic issues surface and hence they are reported less in the forums.

6   Conclusion

In this work, we presented a detailed temporal analysis of two long-running MOOCs from different disciplines and identified the similarities and differences between them. Our methodology and analysis is helpful in determining the stability of these courses and identifying opportunities for improvement. The weakly supervised nature of our approach using a common set of seed words is beneficial in extending our models with minimal effort to analyze newer courses. There are several exciting future research directions. The temporal analysis can potentially be integrated with an automatic feedback mechanism to actively monitor student feedback. This can be especially helpful when large course changes (such as splitting the course, or changing the grading methodology) are deployed. This feedback mechanism can help instructors get notified of abrupt changes in the forums and allow them to address concerns promptly, thus helping in improving students' interaction experience.

References

1. Agrawal, A., Venkatraman, J., Leonard, S., Paepcke, A.: YouEDU: addressing confusion in MOOC discussion forums by recommending instructional video clips. In: Proceedings of the International Conference on Educational Data Mining (EDM) (2015)
2. Anderson, A., Huttenlocher, D., Kleinberg, J., Leskovec, J.: Engaging with massive online courses. In: Proceedings of the International Conference on World Wide Web (WWW) (2014)
3. Blei, D.M., Lafferty, J.D.: Dynamic topic models. In: Proceedings of the International Conference on Machine Learning (ICML) (2006)
4. Chaturvedi, S., Goldwasser, D., Daumé III, H.: Predicting instructor's intervention in MOOC forums. In: Proceedings of the Annual Meeting of the Association for Computational Linguistics (ACL) (2014)
5. Cui, Y., Wise, A.F.: Identifying content-related threads in MOOC discussion forums. In: Proceedings of the ACM Conference on Learning @ Scale (L@S) (2015)
6. Ezen-Can, A., Boyer, K.E., Kellogg, S., Booth, S.: Unsupervised modeling for understanding MOOC discussion forums: a learning analytics approach. In: Proceedings of the International Conference on Learning Analytics And Knowledge (LAK) (2015)
7. Gohr, A., Hinneburg, A., Schult, R., Spiliopoulou, M.: Topic evolution in a stream of documents. In: Proceedings of the SIAM International Conference on Data Mining (SDM) (2009)
8. Hall, D., Jurafsky, D., Manning, C.D.: Studying the history of ideas using topic models. In: Proceedings of the Conference on Empirical Methods in Natural Language Processing (EMNLP) (2008)
9. Jagarlamudi, J., Daumé III, H., Udupa, R.: Incorporating lexical priors into topic models. In: Proceedings of the European Chapter of the Association for Computational Linguistics (EACL) (2012)
10. Ramesh, A., Goldwasser, D., Huang, B., Daume, H., Getoor, L.: Understanding MOOC discussion forums using seeded LDA. In: 9th ACL Workshop on Innovative Use of NLP for Building Educational Applications. ACL (2014)

11. Ramesh, A., Kumar, S.H., Foulds, J., Getoor, L.: Weakly supervised models of aspect-sentiment for online course discussion forums. In: Proceedings of the Annual Meeting of the Association for Computational Linguistics (ACL) (2015)
12. Wang, X., McCallum, A.: Topics over time: a non-Markov continuous-time model of topical trends. In: Proceedings of the ACM SIGKDD International Conference on Knowledge Discovery and Data Mining (KDD) (2006)
13. Wong, J.-S., Pursel, B., Divinsky, A., Jansen, B.J.: An analysis of MOOC discussion forum interactions from the most active users. In: Agarwal, N., Xu, K., Osgood, N. (eds.) SBP 2015. LNCS, vol. 9021, pp. 452–457. Springer, Cham (2015). https://doi.org/10.1007/978-3-319-16268-3_58

Neuroscientific User Models: The Source of Uncertain User Feedback and Potentials for Improving Web Personalisation

Kevin Jasberg[2(✉)] and Sergej Sizov[1]

[1] Web Science Group, University of Duesseldorf,
Universitaetsstr. 1, 40225 Duesseldorf, Germany
sizov@hhu.de
[2] Computational Linguistics, University of Duesseldorf,
Universitaetsstr. 1, 40225 Duesseldorf, Germany
kevin.jasberg@hhu.de
https://jasbergk.wixsite.com/research

Abstract. Recent research revealed a considerable lack of reliability for user feedback when interacting with personalisation engines, often denoted as user noise, user variability or human uncertainty. Whenever research on this topic is done, there is a very strong system-centric view in which user variation is something undesirable and should be modelled with the eye to eliminate. However, the possibilities of extracting additional information were only insufficiently considered so far. In this contribution we consider the neuroscientific theory of the Bayesian brain in order to develop novel user models with the power of turning user variability into additional information for improving web personalisation. This will be exemplified by means of standard collaborative filtering.

Keywords: User noise · Human uncertainty · Collaborative filtering
User models · Bayesian brain · Probabilistic population codes

1 Introduction

Personalisation and recommendation have become indispensable in most systems nowadays and the trend still continues to grow in that direction. During the last decade, the growth of interactions continuously supported innovations in a data-driven fashion. This is advantageous as we need to understand a user along with his preferences, peculiarities and behaviour to adapt recommendation and personalisation in order to provide an appealing user experience. This is done by inventive user models and by injecting information into modern personalisation engines based on techniques of machine learning, but the bedrock of such efforts is a thorough knowledge about the user, either by observation (implicit knowledge) or by questioning (explicit knowledge).

© Springer Nature Switzerland AG 2018
H. Hacid et al. (Eds.): WISE 2018, LNCS 11234, pp. 422–437, 2018.
https://doi.org/10.1007/978-3-030-02925-8_30

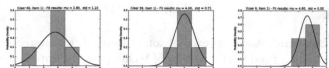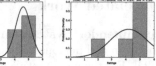

Fig. 1. Visualisation of uncertain user responses for a repeated feedback task.

The strong dependence on user-generated data is curse and blessing at the same time, because the fundamental problem with user feedback is its uncertainty. This means that a considerable fraction of users behave differently or decide otherwise if a task or decision-making has to be repeated within the same context. This phenomenon is often denoted as user noise, user variability or human uncertainty. All these terms refer to the same concept which gives user data the nature of a random variable. As an introductory example we consider the repeated rating of film trailers in a short temporal interval (granting same emotional and cognitive context) with a sufficient number of distractors between rating repetitions. Figure 1 shows the different ratings of four users of this experiment, which will be described in more detail in forthcoming sections. It becomes clear that this user feedback is scattering around a central tendency and hence supports the assumption of a random variable. This feature has recently been in the focus of some research that presented both, induced problems in the evaluation of personalisation approaches as well as attempts for possible solution strategies. The commonality of all these contributions in the field of user modelling, adaptation and personalisation is that there is a very strong system-centric view in which user variation is something undesirable and should be modelled with the eye to eliminate. In this contribution, we want to introduce a new and diametrically opposed paradigm in which we consider uncertainty no more as a mistake or dysfunction with destructive side effects, but rather as an opportunity for gathering additional information. Such an undertaking sensibly has to start with the measurement of user feedback along with its uncertainty and by transmission of this data to a user model, which maps uncertainty into an information space in order to successively supplement a user profile. Since the measurement of user noise or human uncertainty has already been a subject of research, we confine ourselves to the development of a novel user model with special sensitivity for this uncertainty. This naturally leads to the following research questions:

1. What does a possible model look like that considers human decision variability and maps it into the highest possible concentration of additional information?
2. How can this information be integrated into existing recommender systems and personalisation engines?
3. What are the final benefits of this particular user model and what are the benefits of this novel paradigm in general?

2   Related Work

Recommender Systems and Assessment. A lot of research about recommendation and personalisation produced a variety of techniques and approaches [13,24]. For the comparative assessment, different metrics are used to determine the prediction accuracy, such as the root mean squared error (RMSE), the mean absolute error (MAE), along with many others [4,11]. In our contribution, we internalise existing criticism about a lack of understanding human beings in the process of system design [19,22] and develop a user model that is close to the current way of looking at the functionality of the human brain.

Dealing with Uncertainties. The relevance of our contribution arises from the fact that the unavoidable human uncertainty sometimes has a vast influence on the evaluation of different prediction algorithms [1,27]. The idea of uncertainty is not only related to recommender systems but also to measuring sciences such as metrology. Recently, a paradigm shift was initiated on the basis of a so far incomplete theory of error [5,10]. In consequence, measured properties are currently modelled by probability density functions and quantities calculated therefrom are then assigned a distribution by means of a convolution of their argument densities. This model is described in [17]. We transfer this perspective to user feedback by considering it as a single draw from an underlying distribution. This provides us with a probabilistic reference which we can use to verify the predictions based on our own user model.

The Idea of Human Uncertainty. The idea of underlying distributions for user feedback is not far-fetched since the complexity of human perception and cognition can successfully be addressed by means of latent distributions [6]. We adopt the idea of modelling user uncertainty by means of individual Gaussians for constructing our individual response models and thus follow the argumentation in latest research of neuroscience and metrology [18,23]. Probabilistic modelling of cognition processes is also quite common to the field of computational neuroscience. In particular, aspects of human decision-making can be stated as problems of probabilistic inference [9] (often referred to as the "Bayesian Brain" paradigm). There is evidence that populations of neurons provides probability densities over possible unknown states of the world and thus accounts for their uncertainty [23]. However, these estimations slightly differ in each cognition trial due to the volatile concentration of released neurotransmitters, impacting the spiking habits of downstream neurons (neural noise) [8,23]. In other words, human decisions can be seen as uncertain quantities by nature of the underlying cognition mechanisms. In this paper, we adopt the theory of noisy probabilistic population codes (nPPCs) and use them to construct a user model that can naturally represent and explain response uncertainty with neural noise, mapping uncertainty to neural parameters.

Human Uncertainty in Computer Science. User noise or human uncertainty has been mentioned before in computer science. The first reference in the context of user feedback came along with a study where reliability problems have been registered for repeated ratings [12]. The authors, like [11] later on, have already speculated on its impact on personalisation accuracy. This assumption was later confirmed when uncertainty in user ratings (measured by re-rating) and its impact on the RMSE was demonstrated [1, 2]. A more sophisticated analysis is provided by [25], where it could be demonstrated that human uncertainty leads to an offset in a specific metric (magic barrier). This approach was later expanded by [15] and it was shown that this barrier has some uncertainty itself. Additional contributions also revealed that each accuracy metric considering human feedback is naturally biased and that possible rankings built upon these metric scores are subject to probabilities of error [16].

To solve this problem, some strategies have been proposed over the years, just like a pre-processing step that deletes highly deviant values and replaces them by artificial values closer to the mean of a re-rating [2]. Another approach is to provide model-based predictions with uncertainty as well, so that the uncertainties of a rating and a predictor eliminate each other when calculating their difference [20]. Yet another possibility is to compute metrics only with deviations that differ widely from a given predictor and hence can not be explained by human uncertainty [14]. With our contribution, we want to move away from the paradigm of extinction and present a way in which uncertainty can be sensibly used to generate benefits.

3   Theory, Models and Applications

Before we build neurological models, we first need to understand how the human brain processes information and where uncertainty arises in the cognitive process. We will then convert these mechanisms into an adequate user model.

The Single Neuron Model

The response of a single neuron to a stimulus is limited to transmission of electric impulses (spiking) and since each neuron has only got two states of activation, theories of neural coding assume that information is encoded by the spiking frequency (rate) [7]. The functional relationship between responses r of a neuron and the characteristics $s \in S \subset \mathbb{R}$ of a stimulus is given by the so-called tuning curve $r = f(s)$. Besides irregular shapes, tuning curves have frequently been measured to be bell-shaped or sigmoid-shaped respectively. Each tuning curve maximises for a particular value $p := \operatorname{argmax} f$, denoted as the preferred stimulus. For bell-shaped tuning curves, $f : S \to \mathbb{R}$ can be modelled as $f_p(s) := g \cdot h(p, w)(s) + o$, where the shape emerges from the Gaussian density function h with mean p and standard deviation $w \in \mathbb{R}^{>0}$ (tuning curve width). The additional components $g \in \mathbb{R}^{>0}$ and $o \in \mathbb{R}^{>0}$ represent a frequency gain and offset respectively. When measuring tuning curves in reality, one will find that they are somewhat noisy and that even one and the same stimulus never leads to the same response. This fluctuations can be explained by the so-called neural noise [8]. Neuronal responses must therefore be seen as random variables R rather than fixed values determined by tuning curves. It has been found that $R \sim \operatorname{Poi}(\lambda)$ follows a Poisson distribution with expectation $\lambda = f(s)$ [23, 26].

Probabilistic Population Codes

We now consider a population of n neurons, all with the same tuning curve type with (almost) the same neural parameters. The only difference is in the preferred values p_j, which are equidistantly spread across the range of the estimation scale. All parameters determining the population size n, the shape of all tuning curves as well as the assumed stimulus s are summarised in a vector $\xi = (n, g, w, o, s)$ which we will refer to as the cognition vector in the following.

Given a particular fixed s (which is formed from unknown underlying cognitions), each neuron of this population will respond according to its specific tuning curve and interference due to neural noise. Therefore, a response r_j of the j-th neuron must be seen as a realisation of the random variable $R_j \sim \text{Poi}(f_{p_j}(s))$. In order to keep in mind, that these responses are always dependent on the parameters of the cognition vector, we henceforth use the notation $r_j(\xi)$ as realisation of $R_j(\xi)$. The response of the entire population is formed by the response of each neuron and so we denote the n-dimensional random variable $\mathcal{R}(\xi) := (R_1(\xi), \dots, R_n(\xi))$ as the population response for a given ξ with realisation $\varrho(\xi) = (r_1(\xi), \dots, r_n(\xi))$.

This theory of the origin of noisy population responses is illustrated in Fig. 2. In this example, we used $\xi = (11, 10, 0.5, 5, 3)$ as the cognition vector, i.e. we consider $n = 11$ neurons that respond to the assumed stimulus (in this case: cognition result) of $s = 3$ stars where each tuning curve has the offset $o = 5\,\text{Hz}$, the width $w = 1\,\text{Hz}$ and the gain $g = 7$. In the left picture we can see the individual tuning curves, which are distributed equidistantly over the possible range of a rating scale with five stars. For $s = 3$ stars, the responses of each neuron can be fetched from its tuning curve. For a better representation of the population response, it has become a standard to plot the individual responses against the corresponding preferred values, which can be seen in the middle picture. These are the theoretical (static) responses without consideration of neural noise. To add this neural noise, each static response $r_j^{\text{static}}(\xi)$ is replaced by the draw of a random number from the Poisson distribution with parameter $\lambda = r_j^{\text{static}}(\xi)$. This can be seen in the right subfigure. We additionally repeated this sampling once, i.e. the blue and red dots in each case represent a noisy population response and it is obvious that these population responses differ not only from the theoretical reference but also very much from each other. At this point, we see that the same cognition leads to different neural activities on each pass, and that the estimation of a quantity (e.g. product rating) or state of the world is thereby given a natural uncertainty.

Decoder Functions

What we have learned so far is the internal basic cognitive model that allows different neuronal activity for a population of neurons to encode one and the same state of the world. By means of sensory perception, this model can be seen as the a translation of outside reality into inside representation of the external world. By means of cognition, however, this model provides the translation from a cognition black box into internal and measurable representations of thoughts and thinking patterns.

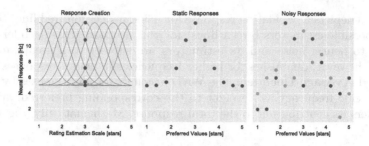

Fig. 2. Genesis of noisy population responses demonstrating the alteration for each cognition trial (red and blue). (Color figure online)

The main question that arises at this point is: How does the human brain translate population activity into estimations for a state of the world or a cognition respectively. Theories assume the use of so-called decoder functions. Mathematically, a decoder function is a mapping $\varphi \colon \mathbb{R}^n \to S$ from population activity onto the estimation scale for a stimulus or cognition. This means that for a particular user-item-pair (u, i), we can obtain an estimation of a single feedback submission directly from the realisation of a population response, i.e. $f_{u,i} = \varphi(\varrho(\xi))$. Hence, noisy user feedback $\mathcal{F}_{u,i}$ can be represented as a random variable given as $\mathcal{F}_{u,i} = (\varphi \circ \mathcal{R})(\xi)$. In neuroscience literature, there are several decoders that have been suggested and frequently used so far [21]. We will give a brief overview of the most frequently discussed decoder functions and will relate them directly to the context of user feedback.

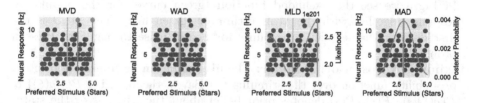

Fig. 3. Visualisation of decoder functions on a population response for a given $\xi = (100, 1, 1, 5, 3)$. The red and green lines show the true and the decoded stimulus. (Color figure online)

Mode Value Decoder: Due to the construction of tuning curves, the MVD assumes that it is exactly the neuron with maximum spiking frequency that is most likely to be addressed by the stimulus or the state of the world. The decoder function is thus given as

$$\varphi_{MVD} \colon \varrho(\xi) \mapsto \underset{p_j \in S}{\operatorname{argmax}} \{r_1(\xi), \ldots, r_n(\xi)\}. \tag{1}$$

Figure 3 depicts a population response for a 3-star-decision (red line) together with possible estimators (green lines) for this decision. This decoder is very prone to neural noise and its estimators are subject to a great ambiguity which, however, diminishes for higher frequencies in neural responses.

Weighted Average Decoder: The WAD accounts for all responses by setting the specific frequency as a weight to the corresponding preferred value and considers its contribution to the total response. Mathematically, the WAD is given by

$$\varphi_{WAD}: \varrho(\xi) \mapsto \frac{\sum_{j=1}^{n} r_j(\xi) \cdot p_j}{\sum_{j=1}^{n} r_j(\xi)}. \tag{2}$$

As to see in Fig. 3, this decoder function does not produce ambiguous estimators and is very stable against neural noise.

Maximum Likelihood Decoder: For a given population response, the MLD chooses the estimator $\hat{s}$ with a view to maximise the corresponding likelihood function:

$$\varphi_{MLD}: \varrho(\xi) \mapsto \underset{s \in S}{\mathrm{argmax}}\ P(\varrho(\xi)|s), \tag{3}$$

where the likelihood itself is given by the i.i.d.-assumption together with the Poisson probability mass function

$$P(\varrho(\xi)|s) = P(r_1(\xi), \ldots, r_1(\xi)|s) = \prod_{j=1}^{n} \frac{f_{p_j}(s)^{r_j(\xi)}}{r_j(\xi)!} \exp\left(-f_{p_j}(s)\right). \tag{4}$$

In Fig. 3 we see the likelihood function (green curve) for the population response together with the MLE estimator (green line). The MLD is the first decoder that explicitly accounts for neural noise through the Poisson probability.

Maximum A Posteriori Decoder: The likelihood can be transformed into a probability function over the stimulus via Bayes' theorem, i.e. $P(s|\varrho(\xi)) \propto P(\varrho(\xi)|\hat{s}) \cdot P(s)$. $P(s)$ denotes prior belief about the stimulus or the states of world that has been learned through former experiences. The estimator is then chosen so that this posterior is maximised, i.e.

$$\varphi_{MAD}: \varrho(\xi) \mapsto \underset{s \in S}{\mathrm{argmax}}\ P(s|\varrho(\xi)) \tag{5}$$

The MAD is much like the MLD but with less variability since the prior works as a stabiliser. In the example of Fig. 3 we arbitrarily used a Gaussian with $\mu = 3$ and $\sigma^2 = 0.75$ as prior belief. The Bayesian brain theory assumes a prominent role of this decoder function, since each population would then naturally represent a probability density over a stimulus or state of the world which can easily be aggregated with other populations' densities by mere addition. For multiple sensory inputs, this decoder function was proven to be a plausible description for the brain's operating principles [3].

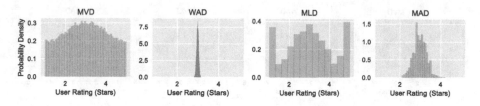

Fig. 4. Feedback distributions by different decoder functions for $\xi = (100, 1, 1, 5, 3)$.

Theoretic Model Properties

As already mentioned, this modelling can explain the genesis of uncertain user feedback $\mathcal{F}_{u,i}$. For the purpose of exemplification, we have computed the resulting feedback distributions for all introduced decoder functions for the cognition vector $\xi = (100, 1, 1, 5, 3)$. The results are depicted in Fig. 4. Already here, certain properties of this model are clearly visible. For the MVD, the vulnerability for neural noise is quite obvious since the corresponding feedback distribution got the largest spread. Even at the boundaries of 1 and 5 stars, there are still high probabilities, hence this distribution is only slightly more informative than a uniform distribution. Using the Bayesian definition of probability (which is interpreted as ones personal confidence), such a user feedback would be provided by users who are not sure about which rating seems appropriate. For the WAD, we notice the robustness to neural noise and the quality of estimation. A user which would utilise this decoder function would surely give constant ratings. Conversely, users with larger uncertainties can probably not be modelled by this decoder. The MLD reveals a remarkable property. Due to the small size of the rating scale $S = [1, 5]$, the likelihood's maximum frequently coincides with the scale boundaries. Therefore, this theory might explain the common user behaviour of giving preference to these boundary ratings. At first glance, the MAD provides the most plausible feedback distributions which seems to strengthen the Bayesian brain theory.

A more thorough analysis of the decoding quality is depicted in Fig. 5. By repeated cognition (population response), computed estimators $\hat{s}$ can be compared with the true stimulus s by means of fractions of the maximum mean squared error (MSE). In this case, the MSE has to be divided by its maximum, because a change of s naturally changes the limits of the MSE which biases analyses (e.g. for $s = 3$ the MSE can only be ≤ 4, but for $s = 1$ the MSE can be up to 16). For all decoders, we see that the estimation quality increases with neural frequency, i.e. the more active the population, the better a cognition can be translated into a numerical estimate.

For the MVD, lower frequencies evoke that the middle of a scale as well as its margins can be estimated slightly worse than the rest. For higher frequencies, it is only the middle of a scale that can be estimated slightly worse. This would inevitably lead to more uncertainty for these values if a rating task is repeated. This decoder thus explains the effect, that margin ratings are much more reliable. For the WAD, we can see the opposite effect. This decoder is suitable for users

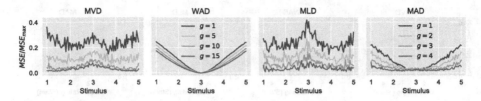

Fig. 5. Reliability analysis – comparing s (input) with possible estimations $\hat{s}$ (output) by means of MSE/MSE_{max} in dependency of the stimulus and tuning curve gain g.

who give reliable ratings for the middle of a scale. Both decoder functions need high frequencies ($g \geq 10$) in order to work with high quality. In contrast, the MAD and the MLD are capable of forming the same quality profile with lower frequencies. This basically means a lower neural energy consumption for a brain while maintaining full functionality (evolutionary advantages). Moreover, the MAD is the only decoder function forming a variety of quality profiles, for which one would otherwise need multiple decoders. Evolutionarily, it is much more reasonable to develop a single mechanism that can be used in all situations than to develop different mechanisms for this task. These arguments can therefore be seen as another indication for the applicability of the Bayesian brain paradigm. This also means that the MAD is again the best candidate for a neuroscientific user model, which is in line with the previous discussion of feedback distributions.

Neuroscientific User Model

The goal of this user model is to find a specific cognition vector $\xi_{u,i}$ for each user-item-pair (u, i) along with a decoder function φ, so that the model-based feedback $\hat{\mathcal{F}}_{u,i}$ minimises the difference to the real user feedback $\mathcal{F}_{u,i}$ by means of an arbitrary disparity metric d. Mathematically, our user model is given by $\mathcal{F}_{u,i} \equiv (\xi_{u,i}, \varphi)$ with

$$(\xi_{u,i}, \varphi) := \underset{(\xi,\varphi)}{\mathrm{argmin}} \ d\big(\mathcal{F}_{u,i}, \hat{\mathcal{F}}_{u,i}\big) = \underset{(\xi,\varphi)}{\mathrm{argmin}} \ d\big(\mathcal{F}_{u,i}, (\varphi \circ \mathcal{R})(\xi_{u,i})\big). \qquad (6)$$

In the case of ambiguity, that is, when several different cognition vectors lead to the same minimum of d, we will select the vector $\xi = (n, g, w, o, s)$ that minimises the population energy $E \propto n \cdot (g + o)$. This reasoning arises from the fact that a human brain always has to work in an energy-efficient manner and thus is most likely to use the cognition vector, in which all n neurons spike as sparsely as possible. The advantage of this model is that each user-item-pair can be mapped into a high-dimensional space that theoretically carries much more information than the consideration of product ratings does alone.

4   Evaluation and Results

The theory of nPPCs has often been confirmed in the context of sensory perception and in our case it can theoretically explain human uncertainty. In this section, we systematically evaluate this ability (goodness of fit) and discuss how content personalisation may benefit from this theory.

Measuring Human Uncertainty (User Study)

We conducted the RETRAIN (Reliability Trailer Rating) study as an online experiment in which 67 participants had watched theatrical trailers of popular movies and television shows and provided ratings in five consecutive repetition trials. User ratings have been recorded for five of ten trailers so that the remaining ones act as distractors, triggering the misinformation effect, i.e. memory is becoming less accurate due to interference from post-event information. The so obtained data set comprises $N = 1\,675$ individual ratings. As mentioned before, we discovered that user responses scattered around a central tendency rather than being constant. From all user ratings, only 35% manifested a consistent response behaviour, while 50% gave two different responses on the same item, and 15% used even three or more different ratings. A detailed breakdown can be found in Fig. 6a. The human uncertainty itself is thereby exponentially distributed as to see in Fig. 6b. In the following, we use this data record to fit individual feedback distributions from all ratings that a user has given to the same item. These will then be compared with our model-based distributions.

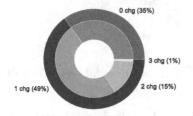

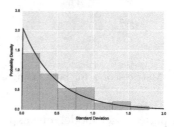

(a) Changes in rating behaviour.   (b) Distribution of rating variances.

Fig. 6. Visualisation of human uncertainty found in the RETRAIN user study.

User Modelling Quality

To assign each user-item-pair its own cognition vector and decoder, we compute the model-based feedback $\hat{\mathcal{F}}_{u,i} = (\varphi \circ \mathcal{R})(\xi)$ for each of the four decoder functions φ and for each cognition vector $\xi \in N \times G \times W \times O \times S$, where each set

$$N := \{1, \dots, 250\} \quad G := \{1, \dots, 100\} \quad W := \{0.1, \dots, 2.0\}$$
$$O := \{1, \dots, 15\} \quad \quad S := \{1, \dots, 5\}$$

contains 100 equidistantly distributed values. Altogether, there are $4 \cdot 10^9$ combinations to be examined brute force. Subsequently, each $\hat{\mathcal{F}}_{u,i}$ will be compared to the real user feedback $\mathcal{F}_{u,i}$ by means of Eq. 6. In doing so, we use two different metrics d, one for a discrete evaluation (close to the original data) and another for a continuous evaluation (more accurate, but on basis of assumptions):

Cohen's Kappa: This metric is intended to evaluate inter-rater-reliability and compares the concurrence of two independent classifications with the probability of reaching this agreement by random guessing. This metric is given by the equation $\kappa = (p_0 - p_c)/(1 - p_c)$, where p_0 is the relative agreement of both raters and p_c denotes the chance of a random agreement.

Jensen-Shannon divergence: We yield probability distributions P_{model} and P_{real} by discretisation of the model-based density as well as for the real user feedback. For those we compute the Jensen-Shannon-Divergence (JSD)

$$\text{JSD}(P_{model}|P_{real}) = \frac{1}{2}D_{KL}(P_{model}|M) + \frac{1}{2}D_{KL}(P_{real}|M), \quad (7)$$

where $D_{KL}(P_1|P_2) = \sum_i P_1(i) \log_2(P_1(i)/P_2(i))$ denotes the Kullback-Leibler-Divergence and $M = \frac{1}{2}(P_{model} + P_{real})$.

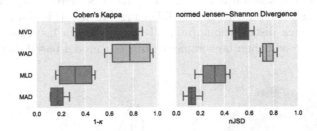

Fig. 7. Metric scores for best fitting cognition vectors from all user-item-pairs

For a perfect user model, one expects that only a single combination of cognition vector and decoder will make the disparity metric d vanish and that all other combinations will maximize d. Therefore, we will not only consider the metric scores themselves, but also their ambiguity. The results show that each decoder function is able to fit constant users when using a sufficiently high frequency gain or sufficiently small tuning curve widths. The average ambiguity is 300, i.e. for about 300 cognition vectors we yield the same minimal metric score. It becomes clear that the strength of our neuroscientific user models is clearly in the modelling of human uncertainty. Therefore, in the following analyses, we will only consider those users who had provided unreliable feedback.

For noisy users, the mean ambiguity is 5, i.e. only five out of 10^9 cognition vectors lead to the same metric minimisation. In Fig. 7 we can see the distribution of metric scores for best fitting cognition vectors. The best decoders are the MLD and the MAD. It is also noteworthy that there are major overlaps when using this metric. This means that there is ambiguity for the decoder function as well. For example, a large proportion of scores for the MLD can also be achieved by the MVD. Moreover, considering the whiskers of the MAD, half of the scores can also be formed by the MLD. Nonetheless, first rankings in model quality can be anticipated. For the normed Jensen-Shannon divergence, this ranking can be verified. In addition, we can notice the increased amount of information when using distributions rather than samples with five draws. So, the scattering of metric scores is much smaller. In summary, the maximum a posteriori decoder can be mentioned as the best decoder function leading to feedback distributions modelling reality with high quality. Therefore, we will explicitly focus on this decoder function for further elaborations.

Information Extraction

Finally, we discuss how personalisation approaches benefit from this new information, provided by a high-dimensional neural space. In doing so, we consider collaborative filtering (CF) in its simplest form: User-item-pairs with corresponding ratings are clustered into user groups in order to recommend new products on the basis of group popularity. To compute a reference for further comparisons, we use a simple k-means approach to find clusters within the samples

$$\text{Sample}_t = \{(u, i, \text{rating}_t) : u = 1, \ldots, 67, i = 1, \ldots, 5\} \tag{8}$$

separately for each rating trial $t = 1, \ldots, 5$. In each sample, we randomly select 30% of the users in each cluster group to delete their ratings for the fifths item (testing-users). We use the mean rating from the remaining 70% of users (learning-users) within each group as the specific group predictor for item 5. This predictor is then compared to the original prediction of the testing-users by means of the RMSE. In this way, we get an RMSE score for each rating trial, and if we repeat the randomised selection of testing-users, we get various scores that form a distribution. This approach will be referred to as *noiseless reference*.

Since this approach does not consider any uncertainty information, we need a second reference to fill this gap. For this purpose, we primarily proceed like above. The only difference is that we execute clustering on the union $\cup_{t=1}^{5}\text{Sample}_t$, i.e. we allow copies of user-item-pairs but with different ratings. Our cluster groups will therefore be much larger and means (predictions) more accurate. Additionally, we do not compare the predictions with ratings of a particular trial, but with the mean rating aggregated from all rating trials. This stochastic approach will be referred to as *noisy reference*.

In contrast, we introduce the following methods, which are based on the additional information of the nPPC user model:

ξ-**Clustering:** We associate $\xi_{u,i}$ to each user-item-pair and use k-means on the neural space $\{(u, i, n, g, w, o, s)\}$. We then proceed with selecting testing-users and learning-users, just as for the references above. Due to the higher dimensional space, user groups may be much more differentiated and more appropriate for testing-user predictions.

Subspace-Clustering: Here, we associate $\xi_{u,i}$ to each user-item-pair and use k-means on the neural subspaces $\{(u, i, n)\}$ (denoted as n-Clustering), $\{(u, i, g)\}$ (g-Clustering), $\{(u, i, w)\}$ (w-Clustering), $\{(u, i, o)\}$ (o-Clustering). We then proceed as above.

Noise-Profiling: We associate $\xi_{u,i}$ to each user-item-pair and aggregate by users to yield sets $S_u := \{\xi_{u,i} : i = 1, \ldots, 4\}$ in which we consider only the first four items. We simply calculate the mean cognition vector $\bar{\bar{\xi}}_{u,i} = (\bar{n}, \bar{g}, \bar{w}, \bar{o}, s)$ for each S_u, where s is left arbitrary. Subsequently, we chose s so that the variance of the model-based feedback distribution $(\varphi \circ \mathcal{R})(\bar{\bar{\xi}}_{u,i})$ is as close as possible to the user's average variance gathered from the rating distributions of the remaining four items.

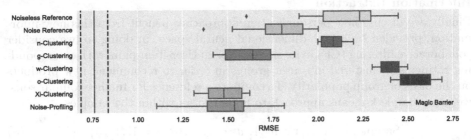

Fig. 8. Feedback distributions from different decoder functions for $\xi^{\mathcal{B}} = (100, 1, 1, 5, 3)$

The results are depicted in Fig. 8. First of all, it has to be noted that the variances of the RMSE distributions are relatively large, which is due to the size of our data record. As a visualisation of the RMSE's offset (which emerges for uncertain user data), we additionally calculated the magic barrier as proposed by [25] together with its 95%-confidence interval. We can see that the noisy reference operates much better than the noiseless reference. Moreover, we see that the w-clustering and the o-clustering behave much worse than both references. This can be explained by the fact that clustering according to user ratings for predicting other ratings can be regarded as sensible since there is a causality. In contrast, the tuning curve width as well as the offset are not causally related to the user ratings. Actually, one would expect the same for the n-clustering and g-clustering respectively. However, the n-clustering performs a little better than the noiseless reference, although both distributions have a complete intersection. The results for the g-clustering is quite surprising since it outperforms the noisy reference. We explain this by a latent causal dependency between a particular rating and neural frequency. As previously mentioned, information is primarily encoded in terms of frequencies within the human brain. Therefore, frequencies might encode ratings and uncertainty simultaneously. For the ξ-clustering as well as for the noise-profiling we can certify an excellent performance result. However, there are some overlaps between all these distributions. For example, the left whisker of the noisy reference reaches the third quartile of the noise-profiling approach. Hence, the noise-profiling does perform doubtlessly better for only 75% of the data whereas the superiority for the other 25% is associated with a certain doubt. Nevertheless, the success of the neuroscientific user models against this stochastic uncertainty model is quite clear, although one should also consider that we have only investigated a very simple approach of collaborative filtering. A focused investigation of more complicated and more sophisticated techniques is therefore needed and will be done in future research.

5 Discussion

In this contribution we have broken with the view that user noise or human uncertainty is something undesirable that only causes trouble in the evaluation of personalisation engines. We explicitly permitted this human property and developed a user model using noisy probabilistic population codes (nPPCs) to reveal and exploit the inherent information. For this purpose, we formulated three research questions at the beginning.

The first question was about how a possible user model could look like that takes into account human uncertainty. For this we consider a population of neurons whose noisy tuning curves are equidistantly allocated over an estimation scale (e.g. rating scale). These tuning curves can be adjusted by various parameters, which we represent in a so-called cognition vector. By this preliminary fixing, the population provides an unreliable response to a stimulus (e.g. a choice of a particular user rating), which can be converted into a real answer through decoder functions. By means of two disparity metrics we can find a cognition vector together with decoder function for each user-item-pair so that measured feedback distributions can be reproduced.

The second research question focused on possible solutions for making the information available to personalisation approaches. For this we have chosen the example of collaborative filtering and introduced three approaches for injecting information by the generation of different underlying data spaces. Even simple clustering algorithms can benefit from these higher-dimensional vector spaces rather than clustering for ratings only. The first results are very promising, i.e. user groups were found to comprise users with a more similar behaviour than yielded for the standard clustering. One might argue that this could have also been achieved by a statistical model. However we revealed that the neuroscientific user models outperform a mere statistical model for representing uncertainty. So these models offer many possibilities for future web engineering: We can use the information about human uncertainty to make predictions more stable and robust against statistical outliers. This would make systems for recommendation and personalisation less susceptible to bad draws of a user's inner feedback distribution.

The third research question referred to the possible benefits of this novel paradigm in general. Every personalisation engine and every recommender system has the goal of being able to map the human being as accurately as possible. The knowledge about the nature of human behaviour, together with a human's peculiarities, is hence crucial. The theory of nPPCs is currently a much-debated theory and is considered by many neuroscientists to be an adequate model of human decision-making which is very close to real structures. The Bayesian brain paradigm is always seen in a prominent role and has been verified many times in neurological experiments. Such a theory about human cognitions is hence a decisive possibility to reach for the goal of personalisation and to map human beings according to their very nature. In this sense, future web personalisation approaches can adapt much closer to human thinking patterns as well as interpreting behaviour more adequately, resulting in improved web experience and increased user satisfaction.

Future Research: In this article, we examined bell-shaped tuning curves. However, sigmoid-shaped tuning curves were also frequently measured in vivo. Further investigations of these with respect to our user model are therefore necessary. Moreover, the present model needs to be extended by many factors and correlates, i.e. there might be dependencies between the cognition vector and the evaluation duration, the testimonial length, but also the weather, acute stress and emotional states. All these information will be used to construct a personalisation approach that is more sophisticated than the clustering technique considered in this introductory technical paper. The resulting engine can then be tested against other systems.

References

1. Amatriain, X., Pujol, J.M., Oliver, N.: I like it... I like it not: evaluating user ratings noise in recommender systems. In: Houben, G.-J., McCalla, G., Pianesi, F., Zancanaro, M. (eds.) UMAP 2009. LNCS, vol. 5535, pp. 247–258. Springer, Heidelberg (2009). https://doi.org/10.1007/978-3-642-02247-0_24
2. Amatriain, X., Pujol, J., Tintarev, N., Oliver, N.: Rate it again: increasing recommendation accuracy by user re-rating. In: RecSys Conference. ACM (2009)
3. Beck, J., Ma, W.J., Latham, P.E., Pouget, A.: Probabilistic population codes and the exponential family of distributions. Prog. Brain Res. **165**, 509–519 (2007)
4. Bobadilla, J., Ortega, F., Hernando, A., Gutiérrez, A.: Recommender systems survey. Knowl.-Based Syst. **46**, 109–132 (2013)
5. Buffler, A., Allie, S., Lubben, F.: The development of first year physics students' ideas about measurement in terms of point and set paradigms. Int. J. Sci. Educ. **23**(11), 1137–1156 (2001)
6. D'Elia, A., Piccolo, D.: A mixture model for preferences data analysis. Comput. Stat. Data Anal. **49**(3), 917–934 (2005)
7. Doya, K., Ishii, S., Pouget, A., Rao, R.P.N.: Bayesian Brain: Probabilistic Approaches to Neural Coding. MIT Press, Cambridge (2006)
8. Faisal, A.A., Selen, L.P., Wolpert, D.M.: Noise in the nervous system. Nat. Rev. Neurosci. **9**(4), 292–303 (2008)
9. Friston, K.: The free-energy principle: a unified brain theory? Nat. Rev. Neurosci. **11**(2), 127–138 (2010)
10. Grabe, M.: Grundriss der Generalisierten Gauß'schen Fehlerrechnung. Springer, Berlin Heidelberg (2011). https://doi.org/10.1007/978-3-642-17822-1
11. Herlocker, J.L., Konstan, J.A., Terveen, L.G., Riedl, J.T.: Evaluating collaborative filtering recommender systems. ACM Trans. Inf. Syst. **22**(1), 5–53 (2004)
12. Hill, W., Stead, L., Rosenstein, M., Furnas, G.: Recommending and evaluating choices. In: SIGCHI Conference (1995)
13. Jannach, D., Zanker, M., Felfernig, A., Friedrich, G.: Recommender Systems: An Introduction. Cambridge University Press, Cambridge (2010)

14. Jasberg, K., Sizov, S.: Assessment of prediction techniques: the impact of human uncertainty. In: Proceedings of WISE (2017)
15. Jasberg, K., Sizov, S.: The magic barrier revisited: Accessing natural limitations of recommender assessment. In: Proceedings of ACM RecSys (2017)
16. Jasberg, K., Sizov, S.: Human uncertainty and ranking error - fallacies in metric-based evaluation of recommender system. In: Proceedings of ACM SAC (2018)
17. JCGM: Guide to the expression of uncertainty in measurement (2008)
18. JCGM: Propagation of distributions using a Monte Carlo method (2008)
19. Knijnenburg, B.P., Willemsen, M.C., Gantner, Z., Soncu, H., Newell, C.: Explaining the user experience of recommender systems. User Model. User-Adapt. Interact. **22**, 441–504 (2012)
20. Koren, Y., Sill, J.: OrdRec: an ordinal model for predicting personalized item rating distributions. In: Proceedings of ACM RecSys (2011)
21. Ma, W., Pouget, A.: Population codes: theoretic aspects. Encycl. Neurosci. **7**, 749–755 (2009)
22. McNee, S.M., Riedl, J., Konstan, J.A.: Being accurate is not enough: how accuracy metrics have hurt recommender systems. In: CHI 2006 Extended Abstracts on Human Factors in Computing Systems, pp. 1097–1101. ACM (2006)
23. Ma, W.J., Beck, J.M., Latham, P.E., Pouget, A.: Bayesian inference with probabilistic population codes. Nat. Neurosci. **9**, 1432–1438 (2006)
24. Ricci, F., Rokach, L., Shapira, B.: Recommender Systems Handbook. Springer, New York (2015). https://doi.org/10.1007/978-1-4899-7637-6
25. Said, A., Jain, B.J., Narr, S., Plumbaum, T.: Users and noise: the magic barrier of recommender systems. In: Masthoff, J., Mobasher, B., Desmarais, M.C., Nkambou, R. (eds.) UMAP 2012. LNCS, vol. 7379, pp. 237–248. Springer, Heidelberg (2012). https://doi.org/10.1007/978-3-642-31454-4_20
26. Tolhurst, D.J., Movshon, J.A., Dean, A.F.: The statistical reliability of signals in single neurons in cat and monkey visual cortex. Vis. Res. **23**, 775–785 (1983)
27. White, J.A., Rubinstein, J.T., Kay, A.R.: Channel noise in neurons. Trends Neurosci. **23**(3), 131–137 (2000)

Modeling New and Old Editors' Behaviors in Different Languages of Wikipedia

Anita Chandra[✉] and Abyayananda Maiti

Department of Computer Science and Engineering, IIT Patna, Patna 801103, India
{anita.pcs15,abyaym}@iitp.ac.in

Abstract. Wikipedia is an open-source multilingual encyclopedia which allows users to edit, create and share their knowledge collaboratively. Size of its contents such as articles, editors, links and language editions grows too fast with time. In this paper, we model the growth of editor-article bipartite network of multilingual Wikipedias to investigate behaviors of editors. In this bipartite network, editors and articles are two disjoint sets and if an editor edits an article then it forms an edge between them. The both editors and articles arrive simultaneously into their respective sets and editing is done by editors. The Wiki networks grow by the creation of external edits performed by new editors and/or internal edits done by old editors. These edits are done with a combination of preferential and/or random attachment mechanism. We consider two different randomness parameters for new and old editors in their attachment procedures. We validate the growth model over 20 largest language editions of Wikipedia and our results show good agreement between model and each of the considered languages. After interpreting the values of parameters we notice contrast in editing behaviors of new and old editors in every language. We also notice this non-uniform behavior of editors varies across all the languages. Thus, we report uncommon growth processes and difference in editing behaviors of editors of different languages of Wikipedia.

Keywords: Growth model · Bipartite network · Wikipedia
Editors' behaviors

1 Introduction

Wikipedia is one of the popular instances of crowd-sourcing, user generated content (UGC) and computer supported collaborative projects (CSCW). This online project is an open-source multilingual encyclopedia which allows any user to create, read, share and edit their articles. Because of this impartial promise of Wikipedia, its contents grow very rapidly. At present, there are about 44M articles, 33M editors, 12K administrators and average of 18.58 edits are contributed[1].

---

[1] https://en.wikipedia.org/wiki/Wikipedia:Size_of_Wikipedia.

© Springer Nature Switzerland AG 2018
H. Hacid et al. (Eds.): WISE 2018, LNCS 11234, pp. 438–453, 2018.
https://doi.org/10.1007/978-3-030-02925-8_31

The immense growth in their articles, editors, edits, links, languages and projects motivates many researchers to comprehend the dynamics of Wikipedia complex topology [6,7,23,24]. Initially, Wikipedia was introduced only in English edition, while at present it is available in 288 different language editions[2]. Multilingual property of Wikipedia influences many researchers to carry out comparative studies across languages from the various aspects [13,16,17,19].

Prior studies on growth of Wikipedia showed an initial exponential growth [2,21], however, further authors found saturation in its contents [12,20] and perhaps plateaued. Bongwon *et al.* [20] explained few factors that cause saturation in growth of Wikipedia: **barrier faced by infrequent editors:** number of reverts-per-edits increased and more pages are protected from infrequent editors; **Running out of easy topics:** editors might have already delivered a lot on common topics and harder topics are left that require more time and efforts. Hence, previous growth models were revised and two distinct models are presented by Wikipedians[3], where the growth is characterized by Gompertz function and its modified form. The Gompertz function model predicts that content grows and ultimately asymptotically approaches zero, and its modified model states that it grows continuously but slows down significantly in their later days.

Barabàsi Albert [1,4] is the first evolving model that incorporates concepts of growth and preferential attachment (PA). PA follows **"rich-get-richer"** mechanism according to which papers with more citations are more probable to be cited again. Several growth models which employed PA have been proposed [6,7,23,24] for the link structure of Wikipedia also known as Wikigraph, where articles/topics are described by nodes and hyperlinks between them represent edges. Gandica *et al.* [11] modeled behaviors of editors, where probability of editing of an editor is proportional to the edits that editor has (PA), editor's fitness [5] and editor's age [9]. Thus, we notice from the above model that authors have considered behaviors of all kinds of editors (e.g., new or old) uniform, which might not be always true. For instance, consider a scenario where a new editor who joins to contribute, may be too naive to handle existing high quality/matured/diverse articles. But, the same articles may be of high interest of old experienced editors. Thus, this clearly shows non-uniformity in behaviors of editors. The proposed growth model tries to explain this scenario and we also compare these non-uniform behaviors of editors across the different languages of Wikipedia.

In this paper, we model the growth of editor-article bipartite network of several Wikipedias to investigate behaviors of editors. In this network, editors and articles are two disjoint sets and if an editor edits an article then it forms an edge. The Wiki network evolves by the simultaneous arrival of editors and articles at different rate while edits arrive only from the editors set (active to contribute). Articles are edited with a combination of preferential and/or random attachment mechanism. To report difference in editing behaviors of new and old

---

[2] https://en.wikipedia.org/wiki/List_of_Wikipedias.
[3] https://en.wikipedia.org/wiki/Wikipedia:Modelling_Wikipedia%27s_growth.

editors we consider two distinct attachment procedures for them. The research questions we want to address here are:

- Does there exist a unique growth process in different language editions of Wikipedia?
- Are the editing behaviors of new and old editors uniform in same or different language editions of Wikipedia?
- Are the behaviors of editors similar in different language editions of Wikipedia?

To the best of our knowledge, this is the first study to model editor-article bipartite network of Wikipedia to compare behaviors of editors belonging to different languages. The rest of the paper is organized as follows: we discuss the related works in the Sect. 2. In Sect. 3, we briefly discuss about theoretical framework and attachment kernel of the growth model. In Sect. 4, we discuss filtration of Wikipedia datasets and these filtered networks are used for calculation of parameters of model. Section 5 highlights interesting results and analysis about editors' behaviors across several language editions of Wikipedia. At last, we conclude and discuss some future works in Sect. 6.

2   Related Work

We model growth of multilingual Wikipedias to understand editing preferences or behaviors of editors toward articles across different language editions. Several research related to various aspects of our proposed work has been done: (i) Wikipedia growth analysis; (ii) Wikipedia cross-language growth analysis; (iii) Wikipedia cross-language editors' behaviors analysis. A few of these studies are underlined as follows:

Wikipedia Growth Analysis: The analysis of statistical properties and growth of the directed Wikigraph is presented in [7]. The authors observed scale-invariant property for in and out hyperlink distributions which was characterized by local rules such as preferential attachment mechanism, although editors are responsible for the evolution of Wikipedia network. Buriol *et al.* [6] report temporal empirical analysis of the Wikigraph through characterizing the growth of users, articles, editions and several statistical properties. A growth model for Wikigraph is given in [24], which incorporates concept of preferential attachment and information exchange via reciprocal arcs. The authors achieved an excellent agreement between in-link distributions of their model and real wiki networks by extracting few parameters of model. They believe their model can be useful for studying feedback network, as reciprocity is typical characteristic of this network. Another growth model was given for Wikigraph by Gandica *et al.* [11], where it was described as editor-editor network. According to it, probability of an editor editing is proportional to number of edits she already has (i.e., Preferential attachment), her fitness [5] and her age [9].

Wikipedia Cross – Language Growth Analysis: Zlactic *et al.* [23] show several statistical properties such as degree distributions, reciprocity, clustering, assortativity, average length path e.t.c are common in several languages of Wikigraph. This uniformity shows that the growth processes of different versions of Wikipedia are universal, even they are in different phase of their development. Voss et al. [21] examine quantitative measurement of contents such as articles, editors, edits, links, bytes, redirects, talk pages on articles and editors in different language versions of Wikipedia. However, authors found similar structural distributions of these contents in different languages while quantitative metrics varied across the languages. They conclude different growth processes among languages after analysing the network in details. Ban et al. [3] study the growth by examining cross link structure of articles in different languages of Wikipedias over the duration of 15 years. They found that there existed six well-defined clusters of Wikipedias which have similar growth patterns. Interestingly, the determined clusters show similarities to culture and information literacy rather than language families and groups. Thus, we conclude from the above works that growth process across languages of Wikipedia is either uniform or non-uniform depends on different aspects.

Wikipedia Cross – Language Editors' Behaviors: Pfeil et al. [18] investigate the relation between editors' behaviors and their cultural backgrounds based on four cultural dimensions proposed by Hofstede [14] (Power Distance, Collectivism versus Individualism, Femininity versus Masculinity and Uncertainty Avoidance). The authors determine correlation among above mentioned cultural dimensions for deletion and creation actions of editors. Their study showed that editors belonging to high distance power countries (e.g., French) feel uncomfortable in deleting others contribution even if the contents are incorrect. While editors from high masculinity countries (e.g., Japan) are more active in adding and clarifying information. Hara et al. [13] also attempt to show cultural diversity in different languages of Wikipedia by analysing its different talk pages such as talk, user talk and wiki talk pages using Hofstedes cultural dimensions [14]. They report differences in cultural behaviors of editors between Eastern and Western Wikipedias. Maddock et al. [15] describe and model the coordination processes of editors across 24 language editions of Wikipedia. They measure coordination by number of posts on user's page. More number of posts represents better coordination and vice versa. Interestingly, they report editors contributing to Portuguese or Turkish Wikipedia maintain less coordination while Hebrew or Farsi editors maintain harmony by reporting more information about their edits.

3 Growth Model

The basic processes of network evolution [10] are (i) addition of nodes and edges (ii) arrival of edges between old nodes (iii) deletion of nodes and edges and (iv) rewiring of edges. In modeling dynamics of Wikipedia networks, we assume that they are always growing due to its prevalence in real-world. Thus, we do not study dynamics situations where edits are reverted, editors leave and rejoin, articles are deleted and rewritten e.t.c.

Theoretical Framework: We follow the growth model presented in [8]. The Wikipedia is described as bipartite network, represented as triplet $G = \langle E, A, E_{dit} \rangle$, where E and A are sets of editors and articles respectively and $E_{dit} \subseteq E \times A$ is a set of edits. Basically, an edit represent an edge in the Wikipedia bipartite network. In this paper, we use the terms "edits" and "edges" interchangeably. To model growth we assume logical time for arrival of a new editor as a new time step. Furthermore, we assume that the new editor who had arrived at previous time step becomes old at each the subsequent time step. Before the network starts to grow, we suppose there are e_0 editors, a_0 articles and $edit_0$ edits between them. From the above initial network, it evolves at each time step as following:

1. One editor and w number of articles arrive into their respective sets.
2. We consider m external edits done by single new editor and n internal edits performed by old editors collectively. We denote the edits done by a new editor as external edits and the edits performed by an old editor as internal edits.
3. Attachment of these external and internal edits is proportional to number of article's previous edits along with an adjustable randomness parameter γ. The attachment kernel for articles is defined as:
   Let $\widetilde{A}(k_{a,t})$ denotes the attachment probability of an edit to any article a at time t where the edits of that article is $k_{a,t}$. Then $\widetilde{A}(k_{a,t})$ defines the attachment kernel that takes the form:

$$\widetilde{A}(k_{a,t}) = \frac{k_{a,t} + \gamma}{\sum_{a=1}^{N}(k_{a,t} + \gamma)} \tag{1}$$

   where N is the total number of articles in set A. Note that a bigger γ indicates that editors edit any random article whereas lower value of γ means editors prefer to edit popular articles.
4. We consider two different randomness parameters for new and old editors: γ_e and γ_i. Hence, their corresponding attachment procedures are described as follows:
   (i) At each time step, m external edits done by a new editor are attached to articles of A with probability $\dfrac{k_a + \gamma_e}{\sum_{a=1}^{a_0+wt}(k_a + \gamma_e)}$.
   (ii) At each time step, n internal edits performed by all old editors collectively and attached to articles of A with probability $\dfrac{k_a + \gamma_i}{\sum_{a=1}^{a_0+wt}(k_a + \gamma_i)}$.

Articles Edit Distribution: A generalised growth model for online emerging user-object bipartite networks is given in [8]. After interpreting the values of parameters for different online networks (e.g., MovieLens, Stack Overflow, Wikibooks (English, French) and so on), authors observe quite opposite trends in selection behaviors of new and old web users even in similar kinds of online networks. This result motivated us to examine editing behaviors of editors across the language editions of Wikipedia. In this paper, we are mainly interested in modeling dynamics of edit distribution of articles as it reflects editing preferences or behaviors of editors towards articles. Thus, for the derivation of edits distribution of articles we refer [8], in which authors have provided derivation of object degree distribution in details.

According to above theoretical framework, the change in edit of an article $a \in A$ is given as:

$$\frac{dk_a}{dt} = m \frac{k_a + \gamma_e}{\sum_{a=1}^{a_0+wt} k_a + \gamma_e} + n \frac{k_a + \gamma_i}{\sum_{a=1}^{a_0+wt} k_a + \gamma_i}. \tag{2}$$

After following all the steps from [8] to solve Eq. 2, we get edit distribution of articles as,

$$p(k) = \frac{r}{s}(k_0)^{\frac{r}{s}}(k + k_0)^{-\left(1+\frac{r}{s}\right)}. \tag{3}$$

Where, $r = (c + w\gamma_e)(c + w\gamma_i)$, $s = mc + mw\gamma_i + nc + nw\gamma_e$, $k_0 = (m\gamma_e c + mw\gamma_e\gamma_i + n\gamma_i c + nw\gamma_e\gamma_i)/s$, $c = m + n$.

From the Eq. 3, we observe that edit distribution of articles follows shifted power law, also known as Mandelbrot Law [22] with an additive shift of k_0 and dynamical exponent of $\left(1 + \frac{r}{s}\right)$.

Further to validate the model, we fit edit distribution of articles from the considered model and edit distribution of articles derived from the real Wiki datasets. For this, we extract values of all the parameters of model: m, n, w, γ_e and γ_i from real Wikis empirically. We again refer [8] for the determining values of parameters for several different languages of Wikipedia. Using the procedure provided in the above paper we calculate all values of parameters for all the mentioned top 20 languages of Wikipedia in the Sect. 4 and these values are stated in Table 2 present in the Sect. 5.

4 Filtration of Wikipedia Networks

In this paper, we evaluate randomness values at each time step for every external edit of new editors and internal edit of old editors. For the very large scale networks, this procedure for extraction of randomness values is quite exhaustive and expensive in terms of time complexity. Thus, instead of determining these values for each of the nodes and edges of network, we consider their subset without losing generality. After filtering out insignificant portion from Wikipedia network we get subset of editors, articles and edits that are used to extract values of parameters and also to validate the model. The procedure for filtering Wikipedia networks is provided below:

Unregistered Editors and Their Edits: Wikipedia permits unregistered or anonymous editors to create and edit articles. For these unregistered editors, IP address of their machines is recorded, but it cannot be considered as an identity of any individual. Because of this ambiguity we do not consider unregistered editors and their contributed edits in our analysis. In Fig. 1(a), we give percentage of unregistered editors and their contributed edits to registered editors and their edits for different languages of Wikipedia. From this plot, we see that percentage of unregistered editors is quite large, but their edits are comparatively less than registered ones. This might be because edits of unregistered editors are generally reverted to prevent vandalism. Thus, discarding huge entries of unregistered editors helps to reduce complexity of our experiment, as now we determine values only for the registered editors.

Multiple Edits: Multiple edits means an editor edits same article multiple times at same or different time stamps. In our study, we concern about the relation of editors and articles not the weight or strength of edits. Consequently, we discard all multiple edits of articles done by any editor. In Fig. 1(b), we give percentage of multiple edits contributed by all editors in different languages of Wikipedia. From this plot, we observe that there exists large fraction of multiple edits in every language. For example, 72% and 86% of edits are multiple in case jawiki and zhwiki respectively. Thus, eliminating large share of multiple edits, also reduce number of external and internal edits that are used for generating synthetic network using the growth model. This also helps to reduce the complexity in evaluation of randomness parameters.

Single Edit Editors and Edited Articles: First edit of an article is actually its creation date[4] and single edit of editors means editors edited only once, which is an insignificant contribution. Thus, we discard all articles that are created but never edited again and editors edited once. In Fig. 1(c), we present percentage of single edited articles and single edit editors for different languages of Wikipedia. From the plot we observe that there is large fraction of single edited articles and single edit editors in every language. Hence, we again step towards reducing complexity of the procedure by filtering these large fractions of articles and editors.

Bots and Their Edits: In Wikipedia, a bot is an automated tool which performs recurrent jobs to maintain articles. We consider list of registered Wikipedia bots from[5]. In Fig. 1(e), we provide percentage of bots and their edits over different editors. From plot we observe bots is too less in all the languages. The share of their edits is average in both the considered cases as bots do several repetitive automated tasks like checking spelling, grammar, proper punctuations e.t.c. We can observe from Fig. 1(d) that editing behavior of bots is similar to the human editors structurally and also its count is too less. Consequently, in our analysis we consider bots and their contributed edits.

We have downloaded complete edit history for 20 largest Wikipedia by the number of articles from Wiki dump[6]. In our study, we have not taken Cebuano,

---

[4] https://en.wikipedia.org/wiki/Wikipedia:Village_pump_(technical)Article_creation_date.

[5] https://en.wikipedia.org/wiki/Category:All_Wikipedia_bots.

[6] https://dumps.wikimedia.org/.

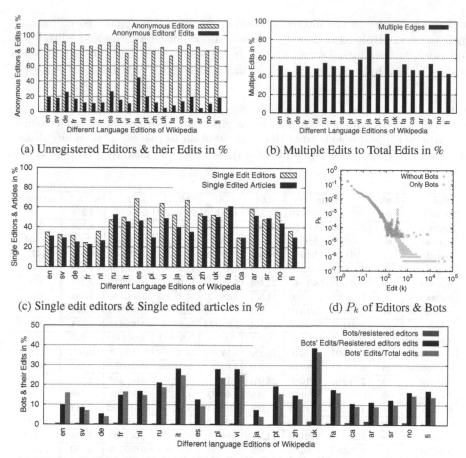

(a) Unregistered Editors & their Edits in %

(b) Multiple Edits to Total Edits in %

(c) Single edit editors & Single edited articles in %

(d) P_k of Editors & Bots

(e) Bots to registered editors, their edits to edits of registered editors and their edits to all edits made by registered as well as unregistered editors in % for top 20 languages of Wikipedia.

Fig. 1. We present percentage of (a) unregistered editors and their edits over total editors and total edits; (b) multiple edits over total edits done by registered editors, unregistered editors and bots; (c) single edited articles over total articles and single edit editors over total editors; (e) bots to registered editors, their edits to registered edits and their edits to total edits. In (d) we give edit distribution of human editors and bots.

Waray and Serbo-Croatian Wikipedia because the set of editors is too small for them. The downloaded XML history files are parsed to extract editor IDs, article IDs and time stamp at which an editor edits any article. We consider 5 years of projects from their publishing date and sorted edit history files for all Wikipedias except English Wikipedia. We have taken 3 years of history information instead of 5 years for English Wikipedia as it grows too fast and has much bigger set of editors. Further, we apply aforementioned filtering procedures to the all considered languages of Wikipedia to achieve filtered networks reported in Table 1. Hence, in the filtered networks, we discard unregistered editors and their edits, multiple edits done by editors, single edited articles and editors who edit once, but we include bots and their edits. In the same table, we also give basic statistics information for all languages of filtered Wikipedia networks.

Table 1. List of languages of Wikipedia, their code, their timespan and basic statistics information of the filtered Wikipedias are given. We provide filtered set of editors $|E|$, articles $|A|$ and unique edits $|E_{dit}|$. In addition to this, we also specify density of network $\rho = \frac{|E_{dit}|}{|E||A|}$, average edits of articles $\langle k_a \rangle$, and maximum edits of articles $a_k(max)$.

Wikis	Codes	Years	$|E|$	$|A|$	$|E_{dit}|$	ρ	$< k_a >$	$a_k(max)$
English	en	09/09/01 - 09/09/04	61,219	4,06,610	29,34,172	0.000117	7.216	888
Swedish	sv	25/05/01 - 25/05/06	7,955	1,67,708	9,76,744	0.000732	5.824	352
German	de	15/09/01 - 15/09/06	1,19,146	6,67,939	83,59,236	0.000105	12.514	2,409
French	fr	13/10/01 - 13/10/06	59,048	5,30,730	45,75,017	0.000145	8.620	1471
Dutch	nl	05/10/02 - 05/10/07	32,023	4,51,076	41,91,949	0.000290	9.293	732
Russian	ru	08/11/02 - 08/11/07	22,468	2,95,284	21,50,477	0.000324	7.285	661
Italian	it	14/07/01 - 14.07.06	14,541	2,05,108	14,36,388	0.000481	7.003	941
Spanish	es	23/10/02 - 23/10/07	71,180	3,75,295	35,58,829	0.000130	9.482	1,484
Polish	pl	26/09/01 - 26/09/06	17,101	2,86,155	20,67,756	0.000422	7.225	694
Vietnamese	vi	16/11/02 - 16/11/07	3,664	53,719	2,69,683	0.00137	5.020	223
Japanese	ja	17/04/02 - 17/04/07	75,558	4,59,298	48,03,119	0.000138	10.457	1,273
Portuguese	pt	30/09/01 - 30/09/06	18,298	2,80,243	1424068	0.000277	5.081	642
Chinese	zh	27/10/02 - 27/10/07	28,051	2,16,177	17,52,109	0.000288	8.104	777
Ukrainian	uk	04/01/03 - 04/01/08	3,566	90,573	474529	0.001469	5.239	164
Persian	fa	13/12/03 - 13.12.08	7,257	76,214	5,43,996	0.0009835	7.137	304
Catalan	ca	17/03/01 - 17/03/06	1,399	28,415	1,50,066	0.003770	5.281	145
Arabic	ar	15/01/01 - 15/01/06	11,187	1,80,195	9,95,419	0.000493	5.524	1,189
Serbian	sr	17/02/03 - 17/02/08	2,982	82,193	5,38,906	0.002198	6.550	208
Norwegian	no	04/10/01 - 04/10/06	6,165	89,710	574558	0.001035	6.382	334
Finnish	fi	28/08/02 - 28/08/07	14,997	1,73,647	15,52,718	0.000596	8.941	800

5   Results and Analysis

In this Section, we present interesting inferences about editors behaviors by interpreting values of parameters given in Table 2 for all considered 20 languages of Wikipedia. We compare average values of m and γ_e of new editors to the corresponding values of n and γ_i of old experienced editors. Some interesting observations are stated below:

Comparison of External and Internal Edits: After analysing values of m and n from Table 2, we notice that m values ($m \approx 1$) are always too small in comparison to the corresponding n values for all the considered languages. This is because m represents external edits done by single new editor whereas n represents internal edits performed by all old editors collectively. Also, we consider single time step for new editors and discards all multiple edits that they have done on articles. This is quite possible that editors keep editing same articles multiple times instead of switching to other articles. In the Wikipedia, the editors should have ample amount of knowledge in diverse topics to contribute adequately which comes with time and efforts.

In Wikipedia editors edit multilingual articles thus here we consider that editors who contribute to Spanish Wikipedia are said to be Spanish editors and similar assumption goes for other Wikis. Further, we observe interesting relative differences of m and n across the languages of Wikipedia. For example, in svwiki, nlwiki, plwiki, urwiki, cawiki and srwiki, ratio of n and m is much larger (marked as magenta color) as compared to other mentioned languages (marked as black color). Thus, this infers that old and expert Swedish, Dutch, Polish, Catalan, Serbian and Finnish editors edit many articles when they grow older in the network as compared to old English, German, French and Spanish editors. Although we notice that frwiki, dewiki, enwiki and jawiki networks grow too fast, the average number of internal edits contributed by old experienced editors is lesser than smaller wikipedias such as ukwiki, srwiki, cawiki, nlwiki, svwiki and plwiki. Similarly, we notice arrival rate of articles per editor in svwiki, cawiki, ukwiki and srwiki (marked as green color) is more as compared to others Wikis (marked as black color). Again these Wikipedias are smaller in size, but the arrival rate of articles is more than that in bigger Wikipedias (e.g., frwiki, enwiki, dewiki and eswiki).

Comparison of Randomness Parameters of Editors: The values of γ_e and γ_i signify randomness in the edit done by new and old editors respectively. In Table 2, we report best average values of γ_e for new editors and γ_i for old editors for all the languages. We can observe from the values that different languages have a different combination of randomness and preferential behaviors of editors in their growth. For example, γ_e is larger than γ_i in dewiki, eswiki, jawiki and frwiki (marked as red), γ_e and γ_i is closer to each other in enwiki, zhwiki and nlwiki (marked as blue) and γ_e is smaller than γ_i in rest of the other Wikis like ruwiki, itwiki, plwiki e.t.c (marked as black). The greater value of γ_e than γ_i shows new editors edit articles more randomly than old experienced editors and vice versa. Consequently, we identify three different classes after comparing average values of randomness γ_e and γ_i of editors.

In dewiki, eswiki, jawiki and frwiki, we notice that value of γ_e is much larger than γ_i which means that new editors performed external edits more randomly than internal edits by old editors. It infers that new German, French, Spanish and Japanese editors edit articles randomly whereas old experienced editors of these Wikis contribute mainly to popular or most frequently edited articles. Conversely, we can say that popular German, French, Spanish and Japanese articles are mainly edited more by old and expert editors than new editors. While in enwiki, zhwiki and nlwiki, values of γ_e and γ_i are almost similar to each other. In other words, editing preferences of these new and old editors are similar. But when we examine the randomness values more precisely, we find that old English, Chinese and dutch editors edit only popular articles after a certain period, bringing the average of γ_i closer to γ_e. On the other hand, in the rest of wikipedias

Table 2. The values of parameters of growth model calculated from all the considered 20 languages of Wikipedia are given. Also, we specify MAD (Mean absolute deviation), MSE (mean square error) and RMSE (root mean square error) for the model.

Wikis	Codes	m	n	w	γ_e	γ_i	MAD	MSE	RMSE
English - enwiki	en	1.001	46.92	6.64	13.395	17.456	7.63e-04	3.59e-05	5.99e-03
Swedish - svwiki	sv	1.112	121.05	20.97	14.589	71.710	1.50e-03	6.15e-05	7.83e-03
German - dewiki	de	1.024	69.13	5.60	28.733	7.485	3.49e-04	4.24e-06	2.05e-03
French - frwiki	fr	1.057	76.42	8.98	25.425	13.565	4.82e-04	1.36e-05	3.69e-03
Dutch - nlwiki	nl	1.004	129.90	14.08	27.339	29.201	5.84e-04	1.26e-05	3.55e-03
Russian - ruwiki	ru	1.041	94.67	13.14	19.442	48.895	8.10e-04	2.25e-05	4.75e-03
Italian - itwiki	it	1.090	97.69	14.10	20.340	62.416	1.38e-03	7.08e-05	8.41e-03
Spanish - eswiki	es	1.000	48.99	5.27	24.069	9.161	4.74e-04	7.55e-06	2.74e-03
Polish - plwiki	pl	1.092	119.82	16.73	19.821	63.359	9.44e-04	3.73e-05	6.10e-03
Vietnamese - viwiki	vi	1.090	72.51	14.66	16.257	67.952	1.71e-03	4.96e-05	7.04e-03
Japanese - jawiki	ja	1.001	62.56	6.07	24.279	10.199	5.10e-04	8.97e-06	2.99e-03
Portuguese - ptwiki	pt	1.029	76.79	15.31	12.936	53.832	7.05e-04	9.50e-06	3.08e-03
Chinese - zhwiki	zh	1.003	61.40	7.69	21.214	29.627	5.87e-04	1.51e-05	3.88e-03
Ukrainian - ukwiki	uk	1.061	132.00	25.39	11.781	72.617	1.65e-03	6.90e-05	8.31e-03
Persian - fawiki	fa	1.105	73.85	10.50	21.361	61.725	1.14e-03	2.75e-05	5.24e-03
Catalan - cawiki	ca	1.853	105.41	20.31	13.292	73.443	3.05e-03	1.17e-04	1.08e-02
Arabic - arwiki	ar	1.063	87.91	16.10	9.442	61.607	8.96e-04	9.63e-06	3.10e-03
Serbian - srwiki	sr	1.161	179.55	27.56	18.105	77.579	1.80e-03	4.20e-05	6.48e-03
Norwegian - nowiki	no	1.110	92.082	14.55	18.120	66.790	2.02e-03	6.84e-05	8.27e-03
Finnish - fiwiki	fi	1.025	102.51	11.57	20.649	60.605	5.99e-04	1.07e-05	3.27e-03

such as svwiki, itwiki, plwiki, viwiki, ptwiki e.t.c, new Swedish, Italian, Polish and Vietnamese editors edit more popular or frequently edited articles whereas the old editors contribute to diverse and random articles. Thus, from the above we summarise that if a new German, French, Spanish and Japanese editor joins she edits any random article whereas if the new editor is Swedish, Polish, Italian, and Vietnamese, she prefers to edit popular or frequently edited articles and vice versa for the old editors.

In Fig. 2 we present plots for edit distribution of articles for top 12 language editions of Wikipedia. Each plot contains the empirical edit distribution of articles from real Wikis and edit distribution of articles from the model after fitting values of parameters. We observe a very good fit between model and all considered languages of Wikipedia. Moreover, in Table 2 we also provide mean absolute deviation (MAD), mean square error (MSE) and root mean square error (RMSE) for model. Thus, good agreement between Wikis and model and lesser values of MAD, MSE and RMSE shows correctness of the growth model. As a result, we observe that the simple model that incorporates trivial concept of preferential attachment [1, 4] can explain dynamics of several Wikipedias described as bipartite network. The aforementioned findings could able to answer all the research

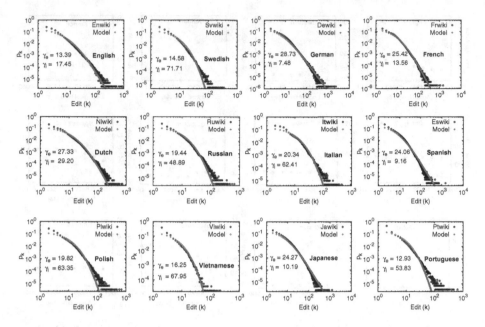

Fig. 2. Plots of articles' edit distribution from real Wiki datasets and model for top 12 language editions of Wikipedia are given. In this, x-axis shows the edit (k) and y-axis depicts the probability of having an article with edit k (p_k).

questions raised in the beginning. We notice non-uniform behaviors in new and old editors in same Wikipedia. For instance, in dewiki, enwiki, frwiki and eswiki, new editors edit articles randomly whereas old editors prefer to edit popular articles. Furthermore, we observe that this non-uniform behaviors of editors varies with the language. For instance, we observe that new German editors edit any random article while new Italian editors edit mostly well known or popular articles. On the contrary, old German editors prefer to edit popular articles while old Italian editors edit less frequently edited articles. Consequently, we notice uncommon growth process and dissimilar editing behaviors and preferences of editors in different language editions of Wikipedia.

Dynamics of Randomness Values: To validate the growth model we derive approximate or average values of all parameters: m, n, w, γ_e and γ_i. On comparing the average values γ_e and γ_i, we can only infer that new editors edit more randomly or preferentially than old editors and vice versa. The randomness in editing behaviors of new and old editors changes over the time. However, we can also have interesting results about editors behaviors by analysing dynamics of randomness values. We notice from the plots that different trends of dynamics exist in different languages of Wikipedia. We can also observe that dynamics in randomness values of new editors extend to longer time steps than old editors. This is because old editors not necessarily edits at each time step. But in case of Wikipedia, this difference is almost insignificant since number of editors are

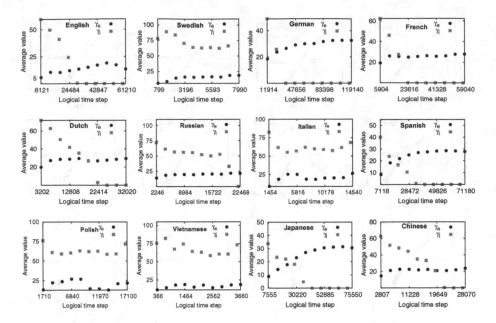

Fig. 3. Evolution of randomness parameters γ_e of new editors and γ_i of old editors in (i) enwiki [bin size: 6121, 5930] (ii) svwiki [bin size: 799, 783] (iii) dewiki [bin size: 11914, 11660] (iv) frwiki [bin size: 5904, 5807] (v) nlwiki [bin size: 2246, 3201] (vi) ruwiki [bin size: 2246, 2202], (vii) itwiki [bin size: 1454, 1425], (viii) eswiki [bin size: 7118, 6748], (ix) plwiki [bin size: 1710, 1677], (x) viwiki [bin size: 366, 353], (xi) jawiki [bin size: 7555, 7164] and (xii) zhwiki [bin size: 2807, 2728]. A bin specifies a time duration. In y-axis we present average randomness values over the time duration given in x-axis.

increasing exponentially with time. So, there is less probability of no show time step. Whereas this difference is quite noticeable in rating or reviewing network of commercial websites (e.g., *amazon.in*), in which several customers buy products but not necessarily rate or review them.

Figure 3 depicts the evolution of γ_e and γ_i for 12 largest languages of Wikipedia. From Fig. 3, we observe that in dewiki, frwiki, nlwiki, ruwiki, eswiki, jawiki and zhwiki, values of γ_e increase slowly and eventually saturated over time. This shows that new editors who join in later stage edit articles more randomly than new editors who had arrived at initial phase of Wiki networks. There is no significant transition in values of γ_e to report in the rest of Wikis. After analysing dynamic values of γ_i given in Fig. 3, we observe different trends across all considered languages. For example, in enwiki, dewiki, frwiki, nlwiki, eswiki, jawiki, ruwiki and zhwiki, we notice that values of γ_i decreases over the time and it approaches to zero after a specific time step. This establishes that after a certain time period old experienced English, German, French, Dutch, Spanish, Japanese, Russian and Chinese editors focus to edit mainly popular or heavily edited articles. We notice this sharp skewness in γ_i values mainly in bigger Wikipedias. In these networks after a certain period, tails become too long (i.e., many articles have lower edits) in edit distribution of articles. There are few heavily edited articles and they become more prominent than tails after a longer time. Thus, every time these heavily edited articles are being selected for editing and making editing behaviors of these editors purely preferential. In addition to this, we can observe in ruwiki, nlwiki and zhwiki, values tends to zero at the later phase of network construction compared to frwiki, eswiki and dewiki. In these networks range of edits of editors is too narrow, this means old editors mainly focus on widely known or frequently edited articles.

On the other hand in svwiki, itwiki, plwiki and viwiki evolution trends of γ_i are quite opposite in which editing preferences of old editors are evenly distributed over the time. This infers that old editors edit articles with wide range (low, high, average edited articles). Again we achieve three different classes of Wikipedia based on evolution patterns of γ_i of old editors. As a result, we notice uncommon growth process and behaviors of editors in different languages of Wikipedia.

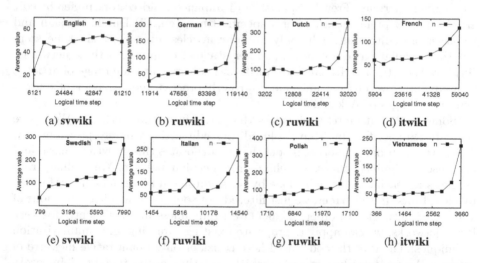

Fig. 4. Evolution of internal edits (n) of old editors of (a) enwiki, (b) dewiki, (c) nlwiki, (d) frwiki, (e) svwiki, (f) itwiki, (g) plwiki, and (h) viwiki is given.

Dynamics of Internal Edits. We have already seen relative differences between average values of m and n, but here we examine evolution of n values. We observe three different classes of Wikipedia based on behaviors of editors from the previous discussions. In the Fig. 4, we consider few languages from each of the groups. Although we consider languages from different classes, we observe similar dynamics trends in all the Wikis except English Wikipedia. For all the networks, these n values are increasing over time, but values are evenly distributed throughout its dynamics in enwiki.

6   Conclusion and Future Works

This paper has model growth of editor-article bipartite networks of Wikipedias
to understand behaviors of editors across the languages. The growth model is
validated over 20 language editions of Wikipedia. We achieve good agreement
between model and all Wikipedias and very less error values of MAD, MSE, and
RMSE for all considered languages. Thus, this signifies that the simple model
that employs general concept of preferential attachment is able to explain the
growth of several Wikipedias.

We identify three different classes based on behaviors of editors across the
languages: (i) German, Spanish, French and Japanese: new editors prefer to
edit any random article as compared to old experienced editors or we can say
that popular articles of dewiki, eswiki, frwiki and jawiki are edited more by old
editors; (ii) English, Dutch and Chinese: choice of editing articles of new as well
as old editors is almost similar; (iii) Swedish, Russian, Italian and Polish: new
editors prefer to edit popular articles than old and expert editors. Similarly,
after exploring dynamics of randomness values we get three different classes:
(i) English, German, French, Spanish and Japanese: old editors prefer to edit
articles purely preferentially after a specific time step; (ii) Dutch, Russian and
Chinese: old editors also edit only popular articles after a certain time step,
but this peculiar pattern is noted at later stage of evolution of these networks;
(iii) Swedish, Russian, Italian and Polish: old editors edit wide range of articles.
Consequently, we observe uncommon growth process and behavior of editors
across languages of Wikipedia.

Some of the future research works that arise from this work are: We notice
that after a specific time step t, old editors edit articles purely preferentially.
Thus, we can propose another model that incorporates this growth scenario and
also discuss significant reasons behind that peculiar pattern. We evaluate randomness for each external edit done by new editors and internal edit performed
by old editors. In this paper, we just filtered out some of insignificant portions of
Wikipedia networks. In order to calculate randomness more efficiently for very
large networks, we can apply appropriate existing sampling and summarisation
techniques. The growth model used here consider only quantitative measure of
articles. We can introduce model in which growth is proportional to joint probability of quantitative and qualitative metrics of the articles. Further to make it
more realistic we can assign different importance to these metrics.

References

1. Albert, R., Jeong, H., Barabási, A.L.: Internet: diameter of the world-wide web.
   Nature **401**(6749), 130–131 (1999)
2. Almeida, R.B., Mozafari, B., Cho, J.: On the evolution of Wikipedia. In: ICWSM.
   Citeseer (2007)
3. Ban, K., Perc, M., Levnajić, Z.: Robust clustering of languages across Wikipedia
   growth. R. Soc. Open Sci. 4(10), 171217 (2017)

4. Barabási, A.L., Albert, R.: Emergence of scaling in random networks. Science **286**(5439), 509–512 (1999)
5. Bianconi, G., Barabási, A.L.: Competition and multiscaling in evolving networks. EPL (Europhysics Lett.) **54**(4), 436 (2001)
6. Buriol, L.S., Castillo, C., Donato, D., Leonardi, S., Millozzi, S.: Temporal analysis of the Wikigraph. In: Proceedings of the 2006 IEEE/WIC/ACM International Conference on Web Intelligence, pp. 45–51. IEEE Computer Society (2006)
7. Capocci, A., et al.: Preferential attachment in the growth of social networks: the internet encyclopedia Wikipedia. Phys. Rev. E **74**(3), 036116 (2006)
8. Chandra, A., Garg, H., Maiti, A.: Investigating selection behavior of new and old users in online emerging user-object networks. In: Proceedings of the 2017 IEEE/ACM International Conference on Advances in Social Networks Analysis and Mining 2017, pp. 455–458. ACM (2017)
9. Dorogovtsev, S.N., Mendes, J.F.F.: Evolution of networks with aging of sites. Phys. Rev. E **62**(2), 1842 (2000)
10. Dorogovtsev, S.N., Mendes, J.F.: Evolution of networks. Adv. Phys. **51**(4), 1079–1187 (2002)
11. Gandica, Y., Carvalho, J., dos Aidos, F.S.: Wikipedia editing dynamics. Phys. Rev. E **91**(1), 012824 (2015)
12. Gibbons, A., Vetrano, D., Biancani, S.: Wikipedia: nowhere to grow (2012)
13. Hara, N., Shachaf, P., Hew, K.F.: Cross-cultural analysis of the Wikipedia community. J. Assoc. Inf. Sci. Technol. **61**(10), 2097–2108 (2010)
14. Hofstede, G.: Cultures and Organizations-software of the Mind. Profile Books, London (2004)
15. Maddock, J., Shaw, A., Gergle, D.: Talking about talk: coordination in large online communities. In: Proceedings of the 2017 CHI Conference Extended Abstracts on Human Factors in Computing Systems, pp. 1869–1876. ACM (2017)
16. Massa, P., Scrinzi, F.: Manypedia: comparing language points of view of Wikipedia communities. In: Proceedings of the Eighth Annual International Symposium on Wikis and Open Collaboration, p. 21. ACM (2012)
17. Ortega, F.: Wikipedia: a quantitative analysis. Ph.D. thesis (2009)
18. Pfeil, U., Zaphiris, P., Ang, C.S.: Cultural differences in collaborative authoring of Wikipedia. J. Comput.-Mediated Commun. **12**(1), 88–113 (2006)
19. Samoilenko, A., Lemmerich, F., Weller, K., Zens, M., Strohmaier, M.: Analysing timelines of national histories across wikipedia editions: a comparative computational approach. In: 11th International Conference on Web and Social Media, ICWSM 2017. AAAI Press (2017)
20. Suh, B., Convertino, G., Chi, E.H., Pirolli, P.: The singularity is not near: slowing growth of Wikipedia. In: Proceedings of the 5th International Symposium on Wikis and Open Collaboration, p. 8. ACM (2009)
21. Voss, J.: Measuring Wikipedia (2005)
22. Xue-Zao, R., Zi-Mo, Y., Bing-Hong, W., Tao, Z.: Mandelbrot law of evolving networks. Chin. Phys. Lett. **29**(3), 038904 (2012)
23. Zlatić, V., Božičević, M., Štefančić, H., Domazet, M.: Wikipedias: collaborative web-based encyclopedias as complex networks. Phys. Rev. E **74**(1), 016115 (2006)
24. Zlatić, V., Štefančić, H.: Model of Wikipedia growth based on information exchange via reciprocal arcs. EPL (Europhysics Lett.) **93**(5), 58005 (2011)

Data Mining Applications

A Novel Incremental Dictionary Learning Method for Low Bit Rate Speech Streaming

Luyao Teng[1,2], Yingxiang Huo[1]([✉]), Huan Song[1], Shaohua Teng[1],
Hua Wang[2], and Yanchun Zhang[2]

[1] Guangdong University of Technology,
Guangzhou 510006, Guangdong, China
yingxiang.huo@qq.com
[2] Victoria University, Melbourne, VIC 3011, Australia

Abstract. Speech streaming, which is widely used nowadays, cost a huge amount of transfer bandwidth and storage space. It is significant to compress them with as few bits as possible, meanwhile keep the voice clear and meaning unchanged. According to speech contexts, the proposed method can dynamically adapt to speech stream of any speaker by appending atoms to dictionary. Furthermore, in order to smoothly represent the amplitude envelopes shifting over the frequency, the dictionary is extended by Hilbert transform. The upper bounds of weights of atoms are constrained, so they can be quantized in practical applications. Experimental results show the advantages of our method. When the minimum reconstruction accuracy is 99.8%, which is applicable to general voice communications, the space saving is over 99%. Our method can adapt to the application with extreme bandwidth/storage limitation and large scale dataset, meanwhile keep reasonable perception quality.

Keywords: Streaming speech · Low bit rate compression
Incremental dictionary learning · Sparse coding

1 Introduction

Speech is the most important way for people to communicate with each other. Digital speech allows us to transfer and storage speech signal with higher quality and lower cost. A huge amount of speech applications that require narrow transmission bandwidth and storage capacity make it important to compress the digital speech signal. For example, extreme environments such as satellite telephone need speech compression because of the narrow and expensive transmission bandwidth. Meanwhile, there are many applications demands on low computation complexity and short latency. Using mobile communications as example, lower bit rate consumed by each user is meaningful for serving more users with limited bandwidth simultaneously. Moreover, computation complexity affects battery life and latency affects the communication quality. So a good encoder should consider all of these factors.

The most common way to obtain lower bit rate meanwhile retains reasonable perception quality is to remove redundancy and irrelevant information from the signal. By training a dictionary and selecting a few atoms from it to make combinations, we

© Springer Nature Switzerland AG 2018
H. Hacid et al. (Eds.): WISE 2018, LNCS 11234, pp. 457–471, 2018.
https://doi.org/10.1007/978-3-030-02925-8_32

can represent long signals in a compressive manner. It has been reported that the dictionary learning has a wide range of applications, such as Speech enhancement [1], audio compressing [2, 3], audio classification [4], music analysis [5], blind source separation [6], image denoising [7, 8], image classification [9–11], image super-resolution [12], and texture synthesis [13]. In these applications, sparse learning models are well adapted to natural signals.

In practical, we may not be able to obtain full signals and construct the dictionary beforehand because they generate in real time, or are too huge to analyze at once. Online dictionary learning, for example [14, 15] enables us to adjust basis vectors to signals dynamically. [16] and [17] aims on optimizing the computation efficiency and memory requirement, they are benefit to circumstances that operating huge datasets.

There are generally linear and non-linear dictionary learning models. Principal component analysis (PCA) [18] is an example of the former. PCA and its variants guarantee that the basis[1] vectors are orthogonal. Their decompositions are easy to evaluate. And they reconstruct signals by linearly combing a few dictionary elements. However, realistic signals may naturally non-linear. So it is meaningful to study non-linear dictionary learning. This goal can be achieved by using neural networks [19], kernel methods [20], or principal geodesic analysis (PGA) [21].

Methods such as PCA and K-SVD [22] generate dictionaries by statistical manner. However, they concentrate too much on significant patterns but neglect small details. By using lest mean square error (MSE) or its equivalent as judgement standard, some methods announce that qualities of results are high. In fact, none of these standards guarantee if the reconstructed results are natural or even legal. This can lead to bad perception quality. According to our research, exchanging of amplitude envelopes recorded with same speaker and same word will not affect much on perception quality despite that they are not exactly the same, but replacing an envelope with their average or linear synthesis can produce worse perception quality. So a clear envelope is more preferable than the blurred ones generated by averaging procedures. Most of the dictionary learning methods seek for highest average quality, but this often leads to poor qualities of several samples. Nevertheless, because human's auditory sense is sensitive to impulse, the bad reconstructed envelopes distributed in full duration of the reconstructed results will be detectable or even significantly affect the perception quality.

This paper concentrates on training the dictionary of frequency domain amplitude envelopes, which is an important task in speech compressing. Our method is different from PCA liked methods. By selecting real envelopes as atoms for constructing the dictionary, meanwhile constraining to use fewer atoms when simulating every envelope, we can produce natural results with less blurring. We first propose a method to evaluate the weight of atoms for representing an envelope. Then, by using the atom and its Hilbert transformed result, we can efficiently match atoms and envelopes with similar shapes but shifting over the frequency. We further explore both one-pass and multi-pass schemes. One pass scheme provides real time coding with slightly higher bit rate, whereas multi-pass schemes encodes in lower bit rate but cause a degree of latency. Different schemes can apply in different scenarios flexibly.

---

[1] In this paper we treat bases and atoms as synonyms.

The organization of this paper is as follows, Sect. 2 describes the proposed method. Section 3 presents brief results. Section 4 is conclusion and future work. Section 5 is references.

2   The Method

In this section, we first propose the atomic matching and dictionary construction method in Subsect. 1. Then, in Subsect. 2, we extend to support shifting matching by using Hilbert transform. In Subsect. 3, we introduce the multi-pass scheme for increasing the flexibility of the algorithm.

2.1   Atomic Matching and Dictionary Construction

The fundamental problem of this subsection is finding reasonable atoms combination within limited time. Let us suppose that each envelope is a n dimension vector, denote the dictionary and its atoms as $D = [d_0, d_1, \ldots, d_{z-1}] \in \mathbf{R}^{n \times z}$, the target envelope as $x \in \mathbf{R}^{1 \times z}$, the selected atoms as $P = [p_0, p_1, \ldots, p_{u-1}] \in \mathbf{R}^{n \times u}$, and the weights of p_i for simulating x as w_i, $W = [w_0, w_1, \ldots w_{u-1}] \in \mathbf{R}^{1 \times u}$, temporary vector as x', temporary weight as u and v, and counter as i, j, and k. The decomposing procedure can be described as procedure 1:

Procedure 1 Decomposing procedure of x

1: $p_0 := \operatorname*{argmax}\limits_{d_i} \|\operatorname{proj}(x, d_i)\|$

2: $w_0 := \|\operatorname{proj}(x, p_0)\| / \|p_0\|$

3: $x' := w_0 p_0$

4: For $j = 1$ to u do

5:     $p_j := \operatorname*{argmax}\limits_{d_i \notin \{d_k, k \in [0, j)\}} \|\operatorname{proj}(x, \{x', d_i\})\|$

6:     $\{u, w_j\} := \operatorname*{argmin}\limits_{\{\dot{u}, v\}} \|\dot{u} x' + v p_j - x\|$

7:     $x' := \operatorname{proj}(x, \{x', p_j\})$

8:     $\forall k \in [0, j), w_k := u w_k$

9:     If x' is similar enough to x, Then break

10: End for

Whereas $\operatorname{proj}(a, b)$ represents the projection of vector a on vector b, and an extension version $\operatorname{proj}(a, \{b_1, b_2\})$ represents the projection of vector a on a (hyper)plane determined by vector b_1 and b_2. Note that u, v, x', and W will be updated and overwritten in each loop. The argmin in line 6 can be evaluated directly and efficiently according to their geometrical relationship.

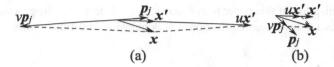

Fig. 1. Ill-posed weights (a) and normal weights (b)

Because the atoms may not be orthogonal, the dictionary is often over completed. As a result, not all the atoms are selected despite that some of them are similar to the target. The selecting procedure is given in procedure 1. Furthermore, for reducing the bit rate, it is not necessary to use n atoms for each envelope. Once the target quality is reached, the decomposing procedure will stop. This is different from PCM liked methods.

In practice, for better quantizing, the upper and lower bound of the weights should be constrained. For example, as shown in Fig. 1, both (a) and (b) are successful to use p_j and x' to represent x. However, because in (a), the scale of weights v and u are very large, and v is even a negative number, they are ill-posed. To solve this problem, we further append constrain for line 5 to filter out the atoms that causing ill posed weights.

When procedure 1 fails to produce result with expected quality, we can simply append the current envelope as an atom to the dictionary. To obtain better efficiency in both argmax procedures, we can further limit the capacity of the dictionary. By using algorithms such as least recently used (LRU) or first in first out (FIFO), we can continue append atom even when the dictionary is full. Furthermore, Parallel computing is implemented in argmax procedures.

Because the shapes of almost all the envelopes are similar to the previous ones, we can further adapt to this character by using the previous envelope as a special atom to represent the current ones. We call this feature as "differential feature" in the following text.

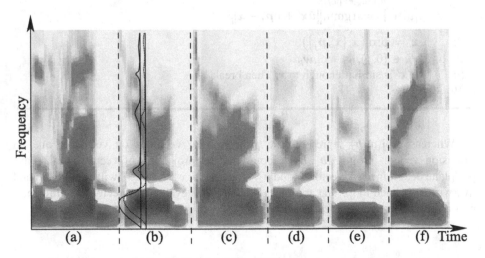

Fig. 2. Example of amplitude envelopes

2.2   Shifting Matching Extensions with Hilbert Transform

For synthesizing the envelops better, we first analyze the character of them varying over the time. As shown in Fig. 2, each column represents an envelope, and the horizontal direction represents the time. Darker colors in Fig. 2 represent higher amplitude, whereas white represents zeros. The envelope spectrum can be divided into 6 voiced sections as the sub charts from (a) to (f) shows. We can obtain that within each section, the envelopes changes over the time gradually. And some of them, for example, the two envelopes plotted in line charts at section (b) of Fig. 2, can approximately represent each other by slightly shifting over the frequency direction.

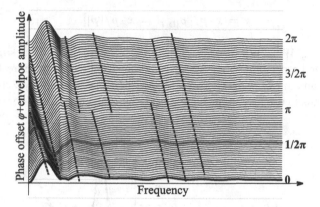

Fig. 3. Shifting characters of linear combining an envelope with its Hilbert transformed result

---

Procedure 2 Extension of procedure 1 for simple shifting mode

---

1: $p_0 := \underset{d_i}{\mathrm{argmax}} \|\mathrm{proj}(x, \{d_i, \widehat{d_i}\})\|$

2: $\{w_0, w_1\} := \underset{\{\dot{u}, v\}}{\mathrm{argmin}} \|\dot{u}p_0 + v\widehat{p_0} - x\|$

3: $x' := w_0 p_0 + w_1 \widehat{p_0}$

4: For j=1 to u do

5:     $p_j := \underset{d_i \notin \{d_k, k \in [0,j)\}}{\mathrm{argmax}} \|\mathrm{proj}(x, \{\mathrm{proj}(x, \{x', d_i\}), \widehat{d_i}\})\|$

6:     $\{u, w_{2j}, w_{2j+1}\} := \underset{\{\dot{u}, v, \dot{\mu}\}}{\mathrm{argmin}} \|\dot{u}x' + v p_j + \dot{\mu}\widehat{p_j} - x\|$

7:     $x' := \mathrm{proj}(x, \{\mathrm{proj}(x, \{x', p_j\}), \widehat{p_j}\})$

8:     $\forall k \in [0, 2j-1), w_k := u w_k$

9:     If x' is similar enough to x, Then break

10: End for

Simple Shifting Mode. In Fig. 3, we plot an atom, its Hilbert transformed result, and some of their linear combinations in one chart. Denote vectors with hat as their Hilbert transformed result. The bold line at bottom of Fig. 3 is the original envelope and is denoted as p, whereas the other bold line is $\hat{p}$. Denoting the phase offset as φ, other linear combinations of p and $\hat{p}$ can be represented as $q(\varphi) = p\cos(\varphi) \times \hat{p}\sin(\varphi)$. Furthermore, for plotting clearer and more institutively, we append a small offset who equals to φ for each linear combinations line at vertical axis. As the dotted lines shown in Fig. 3, the significant peaks and valleys of $q(\varphi)$ are approximately moving linearly over the frequency as the φ increases. So we can simply synthesize an envelope x whose shape is similar to an atom but with slightly offset over the frequency by using:

$$x \approx \alpha p + \beta \hat{p}, \tag{1}$$

$$\alpha = x \cdot p / \|p\|, \ \beta = x \cdot \hat{p} / \|\hat{p}\|. \tag{2}$$

By extending $W = [w_0, w_1, \dots w_{2u-1}] \in \mathbf{R}^{1 \times 2u}$, we propose a simple shifting mode, and extend procedure 1 to procedure 2. Comparing with procedure 1, we need to record double amount of weights for each atom in procedure 2. In practical, because procedure 2 tends to use less atoms for synthesizing each envelope, it consumes less bit rate averagely.

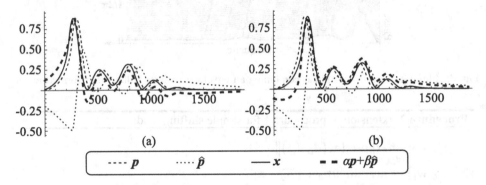

Fig. 4. Target envelopes and their approximation results

For better illustration, we present two examples in Fig. 4. In subplot (a), the target envelope x has same shape with atom p, and has a -30 samples offset to x on frequency, whereas in subplot (b), the offset becomes to 30 samples. Both in (a) and (b), comparing with the p, positions of peaks and valleys of $\alpha p + \beta \hat{p}$ on frequency are closer to the target signal x. Although the amplitudes of $\alpha p + \beta \hat{p}$ are slightly different from x, they affect little on perception quality. Because human's auditory sense is more sensitive to the frequencies of peaks then their amplitudes.

Shifting Mode. We modify simple shifting mode to shifting mode. Shifting mode can shift the atoms directly to match the target envelope. So it can avoid the inaccurate amplitudes' problem. To estimate the offsets efficiently, we need to calculate the Hilbert transform of each atom once it is appended to the dictionary beforehand. Then the offset estimation function can be constructed as:

$$g(p, x) = -\arg(p \cdot x + \hat{\mathbb{i}}\hat{p} \cdot x) \tag{3}$$

Whereas $\hat{\mathbb{i}}$ is the imaginary unit, and function arg(.) retrieve the angle from the positive real axis to the vector in the complex plane. For further explaining the character of $g(p, x)$, function shift(m, s) is defined, and it returns the result that envelope m shifting over the frequency by s samples. A positive or negative s mean that shifting to left or right. By using $x = \text{shift}(p, s)$ to simulate that an envelope x whose shape is same as an atom p shifted over the frequency by offset s, we plot $g(p, \text{shift}(p, s))$ in Fig. 5. As shown in Fig. 5, when the offset is at range between -40 to 40, $g(p, \text{shift}(p, s))$ is a monotonic increasing function of s, whose zero crossing is at $s = 0$. So by selecting a reasonable constant number $c \approx \lim_{s \to 0} g(p, \text{shift}(p, s))/s$ in statistical manner beforehand, we can approximately use Newton's method to estimate the s when at zero crossing of $g(p, \text{shift}(x, s))$.

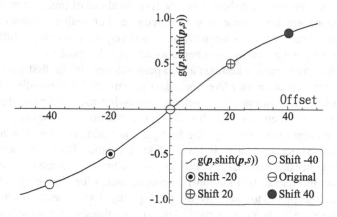

Fig. 5. Shifting estimation function $g(p, \text{shift}(p, s))$ of the example envelops

Practically, We initially set s to 0, then obtain s iteratively by applying equation below multiple times:

$$s := c \times g(p, \text{shift}(x, s)). \tag{4}$$

For applying the shift matching above, the step 1 of procedure 1 should be extent as:

$$\{p_0, s_0\} := \text{argmax}_{\{\text{shift}(d_i, s_0), s_0\}} \|\text{proj}(x, \text{shift}(d_i, s_0))\|. \tag{5}$$

Accordingly, to support shifting matching, step 5 of procedure 1 should be extent as:

$$\{p_j, s_j\} := \text{argmax}_{\{\text{shift}(d_i \notin \{d_k, k \in [0,j)\}, s_0), s_j\}} \left\|\text{proj}\left(x, \left\{x', \text{shift}(d_i, s_j)\right\}\right)\right\|. \tag{6}$$

However, we have not implemented the (7) in this paper because the Newton's method presented in this section cannot be directly applied for this step. We use simple shifting mode in for this step in the experiment, and leave this unsolved problem to our future work. So, for representing each envelope, we need to record the atoms' id in the dictionary, the offset and weight for the first atom, and weights for the next atoms and their Hilbert transform results. Although comparing with general dictionary based methods, more parameters are used when referencing an atom, we need fewer atoms to reconstruct an envelope. This will actually benefit for lowering the bit rate meanwhile increasing the perception quality.

2.3 Multi-pass Schemes

The procedure 1 and its extension described in Subsect. 2.2 are in fact one-pass schemes. They are real time coding, thus do not support updating the dictionary ahead of time. The disadvantage is they could cause more bit rate. For example, by using the current dictionary, we need to use 10 atoms for representing an envelope. But this envelope may need only 1 atom for representing by using the dictionary up-dated after 100 ms. So if we can construct the dictionary 100 ms ahead of time, we may be able to use fewer parameters for representing an envelope. For realistic streaming signals, obtaining them ahead of time is not possible, however, we can delay a little bit when encoding. So we need a multi-pass scheme to achieve this goal.

The simplest multi-pass scheme is a two-pass scheme. In the first pass ahead, the encoder constructs the dictionary like a one-pass scheme, but not recording the weights. Whereas the last pass encoder only selects atoms for reconstructing the envelope, evaluates and records their weights. This allows using some envelopes appeared later to represent the current ones. However, the first pass just adds envelopes as atoms when the testing reconstruction quality is lower than threshold. These operations are not response to whether the dictionary is optimal. So a solution is to use a looser threshold in the first pass in order to find the most important atoms for dictionary. Then the last pass uses a strict threshold to guarantee the final quality. Experimental results show that by using looser thresholds for prior passes and stricter thresholds for the later ones, we can construct a more optimal dictionary.

3   Experimental Results

In this section, we compare the compression ratio between the proposed method and the online dictionary method (ODL) [15]. Then some examples are used to show difference of different parameter in the proposed method. All of the experimental speech signals come from PTDB-TUG database [23]. PTDB-TUG database includes microphone records from 20 speakers, and all of them are recorded with laryngograph signals. By applying STFT to microphone signal and using precise fundamental frequencies extracted from laryngograph signal, we obtain amplitude envelopes. Because the frequency ranges up to 6000 Hz contains significant amplitudes for human's auditory perception, we crop the envelopes within these ranges, and obtain 512 samples in each envelope. Because ODL method requires dictionary scale beforehand, when comparing with ODL, we set the dictionary scale of ODL to be the same as our method.

3.1   Compression Ratio

Supposing that the data consumes A bit originally, and reduces to A' bits after compressing, we call $A'A^{-1}$ as compression ratio (CR). To simplify the experiment, we directly use 32-bit float format to storage each sample in the original envelope and atoms, and use the least bit for representing the atoms ID in our method. Furthermore, by treating each envelope x and its recovered result x' as 512 dimensional vectors, we use $\frac{x'x}{x'x}$ to measure their similarity. And we treat the average similarity of all the envelopes as the overall reconstruction quality (RQ). By randomly selecting the full dataset from 5 males and 5 female speakers, using 3-pass scheme with differential feature and shifting mode in the proposed method, and setting the target RQ = 98% for both methods, we Compare compression ratio between proposed method and ODL and PCA. We noticed that for samples from all speakers, PCA can keep the average RQ at 98% with 6 bases. Which means that the compression ratios of PCA is 0.011719.

Table 1. Compression ratios of female speech

	Speaker 1	Speaker 2	Speaker 3	Speaker 4	Speaker 5
Proposed	0.005627	0.005925	0.007452	0.006117	0.006099
ODL	0.010855	0.011786	0.014018	0.012411	0.011622

Table 2. Compression ratios of male speech

	Speaker 6	Speaker 7	Speaker 8	Speaker 9	Speaker 10
Proposed	0.016553	0.014033	0.014179	0.017093	0.017643
ODL	0.011587	0.009823	0.009925	0.011965	0.012350

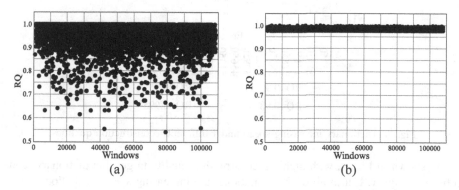

Fig. 6. Reconstruction quality of ODL (a) and proposed method (b) of all reconstructed window from speaker 1. RQ = 98%.

The compression ratios of the proposed method and ODL are shown in Tables 1 and 2. The dictionary sizes of ODL are set to the same as the proposed method in each test. It is shown that the result of ODL is very closed to PCA. When processing with female speech, the proposed method provides lower thus better compression ratio than the ODL and PCA. Whereas for male speeches, the proposed method apparently performed worse. The main reason is that, the meaning of RQ are different between proposed method and the other two. The former treats RQ as minimum threshold, whereas the others treat the RQ as average target. Figure 6 illustrates the difference intuitively. Although the average RQ of ODL is higher than 98%, there are many windows with RQ lower than 98%. And PCA have the similar problem. By contrast, all of the reconstructed window of the proposed method are guarantee to have RQ higher than 98%. The short time reconstruction failures of PCA and ODL above will lead to clicks, pops and cracks of the reconstruction wave, and affects much on the perception quality. So the result of the proposed method is more preferable.

3.2   Comparisons with Different Configurations

By using all of the envelopes of a speaker, we make comparisons on dictionary sizes and numbers of atoms selected in envelopes' reconstruction.

Dictionary Increasing Speed under different Reconstruction Quality. In this experiment, by using 6-passes scheme with differential feature and shifting mode, we compare the incremental speed of dictionaries size under RQ at 0.9, 0.95, 0.98, and 0.995.

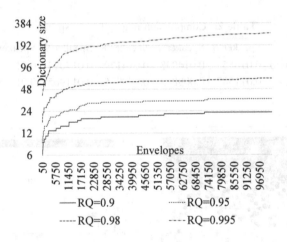

Fig. 7. Dictionary increasing speed under different reconstruction qualities

As shown in Fig. 7, with higher reconstruction quality target, the dictionary tends to become bigger. But in all of the conditions, the increasing speed of dictionary size becomes lower and lower as the input envelopes increases, and finally tends to stop increasing. This means that the dictionary will finally become well adapting to the signal. The transmissions of dictionaries only occur when there are increments on dictionary. For the streaming signal, especially the ones with long duration, dictionary sizes will finally become consistent, thus there are little differences between dictionary transmission band when target RQ changes.

Dictionary Increasing Speed in Different Passes. In our design, we expect that more optimal dictionary is produced by using more passes. In this experiment, by using from 1-pass to 6-passes schemes with differential feature and shifting mode, we compare the incremental speed of dictionary sizes under RQ at 0.98. They are plotted in Fig. 8.

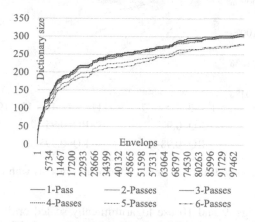

Fig. 8. Dictionary increasing speed of different passes schemes

As Fig. 8 shown, with more passes, the incremental speed of dictionary size becomes slower. The main reason is because we can use the more suitable envelopes to construct the dictionary in the case of using more passes schemes. It is beneficial to reduce the redundancy of the dictionary.

Probability Density of Atoms' Referencing Amounts for Representing Envelops. We have mentioned earlier that it is not necessary to use the maximum number of atoms to represent each envelope. So we estimate the average possibility of number of atoms used for representing an envelope through two experiments. In the first experiment, we use different passes, and plotted as Fig. 9 shown. Whereas by using different RQ, we have Fig. 10. Both of the two experiments are using differential feature and shifting mode. In Fig. 9, RQ = 0.995, whereas Fig. 10 uses 6-passes schemes.

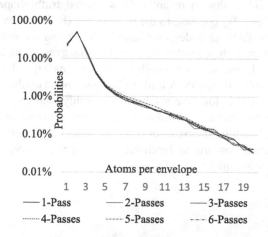

Fig. 9. Probability density of atoms' referencing amounts with different passes

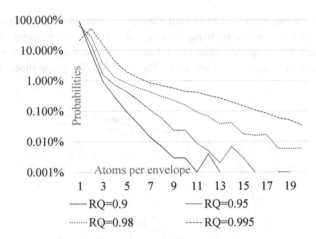

Fig. 10. Probability density of atoms' referencing amounts with different RQs

Note that both Figs. 9 and 10 use logarithmically scaled ordinates for displaying small numbers better. We discovered that for different passes schemes, the probability densities of atom counts are almost identical. And our algorithm tends to use less than 3 atoms for representing an envelope, which means that the results are with good sparse characters. Whereas when using higher RQ, the average number of atoms for representing each envelope becomes just slightly greater. This means that the proposed method is able to produce low bit rate even when RQ target is extremely high.

3.3   Reconstruction Comparison

In this subsection, we illustrate and compare the amplitude envelope spectrums. As shown in sub charts from (a) to (d) of Fig. 11, as the RQ target rises, the spectrums become more closed to the ground truth. Although some sections of spectrums with lower RQ may be different from the ground truth, none of them is obviously illegal. Because we can validate that there are always ground truth shapes similar to any reconstructed envelopes. By contrast, as the rectangle in (f) shows, even though the RQ has been set to 0.98, ODL provides fake amplitudes in this section. From (g) to (i), PCA generates very smooth reconstruction results. However, these results are lack of small texture details. Especially in the high frequency areas, the shapes are very blur. These shortcomings of ODL and PCA leads to bad perception quality. From (a), we can obtain that when the RQ is low, the results are smoothly changing over the time, but still keep clear over the frequency. Furthermore, we can see that our shifting matching method is successful shifting the bases over the frequency to generate smooth shifting results. These characters conform to the character of listening perception, so potentially leads to good perception qualities.

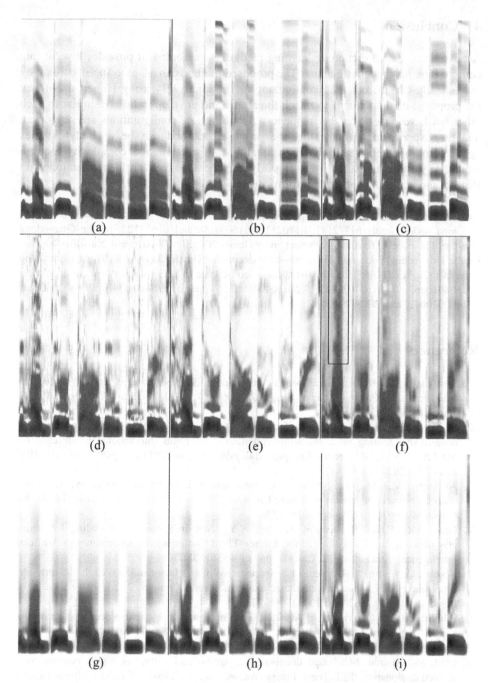

Fig. 11. Reconstructed envelopes and the ground truth. (a) to (d) are the results of the proposed method with target RQ of 0.9, 0.95, 0.98 and 0.995, (e) is the ground truth, (f) is the result of ODL with 0.98 target RQ, and (g) to (i) are the results of PCA with target RQ of 0.95, 0.98, and 0.995.

4   Conclusion

A novel incremental dictionary learning method is presented. It provides low bit rate compression for speech streaming. The proposed method can adapt to speech stream by dynamically and incrementally optimizing its dictionary. By using Hilbert transform, the proposed method is able to match the atoms and envelops who are similar to each other but have offset over the frequency. A wide range configurations are supported, and the experiments validate that the proposed method can be applied to varies applications with different bandwidth, quality, and computation complexity requirements. In further research, we will study for the unsolved problem mentioned in shifting mode, and for phase spectrum compression.

Acknowledgement. This work is supported in part by the National Natural Science Foundation of China under Grants 61772141, 61702110, 61603100, and 61673123, by the Guangdong Provincial Science & Technology Project under Grants 2015B090901016 and 2016B010108007, by the Guangdong Education Department Project under Grants Guangdong Higher Education letter 2015[133], 2014[97], and by the Guangzhou Science & Technology Project under Grants 201508010067, 201604020145, 201604046017, 2016201604030034, 201802030011, 201802010042, and 201802010026.

References

1. Sunnydayal, V., Kumar, T.K.: Speech enhancement using posterior regularized NMF with bases update. Comput. Electr. Eng. **62**, 663–675 (2017). https://doi.org/10.1016/j.compeleceng.2017.02.021
2. Gunawan, T.S., Khalifa, O.O., Shafie, A.A., Ambikairajah, E.: Speech compression using compressive sensing on a multicore system. In: 2011 4th International Conference on Mechatronics (ICOM), Kuala Lumpur, Malaysia, pp. 1–4 (2011). https://doi.org/10.1109/icom.2011.5937130
3. Al-Azawi, M.K.M., Gaze, A.M.: Combined speech compression and encryption using chaotic compressive sensing with large key size. IET Signal Process. **12**(2), 214–218 (2018). https://doi.org/10.1049/iet-spr.2016.0708
4. Grosse, R., Raina, R., Kwong, H., Ng, A.Y.: Shift-invariant sparse coding for audio classification. In: Proceedings of the Twenty-Third Conference on Uncertainty in Artificial Intelligence, Vancouver, BC, Canada, pp. 149–158 (2007)
5. Févotte, C., Bertin, N., Durrieu, J.L.: Nonnegative matrix factorization with the Itakurasaito divergence: with application to music analysis. Neural Comput. **21**(3), 793–830 (2009). https://doi.org/10.1162/neco.2008.04-08-771
6. Zibulevsky, M., Pearlmutter, B.A.: Blind source separation by sparse decomposition in a signal dictionary. Neural Comput. **13**(4), 863–882 (2001). https://doi.org/10.1162/089976601300014385
7. Elad, M., Aharon, M.: Image denoising via sparse and redundant representations over learned dictionaries. IEEE Trans. Image Process. **15**(12), 3736–3745 (2006). https://doi.org/10.1109/TIP.2006.881969
8. Mairal, J., Elad, M., Sapiro, G.: Sparse representation for color image restoration. IEEE Trans. Image Process. **17**(1), 53–69 (2008). https://doi.org/10.1109/TIP.2007.911828

9. Mairal, J., Bach, F., Ponce, J., Sapiro, G., Zisserman, A.: Supervised dictionary learning. Adv. Neural. Inf. Process. Syst. **21**, 1033–1040 (2009)
10. Bradley, D.M., Bagnell, J.A.: Differentiable sparse coding. Adv. Neural. Inf. Process. Syst. **21**, 113–120 (2009)
11. Yang, J., Yu, K., Gong, Y., Huang, T.: Linear spatial pyramid matching using sparse coding for image classification. In: 2009 IEEE Conference on Computer Vision and Pattern Recognition, Miami, FL, USA, pp. 1794–1801 (2009). https://doi.org/10.1109/cvpr.2009. 5206757
12. Lu, X., Wang, D., Shi, W., Deng, D.: Group-based single image super-resolution with online dictionary learning. Geomat. Inf. Sci. Wuhan Univ. **2016**(1), 84 (2016). https://doi.org/10. 1186/s13634-016-0380-9
13. Peyré, G.: Sparse modeling of textures. J. Math. Imaging Vis. **34**(1), 17–31 (2009). https:// doi.org/10.1007/s10851-008-0120-3
14. Warmuth, M.K., Kuzmin, D.: Randomized online PCA algorithms with regret bounds that are logarithmic in the dimension. J. Mach. Learn. Res. **9**, 2287–2320 (2008)
15. Mairal, J., Bach, F., Ponce, J., Sapiro, G.: Online dictionary learning for sparse coding. In: Proceedings of the 26th Annual International Conference on Machine Learning, Montreal, Quebec, Canada, pp. 689–696 (2009). https://doi.org/10.1145/1553374.1553463
16. Mensch, A., Mairal, J., Thirion, B., Varoquaux, G.: Stochastic subsampling for factorizing huge matrices. IEEE Trans. Signal Process. **66**(1), 113–128 (2017). https://doi.org/10.1109/ TSP.2017.2752697
17. Liu, J., Garcia-Cardona, C., Wohlberg, B., Yin, W.: Online convolutional dictionary learning. In: 2017 IEEE International Conference on Image Processing (ICIP), Beijing, China, pp. 1707–1711 (2017). https://doi.org/10.1109/icip.2017.8296573
18. Jolliffe, I.T.: Principal Component Analysis. Springer, New York (2005). https://doi.org/10. 1007/b98835
19. Hinton, G.E., Salakhutdinov, R.R.: Reducing the dimensionality of data with neural networks. Science **313**(5786), 504–507 (2006). https://doi.org/10.1126/science.1127647
20. Schölkopf, B., Smola, A., Müller, K.-R.: Kernel principal component analysis. In: Gerstner, W., Germond, A., Hasler, M., Nicoud, J.-D. (eds.) ICANN 1997. LNCS, vol. 1327, pp. 583–588. Springer, Heidelberg (1997). https://doi.org/10.1007/BFb0020217
21. Schmitz, M.A., et al.: Wasserstein dictionary learning: optimal transport-based unsupervised nonlinear dictionary learning. SIAM J. Imaging Sci. **11**(1), 643–678 (2018). https://doi.org/ 10.1137/17M1140431
22. Aharon, M., Elad, M., Bruckstein, A.: K-SVD: an algorithm for designing of overcomplete dictionaries for sparse representations. IEEE Trans. Signal Process. **54**(11), 4311–4322 (2006). https://doi.org/10.1109/TSP.2006.881199
23. Pirker, G., Wohlmayr, M., Petrik, S., Pernkopf, F.: A pitch tracking corpus with evaluation on multipitch tracking scenario. In: INTERSPEECH 2011, 12th Annual Conference of the International Speech Communication Association, Florence, Italy, pp. 1509–1512 (2011)

Identifying Price Index Classes
for Electricity Consumers via Dynamic
Gradient Boosting

Vanh Khuyen Nguyen[✉], Wei Emma Zhang, and Quan Z. Sheng

Department of Computing, Macquarie University, Sydney, Australia
thi-vanh-khuyen.nguyen@students.mq.edu.au,
{w.zhang,michael.sheng}@mq.edu.au

Abstract. Electricity retailers buy electricity at spot prices and resell energy to their customers at fixed retail prices. However, the electricity market is complex with highly volatile spot prices, and high price events might happen during peak time periods when energy demand significantly increases, leading to the decision of the retail price a challenging task. Understanding consumer price index, a price indicator that is associated with electricity consumption of customers helps energy retailers make critical decisions on pricing strategy. In this work, we apply dynamic gradient boosting model, namely CatBoost, to classify customers into different groups according to their price indices. To benchmark our results, we compare the performance of CatBoost with other baselines, including Random Forest, AdaBoost, XGBoost, and Light-GBM. Our experimental results proved that CatBoost outperformed other algorithms due to its effective overfitting detector and categorical encoding techniques. Besides, the area under the curve of the Receiver Operating Characteristics (ROC), often known as AUC, is used as a standard measure metric to evaluate and compare between classifiers. Hence, CatBoost gained the lowest difference score of 0.02 between train AUC and test AUC scores that successfully competed other models.

Keywords: Classification learning · CatBoost
Gradient boosting model

1 Introduction

The Australian wholesale market is operated as a spot market where retailers buy electricity from the electricity generators and pay at the spot price, and then reselling energy to their customers at the fixed retail price [1,2]. The spot price for each trading interval (half-hour) refers to the current price in the marketplace at which energy can be traded by market participants. However, spot prices in electricity market are highly volatile in regards to supply and demand [2].

Besides this, high price event might happen at any time when energy demand significantly increases. The high price event refers to the spot price impressively

© Springer Nature Switzerland AG 2018
H. Hacid et al. (Eds.): WISE 2018, LNCS 11234, pp. 472–486, 2018.
https://doi.org/10.1007/978-3-030-02925-8_33

increasing at the specific trading interval, e.g. thousands dollar/MWh, and it is sometimes unpredictable. To accurately decide the fixed retailer price and achieve the most profits, an Australian electricity retailer attempts to analyze the price index of each customer.

Price index is defined as the ratio of volume-weighted average price (VWAP) to the mean of spot price at half-hour trading interval over the designated period. Following by that, customers have high price index (greater than 1), i.e., consume more energy during high spot price events or peak time periods, are grouped into group 1. The retailer has less profits from this group of customers as it pays more to the generators but charges a fixed price from the customers. Other customers are grouped into group 0 (lower or equal to 1). By serving those customers in group 0, the company gains more benefits. Therefore, understanding the customer's price index is critical for the decision of the retailer price.

To solve this problem, we formulate it as a binary classification task and data collection is extracted from the retailer's data warehouse. However, the real-world dataset often contains multiple categorical values. Particularly, four variables (age, has solar, network, and tariff types) are defined in categorical formats. Some machine learning algorithms can handle categorical features but many models can not do that. Therefore, data preprocessing is conducted to transform these non-numerical values to numerical forms before training. Ordinal or integer encoding is a simple way to convert categorical features to numbers; but it might produce negative impacts on the accuracies of classifiers [7]. In contrast, one-hot-encoding is a well-known method to improve the effectiveness of categorical transformation even though it can cause the high dimensionality of the problem space [16]. In this study, we take advantages of tree-based methods to perform classification task since they do not often require transformation preprocessing for categorical variables like one-hot-encoding method [3]. But, in most cases, we need to convert categorical features to numeric values before adding input data into the models; therefore, applying ordinal or integer encoding technique is enough. Moreover, tree-based classifiers have been widely-used in many machine learning applications. Decision tree and Random Forest are the classic tree-based models that have applied for solving classification problems. However, these models still have many limitations in terms of accuracy, overfitting detection, and others [3,19]. On the other hand, Gradient Boosting Model (GBM) has become more popular with many advanced techniques proposed in the research field. For instance, XGBoost has gained a lot of success due to its scalability [5] while LightGBM has improved the computations speeds in training process as well as maintain the accuracy of the model [10].

In this research, we will use the recent advanced dynamic gradient boosting, called CatBoost, to address classification problem in energy retailer. The CatBoost algorithm was firstly introduced in the paper of [6]. The experimental results of this research showed that CatBoost was able to reduce bias and improve accuracy of GBM. In particular, it outperformed popular GBMs, including XGBoost and LightGBM. These implementations were conducted based on several public datasets [6]. Similarly, we adopt and employ CatBoost algorithm for solving binary classification problem. We also compare CatBoost over baselines, including Random Forest, AdaBoost, XGBoost, and LightGBM in terms of accuracy and overfitting prevention. Importantly, we perform CatBoost on our real-world dataset that contains multiple categorical variables. Therefore, we can evaluate the robustness of categorical encoding techniques built in CatBoost algorithm while also ensure well-controlling overfitting in the model. Consequently, CatBoost obtained outstanding performance compared to other baselines. Particularly, train AUC score (0.7618) and test AUC score (0.7418) of CatBoost model were very close while its F1 score is also higher than other models, excepting LightGBM.

Our work has not only resolved the specific problems for the Australian electricity retailer, but also addresses the major concerns for participants in electricity market. It is the first work that applies machine learning methods in price index classification for electricity consumers. The work has significant importance for improving investments and revenues of electricity generators and retailers. Furthermore, this paper provides an empirical evaluation on the newest classification algorithm, CatBoost, and proves its effectiveness.

The rest of the paper is organized as follows. Section 2 describes some studies relating to classification problems in literature. Section 3 introduces the real-world dataset and our proposed approach for price index classification. Section 4 presents experiments on real-world dataset and compares CatBoost against other baseline algorithms. Section 5 concludes the paper with our key findings and future research directions.

2   Related Work

Tree-based models are popular techniques for solving classification problems in machine learning. Decision tree is a simple classifier which collects decision nodes arranged in a tree structure [17]. This method has been used widely in practice due to the ease of interpretability [19]. However, if the training data is not large enough and contains noise, it causes to overfitting when the model attempts to fit all of attributes [12]. Random Forest then was introduced by [4] that was also adopted from decision tree model. It is a combination of tree-structured classifiers in which each tree votes for the most popular class at the input vector [4]. Even though this model has gained more popularity in machine learning applications, it is still difficult to analyze. Particularly, its implementation mostly depends on simplifying assumptions and variations of the standard framework [3].

Recently, Gradient Boosting Model (GBM) has become powerful techniques for solving supervised learning problems due to its efficiency, accuracy, and interpretability. Indeed, it has been widely applied in industrial manufacturing systems [13], energy management in buildings [20], solar radiation [18], and etc. Specifically, the study of [13] proposed pattern recognition-based algorithm to classify energy consumption patterns in industrial manufacture. The authors in [20] employed different methods to build an automated classification for energy load profiles in buildings. In the work of [18], GBM was conducted by using

Gradient Boosted Regression and Extreme Gradient Boosting to resolve solar radiation problem. Another study of [11] used Gradient Tree Boosting (GTB) to generate baseline model of electricity usage. Recently, XGBoost and Light-GBM are two representatives of GBM that have gained a lot of success on a wide range of practical tasks. XGBoost is successful due to its scalability resulted from several important systems and algorithmic optimizations [5]. On the other hand, LightGBM tackles the high time-consuming problem of GBM when dealing with high dimensional features and large data size [10]. To do that, they proposed two novel techniques, including Gradient-based One-Side Sampling (GOSS) and Exclusive Feature Bundling (EFB). As the experiments of [10], LightGBM speeded up to over 20 times for the training process while still achieved the same accuracy performance.

However, most of existing GBMs require conversion of dataset to specific format. Particularly, we have to convert all categorical features to numbers before training. Different kinds of categorical encoding techniques were proposed in the research of [7,16]. For example, ordinal encoding method produces unique numeric code to a specific category [7]. Other techniques for encoding are N and $N-1$ encoding methods that are similar to one-hot-encoding or dummy tables [16]. Although these above techniques are widely used in research and practice, they still have many weaknesses; e.g., misleading information, difficult interpretation of original features, and increase of dimensionality of dataset. Therefore, it is necessary to seriously consider the selection of the appropriated categorical encoding in data preprocessing. Besides this, the most challenge in GBMs is to reduce overfitting when training dataset is not large enough. In order to overcome these issues when dealing with categorical data and limited data size, we adopt the most recent gradient boosting method, namely Catboost. In Catboost, a new approach is proposed to handle categorical features more effectively and robustly. It also improves overfitting detector in GBM to achieve better results between training and testing dataset.

3   Method

In this section, we present the problem formulation, our special real data and the way to calculate price index for each customer. After that, we discuss the adoption of CatBoost model in this research.

3.1   Problem Formulation

Given the dataset that contains n available independent objects (customers), for each object i with $i \in \{1, \cdots, n\}$, we aim to classify it to a specific group based on its price index. Let y_i, p_i denote a class and a price index for an object i, respectively, our aim is to predict which class a specific object belongs to, that can be defined as follows:

$$y_i = \begin{cases} 1 & \text{if } p_i > 1 \\ 0 & \text{otherwise} \end{cases}$$

With such transformed formulation, we are able to resolve our problem by using binary classifier with y_i seen as the class label. We define that if $y_i = 1$, the customer has price index greater than 1. If $y_i = 0$, the customer has price index lower or equal to 1. However, we also consider the likelihood of an object assigned to a specific class; i.e., we want to know how is the probability of the predicted class rather than only the label is either 0 or 1.

As each object contains multiple features, each data point denoted as $u_i \triangleq [x_i, y_i]$ is partitioned into a feature vector $\{x_{ik}\}_{k=1}^{m} \in \mathbb{R}^m$ where k is the feature index and a scalar response $y_i \in \mathbb{R}$. For theoretical consideration, we say that the data points u_k are independent and identically distributed according to an unknown probability measure $\mathcal{P}(u)$. Practically speaking, this probability measure should describe which region in the vector space $u \in \mathbb{R}^{m+1}$ that is more likely to contain unforeseen data points.

Our ultimate goal now is to predict the response y associate with any given feature vector x, i.e. to estimate the conditional distribution $\mathcal{P}(y \mid x)$. Many classical approaches to this problem revolve around constructing a statistical model $y = g_\theta(x, \nu)$ where ν can be any form of random variable and θ is the set of parameters that form a concrete definition of $g : (x, \nu) \mapsto y$, [9] for example. In this approach, we can see that for some given feature vector x_i,

$$\mathcal{P}(y \mid x_i) = \mathcal{P}(g_\theta(x_i, \nu) \mid x_i)$$

which, if θ is identified, can certainly be deducted from the distribution of ν.

3.2   Real Data Description

The dataset used in this work is collected from real dataset of an energy company in Australia. It contains the electricity consumption of 658 households from January to August of 2017. For each customer, the information includes network provider, house owner's age, tariff types, and total average of electricity consumption. Table 1 summarizes these information, and we consider them as features in our classification task.

Since our raw data is not labeled, we conduct financial analysis to determine price index for each customer. Firstly, the volume-weighted average price (VWAP) is calculated by aggregating the total cost of supplying electricity for each half hour and dividing the aggregate number by the total consumption (kWh) over the designated time period (Eq. 1). For example, $VWAP_j$ of customer j ($j \in [0, 657]$) is defined as follows:

$$VWAP_j = \frac{\sum_t P_t * Q_t}{\sum_t Q_t} \tag{1}$$

where P_t is the spot price ($/MkW) at the specific 30-min interval, $t \in T$ with T is the designated time period, Q_t is the total energy consumption (kWh) at the time interval t.

Then, the price index, which can be either premium (denoted as 1) or discount (denoted as 0), is derived from the ratio of VWAP to the "time-weighted average

Table 1. Summary of the features of dataset collected from an electricity retailer from January to August of 2017.

Features	Data type	Description
Age	string	age group of customers, including: - group 1 (less than 30 years old), - group 2 (between 30 and 40 years old), - group 3 (between 40 and 50 years old), - group 4 (between 50 and 60 years old), - group 5 (greater than 60 years old)
Has solar	string	household has either solar system or not
Network	string	based on the location of households, we identify their energy network distributors. There are three main network: Ausgrid, Endeavour, and Essential
Tariff type	string	there are two main types of tariff rates, including time of use and single rate. Different tariff types offer different energy prices
Total average usage	float	the average of energy usage during a time period (Jan–Aug 2017)

price" (TWAP) or the mean of the half hourly spot prices over the designated period. The observation is given the value of 1 if its ratio is greater than 1, and 0 otherwise. For the current dataset, there are respectively 328 customers in group 0 and 330 customers in group 1. Thus, our dataset has balance between two groups.

3.3 Price Index Classification

In this session, we will quickly review the theoretical framework for boosting algorithms in machine learning with specific details given to the dynamic boosting approach given in [6]. The overview of classification process is demonstrated in the Fig. 1.

Firstly, input data that contains categorical attributes is going through the preprocessing schemes to convert categorical values into numeric formats. That process depends on requirements of a specific classification model which will be applied, e.g., CatBoost does not require all variables that have to be numerical. Secondly, different kinds of tree-based classifiers are deployed and the final step is to evaluate the performance of classifiers to select the best model for solving problem in our case study.

Preprocessing Schemes for Categorical Features. Real-world data often contains non-numeric values whereas many data mining models require

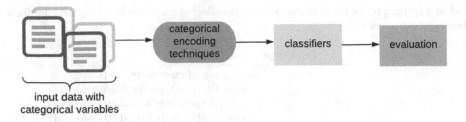

Fig. 1. Price index classification process

numerical data inputs. Thus, it is a serious challenge for data mining applications. Moreover, existing researches indicate that the use of different techniques for handling categorical, nominal, or binary data may significantly impact on the performance of neural networks, linear and logistic regression [16], and the classification accuracy of neural networks [7]. However, there are not many justifications or discussions about categorical data preprocessing in research field. Therefore, in this study, we would like to apply and evaluate some common techniques for dealing with categorical data in different tree-based models.

Table 2. Schemes for encoding categorical inputs variables

Raw value	Ordinal	N			N − 1	
	x_1	x_1	x_2	x_3	x_1	x_2
First	1	1	0	0	1	0
Second	2	0	1	0	0	1
Third	3	0	0	1	0	0

Ordinal encoding or integer encoding is a traditional technique to assign a unique numeric code to each category. This type of technique is easy to interpret as well as avoid the increase of dimensionality of data [7], e.g., this technique might be more appropriated for a feature like a grade variable with the values: "first", "second", and "third". However, for many types of categorical variables, ordinal encoding might lead to misleading information in training models, e.g., the color variable includes values "green", "yellow", and "red" and these categorical values are encoded into 1, 2, and 3 respectively. The problem is that some models might not understand well the relationship between these numerical values. They then assume that number 1 should be lower than number 2 and number 3 is higher than number 2. Therefore, it might produce negative impacts on the results of classifiers.

For solving that problem, another technique can be used that is called 1-out-of-N variable coding (also known as N encoding or dummy tables). In this technique, an attribute with N categories is transformed into N input variables in which an observation is assigned with a value of 1 for a specific category that it represents, 0 otherwise [7]. However, this type of encoding might lead to the increase of dimensionality of the problem space and it is also hard for interpreting the original features. In fact, some researchers also showed that N encoding method can cause to multicollinearity and ill-conditioning [7], i.e., multicollinearity may occur due to the repetition of the same kind of variables in dummy tables. That might lead to ill-conditioning in mathematical problem which means the small change in input (independent variable) causing to the large change in output (dependent variable). Therefore, it will negatively impact on the classification models.

In order to prevent these issues, 1-out-of-$N-1$ encoding (also known as $N-1$ encoding) method was proposed [7]. It is similar to N encoding but only $N-1$ class categories are used to represent for a variable with N categories. As seen in the Table 2, the variable x_3 was removed when applying $N-1$ encoding to avoid multicollinearity. This method is typically used when N is small, says less than ten, but when N grows, the inputs of the model are significantly increased. That may cause to the increase of the number of parameters to be estimated in the model [16]. Thus, with such type of method, we still cannot avoid the increase of dimensionality when dealing with high-cardinality variables containing a large number of distinct values (e.g. hundreds or thousands of distinct values). Overall, any encoding techniques have trade-offs. In this paper, we will explore and apply another method that is an in-built function in CatBoost model to effectively handle categorical features.

CatBoost-Based Categorical Encoding Approach. CatBoost utilizes an effective technique to encode categorical values to numerical values while also prevent overfitting. The process is going through three main steps as described in the Algorithm 1.

Algorithm 1. CatBoost's categorical encoding

1: The set of input observations is permuted in a random order; thereby generating multiple random permutations.
2: The label value with a floating point or category is transformed to an integer.
3: All categorical variable values are converted to numerical values by a given formula:

$$avg\_target = \frac{countInClass + prior}{totalCount + 1}$$

where:
   - *countInClass* is how many times the label value is given to the value of 1 for objects with the current categorical feature value;
   - *prior* is a preliminary value for the numerator;
   - *totalCount* is the total number of objects (up to the current one) that have a categorical feature value matching the current one.

Dynamic Gradient Boosting Method. For most practical machine learning problems, there is really no sensible way to mathematically describe the distribution $\mathcal{P}(u)$, e.g. in term of constructing g_θ in closed-form. Therefore the classical parametric approach is not applicable in these scenarios and some simplifications are necessary. In the case for dynamic boosting, these simplifications can be understood as removing the random perturbation ν and define the function $g(x)$ as the one that minimizes the expected difference between the response y and the predicted value $g(x)$, i.e.

$$g \triangleq \underset{g \in \mathcal{G}}{\operatorname{argmin}} \mathbf{E}_u[l(g, u)], \text{ where } l(g, u) \triangleq d(y, g(x))$$

with $d(\cdot, \cdot)$ being a chosen metric function, e.g. usually the squared difference.

This is now a problem of functional optimization [8,15] where the search domain is defined by $\mathcal{G}$ which is a chosen family of mapping $\mathbb{R}^m \mapsto \mathbb{R}$, e.g. the sum of decision trees where each tree can be described as:

$$h(x) = \sum_i w_i \mathbf{I}_i(x)$$

with $w_i \in \mathbb{R}$ and $\mathbf{I}_i(\cdot)$ are the indicator functions of disjoint subsets covering $\mathbb{R}^m$. The solution to this problem can be approximated in a similar fashion as with numerical optimization, i.e. one can goes about building g in an iterative manner starting from an initial function such as: the constant $g_o(x) = \overline{y_k}$ and adding one tree to the sum at each iteration such that we can write the function g at iteration t as:

$$g_t = g_{t-1} + \alpha_t h_t, \alpha_t \in \mathbb{R} \quad \forall t \in \mathbb{N}/\{0\}$$

As with other optimization methods, the incremental function h_t should be equivalent to the negative gradient of the loss distance d computed at the current functional position g_{t-1}, i.e.:

$$h_t = \underset{h}{\operatorname{argmin}} \mathbf{E}_u[(h(x) - \delta_t(u))^2], \qquad \delta_t(u) \triangleq -\left.\frac{\partial d(y, s)}{\partial s}\right|_{s = g_{t-1}(x)}$$

The above expectation can only be approximate using the dataset S such that:

$$h_t \approx \underset{h}{\operatorname{argmin}} \sum_k (h(x_k) - \delta_t(u_k))^2$$

However, it was shown in [6] that repeated use of the same dataset S through all iterations t will lead to bias in estimation of the gradient and cause overfitting to the resulting model (see Sect. 2 of [6]). What was suggested to mitigate this problem is to compartmentalize the dataset into nested structure that helps minimize sharing data between consecutive iterations. There were two partition scheme suggested in the paper but the more effective one called 'Ordered dynamic boosting' is as follows:

Algorithm 2. CatBoost's ordered dynamic boosting

1: Given a starting model g_o, dataset $S = \{u_k\}_{k=1}^{K}$;
2: In each incremental iteration $t = 1, 2, \ldots K$ and for each index k, we maintain a copy of the model as:

$$g_{t,k} = g_{t-1,k} + h_{t,k}$$

3: Where we have calculated the gradient term $h_{t,k}$ for the kth model as:

$$h_{t,k} \approx \underset{h}{\operatorname{argmin}} \sum_{l=1}^{k} (h(x_l) - \delta_{t,l}(u_l))^2,$$

where we redefine:

$$\delta_{t,l}(u) \triangleq -\left.\frac{\partial d(y, s)}{\partial s}\right|_{s=g_{t-1,l-1}(x)}$$

We can notice that each of the model $g_{t,k}$ will only make use of the first k data points and the residual term for each data point is calculated based only on its preceding data points according to any certain chosen ordering between the data points. This explains why the scheme helps minimize repeated usage of the same data point when calculating the gradient term in each iteration. And it is reasonably followed that we will take $g_{K,K}$ as the final result while discarding the rest of other models.

4 Evaluations

To evaluate the performance of CatBoost, we compare it with other baseline models, including Random Forest, AdaBoost, XGBoost, and LightGBM. The settings for implementation and evaluations will be discussed in the following subsections.

4.1 Setup and Evaluation Metrics

All of models in the experiments were implemented in Python environment with the supports of open-source libraries. For Random Forest and AdaBoost, we use scikit-learn library[1] and most of parameters are set as defaults. On the other hand, XGBoost, LightGBM, and CatBoost have different parameters to tune in order to improve the performance of the models as well as prevent overfitting problems. As shown in the Table 3, we tuned the optimal values for different types of parameters in these models by applying exhaustive grid search. The grid search combines all of possible combinations of parameter values to find the best combination of these values for the classifiers. The results of grid search are displayed in the Table 3.

---

[1] http://scikit-learn.org/.

Furthermore, some models (Random Forest, AdaBoost, XGBoost, and Light-GBM) could not handle categorical values; thereby first encoding these values to numeric values. However, LightGBM and CatBoost have built-in functions for dealing with these categorical features to reduce overfitting and increase accuracy of the models. Thus, we can use them by passing indices of categorical features in our datasets to the models.

Table 3. Tuning hyper-parameters for classifiers

Function	XGBoost	LightGBM	CatBoost
Control overfitting parameters	learning rate: 0.05 max depth: 10 min child weight: 6	learning rate: 0.01 max depth: 25 number leaves: 300	learning rate: 0.15 depth: 10 l2 leaf reg: 9
Parameters for categorical features	not available	categorical features: features indices	categorical features: features indices non-hot-max size: None
Parameters for speed control	estimators: 200	iterations: 200	iterations: 100

Area under the curve of the Receiver Operating Characteristics (ROC) is often denoted as AUC which is used as an alternative measure for machine learning algorithms. Indeed, the research of [14] found that AUC provided the better measure compared to accuracy; especially, using AUC for comparing the performance of different models. Therefore, we use AUC as a standard measure metric in our experiments. Particularly, we calculate separately AUC scores on training and testing datasets. We then compute the difference scores between train AUC and test AUC scores to evaluate the overfitting detector of a specific model. In addition, F1 score is used as an alternative measure to assess the accuracy of the models in this experiment. F1 score measures the accuracy of test dataset and it is described as a weighted average of the precision and recall.

4.2   Results and Discussion

In this section, various methods are used to benchmark our results while also compare the performance of CatBoost model with other existing models in terms of overfitting detection and accuracy. As seen in the Table 4, CatBoost outperformed with highest AUC score on test set (0.74) compared to other methods. It also showed that CatBoost effectively handled overfitting since the AUC scores for both training and testing are approximately close, especially, the difference score is 0.02. LightGBM model has test AUC score of 0.72 that is quite close to the score of CatBoost (0.74), but the difference score of LightGBM (0.123) is sixth time higher than CatBoost's score (0.02). In other words, the overfitting detector of LightGBM did not perform as well as CatBoost in this case.

On the other hand, XGBoost and AdaBoost have low AUC score on test set (approximately 0.63) whereas the difference scores for both models are also high, 0.32 and 0.21 respectively. Similarly, Random Forest performed badly with the difference score of 0.37. Thus, these models did not control well overfitting due to the preprocessing of encoding categorical features into numerical values. Additionally, LightGBM achieved highest F1 score of 0.72, and followed by Cat-Boost with F1 score of 0.66. Besides, the F1 measure of other models (XGBoost, AdaBoost, and Random Forest) achieved over 60%. Generally, the performance of these models are not good in term of accuracy as our dataset is quite small with only 658 customers. For future improvement, we need to collect more data for training process.

Table 4. Comparisons between CatBoost and baseline tree-based algorithms. RDF, ADB, XGB, LGB, and CB stands for Random Forest, AdaBoost, XGBoost, LightGBM, and CatBoost respectively. The difference score is used to evaluate the overfitting detector of a specific model. The lower score is, the better the model is.

	RDF	ADB	XGB	LGB	CB
Training AUC Score (AUC_1)	0.9668	0.8359	0.9575	0.8439	0.7618
Test AUC Score (AUC_2)	0.5981	0.6262	0.6326	0.7210	0.7418
Difference Score ($AUC_1 - AUC_2$)	0.3687	0.2097	0.3249	0.1229	**0.020**
F1 Score	0.6145	0.6180	0.6182	**0.7209**	0.6624

Furthermore, we display the ROC curve in the Fig. 2 to compare the performance of classifiers across the entire range of class distribution and error costs. The true positive rate (TPR) refers to the ratio of the number of correct positive results and all actual positive cases during the test. The false positive rate (FPR), on the other hand, refers to the ratio of the number of incorrect positive results and all actual negative cases during the test. The area under the curve is defined as AUC. The larger the area is, the better the performance of the model is. As shown in the Fig. 2, the ROC curve of CatBoost and LightGBM clearly dominate XGBoost, AdaBoost, and Random Forest. Additionally, the curves of CatBoost and LightGBM are not dominating each other in the whole range, but CatBoost has smoother and larger curve than LightGBM in the entire range.

Consequently, the performance of CatBoost on dataset with multiple categorical features is better than other existing gradient boosting and tree-based methods. If we consider to measure the likelihood of the level of energy consumption rather than only the class label (0 for low price index and 1 for high price index), CatBoost will be the best choice based on our experimental results. However, we will consider to improve the accuracy for CatBoost by collecting more data for training model in our future works.

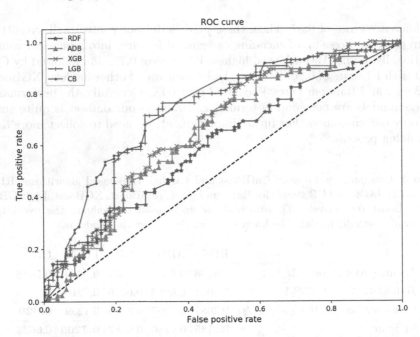

Fig. 2. Receiver operation characteristics of different methods. The area under the curve is denoted as the AUC. The black dot line refers to random guess. Data points are greater than random points, that represents good classification; otherwise the classifier produces poor results. RDF, ADB, XGB, LGB, and CB stands for Random Forest, AdaBoost, XGBoost, LightGBM, and CatBoost respectively

5   Conclusion and Future Work

In this work, we introduced and applied the most recent dynamic gradient boosting model, named CatBoost, to solve real-world classification problems in energy industry. To benchmark our results, the comparison between CatBoost and other baseline algorithms (LightGBM, XGBoost, AdaBoost, and Random Forest) was conducted. The experimental results showed that CatBoost outperformed other classifiers in terms of overfitting detection with the small difference score (0.02) between training and test AUC scores. Even though F1 score of CatBoost was a bit lower than LightGBM but it still performed better than other models, especially its F1 score is higher than the scores of XGBoost, AdaBoost, and Random Forest. Therefore, it proves that CatBoost can provide robust and effective categorical encoding system while also maintain accuracy and reduce overfitting.

For future works, we would like to collect a larger dataset for training to improve the accuracy of the model. Possibly, we might explore the impacts of other qualitative features on the energy consumer price indices, such as solar system size, house size, number of rooms, number of adults/ kids, and the like. By doing that, we can add new input features to improve the performance of classifiers and evaluate the importance of each feature for classification task.

Also, we want to assess the computation speeds of CatBoost when handling multiple categorical variables in big data.

Acknowledgement. This study was funded by Capital Markets Cooperative Research Centre (CMCRC) (https://www.cmcrc.com) and supported for data collection by Mojo Power, Australia.

References

1. AEMC. Fact sheet: How the spot market works. p. 4 (2017)
2. Anderson, E.J., Hu, X., Winchester, D.: Forward contracts in electricity markets: the Australian experience. Energy Policy **35**(5), 3089–3103 (2007)
3. Au, T.C.: Random Forests, Decision Trees, and Categorical Predictors: The "Absent Levels" Problem (2017)
4. Breiman, L.: Random forests. Mach. Learn. **45**(1), 5–32 (2001)
5. Chen, T., Guestrin, C.: XGBoost: a scalable tree boosting system. In: Proceedings of the 22nd ACM SIGKDD International Conference on Knowledge Discovery and Data Mining, pp. 785–794 (2016)
6. Dorogush, A.V., Gulin, A., Gusev, G., Kazeev, N., Prokhorenkova, L.O., Vorobev A.: Fighting biases with dynamic boosting. CoRR, arXiv:abs/1706.0 (2017)
7. Fitkov-Norris, E., Vahid, S., Hand, C.: Evaluating the impact of categorical data encoding and scaling on neural network classification performance: the case of repeat consumption of identical cultural goods. In: Jayne, C., Yue, S., Iliadis, L. (eds.) EANN 2012. CCIS, vol. 311, pp. 343–352. Springer, Heidelberg (2012). https://doi.org/10.1007/978-3-642-32909-8_35
8. Friedman, J., Hastie, T., Tibshirani, R.: Additive logistic regression: a statistical view of boosting (With discussion and a rejoinder by the authors). Ann. Stat. **28**(2), 337–407 (2000)
9. Gelman, A., Carlin, J.B., Stern, H.S., Dunson, D.B., Vehtari, A., Rubin, D.B.: Bayesian Data Analysis, 3rd edn. Chapman & Hall/CRC Texts in Statistical Science. Taylor & Francis (2013)
10. Ke, G., et al.: LightGBM: a highly efficient gradient boosting decision tree. In: Advances in Neural Information Processing Systems 30: Annual Conference on Neural Information Processing Systems 2017, pp. 3149–3157 (2017)
11. Kim, T., et al.: Extracting baseline electricity usage using gradient tree boosting. In: 2015 IEEE International Conference on Smart City/SocialCom/SustainCom (SmartCity), pp. 734–741 (2015)
12. Kokol, P., Zorman, M., Stiglic, M.M., Malèiæ, I.: The limitations of decision trees and automatic learning in real world medical decision making. In: 9th World Congress on Medical Informatics, MEDINFO 1998, pp. 529–533 (1998)
13. Le, C.V., et al.: Classification of energy consumption patterns for energy audit and machine scheduling in industrial manufacturing systems. Trans. Inst. Measur. Control **35**(5), 583–592 (2013)
14. Ling, C.X., Huang, J., Zhang, H.: AUC: a better measure than accuracy in comparing learning algorithms. In: Xiang, Y., Chaib-draa, B. (eds.) AI 2003. LNCS, vol. 2671, pp. 329–341. Springer, Heidelberg (2003). https://doi.org/10.1007/3-540-44886-1_25
15. Mason, L., Baxter, J., Bartlett, P., Frean, M.: Boosting algorithms as gradient descent. In: Proceedings of the 12th International Conference on Neural Information Processing Systems, NIPS 1999, pp. 512–518. MIT Press, Cambridge (1999)

16. Micci-Barreca, D.: A preprocessing scheme for high-cardinality categorical attributes in classification and prediction problems. SIGKDD Explor. **3**(1), 27–32 (2001)
17. Hastie, T., Tibshirani, R., Friedman, J.: The elements of statistical learning: data mining, inference and prediction (2001)
18. Torres-Barrán, A., Alonso, Dorronsoro, J.R.: Regression tree ensembles for wind energy and solar radiation prediction. Neurocomputing (2017)
19. Wu, D.J., Feng, T., Naehrig, M., Lauter, K.E.: Privately evaluating decision trees and random forests. PoPETs **2016**(4), 335–355 (2016)
20. Li, X., Bowers, C.P., Schnier, T.: Classification of energy consumption in buildings with outlier detection. IEEE Trans. Ind. Electron. **57**(11), 3639–3644 (2010)

Big Data Exploration for Smart Manufacturing Applications

Ada Bagozi[✉], Devis Bianchini, Valeria De Antonellis, and Alessandro Marini

Department of Information Engineering, University of Brescia,
Via Branze, 38, 25123 Brescia, Italy
{a.bagozi,devis.bianchini,valeria.deantonellis}@unibs.it,
alessandro@marinistudio.com

Abstract. Industrial Big Data management is gaining momentum as a relevant research topic for the development of innovative smart manufacturing applications. Big data technologies enable the collection, management and analysis of large amount of data from Cyber Physical Systems. In this context, data exploration is becoming a fundamental facility to let users/operators learn from collected data and take decisions. Exploration has to be performed according to different perspectives, spreading over all the hierarchy levels of the smart factory asset (from each device up to the fully connected enterprise and its products) and covering the entire life cycle value stream, from development to production stages. In this paper, we propose a model-based approach to represent data exploration scenarios, by abstracting from implementation details and taking into account different perspectives of the Reference Architectural Model for Industry 4.0 (RAMI 4.0). In particular, each scenario is related to the relevance of data to be explored and different user/operator requirements. A framework based on the approach and experiments in a real Industry 4.0 case study are also described.

Keywords: Big data exploration · Data summarization
Data relevance · Industry 4.0 · Smart manufacturing

1 Introduction

Nowadays, concepts and benefits of digital enabled manufacturing are under the light of a large media interest, involving all the actors of the entrepreneurial, political and scientific communities [1]. Specifically, after the launch of the Industry 4.0 program by the German government, expectations about the advantages of a wide adoption of big data technologies for the development of smart manufacturing applications [2] have been growing faster and faster. The Reference Architectural Model for Industry 4.0 (RAMI 4.0 [3]) suggests an holistic view over the smart factory, where new technologies can be integrated at different factory *hierarchy levels* (from each device up to the fully connected enterprise and its products), act over the entire *life cycle value stream*, from the development to

© Springer Nature Switzerland AG 2018
H. Hacid et al. (Eds.): WISE 2018, LNCS 11234, pp. 487–501, 2018.
https://doi.org/10.1007/978-3-030-02925-8_34

the production stage, and span on all functional layers of the factory, including asset management and health assessment of machines and plants, up to advanced business layer functions. Furthermore, data is emerging as a new industrial asset, to implement advanced functions like state detection, health assessment, as well as manufacturing servitization [4]. Big data and related technologies enable to collect, manage and analyze data at the highest level of details, at the cost of facing its disruptive characteristics, namely, volume, velocity and variety. In the last years a number of different technologies has been developed and several information architectures and techniques are available in order to analyze the huge amount of data collected from the field and to support logics for smart manufacturing applications. In particular, data exploration is becoming a key facility to let users/operators learn from collected data and take decisions. In this context, human operators still play a crucial role to recognize critical situations that have not encountered before, based on their long-term experience, but they must be supported in the identification of relevant data without being overwhelmed by the huge amount of information. In the so-called "Human in the Loop Cyber Physical System (CPS)" [5], human actions and machine actuation go hand-by-hand and can often complement each other.

Currently, big data management in industrial applications mainly focuses on innovative solutions at machine and shop floor levels, such as anomaly detection [6] or communication performances [7]. In this context, one of the open issues concerns the identification of a global information model able to encompass the aspects addressed at the different functional layers of the smart factory.

In this paper, we propose a model-based approach to describe big data exploration scenarios for smart manufacturing applications. Each scenario is related to the relevance of data to be explored and different users'/operators' requirements, spreading over all the hierarchy levels of the smart factory and covering the entire life cycle value stream. To this purpose, we rely on a multi-dimensional model, that organizes big data collected from CPS according to multiple dimensions, and on relevance evaluation techniques to focus the attention of users/operators on relevant data only. This model aims to make large data volumes searchable [8] and data exploration affordable. In [9] summarization and relevance evaluation techniques have been suggested as ingredients to perform exploration of real time data in a dynamic context of interconnected systems. These techniques have been applied to anomaly detection on single machines in [10]. Here we extend this research being inspired by the Reference Architectural Model for Industry 4.0. In particular, the model-based approach is used to define multiple data exploration scenarios, that enable to identify relevant data for exploration requirements of specific categories of users/operators. A framework based on the approach and experiments in a real Industry 4.0 case study are also described.

The paper is organized as follows: in Sect. 2 motivations and approach overview are introduced; in Sect. 3 we present the multi-dimensional model; Sect. 4 describes the model-based data exploration approach; in Sect. 5 implementation and experimentation issues are discussed; in Sect. 6 we highlight cutting-edge features of our research compared to state of the art; finally, Sect. 7 closes the paper with final remarks and future work.

2 Context and Motivations

2.1 Running Example

Let's consider an Original Equipment Manufacturer (OEM), producing multi-spindle machines for various industrial sectors: automotive, aviation, water industry, etc. A multi-spindle machine is a turning machine, composed of three to five spindles, that use multiple tools to cut pieces of material simultaneously and operate independently each other. Each spindle is mounted on a unit moved by an electrical engine to perform X, Y and Z movements. The spindles are carried in a rotating drum where raw material is positioned. On such multi-spindle machines we had the opportunity of testing our approach.

Asset Hierarchy and Data Collection. Asset hierarchy in this example is given by spindles (equipped with cutting tools), aggregated into multi-spindle machines, in turn organized into plants. Real time data collected from the multi-spindle machines concerns, for each spindle, the velocity of the three axes (X, Y and Z) and the electrical current absorbed by each of the engines, the value of rpm for the spindle rotation, the percentage of power absorbed by the spindle engine (charge coefficient). Hereafter, we will refer to the measured quantities as *features*.

Data Exploration Requirements. Spindle precision, working performances and reduction of machine downtime are critical factors to be monitored. One of the problems that is observed on multi-spindle machines is the spindle rolling friction torque increase, that is, a specific behavior of the spindle shaft that turns hard more and more due to different possible reasons: lack of lubrication and bearing wear that may lead to possible bearing failures. Specifically, spindle rolling friction torque increase can be monitored by observing the spindle power absorption for similar rpm values. If, exploring data, an increased power absorption is detected, the spindle rolling friction torque increase is identified as the possible anomaly that increases the energy request to perform the manufacturing operations. We will refer to the set of features used to monitor a phenomenon of interest as *feature space*.

Target Users/Operators. Monitoring phenomena, like the spindle rolling friction torque increase, may serve needs of different users according to their different perspectives and objectives. The aim of the OEM's clients, who use the machines in their manufacturing processes, is to promptly detect and even prevent spindle rolling friction torque increase. Given the hierarchical organization of industrial assets, propagation of anomalies over hierarchy levels from each spindle to the whole plant can help to identify critical areas in the factory and to properly plan supply chain activities (e.g., by ordering new spare parts for machines or revising raw material orders). On the other hand, the OEM aims to provide remote maintenance services or to plan innovation changes at machine design stage. The OEM can also be interested in comparing performances of multiple installations of the same machine model in different plants, to perform configuration improvements on single assets. Insights extracted from measures can also suggest to collect new data, because the available ones are not able to detect the problems directly observed on the machine, thus impacting on feature engineering phase.

The example above rises the need of making possible the definition of distinct data exploration scenarios according to different users' needs, considering both an enterprise-wide organization of industrial assets and different target users/operators to meet multiple data exploration requirements.

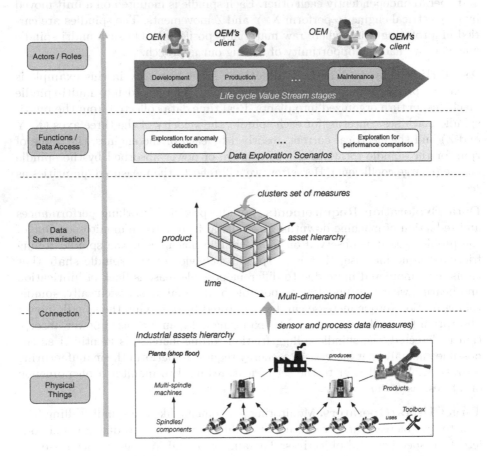

Fig. 1. Overview of the model-based data exploration approach.

2.2 Approach Overview

Figure 1 presents an overview of the proposed approach. Process data is gradually collected from the physical systems, that present a hierarchical or interconnected organization of assets as introduced in the motivating example.

Collected measures are incrementally processed and organized within a multi-dimensional model. Dimensions are related to the industrial assets: the monitored components (e.g., spindles, multi-spindle machines, plants), the tools which spindles are equipped with, the manufacturing product that is being produced. A data summarization process aggregates the huge amount of collected data, in order to face high volumes, instead of considering each single data record, that might be not relevant given the high level of noise in the working environment (slight variations in the measured variables). Data summarization is based on a clustering procedure whose output is a set of clusters, updated over time, that are used to represent the behavior of the monitored assets (see Sect. 3.1). On top of the multi-dimensional model, data exploration scenarios can be defined, to serve different users' needs, e.g., anomaly detection on a single machine or performances comparison across different machines (see Sect. 4). Each scenario focuses on a set of clusters and leverages (possibly) different dimensions of the clustered data. Firstly, data relevance evaluation techniques help to restrict the exploration to the relevant clusters only (e.g., those corresponding to anomalies for state detection purposes). Therefore, specific dimensions are selected. For instance, anomaly detection is focused on asset hierarchy from spindles to machines and plants that aggregate them, while performances comparison focuses on shop floor level in the same plant or across multiple plants in distinct enterprises.

Depending on the data exploration scenario, users/operators interact during different stages of the life cycle values stream (development, production, maintenance).

3 Clustering-Based Multi-dimensional Model

The characteristics of big data, namely, volume, velocity and variety, pose additional issues for data collection and organization. High volume calls for techniques and tools to provide a compact view over the large amount of collected data. Furthermore, when dealing with real time data, collected in CPS, data streams must be considered, where not all data is available since the beginning, but is collected in a fast and incremental way. To face these issues, data summarization techniques are shortly described in the following.

3.1 Clustering-Based Data Summarization

In our approach, data summarization is based on clustering-based techniques. Clusters offer a two-fold advantage: (a) they give an overall view over a set of measures, using a reduced amount of information; (b) they allow to depict the behavior of the system better than single records, that might be affected by noise and false outliers, in order to observe a given physical phenomenon.

The clustering algorithm is performed in two steps: (i) in the first one, a variant of Clustream algorithm [11] is applied, that incrementally processes incoming data to obtain a *set of syntheses*, that provide a loss-less representation of measures; (ii) in the second step, X-means algorithm is applied [12] in order to cluster syntheses obtained in the previous step. Clusters give a balanced view of the observed physical phenomenon, grouping together syntheses corresponding to the same working status. Syntheses and clusters are defined as follows.

Definition 1 (Synthesis). *We define a synthesis S_j in a feature space fs_j as a tuple $S_j = \langle N_j, \boldsymbol{LS}_j, SS_j, \boldsymbol{X}_j^0, R_j \rangle$, where: (i) N_j is the number of points included into the synthesis; (ii) $\boldsymbol{LS}_j$ is a vector representing the linear sum of points; (iii) SS_j is a scalar representing the quadratic sum of points; (iv) $\boldsymbol{X}_j^0$ is a vector representing the centroid of S_j; (v) R_j is the radius of the synthesis.*

Definition 2 (Cluster). *A cluster C_i is defined as follows: $C_i = \langle \boldsymbol{C}_i^0, \mathcal{S}_{set}^{C_i} \rangle$, where $\boldsymbol{C}_i^0$ is the cluster centroid, $\mathcal{S}_{set}^{C_i}$ is the set of syntheses belonging to the cluster. We denote with SC the set of identified clusters.*

The two-steps clustering algorithm enables an incremental procedure specifically developed to face data streams. Considering Δt as the time interval in which feature values are grouped in syntheses, that in turn are clustered, every Δt seconds the clustering algorithm outputs a new cluster set SC built on top of the previous iterations.

3.2   Multi-dimensional Model

As remarked in Sect. 2.2, several dimensions, related to the characteristics of industrial assets, are associated to measured features, grouping them according to "facets", such as the tool used during manufacturing, the monitored system, the manufacturing product. Dimensions can be organized through hierarchies: for example, tools can be aggregated into tool types, monitored system follows the asset hierarchy. Beyond these dimensions, we also consider other aspects that can be exploited to guide data exploration. Among them, we mention feature spaces, time and operational parameters. Feature space and time are of paramount importance, since the data summarization procedure relies on these dimensions. As we will discuss in the next section, also data relevance techniques are strictly related to the time dimension. Operational parameters represent domain-specific settings or variables, set during manufacturing: for example, we mention the working mode (G0, fast movement of the spindle to catch the tool, or G1, slow movement of the spindle during the manufacturing). Data comparison across different values of these parameters usually does not make sense. The resulting multi-dimensional model can be viewed as a hypercube, where axes represent exploration dimensions (time, asset hierarchy, products, feature spaces, operational parameters) and nodes represent clusters sets computed on measures collected at given exploration dimensions.

4   Model-Based Data Exploration Approach

Data exploration is performed on top of the multi-dimensional model through the definition of several data exploration scenarios in order to pursue different requirements. Each scenario is defined by specifying: (i) the dimensions over which exploration is performed, according to the multi-dimensional model, and their granularity level (e.g., for the monitored system, either machine or shop floor level); (ii) the relevance techniques used to focus exploration on relevant clusters only; (iii) the target users/operators.

In the following, we will discuss two possible model-based data exploration scenarios. We remark that the list is not exhaustive. Our aim is to demonstrate how this model is able to describe scenarios by abstracting from implementation details. In particular, we consider:

- *exploration for anomaly detection*, to detect anomalies by observing if collected data overtakes or gets closer to physical limits of breakage for the monitored system;
- *exploration for performance comparison*, to compare the performances of different machines in the same plant or across different plants.

4.1   Data Relevance Evaluation

Relevance-based techniques are used to focus exploration on relevant clusters only. In literature, data relevance is defined as the distance from an expected status. The point here is to define the expected status and how to compute such a distance. In the following, we will describe the expected status and distance computation metrics for each monitored system at machine or single component level (e.g., the spindle) and for composite plants at shop floor level.

Data Relevance at Machine/Component Level. At this level, the expected status is identified through the set $\hat{SC} = \{\hat{C}_1, \hat{C}_2, \dots \hat{C}_n\}$ of clusters computed during normal working conditions of the monitored machine or component. Let's denote with $SC = \{C_1, C_2, \dots, C_m\}$ the current clusters set, where n and m do not necessarily coincide. Relevant data are recognized when SC differs from $\hat{SC}$. Therefore, the proposed relevance techniques are based on clusters set distance between SC and $\hat{SC}$, denoted with $\Delta(SC, \hat{SC})$, and enable to detect over time clusters movements, clusters contraction/expansion, changes in the number of clusters. We refer to [9] for details on $\Delta(SC, \hat{SC})$ computation.

Data Relevance at Shop Floor Level. In this case, expected status is derived by considering status of each involved machine or component in the same plant or across different plants. We denote this as *baseline distance*, occurring when all the machines are working in normal conditions; in case of two machines $\mathcal{M}_1$ and $\mathcal{M}_2$, it is computed as follows:

$$\Delta_{baseline}(\mathcal{M}_1, \mathcal{M}_2) = \Delta(\hat{SC}_1, \hat{SC}_2) \tag{1}$$

where $\hat{SC}_1$ (resp., $\hat{SC}_2$) is the clusters set obtained for machine $\mathcal{M}_1$ (resp., $\mathcal{M}_2$) during normal working conditions. This distance is usually different from 0 since the two machines work in different environments and the likelihood of having exactly the same measures for considered features over the compared physical systems is very low. When we are dealing with more than two machines, Eq. (1) is extended by computing $\Delta_{baseline}(\cdot)$ for each pair $\mathcal{M}_i$ and $\mathcal{M}_j$ of machines and considering the average value. The metric of relevance, in this case, aims at highlighting the difference with respect to the baseline distance, that is:

$$\Delta_{t'}(\mathcal{M}_1, \mathcal{M}_2) = \Delta(SC_1, SC_2) - \Delta_{baseline}(\mathcal{M}_1, \mathcal{M}_2) \tag{2}$$

where SC_1 (resp., SC_2) is the clusters set obtained at the current time t' for machine $\mathcal{M}_1$ (resp., $\mathcal{M}_2$). Also in this case, for more than two machines, the average value of $\Delta(SC_1, SC_2)$ and $\Delta_{t'}(\mathcal{M}_i, \mathcal{M}_j)$ for each pair $\mathcal{M}_i$ and $\mathcal{M}_j$ is computed.

4.2 Exploration for Anomaly Detection

This scenario may be implemented in a state detection service by OEMs to prevent downtime of monitored systems, and by OEM's clients to plan supply chain activities. Figure 2 shows an example of clusters changes for the detection of anomalous working conditions on spindles considered in the running example: changes in clusters set over time are detected due to spindle rolling friction torque increase, causing a decrease of rpm and an increase of the percentage of absorbed power. In this scenario, evolution of measures over time is observed within the feature space composed of rpm and absorbed power features. The exploration is performed by fixing the observed system (e.g., the spindle). Moreover, the status of the spindle can be propagated up to the machine and plant the component belongs to.

The relevance techniques at machine/component level allow to identify what are the clusters that changed over time, getting closer to the physical limits (bounds) of breakage of the monitored system. Let's denote with $\{\overline{C_i}\}$ the set of such clusters. We distinguish among *warning* and *error* bounds: (a) warnings identify anomalous conditions that may lead to breakdown or damage of the monitored system; (b) errors identify unacceptable conditions in which the machine or component can not operate. Besides defining features bounds, we introduced the notion of *context* to specify contextual bounds. A contextual bound represents the limit of a feature within a specific context, where the feature is measured. The context is specified by fixing operational parameters (e.g., the working mode G0/G1). The rationale is that, in a specific context, when the CPS works normally, a feature can assume values within a specific range, that might be different from the overall physical limits for the same feature, disregarding the context.

The warning bound is used to perform an early detection of a potential deviation towards an error state. For each cluster in $\{\overline{C_i}\}$, the distance of cluster centroid from the warning and error bounds is computed. If this distance is equal

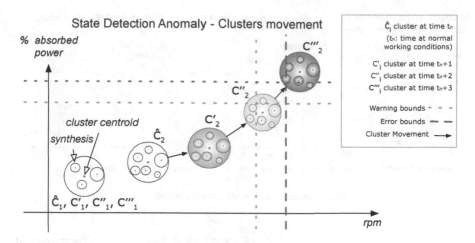

Fig. 2. Illustration of clusters sets changes over time, in data exploration scenario for anomaly detection. Clusters set $\hat{SC}$ is generated at time t_n at normal working conditions and consists of clusters $\hat{C}_1$ and $\hat{C}_2$.

or lower than the cluster radius, this means that a warning or error status has to be detected. Note that distance also helps to detect *potential* state changes. Consider for example Fig. 2, that shows an example of cluster evolution over time. The figure shows how the cluster C_1 does not changed its position, as well as its size, from time t_n to t_{n+3}. On the other hand, cluster C_2 evolves from the wealth zone to the warning and error zones. At time t_{n+2} cluster C_2'' crosses the warning bound of rpm feature causing a warning alert, at time t_{n+3} cluster C_2''' moves into the error zone, crossing error bounds of both the considered features. At time t_{n+1} cluster C_2' still remains inside the wealth zone, however relevance techniques detected its change. Therefore, cluster C_2' is recognized as relevant and monitored to detect warning or error state changes.

The OEM can remotely monitor state changes and cooperate with OEM's clients in order to revise features and contextual bounds for specific working conditions. This can be also used by OEM to revise machine design at development stage and to aggregate assets within the asset hierarchy during plant design. These considerations deserve further research and will be addressed in future work.

4.3   Exploration for Performance Comparison

In this scenario OEM compares performances across different systems, in order to remotely adjust the machine configuration parameters or to plan maintenance activities spanning across different plants. In this case, clusters sets are compared over the monitored system dimension, at plant or shop floor level. Let's consider the situation depicted in Fig. 3. In the figure, two spindles are compared considering how the clusters sets distance between two spindles evolves over time. At

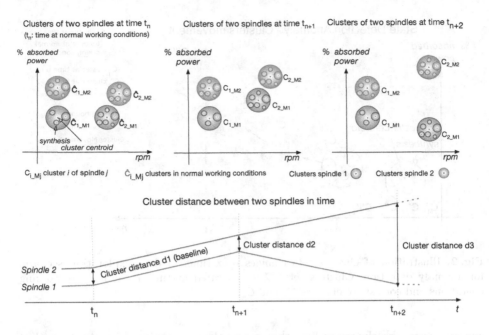

Fig. 3. Illustration of clusters sets changes over time, in data exploration scenario for performance comparison. $\hat{C}_{i\_\mathcal{M}j}$ represents a cluster for spindle $\mathcal{M}_j$ while working in normal conditions ($j = 1, 2$).

time t_1 the two spindles present a cluster set distance equal to d_1, corresponding to the baseline distance (see Sect. 4.1). At time t_{n+1} the two spindles present a distance $d_2 = d_1$. This means that the relative distance in terms of clusters sets between the two spindles is not changed. We remark here that, as shown in Fig. 3, the condition $d_2 = d_1$ holds also if the two spindles changed their behaviors, and their corresponding clusters sets evolved accordingly. In this case, data exploration for anomaly detection explained above can help to identify anomalous conditions on one of the considered spindles. On the other hand, at time t_{n+2} the distance d_3 is changed compared to d_1 and d_2, meaning that the two spindles started behaving differently each other.

5   Implementation and Experimental Validation

The model-based approach presented in this paper guided the implementation of the IDEAaS (Interactive Data Exploration As a Service) framework, whose architecture is shown in Fig. 4. IDEAaS Core Modules include Data Acquisition, Data Summarization and Data Relevance Evaluation. Model-based data exploration scenarios are implemented as services, such as CPS State Detection and CPS Performance Comparison. As shown in Fig. 4, data coming from the physical system, properly collected through sensors and IoT technologies, is sent to the Data Acquisition Module to be processed.

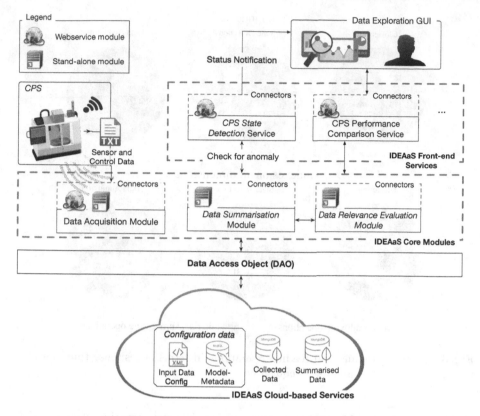

Fig. 4. The IDEAaS framework architecture.

The Data Acquisition Module is in charge of storing data in the cloud. This module operates in order to minimize time spent for data acquisition. Specifically, data collected is first saved in a NoSQL database (*Collected Data*), minimizing any other operation, namely data control and cleaning and data analysis that are performed in parallel. Dimensions are stored as metadata of the multidimensional model into the *Model-MetaData* relational database. On top of the IDEAaS front-end services, a Data Exploration GUI will be developed to allow users/operators to interact with the system and explore collected data. Also data visualization issues within the Data Exploration GUI will be investigated with reference to the data organization within the multi-dimensional model.

IDEAaS architecture is implemented in Java, on top of a Glassfish Server Open Source Edition 4. NoSQL MongoDB technology and MySQL DBMS have been used within the cloud-based services.

5.1 Efficiency Experiments

We collected ~140 millions of records from three multi-spindle machines, each one equipped with three spindles and different tools. These multi-spindle

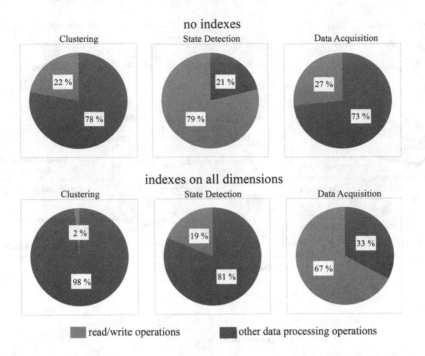

read/write operations   other data processing operations

Fig. 5. Impact of writing and reading operations in terms of response times on clustering, state detection and data acquisition.

machines generate 144 measures every second. We run experiments on a VPS Ubuntu Server 16.04, 1 vCore CPU, RAM 4 GB, hard disk 20 GB (SSD). In order to speed up reading and writing operations, we introduced indexes on MongoDB documents.

Figure 5 shows the percentage of reading/writing operations in terms of response time during incremental clustering, data acquisition and state detection. Average response time values are listed in Table 1, considering 8.000 measures. These values also demonstrate how the incremental approach is able to face the measures input rate (144 measures/sec).

If indexes are defined, reading/writing operations negatively impact on clustering and state detection response time, while defining indexes slows down data acquisition processing.

Figure 6 shows how indexes impact on reading/writing response time. When no indexes are applied, as expected, reading operations are the most expensive ones. On the other hand, setting indexes on all the dimensions negatively impacts on the acquisition rate. An acceptable trade-off between these two situations can be reached by establishing which and how many indexes to set thanks to the designed data exploration scenarios. In particular, in this paper we presented two kinds of scenarios, where data exploration is performed over the asset dimension: in the exploration for anomaly detection the whole hierarchy is considered, while in the exploration for performance comparison multi-spindle machines are

Table 1. Average response time of clustering, data acquisition and state detection, including read/write operations (on 8.000 measures).

	Clustering	State detection	Data acquisition
Read/write operations without indexes	1.219 ms	1.219 ms	54 ms
Read/write operations with indexes on all dimensions	76 ms	76 ms	306 ms
Other data processing operations	4.300 ms	330 ms	150 ms

compared at shop floor level. According to these scenarios, we tested reading/writing response time by setting indexes only on timestamp and asset hierarchy. Finally, given the extensible nature of our model-based approach, we performed further tests considering indexes set on features and feature spaces as well. Experimental evaluation shows how data exploration scenarios help to define dimensions of interest, on which indexes are set.

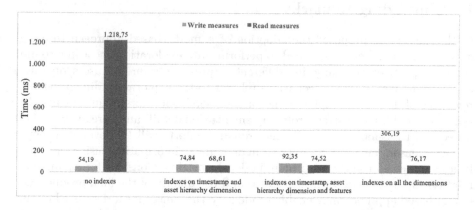

Fig. 6. Response time for writing and reading operations over 8.000 measures by applying indexes on the documents stored in the MongoDB database.

6   Related Work

In literature, recent approaches agreed upon the need of holistic solutions to make production systems Industry 4.0-ready. However, considering the Reference Architectural Model for Industry 4.0 (RAMI 4.0), existing approaches focused on specific layers or aspects, such as communication issues [1], state detection [6, 13,14] and big data prognostics [15] at machine level. In particular, authors in [1] propose a CPS-enabler that is able to discover and connect production systems in a dynamic way through the interaction with an Information Server, that sends notifications about detected connection points. The work in [16] addresses big data analytics for Industry 4.0, proposing an architecture that deals with big data collection, analysis and distribution. Similarly, the work in [17] is focused on big data analytics for asset management (namely, equipment health, reliability, effects of unplanned breakdown). Approaches in [16,17] do not consider multiple perspectives on explored data, making difficult to quickly adapt data exploration to different requirements, expressed for specific categories of users/operators in order to better focus the attention on relevant data.

For what concerns anomaly detection and prognostics approaches, existing solutions operate at machine level, without considering the smart factory as a set of interconnected layers, where data exploration can be declined over different aspects. We refer to [10] for a survey on this kind of approaches. Data exploration has been studied in [18–20]. Nevertheless, these approaches have not been specifically applied to the Industry 4.0 context and its reference architectural model. Data exploration scenarios we defined here can be seen as building blocks to create flexible exploration strategies, by abstracting from implementation details.

7   Concluding Remarks

In this paper, we proposed the adoption of a model-based approach to design data exploration scenarios, aimed to perform data exploration according to multiple perspectives and targeting different requirements for the users/operators of the smart factory. Each scenario is related to the exploration dimensions, the relevance of data to be explored and target users'/operators' requirements. The approach enables to abstract from implementation details and spread over all the aspects of the smart factory. Further research efforts will be devoted to better connect sensors and process data to management information, using data analytics solutions for performing more in depth insights. Consider for example the relationship between asset health status and the quality of the final product. Our model-based approach may enable to identify the causes for a decreased quality observed on products, from machine level up to an enterprise-wide scope. On the other hand, behavior changes at machine or shop floor levels may be promptly detected to prevent undesirable deterioration of the quality of the product at the highest levels. Finally, the absence of any correlation between sensors and process data and management information might suggest new data collection campaigns to refine and improve the knowledge extraction process.

References

1. Schlechtendahl, J., Keinert, M., Kretschmer, F., Lechler, A., Verl, A.: Making existing production systems industry 4.0-ready. Prod. Eng. 9(1), 143–148 (2015)
2. Lee, J., Ardakani, H., Yang, S., Bagheri, B.: Industrial big data analytics and cyber-physical systems for future maintenance and service innovation. Proc. Conf. Intell. Comput. Manuf. Eng. (CIRP) 38, 3–7 (2015)
3. Hankel, M., Rexroth, B.: The reference architectural model industrie 4.0 (RAMI 4.0). In: ZVEI (2015)

4. Lee, J., Bagheri, B., Kao, H.: A cyber-physical systems architecture for industry 4.0-based manufacturing systems, Manufacturing Letters. **3**, 18–23 (2015)
5. Nunes, D., Silva, J.S., Boavida, F.: A Practical Introduction to Human-in-the-Loop Cyber-Physical Systems. Wiley IEEE Press, Hoboken (2018)
6. Huber, M.F., Voigt, M., Ngomo, A.-C.N.: Big data architecture for the semantic analysis of complex events in manufacturing. In: Informatik 2016. Gesellschaft für Informatik eV (2016)
7. Iglesias-Urkia, M., Orive, A., Barcelo, M.: Towards a lightweight protocol for industry 4.0: an implementation based benchmark. In: IEEE International Workshop on Electronics, Control, Measurement, Signals and their Application to Mechatronics (2017)
8. Embley, D.W., Liddle, S.W.: Big data—conceptual modeling to the rescue. In: Ng, W., Storey, V.C., Trujillo, J.C. (eds.) ER 2013. LNCS, vol. 8217, pp. 1–8. Springer, Heidelberg (2013). https://doi.org/10.1007/978-3-642-41924-9_1
9. Bagozi, A., Bianchini, D., De Antonellis, V., Marini, A., Ragazzi, D.: Summarisation and relevance evaluation techniques for big data exploration: the smart factory case study. In: Dubois, E., Pohl, K. (eds.) CAiSE 2017. LNCS, vol. 10253, pp. 264–279. Springer, Cham (2017). https://doi.org/10.1007/978-3-319-59536-8_17
10. Bagozi, A., Bianchini, D., De Antonellis, V., Marini, A., Ragazzi, D.: Big data summarisation and relevance evaluation for anomaly detection in cyber physical systems. In: Panetto, H. (ed.) On the Move to Meaningful Internet Systems. LNCS, vol. 10573, pp. 429–447. Springer, Cham (2017). https://doi.org/10.1007/978-3-319-69462-7_28
11. Aggarwal, C., Han, J., Wang, J., Yu, P.: A framework for clustering evolving data streams. In: Proceedings of 29th International Conference on Very Large Data Bases (VLDB 2003), pp. 81–92 (2003)
12. Pelleg, D., Moore, A.: X-means: extending k-means with efficient estimation of the number of clusters. In: Proceedings of 17th International Conference on Machine Learning (ICML), pp. 727–734 (2000)
13. Stojanovic, L., Dinic, M., Stojanovic, N., Stojadinovic, A.: Big data-driven anomaly detection in industry (4.0): an approach and a case study. In: 2016 IEEE International Conference on Big Data (Big Data), pp. 1647–1652 (2016)
14. Hanamori, T., Nishimura, T.: Real-time monitoring solution to detect symptoms of system anomalies. FUJITSU Sci. Tech. J. **52**, 23–27 (2016)
15. Yan, J., Meng, Y., Guo, C.: Big data-driven based intelligent prognostics scheme in industry 4.0 environment. In: IEEE Prognostics and System Health Management Conference (PHM-Harbin) (2017)
16. Santos, M.Y., et al.: A big data analytics architecture for industry 4.0. In: Rocha, Á., Correia, A.M., Adeli, H., Reis, L.P., Costanzo, S. (eds.) WorldCIST 2017. AISC, vol. 570, pp. 175–184. Springer, Cham (2017). https://doi.org/10.1007/978-3-319-56538-5_19
17. Campos, J., Sharma, P., Gabiria, U., Jantuen, E., Baglee, D.: A big data analytical architecture for the asset management. Procedia CIRP **64**, 369–374 (2017)
18. Kalinin, A., Cetintemel, U., Zdonik, S.: Interactive data exploration using semantic windows. In: Proceedings of the ACM SIGMOD International Conference on Management of Data, pp. 505–516 (2014)
19. Dimitriadou, K., Papaemmanouil, O., Diao, Y.: Aide: an active learning-based approach for interactive data exploration. IEEE Trans. Knowl. Data Eng. **28**(11), 2842–2856 (2016)
20. Costa, C., Chatzimilioudis, G., Zeinalipour-Yazti, D., Mokbel, M.F.: Efficient exploration of telco big data with compression and decaying. In: 2017 IEEE 33rd International Conference on Data Engineering (ICDE 2017), pp. 1332–1343 (2017)

Dark Web Markets: Turning the Lights on AlphaBay

Andres Baravalle[(⊠)] and Sin Wee Lee

University of East London, London, UK
andres@baravalle.it

Abstract. Dark web markets have been, for a while now, at the centre of the attention of governmental bodies. After the rise and fall of The Silk Road, more and more marketplaces have followed the same path. Alphabay is the last of these markets to come under the spotlight, for its staggering amount of products on sale (estimated value of around $88M) during our analysis, and $590M of successful transactions during its existence.

In this paper we describe the spider that we developed for this project, our data analysis pipeline and the key findings of our data analysis.

The data analysis focuses on quantitative research, covering Alphabay's adverts, its sellers and its categories.

The outcomes of the research provide an insight on Alphabay, its general trend and the nature of its ads, which appear being a majority of drug products. United States and the United Kingdom are the most active countries on the website.

Keywords: Dark web · Dark web markets · Data wrangling · Text mining
Spidering

1 Introduction

Since the beginning of the web, illegal activities have been widespread. Popular software such as Napster, Kazaa, Gnutella, uTorrent and more recently Megaupload [1] have often being used for illegal file sharing. On-line auctions web sites and marketplaces are rife with fake goods and goods of questionable source. Web sites selling (and not delivering) tickets for event, planes or holiday packages is a recurring feature in newspapers.

In the modern world security and privacy are playing an increasingly important role, with numerous privacy related issues, such as governmental surveillance and corporate tracking facing modern Internet users. Such occurrences have led Internet users and software developers to deploy technology in order to protect their freedom and anonymity online.

The most widely known examples of such technology are Tor (The Onion Routing) and I2P (Invisible Internet Project), both volunteer maintained open source project which aims to protect end users by anonymising their Internet traffic by forwarding user data through a series of anonymous nodes.

© Springer Nature Switzerland AG 2018
H. Hacid et al. (Eds.): WISE 2018, LNCS 11234, pp. 502–514, 2018.
https://doi.org/10.1007/978-3-030-02925-8_35

Sites hosted on Tor (or I2P) can be visited by users, but it is hard to identify where they are hosted and who hosts them. This is what we call the dark web.

Anonymity and privacy are key characteristics of the dark web. Although dark websites are publicly visible, the IP addresses of the servers that run them are hidden (Egan n.d.). As a result, these websites can be visited by users, but it is hard to identify where they are hosted and who hosts them, and who are its users.

This is in contrast with the "surface" web and on the "deep web", where both users and servers are easily identified by their IP address. Content on the surface web is easily reachable with search engines; content on the deep web is not (e.g. password protected fora), but there is no anonymity in either case.

Dark web markets operate in anonymous networks, protecting the identity of all participants, buyers and sellers.

In the last years, dark web markets have quickly come to prominence as a safe haven where to carry out illegal transactions. Due to the difficulty of the Internet regulation and the convenience of online transaction, more and more illegal activities occur on dark web [20]. However, due to the initiative of the enforcement, these dark web sites, especially dark web markets, frequently shut down, and players come to prominence.

The size of the dark web is currently estimated to be between 7,000 and 13,000 websites "up and running" [17, 19]. The research carried out have identified about 30,000 websites, but most of them are short-lived (hence the lower number of web sites "up and running").

However, the estimation of the size of the dark web can be varied, depending on the method used, and researchers typically distinguish between websites online consistently enough to be investigated and short-lived websites.

In comparison, ISC reports 1.07 billion hosts in July 2017 (http://ftp.isc.org/www/survey/reports/current/) on the surface web.

Regarding the type of content hosted, according to INTELLIAGG's analysis, just under 70% of the dark web content is illegal (under UK and US law), while the remaining content would be legal. 30% of the illegal dark web are dark web markets.

The first dark web marketplace went online in 2011, and several dark web markets have risen to fame and have fallen later (notably: The Silk Road, The Silk Road 2.0, Evolution, Hydra, Agora).

Alphabay and Hansa were the largest dark web market places in 2017. Both have been shut down at the same time (July 2017) as a part of a law enforcement operation by the Federal Bureau of Investigation, the Drugs Enforcement Administration and European law enforcement agencies acting through Europol [14, 18]. On July 12, 2017, Alexandre Cazes, the founder of AlphaBay, allegedly committed suicide in the prison.

This paper presents research carried out on the data collected on the last weeks of life of AlphaBay, between June and September 2017.

The closure of Alphabay effectively ended the data collection.

Based on past experiences, it is widely believed that other web markets will take the place of AlphaBay (although that is not yet the case as we write in August 2018).

2 Experimental Environment

The first step in data collection from the dark web markets involved spidering the dark web markets [8]. The first proof-of-concept spider was developed with a few lines of code to simulate human authentication on the market. This was followed by a full spider, developed using Python and Amazon Web Services (AWS).

Building on top of lessons learnt and on expectations of an evolved landscape, we decided to:

- Implement a Wizard of Oz approach: non-headless browser (Selenium); to speed up the development loop and to reduce detection risks
- Agnostic software (doesn't care on what operating system runs), building on top of cloud technologies. Amazon Simple Storage was used to store (some) binary data, and MySQL was used to store textual data
- Use of Computer-to-Human interfaces
- A higher degree of anonymisation: disposable Alphabay accounts were created and used for the spidering, with a new account created after each spidering session.

Regarding architecture, we have developed a focused, vertical spider.

The spider run over a period of 6 weeks, with incremental improvements. Screenshots were taken automatically, which greatly helped in the ex-post examination, once Alphabay was down.

Critical issues around performance were immediately evident. Multi-thread execution and horizontal scaling are typically options used to increase the performance of web spiders; one of the shortcomings of our system was its limited possibilities for horizontal scaling (Fig. 1).

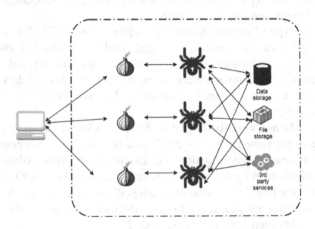

Fig. 1. Spider architecture

A single spider in our development box collected about 2,000 ads per day (using a single instance; more using multiple instances), which means that a full spidering if no new products added, would have taken about six months. Surface web spiders using multiple threads and multiples proxies can easily reach 200 pages per minute, with just 5 proxies [1].

Spidering dark web markets requires researchers to heavily invest time to avoid detection during the spidering. Anti-DOS protections and anti-spidering protections (e.g. CAPTCHAs) limited heavily the speed of spidering.

Disposable accounts were created for each spidering session, and the IP address was always refreshed. In our project, each spider used a different Tor connection, and a different disposable user.

In order to address performance issues, multiple spiders were running at the same time as soon as the spider stabilised. No local storage was used; all data was stored on AWS in order support quick deployment and data collection.

2.1 Data Analysis Tools

Different technologies and programming languages that can be used for data analysis. Python, SAS and R (Anon 2017) are all valid and popular solutions, typically attracting support from different communities.

The data analysis presented in this paper has been performed using a combination of Python and R, with some exploratory data analysis done in SAS and Microsoft Access.

R and Jupiter (Python) Notebooks have been used through the data analysis to discuss the results and draft this paper (Anon n.d.).

2.2 Data Retrieval

It is important to highlight that Alphabay had a limitation on the number of pages that can be visualised per category (50). Essentially, this means that older ads are typically hidden (unless the user would perform a keyword search rather than browse the side).

Just over 380,000 ads were recorded to be present on Alphabay, both from screenshots and from other sources. Given the technical constraints, the number of ads that could be spidered at the time was just over 180,000. In most categories, the spidering could not retrieve ads before 2015 (Table 1).

The coverage of the data extracted is well over 10% in most categories, with few exceptions. One is "Other listings" (very small category, 3% coverage) and the others are Drugs and Chemicals (6%) and Fraud (7%). This coverage is based on the theoretical maximum (380,000). However, in reality, the users could not access more than 180,000 ads, which would give nearly double coverage.

For each advert, title, description, price (in USD), URL, seller, payment, origin, destination, category, collection timestamp, date of posting and number of products sold were collected.

Before analysis, data was prepared, by:

1. Removing special characters and switching to lowercase
2. Finding in the title or description of the ads the amount (number and mass) of the product

Table 1. Coverage per category

Category	Sample	Products	Coverage
Carded items	504.00	3,997.00	12.61%
Counterfeit items	2,658.00	10,098.00	26.32%
Digital products	2,908.00	18,941.00	15.35%
Drugs & Chemicals	16,536.00	259,399.00	6.37%
Fraud	3,725.00	48,046.00	7.75%
Guides & Tutorials	3,531.00	16,802.00	21.02%
Jewels & Gold	524.00	1,894.00	27.67%
Other listings	138.00	4,344.00	3.18%
Security & Hosting	264.00	903.00	29.24%
Services	1,376.00	8,182.00	16.82%
Software & Malware	1,205.00	3,653.00	32.99%
Weapons	614.00	5,562.00	11.04%

3. Calculating the price of one unit of one dose (1 g) for each transaction for the "Drugs and Chemicals" category.

A small number of outliers had drastically affected our initial analysis. To start, a "manual" review of the worst outliers showed some issues around data quality.

First of all, sellers wanting to retire a product typically increased the price to absurd, non-market values (an indicated in the ad that the product was not on sale any more). Manually analysing all products on sale for over 8,000 USD resulted in identifying 535,000 USD worth of retired products. Secondly, a number of "real" products where on sale at 0 USD – possibly as "feedback factories" and this had to be considered when deciding what to do with null values. Finally, a number of extreme values where actually genuine products.

While deciding to identify and remove extreme values, likely to be "dirty" data, we had to deal with the fact that the distribution of the data across most subcategories in the dataset is very long-tailed.

Different techniques have been tried on the data: z-score, boxplot (using 2 and 3 times the interquartile range) and percentiles.

For this analysis, we have decided to remove as little data as possible, slicing (by price) at the top 0.5^{th} percentile and at the bottom 99.5^{th} percentile.

3   Alphabay

At the time it was shut down, AlphaBay was the largest dark web marketplace. Its reputation can be confirmed by referring to the Google search statistics with the keywords AlphaBay and Dream Market (another popular dark web market) between January 2015 until June 2017 (Anon n.d.) (Fig. 2).

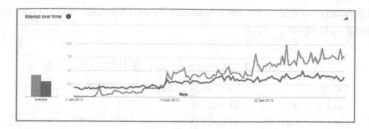

Fig. 2. Evolution of AlphaBay and Dream Market Google researches (Color figure online)

AlphaBay (in red in the graph) has become more and more popular since the demise of Agora, and before being shut down, it was the most popular dark web market (Anon 2017).

This assumption can be reinforced by looking at the data collected on AlphaBay. The number of ads per month (in the samples collected) from January 2015 until June 2017 shows the growing popularity of Alphabay (Fig. 3).

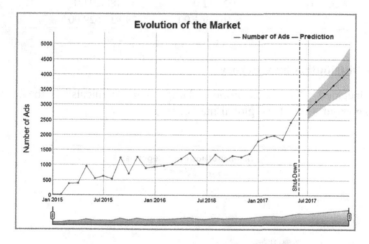

Fig. 3. Evolution of AlphaBay web market

Between 2015 and 2016, there was a significant jump, the number of ads (in our sample) rose from 7,712 up to 14,161. The number of ads that have been posted during the six first months of 2017 (before the closing) is 12,878, which is almost the same that in the whole 2016. If AlphaBay would have stayed online, the number of ads would reached a peak of 4,000/month by the end of the year.

Estimating the value of the products posted in the market and of each of the categories is key to a better understanding of dark web markets

As it can be seen in Table 2 we have mined products for a total value of over 6M USD (after removing extreme values); our estimates put the total value of the individual products on sale at the time between 79 and 88 million USD (depending on the modelling technique).

Table 2. Market estimates (in USD)

Category	SumOfPriceSample	SumOfPRiceEstimate
Carded items	24,940.00	197,788.00
Counterfeit items	431,625.00	1,639,785.00
Digital Products	30,404.00	198,034.00
Drugs & Chemicals	4,413,189.00	69,229,367.00
Fraud	134,297.00	1,732,197.00
Guides & Tutorials	44,168.00	210,170.00
Jewels & Gold	197,899.00	715,307.00
Other listings	3,986.00	125,472.00
Security & Hosting	10,522.00	35,990.00
Services	227,700.00	1,353,955.00
Software & Malware	246,919.00	748,544.00
Weapons	402,712.00	3,648,020.00
	6,168,361.00	79,834,629.00

3.1 Ads Distribution by Category

Alphabay included 12 main categories; "Drugs and Chemicals" is the largest one, representing about 45% of the global market (Fig. 4).

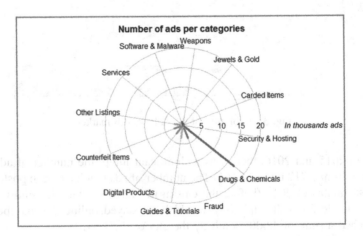

Fig. 4. Distribution of the market

The second most popular category is "Fraud", including all the ads regarding impersonation, fake ids and accounts, and representing about 13% of the market.

Finally, all other items (digital product, weapons, jewellery etc.) represent a small portion of the marketplace. It is worth mentioning that although weapons contributed to just a small portion of the listings, the total value of the listings is about $3.6M, with the United States as the main sellers, followed by UK. These two countries together, representing 97% of the global market.

3.2 Ads Distribution by Country

Figure 5 represents the 10 main countries in the world based on the number of adverts. As we can see, United States is the largest country by number of adverts, with more than twice as the number of ads than the second one, United Kingdom.

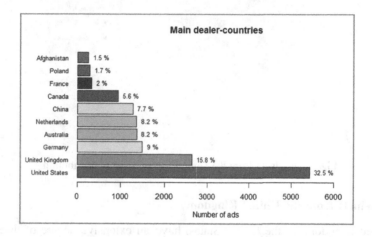

Fig. 5. Countries by number of adverts

Moreover, it is noticeable that most of these countries have strong economies. Five of the top 10 countries belong to the Group of Seven (G7), only Japan and Italy are not present. Top markets

In the next section, a more in-depth products analysis is presented on the top three sellers countries, United States, United Kingdom and Germany, plus a country of interest, Afghanistan.

A high-level analysis of some peculiarities of the specific markets is also included at the end of the section.

3.3 Products from the United States

The United States have the most diversity in terms of the products listings on AlphaBay. Its sellers trade products ranging from Drugs and Chemical (74%), Jewels and Gold (10%), Services (4%), counterfeit items (8%) and Weapons (2%).

Comparing with other countries, United States have the one of the highest concentration of Drugs and Chemicals. 29% of this are Cannabis and Hashish, followed by Psychedelic drugs.

The United States are also the largest market for weapons (Fig. 6).

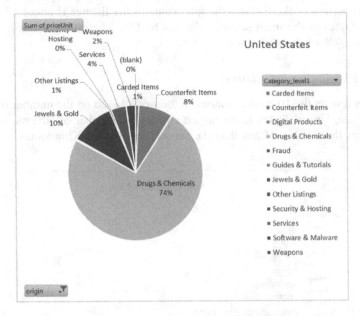

Fig. 6. Products from the United States, market distribution

3.4   Products from the United Kingdom

The United Kingdom, as the United States, have an extensive degree of diversity in terms of products listings. A huge portion of the products are in the "Cannabis & Hashish" category, but stimulants and other illegal drugs are significantly present as well (Fig. 7).

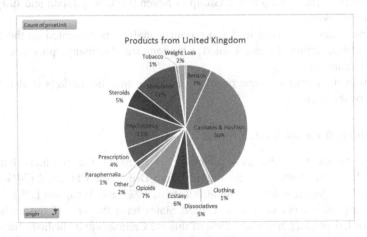

Fig. 7. Products from the United Kingdom, market distribution

3.5 Products from Germany

As for Germany, the country has very large number of products on sale in the category "Cannabis & Hashish", as the United Kingdom. Most of the European countries follow a similar pattern.

In Germany, the other most popular products are in the "Disassociatives" and "Psychedelics" categories (Fig. 8).

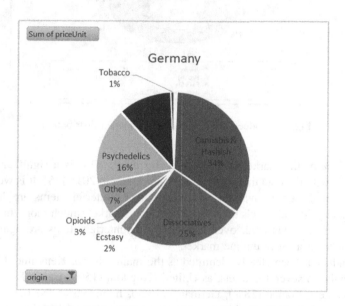

Fig. 8. Products from Germany, market distribution

3.6 Products from Afghanistan

It is interesting to notice that, unlike most of the countries, Afghanistan does not retail drugs on AlphaBay. The vast majority of the products advertised are false identity documents and fake accounts. Sellers from Afghanistan are also dealing in electronic devices and software (Fig. 9).

3.7 Other Markets

Here are some of the key findings and some of the peculiarities of other individual markets:

- **Netherland:** 100% of the products advertised by sellers in this country are Drugs and Chemicals. Two type of drugs, Ketamine and Crack are the top products advertised, with 36% and 38% respectively. This is similar for **Germany** (95%), **UK** (87%) and **Australia** (90%), **Canada** (97%), **France** (95%), **India** (nearly 100%).

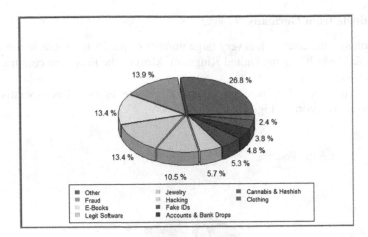

Fig. 9. Products from Afghanistan, market distribution

- **China:** 74% of the market is represented by drugs. This is a significant increase comparing with the data that was collected in Agora in 2015 [15]. It is worth noting that most of the drugs available are chemical. Counterfeit items are 18% of the market, and gold and jewels represent 16% of the market (much more than in all the other countries). That said, over 16% of the counterfeit items are again jewelry, which makes it a fairly unique market.
- **India:** India has been clearly identified as the main seller of Ketamine. Ketamine is very popular in several markets, as United Kingdom (15%), Czech Republic (68% of the market), France (58%), Germany (25%), India (96%), Netherlands (36%), Switzerland (36%). Worldwide, ketamine represents about 10% of the market. It is worth noting the presence of India – both a producer and consumer of Ketamine (while in most other countries is likely to be imported). It is also worth mentioning that Ketamine can and is also used as a date rape drug. In India's case it seems to be supply driven rather than just demand driven.
- **Japan:** Japan's market stands out; sellers mainly distribute digital products and online services. Drugs and Chemicals only contribute to 2% of the market. Examples of digital goods on sales include game keys, fraud softwares, malwares and security software.

4   Conclusion

The work presented in this paper is a follow up research from [15], reporting on the analysis of the largest dark web marketplace at the time.

The conclusion drawn from our research are alarming.

Dark web markets keep resurfacing regardless of the efforts made to shut them down, and Alphabay was much larger than any other marketplaces before.

Around \$88M of individual products are estimated to have been on sale on AlphaBay at the time of our research. The total value for the transactions during the existence of the marketplace is in the region of \$590M.

Drugs, fake IDs and weapons were readily available in a trans-national marketplace, just one click away and (fairly) anonymously. When it comes to counterfeit documents, any EU ID card would allow potential buyers to travel through any country in the EU, open bank accounts and in general create a new identity for himself/herself.

A new development is that dark web markets are now working very cautiously, implementing security measures and hacker avoidance updates regularly. The architectures are clearly more secure.

The understanding of illegal dark web markets is crucial. Information gathered on dark web markets can provide police forces and policymakers with greatly needed data to make informed decisions.

Although this research presented in this paper provides a detailed analysis of AlphaBay market, it can be extended to other dark web markets as well.

Acknowledgements. The authors wish to thank Simon Delecourt and Edouard Donze, who spent a summer internship at UEL in June–August 2017 and contributed to the preliminary data analysis for this paper.

References

1. Achsan, H.T.Y., Wibowo, W.C.: A fast distributed focused-web crawling. Proc. Eng. **69**, 492–499 (2014)
2. Anon: Megaupload file-sharing Site Shut Down (2012). http://www.bbc.co.uk/news/technology-16642369. Accessed 07 Aug 2017
3. Anon: 2017 SAS, R, or Python flash survey results (2017). http://www.burtchworks.com/2017/06/19/2017-sas-r-python-flash-survey-results/. Accessed 08 July 2017
4. Anon: Forget silk road, cops just scored their biggest victory against the dark web drug trade (2017). https://www.forbes.com/sites/thomasbrewster/2017/07/20/alphabay-hansa-dark-web-markets-taken-down-in-massive-drug-bust-operation/#480a3cb05b4b. Accessed 07 Aug 2017
5. Anon: Cocaine – Trafficking and supply (n.d.). http://www.emcdda.europa.eu/publications/eu-drug-markets/2016/online/cocaine/trafficking-and-supply. Accessed 24 July 2017
6. Anon: Explore: AlphaBay (n.d.). https://trends.google.co.uk/trends/explore?date=2014-12-01%202017-07-01&q=alphabay. Accessed 22 July 2017
7. Anon: Explore: AlphaBay, Dream Market (n.d.). https://trends.google.co.uk/trends/explore?date=2015-01-01%202017-07-01&q=alphabay,Dream%20Market. Accessed 22 July 2017
8. Anon: Facts about drugs (n.d.). http://www.thestudentpocketguide.com/2012/01/student-life/health-and-relationships/facts-about-drugs/. Accessed 12 July 2017
9. Anon: How much do drugs cost? (n.d.). http://www.drugwise.org.uk/how-much-do-drugs-cost/. Accessed 12 July 2017
10. Anon: R Markdown tutorial (n.d.). http://rmarkdown.rstudio.com/lesson-1.html. Accessed 29 June 2017
11. Anon: The average cost of illegal street drugs (n.d.). http://www.rehabcenter.net/the-average-cost-of-illegal-drugs-on-the-street/. Accessed 12 July 2017

12. Anon: The cost of street drugs in Britain (n.d.). http://www.telegraph.co.uk/news/uknews/crime/11346133/The-cost-of-street-drugs-in-Britain.html. Accessed 12 July 2017
13. Anon: What illegal drugs cost on the street around the world (n.d.). http://o.canada.com/business/interactive-what-illegal-drugs-cost-on-the-street-around-the-world. Accessed 12 July 2017
14. Baraniuk, C.: Alphabay and Hansa Dark Web Markets shut down (2017). http://www.bbc.co.uk/news/technology-40670010. Accessed 24 July 2017
15. Baravalle, A., Lopez, M. M., Lee, S.W.: Mining the dark web: drugs and fake IDs. IEEE, Barcelona (2017)
16. Egan, M.: What is the Dark Web? How to access the Dark Web. What's the difference between the Dark Web and the Deep Web? (n.d.). http://www.pcadvisor.co.uk/how-to/internet/what-is-dark-web-how-access-dark-web-deep-joc-beautfiulpeople-3593569/
17. INTELLIAGG: Deeplight. (Shining A Light On The Dark Web), s.l., s.n (2016)
18. Kopan, T.: DOJ announces takedown of dark web market Alphabay (2017). http://edition.cnn.com/2017/07/20/politics/doj-takes-down-dark-web-marketplace-alphabay/index.html. Accessed 24 July 2017
19. Lewis, S.J.: OnionScan Report: July 2016 - HTTPS Somewhere, Sometimes, s.l., Mascherari Press (2016)
20. Zulkarnine, A., Frank, R., Monk, B., Mitchell, J.: Surfacing collaborated networks in dark web to find illicit and criminal. s.l., s.n (2016)

Author Index

Printed in the United States
By Bookmasters